CORONAL DISTURBANCES

INTERNATIONAL ASTRONOMICAL UNION

UNION ASTRONOMIQUE INTERNATIONALE

SYMPOSIUM No. 57

HELD AT SURFERS PARADISE, QUEENSLAND, AUSTRALIA, 7–11 SEPTEMBER, 1973

CORONAL DISTURBANCES

EDITED BY

GORDON NEWKIRK, JR.

High Altitude Observatory, National Center for Atmospheric Research, Boulder, Colo., U.S.A.

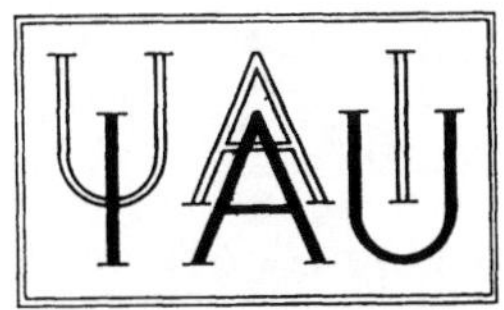

D. REIDEL PUBLISHING COMPANY

DORDRECHT-HOLLAND / BOSTON-U.S.A.

1974

Published on behalf of
the International Astronomical Union
by
D. Reidel Publishing Company, P.O. Box 17, Dordrecht, Holland

Sold and distributed in the U.S.A., Canada, and Mexico
by D. Reidel Publishing Company, Inc.
306 Dartmouth Street, Boston,
Mass. 02116, U.S.A.

Library of Congress Catalog Card Number 74–80521
ISBN-13: 978-90-277-0492-4 e-ISBN-13: 978-94-010-2257-6
DOI: 10.1007/978-94-010-2257-6

TABLE OF CONTENTS

PART III / SHOCK WAVES AND PLASMA EJECTION

Chairman: *J. T. Jefferies*

Chairman: *F. B. McDonald*

PART IV / ACCELERATION, CONTAINEMENT AND EMISSION OF HIGH-ENERGY FLARE PARTICLES

Chairman: *R. N. Bracewell*

Chairman: *S. F. Smerd*

PART V / REPORTS ON SPECIAL OBSERVATIONS

Chairman: *R. G. Athay*

Editor's Note: The authors intend that papers presented only as abstracts in this volume will appear in full form in the journals indicated in parentheses at the beginning of the abstract.

PREFACE

This Symposium was held at Surfer's Paradise, Queensland, Australia, from 7 to 11 September 1973. The Organizing Committee, chaired by J. P. Wild, consisted of A. Boischot, A. Bruzek, J. T. Jefferies, G. Newkirk, T. Takakura, and V. V. Zheleznyakov. We are indebted to the Local Organizing Commettee, chaired by S. F. Smerd and including R. G. Giovanelli, R. E. Loughhead, N. G. Seddon, K. V. Sheridan, and J. P. Wild, for advice in preparing this volume as well as for the smooth arrangement of the sessions. In addition, the session chairmen and reporters are to be thanked for their assistance in preparing the recorded discussions. It is a pleasure to thank Mrs R. Toevs and Mr A. Csoeke-Poeckh of High Altitude Observatory for assistance in editing these Proceedings. The financial aid for the Symposium afforded by the International Astronomical Union, the Ian Potter Foundation of Melbourne, and the Sunshine Foundation of Melbourne, as well as generous assistance of the CSIRO Divisions of Physics and Radiophysics is gratefully acknowledged.

That the solar corona is not a quiescent plasma was first fully appreciated through the discovery of solar radio bursts thirty years ago. Since that time intensive research has uncovered a vast variety of coronal disturbances and revised our concept of this region of the solar atmosphere to that of a dynamic medium undergoing continuous expansion, constantly evolving under the influence of underlying photospheric activity, and frequently traversed by transient phenomena. Such transients were the subject of this Symposium. Many, but not all, coronal disturbances are initiated by flares or sub-flares; in fact, we have begun to realize that an important class of such disturbances accompany eruptive prominences and may represent a significant factor in the evolution of the corona. The development of those disturbances initiated by flares may conveniently be discussed in terms of two stages of development; the flash phase, characterized by the cataclysmic release of energy, acceleration of particles to energies of a few keV to BeV per nucleon, X-ray, microwave, and type III radio bursts, and the onset of thermalization; and the post flash phase, characterized by coronal shock waves, type II and IV radio bursts, further particle acceleration, coronal depletions, and interplanetary shocks. The Symposium brought out several new perspectives and raised new questions in each of these areas.

Many puzzles remain concerning the fundamentals of flare energy release. One in particular is the acceleration of particles to high energy. Although it cannot yet be stated that we have a complete picture of this process, a possible view has emerged – that the acceleration occurs by two comparatively distinct processes. The first step appears to be direct acceleration predominantly of electrons to moderate energy by the Fermi or betatron mechanisms in rapidly changing magnetic configurations. The stage is then set for subsequent acceleration of these electrons and resident ions to higher energy in the turbulent shocks which race through the corona. There appears to be no question that the source of this acceleration is low in the corona rather than in the chromosphere or photosphere. In addition, as spacecraft observations are pursued to lower and lower particle energies, it becomes apparent that a major fraction of the total flare energy may reside initially in these fast particles.

Diagnosing such coronal disturbances as type III bursts has presented many problems since it could not be ascertained until recently whether the exciting particles were electrons or protons, how the beam of particles propagates through the corona, and where the beam moves in relation to coronal magnetic and density structures. Simultaneous *in situ* detection of low frequency type III bursts and electrons at 1 AU appears to have resolved the first question – the exciter is a beam of electrons having energies 10–100 keV. However, just how the excitation occurs remains a problem. A complete explanation of how such a beam manages to propagate through the corona out into interplanetary space without complete dissipation is particularly elusive. However, one question concerning the generation of the radio burst – whether interplanetary type III's radiate at the fundamental or first harmonic – appears to have been resolved in favor of dominance of the harmonic for most bursts. Thus, the electron densities inferred from these bursts are now in quite good agreement with those inferred optically or measured directly from space probes.

High resolution radio as well as synoptic coronal observations appear on the threshold of determining just where the electrons exciting type III bursts traverse the corona and the long accepted dictum – that type III's propagate out along the dense cores of coronal streamers – seems to be overthrown. It now appears that the favored path for the type III electron beam is outwards along open magnetic field lines emerging from active regions containing the flare source but not necessarily along coronal streamers. Other electrons, temporarily trapped in closed magnetic arches where the coronal density is high, give rise to the type V radio continuum.

During and following the flash phase, those accelerated particles which have not escaped the sun give up their energy to the surrounding corona or impact on the chromosphere. Observation of the resulting X, EUV, microwave, and visible radiation provides a powerful potential tool for the diagnosis of this complex process. Hopefully, the new observations of high temporal and spacial resolution, when combined with new theoretical methods, will allow a determination of which parts of this rich spectrum arise from impact on the chromosphere, bremsstrahlung in the corona, thermal radiation from the corona, and conduction from a superheated corona to the chromosphere.

Among the most spectacular post flash phase coronal disturbances are the coronal shocks, type II radio bursts, and interplanetary shocks. We have come to recognize that these are all various ramifications of the same type of disturbance although many aspects of the phenomena, such as the patchy spacial appearance of type II bursts, have remained perplexing. Interpretation of these bursts as fast-mode MHD shocks appears to offer a convincing explanation since such waves are refracted away from regions of high Alfvén speed and intensify into shocks in the low Alfvén speed regions, which tend to reflect the intricate density structure of the corona. At long last, the optical counterpart of such shocks have been observed by space-borne coronagraphs. In some cases, such disturbances appear to accompany realignments of the coronal magnetic field leading to a subsequent depletion of the inner corona and complete restructuring of its form. In others, magnetic arches are expelled from the Sun carrying with them a significant fraction of the coronal plasma. These dramatic new observations are bound to produce new perspectives on the nature of coronal disturbances.

Satellite probes have provided a wealth of definitive data on these shocks as they propagate into interplanetary space and much of our knowledge of the nature of shocks as they exist in the corona is derived from such observations. Interpreted in the simplest fashion, interplanetary shocks appear to invove $\sim 10^{16}$ g and $\sim 10^{32}$ erg and thus comprise a substantial fraction of the mass and energy ascribed to a flare. However, such coronal disturbances are complex, and the question of distinguishing between the primary flare-induced shock and a subsequent (or coincident) modification of the coronal magnetic field leading to an escape of plasma remains unresolved.

GORDON NEWKIRK, JR.

SCIENTIFIC ORGANIZING COMMITTEE

Wild, Dr J. P. (Chairman), CSIRO Division of Radiophysics, Epping, N.S.W., Australia
Boischot, Dr A., Observatoire de Meudon, Meudon, France
Bruzek, Dr A., Fraunhofer Institut, Freiburg, F.R.G.
Jefferies, Dr J. T., University of Hawaii, Honolulu, HI, U.S.A.
Newkirk, Dr G., Jr., High Altitude Observatory, National Center for Atmospheric Research, Boulder, Colo., U.S.A.
Takakura, Dr T., University of Tokyo, Bunkyo-ku, Tokyo, Japan
Zheleznyakov, Prof. V. V., Radiophysical Research Institute, Gorkii, U.S.S.R.

JOINT LOCAL ORGANIZING COMMITTEE

Smerd, Dr S. F. (Chairman), CSIRO Division of Radiophysics, Epping, N.S.W., Australia
Giovanelli, Dr R. G., CSIRO Division of Radiophysics, Chippendale, N.S.W., Australia
Loughhead, Dr R. E., CSIRO Division of Radiophysics, Chippendale, N.S.W., Australia
Seddon, N. G. (Mr), CSIRO Division of Radiophysics, Epping, N.S.W., Australia
Sheridan, Dr K. V., CSIRO Division of Radiophysics, Epping, N.S.W., Australia
Wild, Dr J. P., CSIRO Division of Radiophysics, Epping, N.S.W., Australia

LIST OF PARTICIPANTS

Acton, Dr L. W., Lockheed Palo Alto Res. Labs., Palo Alto, Calif., U.S.A.
Aller, Prof. L. H., University of California, Los Angeles, Calif., U.S.A.
Altschuler, Dr M. D., High Altitude Observatory/National Center for Atmospheric Research, Boulder, Colo., U.S.A.
Athay, Dr R. G., High Altitude Observatory/National Center for Atmospheric Research, Boulder, Colo., U.S.A.
Bappu, Dr M. K. V., Indian Institute of Astrophysics, Kodaikanal, India
Beckers, Dr J. M., Sacramento Peak Observatory, Sunspot, N.M., U.S.A.
Bhattacharyya, Prof. J. C., Indian Institute of Astrophysics, Hebbal, Bangalore, India
Bhavilai, Prof. R., Chulalongkorn University, Bangkok, Thailand
Boischot, Dr A., Observatoire de Meudon, Meudon, France
Bonnet, Dr R. M., Laboratoire de Physique Stellaire et Planétaire, Verrières-le-Buisson, France
Bracewell, Prof. R. M., Stanford University, Stanford, Calif., U.S.A.
Brandt, Dr J. C., NASA Goddard Space Flight Center, Greenbelt, Md., U.S.A.
Brault, Dr J. W., Kitt Peak National Observatory, Tucson, Ariz., U.S.A.
Bray, Dr R. J., CSIRO Division of Physics, Chippendale, N.S.W., Australia
Brown, Dr J. C., University of Glasgow, Glasgow, Scotland
Brown, Dr N., CSIRO Solar Observatory, Narrabri, N.S.W., Australia
Brueckner, Dr G. E., U.S. Naval Research Laboratory, Washington, D.C., U.S.A.
Bruzek, Dr A., Fraunhofer Institut, Freiburg, i. Br., F.R.G.
Bumba, Dr V., Astronomical Institute of the Czechoslovak Academy of Sciences, Ondrejov, Czechoslovakia
Cannon, Dr C. J., University of Sydney, Sydney, N.S.W., Australia
Caroubalos, Dr C., Observatoire de Meudon, Meudon, France
Castelli, Dr J. P., Air Force Cambridge Research Labs., Bedford, Mass., U.S.A.
Catura, Dr R. C., Lockheed Palo Alto Research Labs., Palo Alto, Calif., U.S.A.
Cole, Dr T. W., CSIRO Division of Radiophysics, Epping, N.S.W., Australia
Casanovas Corderroure, Dr J., Instituto Universitario de Astrofisica, Tenerife, Spain
Crapps, Dr G. W., CSIRO Solar Observatory, Narrabri, N.S.W., Australia
Delache, Dr P., Observatoire de Nice, Nice, France
Deubner, Dr F.-L., Fraunhofer Institut, Freiburg, F.R.G.
Dravins, Dr D., Lund Observatory, Lund, Sweden
Dryer, Dr M., Space Environment Laboratory, Boulder, Colo., U.S.A.
Dulk, Dr G. A., University of Colorado, Boulder, Colo., U.S.A.
Elgarøy, Dr Ø., Oslo University, Oslo, Norway

Énomé, Dr S., Nagoya University, Toyokawa, Japan
Erickson, Dr W. C., University of Maryland, College Park, Md., U.S.A.
Fainberg, Dr J., NASA Goddard Space Flight Center, Greenbelt, Md., U.S.A.
Frazier, Dr E. N., Aerospace Corporation, Los Angeles, Calif., U.S.A.
Fredga, Dr K., Royal Institute of Technology, Stockholm, Sweden
Frost, Dr K. J., NASA Goddard Space Flight Center, Greenbelt, Md., U.S.A.
Gabriel, Dr A. H., Culham Laboratory, Abingdon, Berks., England
Gebbie, Dr K. B., University of Colorado, Boulder, Colo., U.S.A.
Giovanelli, Dr R. G., CSIRO Division of Physics, Chippendale, N.S.W., Australia
Godoli, Dr G., Osservatorio Astrofisico, Catania, Italy
Gotwols, Dr B. L., The Johns Hopkins University, Silver Spring, Md. U.S.A.
Grossman-Doerth, Dr U., Fraunhofer Institut, Freiburg, i. Br., F.R.G.
Hachenberg, Prof. O., Max-Planck-Institut for Radioastronomie, Bonn 1, F.R.G.
Hartz, Dr T. R., Communications Research Centre, Ottawa, Ontario, Canada
Heisler, Dr L. H., CSIRO Division of Radiophysics, Epping, N.S.W., Australia
Hirabayashi, Dr H., Tokyo Astronomical Observatory, Nagano-ken, Japan
Holt, Dr H., University of Sydney, Sydney, N.S.W., Australia
Jefferies, Dr J. T., University of Hawaii, Honolulu, HI, U.S.A.
Jensen, Prof. E., University of Oslo, Blindern, Norway
Jordan, Dr C., Culham Laboratory, Abingdon, Berks., England
Kai, Dr K., Tokyo Astronomical Observatory, Mitaka, Tokyo, Japan
Kane, Dr S. R., University of California, Berkeley, Calif., U.S.A.
Kaufmann, Prof. P., Universidade MacKenzie, Sao Paulo, Brazil
Kiepenheuer, Prof. K. O., Fraunhofer Institut, Freiburg, i. Br., F.R.G.
Krall, Dr N. A., University of Maryland, College Park, Md., U.S.A.
Kubota, Dr J., Kyoto University, Kyoto City, Japan
Kundu, Prof. M. R., University of Maryland, College Park, Md., U.S.A.
Labrum, Dr N. R., CSIRO Division of Radiophysics, Epping, N.S.W., Australia
Lantos, Dr P., Observatoire de Meudon, Meudon, France
Leblanc, Dr Y., Observatoire de Meudon, Meudon, France
Leibacher, Dr J. W., Laboratoire de Physique Stellaire et Planétaire, Verrières-le-Buisson, France
Liebenberg, Dr D. H., Los Alamos Scientific Laboratory, Los Alamos, N.M., U.S.A.
Lin, Dr R. P., University of California, Berkeley, Calif., U.S.A.
Loughhead, Dr R. E., CSIRO Division of Physics, Chippendale, N.S.W., Australia
McCabe, Dr M., University of Hawaii, Honolulu, HI, U.S.A.
McDonald, Dr F. B., NASA Goddard Space Flight Center, Greenbelt, Md., U.S.A.
McKenna-Lawlor, Dr S., Maynooth University, Dunboyne, Co., Meath, Ireland
McLean, Dr D. J., CSIRO Division of Radiophysics, Epping, N.S.W., Australia
MacQueen, Dr R. M., High Altitude Observatory, National Center for Atmospheric Research, Boulder, Colo., U.S.A.
Malitson, Ms H. H., NASA Goddard Space Flight Center, Greenbelt, Md., U.S.A.
Mangeney, A., Observatoire de Meudon, Meudon, France

Martin, Mrs S. F., Lockheed Solar Observatory, Burbank, Calif., U.S.A.
Maxwell, Dr A., Harvard College Observatory, Cambridge, Mass., U.S.A.
Mayfield, Dr E. B., Aerospace Corporation, Los Angeles, Calif., U.S.A.
Melrose, Dr D. B., Australian National University, Canberra, A.C.T.
Meyer, Dr F., Max-Planck-Institut für Physik, Munchen, F.R.G.
Molnar, Dr H., Observatorio de Fisica Cosmica de San Miguel, San Miguel, Argentina
Morimoto, Dr M., Tokyo Astronomical Observatory, Mitaka, Tokyo, Japan
Mugglestone, Prof. D., University of Queensland, St. Lucia, Qld., Australia
Mullaly, Dr R. F., University of Sydney, Sydney, N.S.W., Australia
Müller, Prof. E. A., Observatoire de Genève, Geneva, Switzerland
Nakagawa, Dr Y., High Altitude Observatory/National Center for Atmospheric Research, Boulder, Colo., U.S.A.
Nelson, Dr G. J., CSIRO Solar Observatory, Narrabri, N.S.W., Australia
Newkirk, Dr G. A., High Altitude Observatory/National Center for Atmospheric Research, Boulder, Colo., U.S.A.
Pande, Dr M. C., Uttar Pradesh State Observatory, Manora Peak, Naini Tal, India
Pasachoff, Prof. J. M., Williams College Observatory, Williamstown, Mass., U.S.A.
Payten, W. J., CSIRO Solar Observatory, Narrabri, N.S.W., Australia
Pecker, Prof. J. C., Institut d'Astrophysique de CNRS, Paris, France
Pick, Mrs M., Observatoire de Meudon, Meudon, France
Piddington, Dr J. H., University Grounds, Chippendale, N.S.W., Australia
Pierce, Dr A. K., Kitt Peak National Observatory, Tucson, Ariz., U.S.A.
Pneuman, Dr G. W., High Altitude Observatory/National Center for Atmospheric Research, Boulder, Colo., U.S.A.
Prokakis, Dr T., National Observatory of Athens, Athens, Greece
Rees, D. E., University of Sydney, Sydney, N.S.W., Australia
Ribes, Mrs E., Observatoire de Meudon, Meudon, France
Riddle, Dr A. C., CSIRO Division of Radiophysics, Epping, N.S.W., Australia
Robinson, R., University of Colorado, Boulder, Colo., U.S.A.
Roddier, Dr F., Université de Nice, Nice, France
Rösch, Prof. J., Observatoires du Pic-du-Midi et de Toulouse, Bagneres-de-Bigorre, France
Rosenberg, Dr J., Observatory 'Sonnenborgh', Utrecht, The Netherlands
Rutten, R. J., Observatory 'Sonnenborgh', Utrecht, The Netherlands
Sakurai, Dr K., NASA Goddard Space Flight Center, Greenbelt, Md., U.S.A.
Scalise, Prof. E., Jr., Universidade MacKenzie, Sao Paulo, Brazil
Schatten, Dr K. H., Victoria University, Wellington, New Zealand
Schmidt, Dr H. U., Max-Planck-Institut für Physik und Astrophysik, Munchen, F.R.G.
Sheridan, K. V., CSIRO Division of Radiophysics, Epping, N.S.W., Australia
Simon, Dr P., Observatoire de Meudon, Meudon, France
Sivaraman, Dr K. R., Indian Institute of Astrophysics, Hebbal, Bangalore, India

Smerd, Dr S. F., CSIRO Division of Radiophysics, Epping, N.S.W., Australia
Smith, Dr D. F., High Altitude Observatory/National Center for Atmospheric Research, Boulder, Colo., U.S.A.
Souffrin, Dr P., Observatoire de Nice, Nice, France
Speer, Dr R. J., Imperial College of Science and Technology, London, England
Steinberg, Dr J. L., Observatoire de Meudon, Meudon, France
Stewart, R. T., CSIRO Division of Radiophysics, Epping, N.S.W., Australia
Stix, Dr M., Universitäts-Sternwarte, Göttingen, F.R.G.
Sturrock, Prof. P. A., Stanford University, Stanford, Calif., U.S.A.
Sy, Dr W. N.-C., University of Papua and New Guinea, Boroko, P.N.G.
Takakura, Prof. T., University of Tokyo, Bunkyo-ku, Tokyo, Japan
Tanaka, Dr K., Tokyo Astronomical Observatory, Mitaka, Tokyo, Japan
Thomas, Prof. R. N., University of Colorado, Boulder, Colo., U.S.A.
Uchida, Dr Y., Tokyo Astronomical Observatory, Mitaka, Tokyo, Japan
van Nieuwkoop, Dr J., Sterrewacht 'Sonnenborgh', Utrecht, The Netherlands
Vorpahl, Dr J., Sacramento City College, Sacramento, Calif., U.S.A.
Vrabec, D., The Aerospace Corporation, Los Angeles, Calif., U.S.A.
Waldmeier, Prof. M., Swiss Federal Observatory, Zürich, Switzerland
Wentzel, Dr D. G., University of Maryland, College Park, Md., U.S.A.
Westfold, Prof. K. C., Monash University, Clayton, Vic., Australia
Wiehr, Dr. E., Universitäts-Sternwarte, Göttingen, F.R.G.
Wild, Dr J. P., CSIRO Division of Radiophysics, Epping, N.S.W., Australia
Wilson, Prof. P. R., University of Sydney, Sydney, N.S.W., Australia
Wolf, Dr B., European Southern Observatory, Hamburg, F.R.G.
Yousef, Dr S. M. A., Cairo University, Giza, Cairo, Egypt
Zirin, Prof. H., California Institute of Technology, Pasadena, Calif., U.S.A.
Zwaan, Dr C., Observatory 'Sonnenborgh', Utrecht, The Netherlands

PART I

MAGNETIC STRUCTURE RESPONSIBLE FOR CORONAL DISTURBANCES

MAGNETIC STRUCTURE RESPONSIBLE FOR CORONAL DISTURBANCES: OBSERVATIONS

MARTIN D. ALTSCHULER

High Altitude Observatory, National Center for Atmospheric Research, Boulder, Colo., U.S.A.*

Abstract. Coronal disturbances are considered as consequences of the ejection of electric currents (or non-potential magnetic fields) from the photosphere and chromosphere into the corona. It may be that electric currents are generated near neutral lines in the photosphere and are later ejected into the corona.

1. Introduction and Point of View

The corona of the Sun is a tenuous fully-ionized plasma, and therefore extremely responsive to magnetic fields. Indeed, the large-scale inhomogeneous structure of the corona, with streamers, condensations, holes, polar plumes, and helmets, is largely a consequence of the distribution of magnetic field throughout the solar atmosphere.

On occasion, the corona or some part of it becomes disturbed over short time scales. Mass motions, particle accelerations, and changes which affect the density and temperature of the corona are observed over intervals ranging from a few seconds in the case of certain radio and hard X-ray bursts to an hour or so for increased emission in the visible and radio continua and in the soft X-ray bands.

For the purpose of this talk, I will divide coronal disturbances into three different categories:

(1) long-period, or evolutionary, disturbances which persist for several days or more and which are undoubtedly controlled by persistent coronal magnetic fields rooted in the photosphere.

(2) fast disturbances which occur over times ranging from minutes to hours and which probably involve hydromagnetic processes.

(3) impulsive disturbances which occur in a few seconds or less and which are possibly a consequence of particle acceleration processes in certain coronal regions.

Fast and impulsive coronal disturbances are closely related to flare processes and eruptive prominences, thus to changing magnetic fields. Long-period coronal disturbances reflect the large-scale photospheric magnetic field.

These different types of coronal disturbances can affect the Earth's magnetic environment in different ways. Long-period coronal disturbances control the fast streams of solar wind and the interplanetary magnetic sector structure. Fast coronal disturbances may cause strong interplanetary shocks. Impulsive coronal disturbances are an important source of energetic particles and X-rays.

Thus it is difficult and misleading to study coronal disturbances apart from other solar and interplanetary activity. Indeed, all forms of solar and interplanetary activity are consequences of magnetic fields generated initially in the subphotosphere. (In the corona, concentrations of thermal energy from causes unrelated to magnetic fields

* The National Center for Atmospheric Research is sponsored by the National Science Foundation.

Gordon Newkirk, Jr. (ed.), Coronal Disturbances, 3–33. *All Rights Reserved.*

are probably of little importance and will not be discussed.) What we are concerned with basically is the transport of magnetic energy from the subphotosphere to interplanetary space either directly by convection of magnetic field or indirectly in the form of fast particles and mechanical energy. Coronal disturbances are a key link in that chain of physical processes.

The purpose of this conference is to trace in as much detail as possible the emergence of magnetic energy from the photosphere or chromosphere into the corona, and the partition of this energy into mass motions, hydromagnetic waves, shocks, fast particles, and heat. Viewed in this way, coronal disturbances should eventually be useful diagnostic tools to help understand both terrestrial magnetic disturbances and the nature of photospheric activity.

Let me now ask a specific question. Are present observations of coronal disturbances and magnetic fields sufficient to allow us to construct a physical model of a given coronal disturbance? To construct even a crude hydromagnetic model, we need to know the spatial and temporal distributions of the mass density, momentum density, magnetic field, and temperature of the disturbance. Even this information may not be sufficient to understand impulsive bursts, because a hydromagnetic description averages over velocity space and cannot describe non-Maxwellian particle acceleration processes. Since at present we do not have complete observations even of the three-dimensional time-changing magnetic field of a coronal disturbance, we cannot construct a unique physical model.

Consequently, we must make inferences from imperfect data. This means we must use observations together with established physical principles to guess how the magnetic field and the plasma are interacting. In general, we cannot expect a unique model of a coronal disturbance to emerge from this approach. Probably the best thing to do is to classify the observed coronal disturbances and the different physical processes that are likely to be present, and then to see if we can guess at the correspondences between the observations and the physical processes.

Immediately, however, we encounter two difficulties. First we must agree on which observations should be explained, and second we must decide which physical processes are important. These are not trivial problems.

Our ability to observe coronal and solar activity has vastly improved over the past ten years. With satellites, space probes, and orbiting laboratories, we are no longer limited by the Earth's atmosphere. We have observed virtually the entire electromagnetic spectrum from gamma rays to hektometer wavelengths as well as charged particles of all types over a wide energy range. Good spatial resolution in the plane of the sky has become available at wavelengths previously undetectable. At present, we have the immense task of correlating in space and time all the different kinds of observations. Thus there is now so much data and so much detail that a significant problem is to decide which observations to try to explain. On the other hand, just listing the different models that have been proposed for solar flares and prominences will show that agreement among theoreticians, even about the dominant physical processes, is not always present.

In fact, the difficulties faced by both observers and theoreticians derive from the same underlying cause, which is both the curse and the charm of solar physics. What we are concerned with in solar activity is a complex system of interacting fields and particles which can be viewed from many levels of sophistication.

To be rigorous, we must picture a coronal disturbance as a plasma of electrons, several different kinds of ions, and perhaps some neutrals, together with a magnetic field which affects the dynamics of these various plasma particles and is in turn perturbed by their motions. Any given ion may continually change its excitation, ionization, position in space, and velocity. Time changes may occur over the entire plasma and in local regions. The plasma distributions in space and velocity are extremely anisotropic because of gravity, magnetic fields, and boundary conditions on the plasma. Billings (1966) describes how complex a single cubic millimeter of coronal plasma really is.

However, coronal disturbances cannot be as completely chaotic as a microscopic description would imply. If they were, there could not be such easily recognizable coronal events as meter wavelength bursts of types II, III, and IV, eruptive prominences, flare surges, and so on. The very fact that it is possible to classify or define different kinds of coronal events on the basis of observation means that there are certain patterns of interaction between the magnetic field and the plasma that justify a macroscopic or fluid description. This is the reasoning that encourages us to attempt simple models of coronal disturbances on the basis of a few selected observational features.

Of course, a fluid description can also be extremely complicated. If we treat each plasma component as a separate fluid with a mean density, mean velocity, mean temperature, etc., the number of nonlinear partial differential equations that we must solve as a system becomes unmanageable. If we lump all the plasma components together and consider only one conducting fluid with an imbedded magnetic field as in magnetohydrodynamics, we still have several dependent variables such as the magnetic field, the fluid velocity, the density, and the temperature, which are functions of space and time, and constitutive parameters such as viscosity, electrical conductivity, and thermal conductivity, which are often treated as constants but which in fact could be functions of the dependent variables. Boundary conditions are also a problem because a coronal disturbance is not a closed system; it is imbedded in a plasma through which other electromagnetic, plasma, and acoustic waves (or disturbances) are continually propagating. Therefore, even a single-fluid hydromagnetic description of a coronal disturbance is usually so complex that we can obtain neither a numerical solution nor a conceptual picture to compare with observations.

Thus understanding the microscopic physics and being able to write down the equations is not enough in the case of coronal disturbances. To interpret observations we must understand the macroscopic interactive system of field and plasma. This means we must be able to write, solve, and interpret some rather complex closed sets of partial differential equations. At present we cannot properly solve such sets of equations and therefore must either do computations with a few terms or else solve the linearized (small-amplitude) approximations. As a result, except for the simplest cases, we do not yet know all the macroscopic processes and nonlinear feedback chains that may be contained in the plasma equations.

I have just painted a bleak picture primarily to emphasize that interactive plasma systems are sufficiently complex that deciding what physical processes and what observations should be emphasized in a model of a coronal disturbance is not a completely trivial matter. With regard to both the theory and the observations of a coronal disturbance, we must at present be satisfied with incomplete descriptions. An incomplete description, however, is largely a matter of judgement. It runs the risk of over-interpreting some observations while ignoring others, and of imagining physical processes that might not actually occur while neglecting those that are crucial. Moreover, to be interesting it should provide a picture broad enough to incorporate several diverse phenomena from only a few assumptions.

Granted that we must settle for an incomplete description and that the game is risky, I will now assume an optimistic attitude and return to my original goal of classifying the observations of coronal disturbances and a few hopefully relevant physical processes, to see if we can distil some useful concepts from this enormous complexity. There are a large number of ways to classify observations and plasma processes, and I will now commit myself to those at the very lowest level of sophistication.

2. Coronal Disturbances and the Electric Current Picture

If a magnetic field supplies energy which affects the temperature, density, or flow field of the surrounding medium, then the magnetic field must be non-potential, that is, it must have a twist or curl, therefore an electric current. Probably any observable solar phenomenon which can be classified as solar activity (whether in the photosphere, chromosphere, or corona) is a consequence of a non-potential magnetic field, or equivalently, an electric current.

In simple hydromagnetic theory, it does not matter whether phenomena are described in terms of the magnetic field or in terms of the electric current (Gold, 1968). If we emphasize the magnetic field, we can determine the electric current by taking the curl (or rot) of the magnetic field; if we emphasize the electric current, we can determine the vector potential and then the magnetic field by solving a Poisson-type vector equation with suitable boundary conditions.

If, however, the regions of electric current are sufficiently localized, then there is a decided advantage in trying to map the electric current rather than (or in addition to) the magnetic field. This is simply because the electric current regions of the solar atmosphere are the regions where magnetic energy is available for conversion into various kinds of kinetic energy. In fact, the neutral lines of the photospheric magnetic field must parallel the major large-scale electric currents of the Sun's surface. Whenever a neutral line is sharply defined in the photosphere, the photospheric electric current is probably strong and localized. Coronal activity, then, is most likely to originate over or near the photospheric neutral lines.

On the other hand, many coronal disturbances seem to follow the lines of potential (or current-free) magnetic field. Of course, any perturbation or kink in the potential field is equivalent to an electric current in the hydromagnetic sense. Nevertheless, for such coronal disturbances there is no clear advantage in using a current description.

There are also many cases where the magnetic field (**B**) is nearly parallel to the electric current (curl **B**). In this situation there usually is no discernible geometrical symmetry and the problem is difficult both mathematically and conceptually. The force-free field is an example.

The situation may also become quite complex if a broader view of hydromagnetic theory is taken, for example if we use a general Ohm's law, or equivalently, treat the electrons and ions as two separate fluids. Then non-parallel gradients of electron pressure and density may generate electric current. In practice, however, the electric current picture is interchangeable with the magnetic field picture at low frequencies until charge separation becomes important in the plasma.

Here I will emphasize the electric current picture and try to interpret observations of coronal disturbances accordingly. At low frequencies, the electric current picture does not introduce new physics, but hopefully follows a less familiar approach.

Let us now discuss the different ways an electric current may be generated in the corona. Since an electric current (that is, a non-potential magnetic field) in the corona is presumably the cause, effect, or kernel of a coronal disturbance, we are in fact classifying physical processes involved in a coronal disturbance.

Suppose that initially electric currents exist only in the photosphere and that the coronal magnetic field is everywhere current-free or potential. How can we generate an electric current in the corona? Four general ways are listed below.

(1) Perturb the potential field of the corona with coronal forces. Examples of this process are:

(a) Drop or condense matter at the top of a closed potential field line thereby bending it to create an electric current. This is equivalent to the magnetic buoyancy of plasma due to the tension forces of the magnetic field (Kippenhahn and Schlüter, 1957; Brown, 1958; Nakagawa and Malville, 1969; Anzer and Tandberg-Hanssen, 1971; Hildner, 1971; Raadu and Kuperus, 1973).

(b) Pull the potential field outward by forces of the solar wind expansion as in helmet streamers; create a current sheet thereby (Sturrock, 1968; Pneuman and Kopp, 1971).

(c) Set up shear flows in the corona which distort or twist the potential magnetic field.

(d) Create a shock or violent mass motion either parallel or perpendicular to a loop of potential magnetic field thereby causing a large-amplitude perturbation of the potential field lines (Uchida, 1970; Pneuman, 1967; Meyer and Schmidt, 1968; Schatten, 1970).

(2) Perturb the potential field of the corona with photospheric or chromospheric

motions. Examples of this process might be:

(a) Change the electric current density in the photosphere by expanding or collapsing the electric current cross-section; kinks in the magnetic field (hence electric currents) propagate out at about the Alfvén speed to readjust the field configuration; changes occur in the strength of the potential magnetic field, or equivalently, in the localization of the magnetic flux.

(b) Change the configuration of the electric current in the photosphere by convective motions; meanders or shears in the photospheric electric current region can twist the potential field lines of the corona, thereby generating an electric current (Sturrock and Coppi, 1966; Levine and Nakagawa, 1974).

(c) Move the footpoints of the potential magnetic field by displacement motions or by vortical mass motions; the kinks or twists propagate into the corona at the Alfvén speed; the twisted field (or current) may continue to build up and store energy, or act as a force-free field (Gold, 1964; Anzer, 1968; Stenflo, 1969; Nakagawa and Raadu, 1972); this situation is similar to that of (b).

(d) Create an electric current in the photosphere by means of non-parallel gradients of electron temperature and electron pressure (Kopecky and Kuklin, 1971), and thus cause readjustment of the coronal field.

(3) Create an electric current in the corona by various plasma processes.

(4) Eject a photospheric electric current or electric current filament upward into the corona. Possible methods of doing this are:

(a) Concentrate the photospheric electric current; then magnetic buoyancy should be effective (Parker, 1955).

(b) Concentrate the photospheric electric current into a thin filament; then the region surrounding the current is heated by magnetic diffusion; current becomes buoyant.

(c) Create magnetic forces (that is, antiparallel electric currents) either by reconnection of field lines in the photosphere (Sweet, 1958; Petschek, 1964; Coppi and Friedland, 1971) or by meandering the photospheric electric current thus creating a small area of opposite magnetic polarity in a unipolar photospheric region (Altschuler *et al.*, 1968).

These four general methods of producing a coronal electric current provide a conceptual scheme to describe the prerequisite conditions for a coronal disturbance. Of course, in reality the fluid flow cannot be merely assumed as we have done, but must be considered self-consistently with the magnetic field and other forces. Now let us take a brief panoramic view of the observations of coronal disturbances.

3. Classification of Observations of Coronal Disturbances

Classifying a coronal disturbance by where it appears in the electromagnetic spectrum is probably safest (1) because each spectral region reveals a different parameter domain of the solar plasma and (2) because the sophistication, sensitivity, and resolution of our detection equipment varies greatly over the spectrum. Thus if radiation enhancements observed in different spectral regions are considered different types of coronal disturbances for classification purposes, we need not decide *a priori* whether we are observing (1) a single coronal region in which several different physical processes are operating over a wide range of energy, or (2) separated coronal regions emitting at the same time under different ambient conditions (such as inside or outside a coronal streamer).

Let us list coronal disturbances and associated phenomena according to the spectral range in which they are observed. No attempt is made for completeness, and fast disturbances are emphasized.

(1) Hα measurements (and other strong hydrogen lines):
- (a) brightenings on disk and limb (flares)
- (b) surges, sprays, other ejecta
- (c) active loops, coronal rain
- (d) flare waves (Moreton disturbances)
- (e) disappearing or winking filaments on the disk
- (f) large erupting prominences on the limb (particularly hedgerow)

(2) Monochromatic measurements of coronal emission lines in the visible spectrum:
- (a) expansion of coronal arches: slowly, rapidly, or explosively
- (b) whips: opening of coronal arches
- (c) hot plasma regions at tops of flare loops

(3) White Light Measurements:
- (a) coronal changes over eclipse path
- (b) thin coronal rays or sheets (possibly electric current sheets)
- (c) electron density changes
- (d) moving blobs, mass motions

(4) Measurements at X-ray wavelengths:
- (a) impulsive brightenings
- (b) small hot emission cores in coronal loops or filaments
- (c) EUV flares and ejecta

(5) Radio measurements (millimeter to hektometer wavelengths):
- (a) sharply defined frequency drifts at decimeter and longer wavelengths
- (b) impulsive microwave bursts
- (c) continuum emission
- (d) enhanced emission and proper motions (two-dimensions in plane of sky) at a single frequency (for example 80 MHz)

(6) Non-Electromagnetic Measurements:
  (a) terrestrial ionospheric disturbances
  (b) solar wind enhancements or (shock) discontinuities in speed, density, and magnetic field
  (c) enhancements in number, flux, and energy of fast charged particles (such as protons, electrons, solar cosmic rays)

At this conference these phenomena will be reviewed in detail. Here I will confine my remarks to aspects of these coronal events which concern magnetic fields. Let us now look at the observations, deductions, and inferences regarding the solar magnetic field.

4. Determining the Coronal Magnetic Field (Long-Period Disturbances)

4.1. Coronal emission line polarization

The coronal magnetic field configuration can be inferred (at least in projection over the limb) if we can observe the monochromatic emission from certain magnetically-sensitive coronal lines and determine the distribution of polarization in the plane of the sky. The degree of polarization together with the angle of maximum polarization provide information on the direction (but not the magnitude) of the coronal magnetic field at the position where the emission line radiation originates. Such observations have been made with a coronameter (Charvin, 1965, 1971) and during eclipses (Hyder, 1966; Eddy and Malville, 1967; Hyder *et al.*, 1968; Beckers and Wagner, 1971; Eddy *et al.*, 1973) for various coronal emission lines. To measure the Stokes parameters, new coronameter-type instruments have been built at Meudon (Charvin, 1971), the University of Hawaii (Orrall, 1971), and at HAO (Querfeld, 1973).

The theory of coronal emission line polarization is quite involved (Charvin, 1965; Hyder, 1965; House, 1972). If the three-dimensional coronal magnetic field is known, House (1972) can determine the polarization that should be observed in the plane of the sky. That was in itself a difficult problem. House, Querfeld, and I are now working on a method which we hope will solve the converse problem of determining the coronal magnetic field geometry in three dimensions from daily polarimeter observations. Our plan is to observe the Stokes parameters of a coronal emission line in the plane of the sky over several days and then use regression analysis together with a few assumptions to find the three-dimensional coronal magnetic field configuration that best fits the plane-of-the-sky observations. Some information about the non-static magnetic fields in fast coronal disturbances might also be inferred with such a method.

4.2. Limb prominence fields

Magnetic fields of limb prominences have been determined from measurements of the Zeeman splitting in several strong spectral lines (Tandberg-Hanssen, 1971). Of the quiescent prominences observed, more than half have a mean line-of-sight field strength between 3 and 8 G. The magnetic field appears to enter and leave at the sides, but in the quiescent prominence itself there is a component of the field parallel to the prominence axis. Rust (1966, 1967) and Harvey (1969) found some evidence that stronger fields occur higher in the prominence. Thus from the available measurements, a quiescent prominence appears to illustrate how a magnetic field may support matter. However, the limited spatial resolution of $10'' \times 10''$, or 7.5 Mm in distance on the Sun, does not allow an estimate of the magnetic fields in prominence fine structures. The fine structures of quiescent prominences may indicate a circulation of matter (Dunn, 1960; Engvold, 1972; Tandberg-Hanssen, 1974). Tandberg-Hanssen and Malville are now studying the Climax measurements of magnetic fields in active limb prominences. A new instrument to measure the four Stokes parameters of spectral lines (and hence the magnetic field) in limb prominences is under construction at HAO.

4.3. Other coronal measurements pertaining to magnetic fields

In addition to measurements of the coronal emission line polarization and the Zeeman splitting of prominence lines, other direct information concerning the general configuration of the coronal magnetic field may be obtained from studies of the X-ray loops and structures (Krieger *et al.*, 1971) and from radio measurements (Daigne *et al.*, 1971; Kundu, 1971).

4.4. Current-free fields: small scale (no surface curvature)

At present, however, the coronal magnetic field cannot be determined on a routine basis from measurements of coronal phenomena. Instead, we must calculate the coronal field from measurements of the photospheric field. One way of doing this is to assume that the magnetic field is current-free (or potential) above the photosphere and then to solve a Laplace equation with the measured photospheric magnetic field distribution providing the boundary condition. Since only the line-of-sight photospheric field component can be accurately measured, observations are usually taken as near as possible to disk center so that the measured field is normal to the surface. The current-free approximation provides a mathematically unique solution for the three-dimensional coronal magnetic field. Any observed deviation from the calculated field geometry is an indication of coronal electric currents.

Schmidt (1964) was the first to use detailed measurements of the photospheric magnetic field to trace the current-free coronal field configuration. His program was designed to represent a limited region not exceeding about 200 Mm on a side; therefore, the curvature of the solar surface was not included. The potential magnetic field of the corona calculated by this method has been compared with active and quiescent prominence features above the limb (Rust, 1966; Harvey, 1969; Rust, 1970; Rust and Roy, 1971; Roy, 1972) and with chromospheric Hα filaments (Rayrole and Semel, 1968; Harvey *et al.*, 1971).

Above strong but reasonably static photospheric fields, the predicted coronal potential field is consistent with the coronal loops observed in monochromatic emission as far as 150 Mm from the limb (Rust and Roy, 1971); this agreement appears in spite of the fact that the photospheric magnetic fields of an active region are measured about a week before or after limb passage. Surprisingly, coronal loops formed after a large flare also agree with the potential field configuration (Roy, 1972). This might mean that non-potential fields in the corona can relax rapidly to potential fields after releasing energy which heats or disturbs the plasma. However, there is also the possibility that the flare-loop magnetic fields have highly twisted fine structure and therefore contain electric current.

In the chromosphere, the agreement between the potential field and the direction of the Hα fine structure is often poor (Rayrole and Semel, 1968; Harvey *et al.*, 1971). This indicates that the fine structure of the chromosphere is associated with non-potential or twisted magnetic fields. In fact the very existence of a filamentary structure is good evidence for complex plasma processes and non-potential magnetic fields. Photospheric fields also appear to be filamentary (Howard and Stenflo, 1972; Frazier and Stenflo, 1972) and therefore non-potential on a fine scale.

Programs have recently been devised which use photospheric field measurements to calculate force-free magnetic fields above active regions (Nakagawa and Raadu, 1972). In these calculations, the electric current and the magnetic field are everywhere parallel and have a constant ratio of magnitudes. The derived force-free magnetic fields sometimes are aligned with active filaments of the chromosphere. Again this indicates that chromospheric fields are often twisted, and that the magnetic field and the electric current are not always perpendicular. However, large active-region filaments do seem to lie along the boundary (neutral-line) which separates photospheric regions of opposite magnetic polarity (Howard and Harvey, 1964).

I do not wish here to enter the controversies concerning the orientation of chromospheric features with respect to the magnetic field (Veeder and Zirin, 1970; Frazier, 1972a, b; Zirin, 1972; Foukal and Zirin, 1972; Cheng *et al.*, 1973) except to emphasize that this is an extremely important problem for our purposes because we want to understand how electric currents (or non-potential fields) are created in the photosphere and chromosphere and how they generate coronal disturbances. Undoubtedly changes in the opacity and orientation of chromospheric filaments during the flare process are associated with changes in magnetic fields and electric currents (Zirin and Tanaka, 1973) although the precise mechanism is still not clear. In any case, the measurement of magnetic fields in the photosphere and chromosphere, particularly for the fine scale, is difficult both observationally (Beckers, 1971; Harvey, 1972) and theoretically (Stenflo, 1971). (See also the other related articles in *IAU Symp.* **43**.)

Since the calculated potential field agrees better with large coronal structures than with fine-scale chromospheric features, maps of the potential magnetic field on a global scale should be useful for the study of those coronal disturbances which are guided over long distances by the general field structure. Let us now discuss the potential field of the solar corona on the global scale.

4.5. Current-free-fields: global scale

Methods have been developed to calculate the current-free coronal magnetic field on a global scale using as data only the measured line-of-sight component of the photospheric magnetic field. In recent years, the mathematical techniques and limitations for such global maps have been discussed in detail in the literature (Newkirk *et al.*, 1968; Schatten *et al.*, 1969; Altschuler and Newkirk, 1969; Schatten, 1971a). Here I will merely make a few general remarks and then discuss applications relevant to coronal disturbances.

The Mt. Wilson data are the only full solar disk magnetic measurements continual over a long period of time. The equipment, the observational techniques, and the method of reduction were described by Howard *et al.* (1967). To obtain the global coronal field in the current-free approximation, the photosphere is first divided into 1080 surface elements of equal area, with 30 zones ($\Delta \sin\lambda = 1/30$ in latitude λ) and 36 sectors ($\Delta\phi = 10°$ in longitude ϕ). For each surface element an average line-of-sight magnetic field is found from the Mt. Wilson data. Corrections for magnetograph saturation are added to those surface elements where strong sunspot fields are present. The average line-of-sight fields of the 1080 equal surface elements are then used to calculate the Legendre coefficients of the harmonic series which solves the Laplace equation and best fits the global photospheric magnetic data (Altschuler and Newkirk, 1969). Once the Legendre coefficients are known, the magnitude and direction of the current-free (potential) coronal magnetic field can be determined at any point in space within about $r = 2.5\ R_0$, beyond which the solar wind dominates.

There are several limitations of this procedure which must be kept in mind. The Mt. Wilson data are restricted to one magnetic component (line-of-sight), to one atmospheric level (the photosphere), and to a relatively small intensity range (0.5 to 100 G). Because of foreshortening effects, the magnetograph measurements are representative of actual fields only near the center of the visible solar disk. Thus photospheric magnetic data for the polar regions are of limited accuracy, and data covering the entire Sun must be collected over at least one complete solar rotation. As a result, any magnetic field fluctuations in the photosphere can be detected only at three to four week intervals. The unavoidable errors in correcting for strong fields and in measuring the photospheric field over an entire solar rotation cause a spurious net monopole component for the global solar field. This spurious monopole contribution is removed by adding a constant to all the line-of-sight field measurements (Altschuler and Newkirk, 1969). The resulting field has no monopole component larger than one part in 10^8. A zero potential surface is also included to make the coronal field radial at $r = 2.5\ R_0$ and thereby simulate the effects of the solar wind.

From a set of Legendre coefficients, we can at present draw four different kinds of maps to help visualize the coronal potential magnetic field.

The first type of map traces the lines of coronal magnetic field from footpoints which are distributed geometrically over the photosphere. One coronal magnetic field line is drawn from each of 648 elements of equal photospheric area (that is, 27 equal divisions in longitude and 24 equal divisions of the north-south axis). Thus this map shows the overall geometry of the coronal magnetic field but does not distinguish strong from weak fields either in the corona or in the photosphere.

The second type of map shows a particular subset of the field lines which appear in the map of the first type. The photosphere is first partitioned into regions of similar magnetic polarity (unipolar regions). The total number of field lines (a number chosen in advance) is then distributed among the unipolar regions in proportion to the amount of radial magnetic flux. Thus this map shows the coronal distribution of the largest amounts of magnetic flux. Strong fields from small photospheric areas and weak fields from large photospheric areas can appear in this map provided a sufficient amount of flux passes through the unipolar photospheric region.

The third type of map shows the field lines which originate from the photospheric regions of strong magnetic intensity. A grid four times finer with $648 \times 4 = 2592$ elements of equal photospheric area is used. The field strength at the center of each area element is calculated and ranked. Field lines are drawn from the 400 area elements with the strongest calculated magnetic field. Thus this map plots only 15% of the possible field lines and shows how photospheric regions of strong magnetic field influence the solar corona. Strong fields correlate with active regions in the corona such as those appearing in X-ray rocket photographs.

The fourth type of map draws a continuous intensity distribution so that the coronal regions with largest $|\mathbf{B}|$ appear brightest. This map is being used to compare the three-dimensional magnetic field distribution with the three-dimensional density distribution as calculated by Altschuler and Perry (1972) and Perry and Altschuler (1973).

In Figure 1, the first three types of maps are shown for the November 1966 eclipse together with an Hα disk picture. Figure 2 shows these types of maps for the March 1970 eclipse together with the X-ray picture taken by American Science and Engineering (Krieger *et al.*, 1971). Figure 3 is a map of the fourth type for the November 1966 eclipse (devised by R. M. Perry) to show the absolute magnitude of the magnetic field strength. The calculated coronal fields for the November 1966 eclipse correspond well with the global density structure and the strong Hα emission regions (Newkirk and Altschuler, 1970; Newkirk, 1971). Around the time period of the March 1970 eclipse, the Sun's photospheric field changed considerably. Even so, there is some agreement between the strong field map (type 3) and the X-ray emitting regions. It is likely therefore that the X-ray emission occurs where the coronal field is strong. The direct comparison of calculated coronal fields with eclipse photographs (Newkirk, 1971; Schatten, 1971b; Altschuler, 1971) has shown that the global potential field is useful for tracing the inhomogeneous coronal structure, thus for long-period (or evolutionary) disturbances.

5. Fast Coronal Disturbances and Coronal Magnetic Fields

Fast coronal disturbances usually occur in less than an hour and often in a few minutes. They probably involve complex hydromagnetic processes. When we try to conceptualize such processes we generally think of 'static' and 'dynamic' magnetic fields. When static, magnetic fields may (1) store energy and fast particles, (2) guide

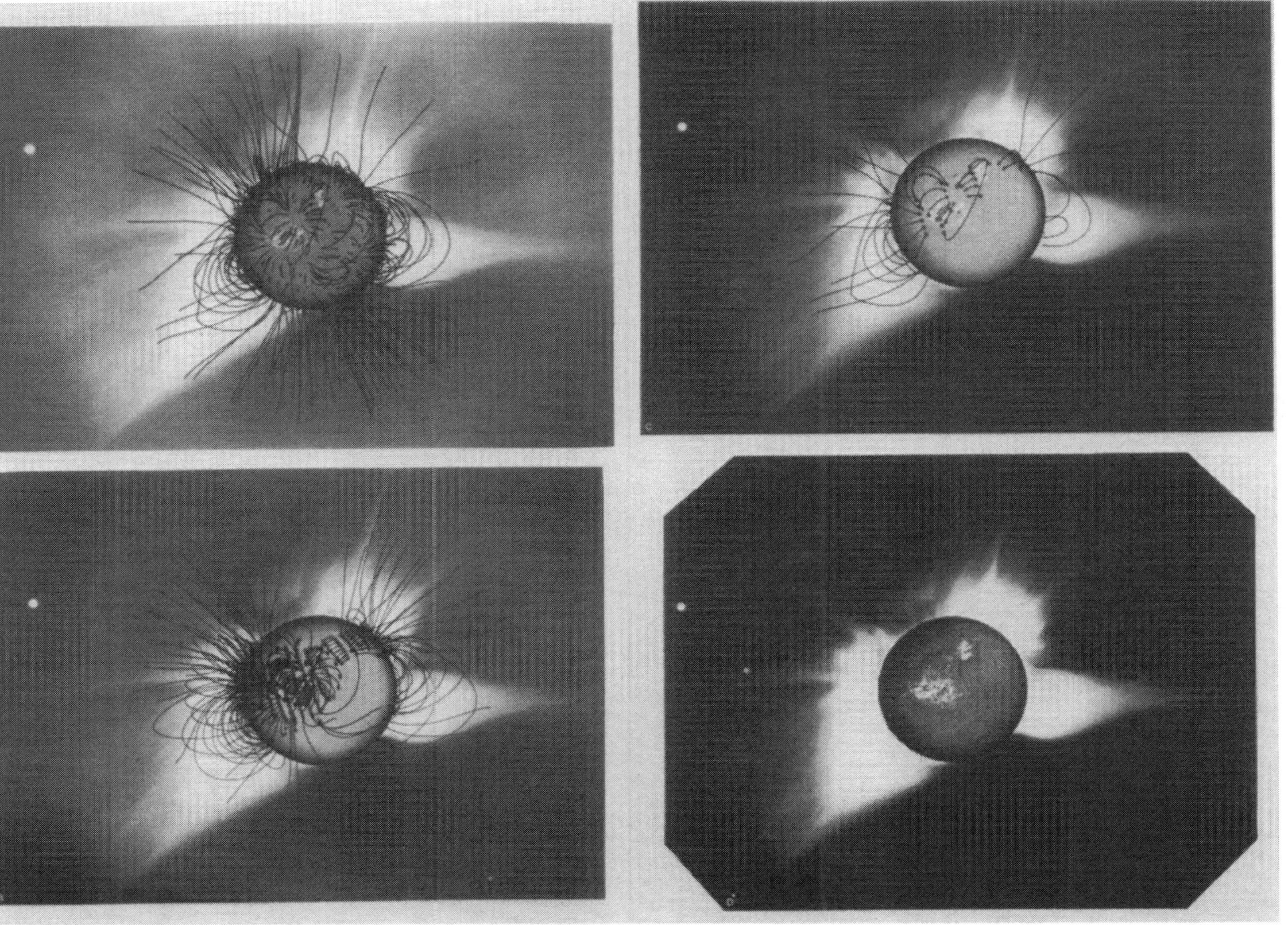

Fig. 1. Global potential magnetic field maps for the November 1966 eclipse. *Upper left*: general field map (type 1). *Upper right*: flux map (type 2). *Lower left*: strong field map (type 3). *Lower right*: eclipse photograph with Hα picture superposed over lunar disk. North is at upper left.

Fig. 2. Global potential magnetic field maps for the March 1970 eclipse. *Upper left*: general field map (type 1). *Upper right*: flux map (type 2). *Lower left*: strong field map (type 3). *Lower right*: eclipse photograph with AS & E rocket X-ray picture superposed over lunar disk. North is at upper right.

disturbances, waves, and heat flow, and (3) support matter. When dynamic, magnetic fields may (1) accelerate particles, (2) compress matter, and (3) set up waves, shocks, and mass motions which in turn feed back changes to the magnetic field configuration.

In terms of the electric current description of Section 2, the 'static' fields correspond to relatively small perturbations of the pre-existing coronal potential field, whereas the 'dynamic' fields usually correspond to the transport of large non-potential mag-

Fig. 3. Absolute magnitude of global potential magnetic field for November 1966 eclipse. Brightest features have most intense magnetic fields. North is up.

netic fields (or electric currents) through the corona. Electric currents generated directly in the corona by plasma processes have not been discussed much in the literature and will be neglected here.

The global current-free field approximation which we have just discussed in Section 4.5 can be used to study fast coronal disturbances of two extreme types. The first type includes static magnetic fields which guide disturbances over global distances. No permanent changes in the photospheric or coronal field geometries are obvious. The second type includes major dynamic disturbances which alter the large-scale photospheric field and therefore the global coronal field.

Let us first see what can be learned about static (or quasi-static) coronal magnetic fields and the role they play in guiding, focusing, and otherwise controlling coronal disturbances.

5.1. Disturbances guided by magnetic fields

There are several transient phenomena which appear to be guided or controlled by quasi-static coronal fields. These include chromospheric flare waves, certain radio disturbances, and probably fast streams of plasma in the interplanetary medium.

On occasion, a fast (up to 1 Mm s^{-1}) wave pulse can be seen in the Hα chromosphere moving away from a flare region (Moreton and Ramsey, 1960; Smith and Harvey, 1971). The pulse usually remains within some angle centered at the flare, and propagates over a significant fraction of the solar circumference. Although this transient is observed at the chromospheric level, its energy source must propagate as a hydromagnetic disturbance through the corona; in the chromosphere, disturbances are slower and are damped over shorter distances (Anderson, 1966; Meyer, 1968; Uchida, 1968). Apparently the flare emits an MHD fast-mode wavefront which expands into the corona. The intersection of this MHD fast-mode wavefront with the chromosphere then causes the observed wave pulse. Recently Uchida *et al.* (1973) have traced the propagation of MHD fast-mode wavefronts from flare regions by means of (global) potential field configurations derived from magnetograph data and electron density distributions derived from K-coronameter data. An isotropic wavefront was assumed at the source. Figures 4 and 5 show for different flares (1) the

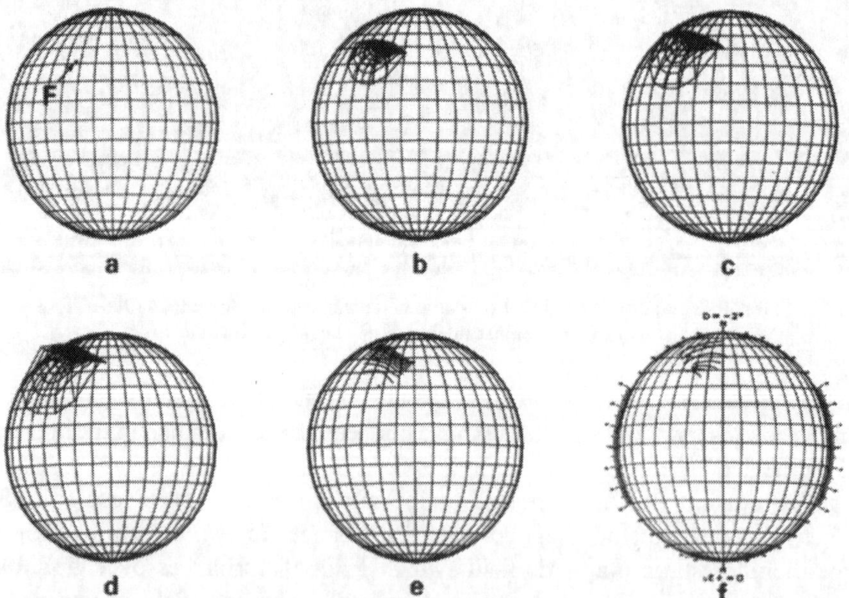

Fig. 4. Calculations for flare wave of 23 May 1967. Diagrams 4a–d show development of coronal wavefront and regions of energy concentration. Diagram 4e shows calculated intersections of the coronal wavefront with the chromosphere at different times. Diagram 4f shows observed positions of the flare wave at different times.

calculated time development of the coronal wavefront, (2) the calculated intersection of the wavefront with the chromosphere at different times, and (3) the observed chromospheric flare wave at different times. The agreement is remarkable. Thus a chromospheric flare wave is caused by a hydromagnetic fast-mode disturbance which propagates into the corona and concentrates in coronal or chromospheric regions

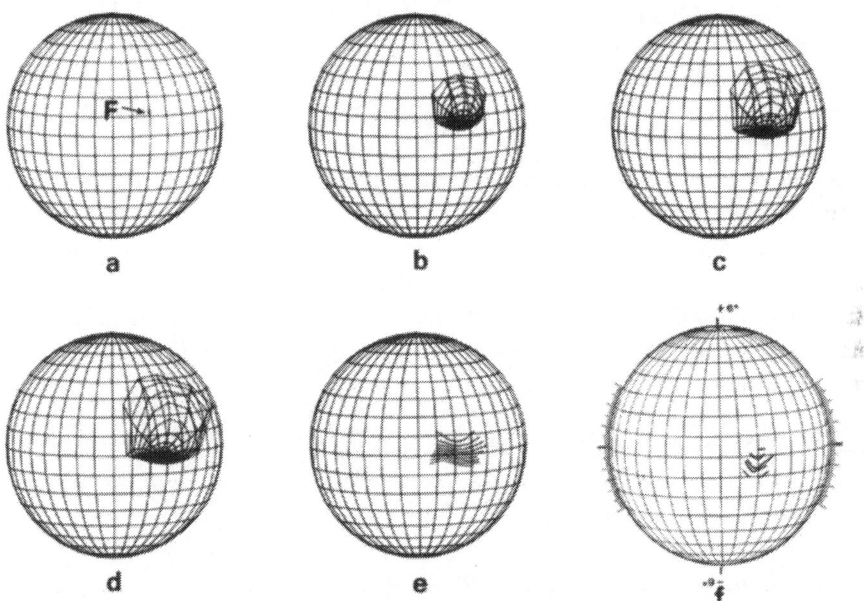

Fig. 5. Calculations for flare wave of 31 July 1967. Diagrams 5a–d show development of coronal wavefront and regions of energy concentration. Diagram 5e shows calculated intersection of the coronal wavefront with the chromosphere at different times. Diagram 5f shows observed positions of the flare waves at different times.

where the Alfvén speed is low. There seems to be some correlation between flare waves and type II radio bursts (Smith and Harvey, 1971; Uchida *et al.*, 1973). Perhaps the type II burst is itself a large-amplitude MHD-fast-mode shock (McLean, 1967), or else a disturbance which originates in a coronal region of low Alfvén speed where MHD fast-mode energy is concentrated.

To determine the magnetic field geometry associated with a radio disturbance, we must accurately locate the radio disturbance at least in the plane of the sky. In general this can be done with interferometry (Wild, 1970). Global maps of the potential magnetic field have been compared with radio data from Culgoora, the University of Maryland, and several other observatories. In general, the results show that fast outwardly-moving radio bursts such as type II, type III, and moving type IV are guided by open field lines (Smerd and Dulk, 1971; Dulk *et al.*, 1971; Dulk and Altschuler, 1971; Kuiper, 1973). Such comparisons are not completely conclusive because we do not know the three-dimensional positions of the radio sources. Moreover, since density gradients are generally small in the corona, their effect on the propagation of radio disturbances is not easy to discern.

The farther we go from the photosphere, the simpler the potential field configuration becomes. Higher harmonics of the photospheric field drop off at higher powers of the radial distance. At $r=2.5\ R_0$, only the dipole, quadrupole, and sometimes the octupole components are influential. These low harmonic components dominate the magnetic field in interplanetary space (Wilcox and Ness, 1965; Schatten, 1971b; Scherrer *et al.*, 1972).

So far we have shown that several coronal phenomena, including the inhomogeneous coronal density distribution, flare-emitted MHD fast-mode disturbances, and certain radio emitting sources, appear to be guided or influenced by the quasi-static magnetic field of the solar corona as determined by the current-free approximation. Thus theory and observation are beginning to find some common ground in the study of coronal activity, at least on the coarse scale. However, the coronal disturbances we have examined so far do not obviously alter the coronal field. They probably involve electric currents formed from kinks or twists in the coronal potential field. Now let's look at disturbances which are associated with major changes in the photospheric and coronal magnetic fields.

5.2. Eruption of Photospheric Electric Currents

With global potential field maps the time resolution is poor. We can only see the magnetic configuration before and after a disturbance with 28 days in between. The most violent event on the Sun is a proton flare. Some years ago, Valdez and Altschuler (1970) found that after proton flares the surrounding coronal magnetic field seems to decrease in flux and to change from a closed-loop (arcade) structure to an open or diverging field. At that time, we had only (the type 1) maps which plot the general coronal field but do not distinguish strong from weak fields, and (the type 2) maps which give the major flux connections. Now we have a microfilm atlas of the coronal field for the period 1959–1970 which contains maps of both the general field (type 1) and the strong field (type 3) (Newkirk *et al.*, 1972). So in preparing this talk I thought it would be worthwhile to look again at the problem. Figures 6 through 11 show the changes of the global coronal field associated with proton flares. At the top are the strong field maps; at the bottom are the general field maps. As a rule, the magnetic field changes drastically in strength and geometry around the flare region. Low magnetic arcades disappear, or decrease significantly in field strength. Since the low magnetic arcades seen in the strong field maps are caused by strong electric currents flowing in the underlying photosphere, it appears that photospheric electric currents disappear or disintegrate at the time of large flares or shortly thereafter. There are only a few ways that this can be done. The currents can disappear by some very efficient magnetic diffusion process; they can disperse if the electric current expands in cross-sectional area, or branches into many small filaments; they can be pulled below the photosphere, or they can be ejected out of the photosphere into the corona. For changes in the time scale of one solar rotation or less, I am willing to wager that strong and extensive photospheric electric currents can disappear so completely only by being ejected upward from the photosphere. Of course, I do not mean that the electric current must be ejected all at once. It could rise gradually, interact with the chromosphere in some complicated way, and be ejected bit by bit. But somehow strong electric currents do disappear rapidly over rather extensive photospheric regions.

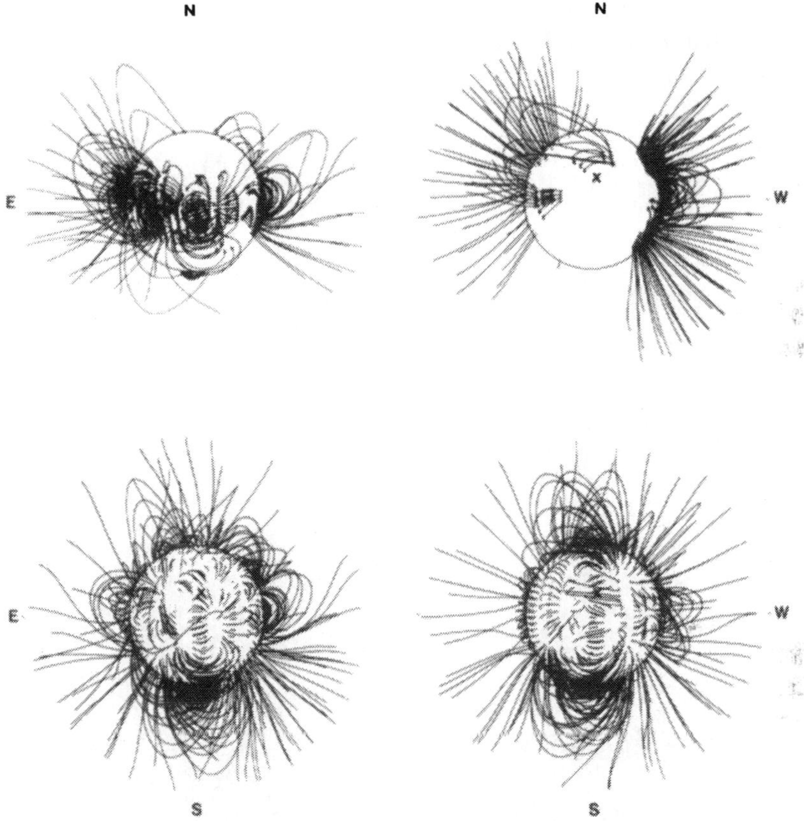

Fig. 6. Changes in calculated coronal magnetic field before and after flare of 29 April 1960. Strong field maps are at top; general field maps are below.

Do we have other evidence that photospheric or chromospheric electric currents are being ejected into the corona? I think we do, and I will try to argue the case. In doing so, I will discuss observations of some of the coronal disturbances listed in Section 3.

If well-defined or localized electric currents are ejected into the corona, we would expect that the accompanying plasma is either hot and dense because of the current pinch effect, or in violent motion because of unbalanced $\mathbf{J} \times \mathbf{B}$ forces.

The hottest and densest plasma regions in the corona are associated with the X-ray filaments or emission cores. Temperatures in such filaments have been put at upwards of 10^7 K. Estimates for the electron density range from 10^{11} to 10^{14} cm^{-3} depending on the assumed volume of the emitting region. Neupert (1971) finds 10^{13} cm^{-3} is sometimes possible. Such dense hot filaments may occur 5 to 50 Mm above the photosphere according to the X-ray pictures of Vaiana and Giacconi (1969) and Krieger *et al.* (1971). During the March 1970 eclipse, Thomas and Neupert (1971) observed that the de-occultation of X-ray emitting regions by the Moon's limb occurred in 0.3 s, corresponding to about 400 km on the Sun. Neupert (1971) believes that the X-ray filaments could be as thin as 16 km.

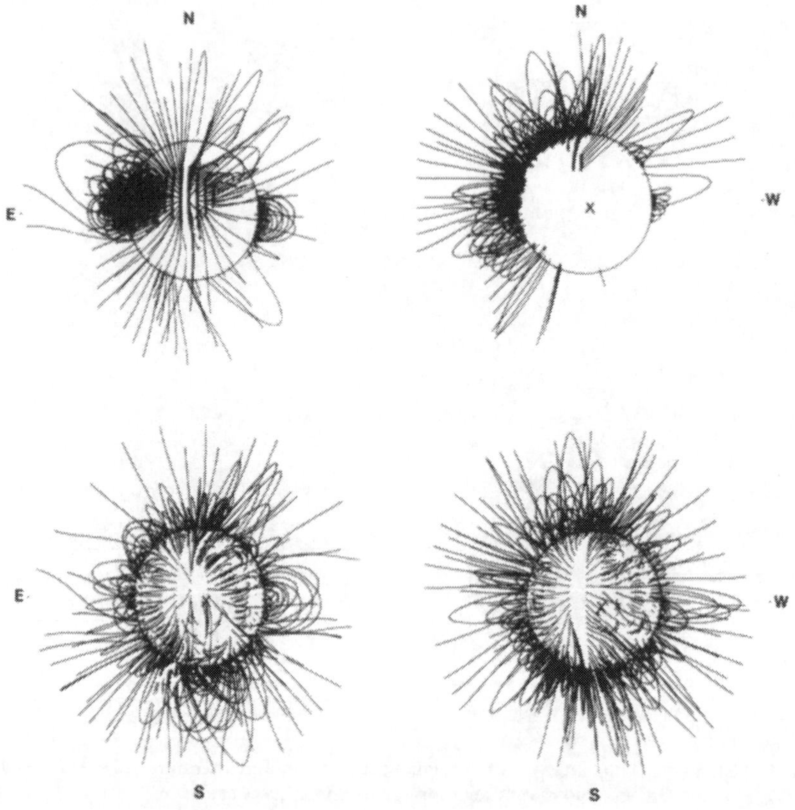

Fig. 7. Changes in calculated coronal magnetic field before and after flares of 4, 6 May 1960. Strong field maps are at top; general field maps are below.

To contain a plasma of 10^{13} cm^{-3} at 10^7 K in the tenuous solar corona requires magnetic fields of about 500 G. If the plasma is held together in a cylinder 100 km in radius, the electric current is about 3×10^{10} A and the current density is about 1 A m^{-2}. Such conditions of temperature, density, and magnetic field permit nuclear reactions to occur. If the electron density were wrong by a factor of 100, the magnetic field and electric current would be wrong only by a factor of 10. In any case, a hot dense plasma cannot be contained for several minutes as a thin filament unless

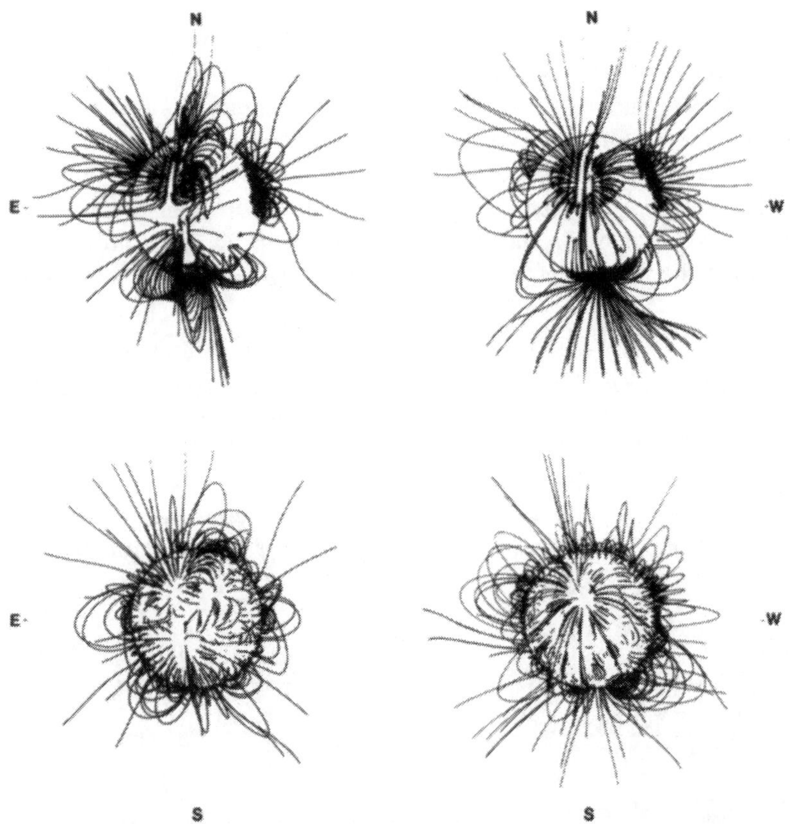

Fig. 8. Changes in calculated coronal magnetic field before and after flare of 24 March 1966. Strong field maps are at top; general field maps are below.

there is a strong electric current along the filament axis. Where could such an electric current originate?

Neupert *et al.* (1974) observed an X-ray loop associated with an importance 1B flare, and found no detectable coronal feature at the flare site before the event. Thus pre-existing coronal material cannot account for the X-ray emission. They found two distinct structures in the X-ray emission: a cooler region (2×10^6–10^7 K) which formed over the Hα flare above the neutral line, and a hot (3×10^7 K) arch about 35 Mm above the Hα flare. The high temperature arch appeared to be more stable in position and lasted about 6 min. They conclude that ionization and heating of chromospheric material must have occurred as the matter moved upward, and that the magnetic field lines must have been closed to provide thermal insulation. Clearly, X-ray flare observations are crucial if we are to observe coronal plasma at temperatures above 5×10^6 K.

Let us now look at a few of the events associated with large flares and prominences that might indicate the ejection into the corona of large electric currents from the photosphere or chromosphere. Most of the flare observations have been made in Hα, a line which shows only the cooler parts of the flare (10^4–10^5 K). Thus in Hα, we would not expect to see hot X-ray emitting plasma but rather mass motions from $\mathbf{J} \times \mathbf{B}$ forces. A few years ago, a special Nobel Symposium was held to discuss the observations of mass motions in solar flares (Öhman, 1968).

Different parts of the flare process are observed on the disk and above the limb (Smith and Smith, 1963; Švestka, 1969). The Hα disk flare primarily shows enhanced densities of active filaments in the chromosphere with little mass motion, while the limb flare shows the more tenuous matter ejected into the corona with large mass motions.

Traditionally, flares have been classified in terms of Hα brightenings (or emission regions) on the solar disk. The brightenings occur first at a number of small points in an active region and then spread along filamentary structures which are at or near the neutral line (which separates photospheric regions of opposite magnetic polarity). According to Severny (1969), the flare brightenings occur over regions where the photospheric electric current has either a large radial component or a large component in the photospheric surface. Thus the activation of a flare filament may corre-

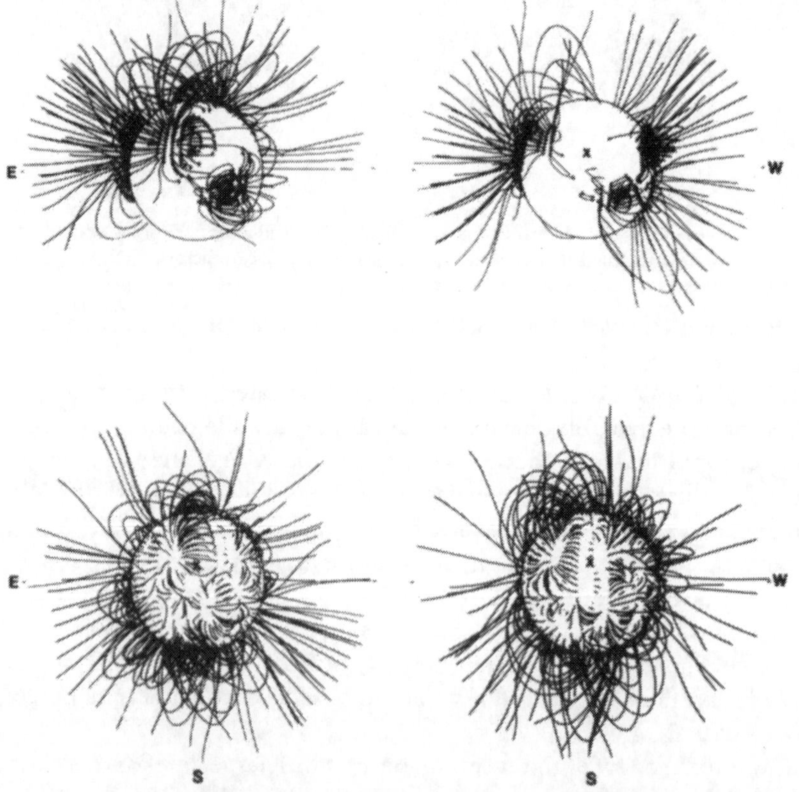

Fig. 9. Changes in calculated coronal magnetic field before and after flare of 18 November 1968. Strong field maps are at top; general field maps are below.

spond to the eruption of an electric current from the photosphere. The dynamical effects that would accompany the rapid elevation of a strong photospheric electric current are complex and violent. Certainly shock waves, adiabatic compression, ionization, and rapid mass motions can be expected. In any case, it is clear that at least chromospheric electric currents are involved in the flare process. The large

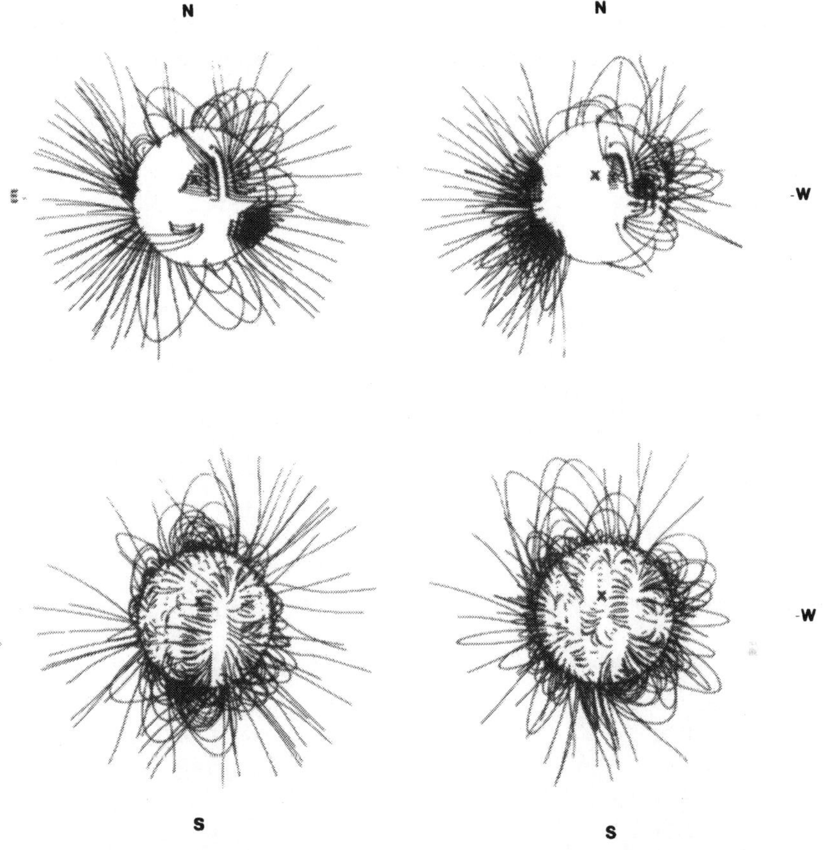

Fig. 10. Changes in calculated coronal magnetic field before and after flare of 29 March 1970. Strong field maps are at top; general field maps are below.

August 1972 flares occurred along a neutral-line filament which was located in a region of strong shear flow (Zirin and Tanaka, 1973). Considerable twists in the smaller filaments, hence presumably electric currents, were also observed.

In addition to the active-region filaments, there are also occasional ejecta or surges into the corona observed in Hα (Macris, 1971). By looking off-center in the Hα line during a disk flare, we can often see evidence for material ejected upward into the corona. Limb observations show flare-associated surges and sprays which correspond to the ejection of matter at about the Alfvén speed. Surges originate in the low chromosphere or photosphere and seem to be associated with small regions of opposite magnetic polarity to that of the nearby sunspot or active surroundings (Rust, 1968; Roy, 1973). Now one of the simplest ways to create antiparallel electric currents with $\mathbf{J} \times \mathbf{B}$ forces directed upward is to take a straight photospheric electric current and create an almost circular bend or meander at some point in the current. This is equivalent to the intrusion of opposite magnetic polarity into a larger unipolar region. The forces on such a bent electric current can be directed upward (Altschuler *et al.*, 1968; Piddington, 1972). I suspect that a surge is the visible manifestation in Hα of the ejection of a ring of electric current and its accompanying

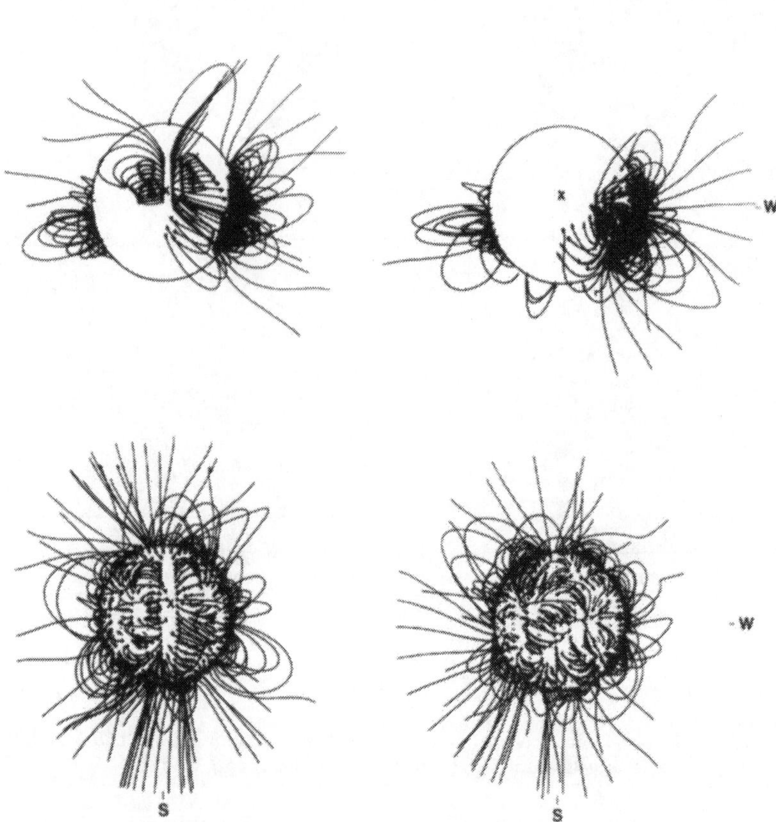

Fig. 11. Changes in calculated coronal magnetic field before and after flares of 2, 4, 7 August 1972. Strong field maps are at top; general field maps are below.

plasma into the corona. What we see in Hα is the mass that recombines and leaks out of the current ring.

The most energetic coronal disturbances are undoubtedly the flare loops which are generally associated with large two-ribbon (or proton) flares (Bruzek, 1964). At about flare maximum, coronal loops appear which connect the two chromospheric flare ribbons on the opposite sides of the neutral line. Seen on the limb in Hα light, loops appear above previous ones as downstreaming matter flows from Hα knots formed at increasing heights. Thus the loop system grows, reaching heights of 60 Mm or more. Wide Hα profiles indicate internal motions of the order of 1 Mm s^{-1} characteristic of Alfvén speeds. Flare loops are also seen in coronal lines of highly ionized metals, indicating very high temperatures. The most interesting observation is that the high temperature coronal yellow line (λ569.4 nm) of Ca XV is characteristically seen in emission at the top of these loops (Billings, 1966). Moreover, several coronal emission lines of highly ionized metals are all simultaneously enhanced in flare loops, indicating that the coronal density is also significantly enhanced. Billings writes "Thus we are confronted with the paradox that the type of prominence that appears to deplete matter from the corona most vigorously is the one in whose vicinity the corona remains most dense. This paradox strongly indicates that the source of descending material in the prominence is not the corona but the same source that enhances the coronal density." Kleczek (1964) writes "A powerful mechanism must exist for transporting material from lower, denser atmospheric layers during and after some flares."

The mass required for a flare loop system is about 10^{16} gm (Jefferies and Orrall, 1964; Kleczek, 1964) which at 50 Mm height is more than the mass of the surrounding corona. To obtain such a mass we can (1) condense coronal matter over a large volume using magnetic compression, (2) eject the mass from the chromosphere in the form of high energy particles (either at once or continuously) and then allow recombination processes to occur in a small volume at the top of the loop, or (3) carry up hot dense plasma confined in a filament of electric current. Here I will not argue the pros and cons of the different theories, but merely talk about the last. After seeing the results of Neupert *et al.* (1973) on the X-ray flare loop, I have become less cautious about suggesting the eruption from below of electric current filaments which contain hot dense plasma.

A current filament provides the proper magnetic forces to contain and elevate hot dense plasma. As the current filament rises, mass motions induced by the hydromagnetic forces will disperse the twisted magnetic flux into a larger volume. Eventually the electric current of the filament will become too weak either to contain the hot dense plasma or to supply sufficient energy to maintain ionization. As a result, the current filament will dissipate, and the dense plasma will cool by free-free emission, recombine, and flow down along the pre-existing potential field lines, creating the flare loops. The impact of this downfalling material will produce flare ribbons in the chromosphere (see for example, Hyder, 1967).

To lift 10^{15} to 10^{16} gm of ionized plasma to a height of 60 Mm in the corona and then to contain it for some minutes or longer, we could use an electric current of 3×10^{11} A flowing through a filament 1 Mm in radius and about 60 Mm in length. This situation corresponds to a twisted magnetic field of 500 G, a hydrogen density of about 10^{13} cm^{-3}, and a temperature of 10^7 K. If we assume a filament 10 Mm in radius, we require only 50 G and 10^{11} cm^{-3} for the same temperature, mass, and electric current.

Now let us briefly discuss other evidence for the ejection of electric currents into the corona.

The eruption or disintegration of large quiescent prominences is clearly associated with twisted magnetic fields even on a macroscopic scale (Valnicek, 1968; Dodson *et al.*, 1972). Beautiful cases of twisting or untwisting erupting prominences have appeared in the literature (Figure 12).

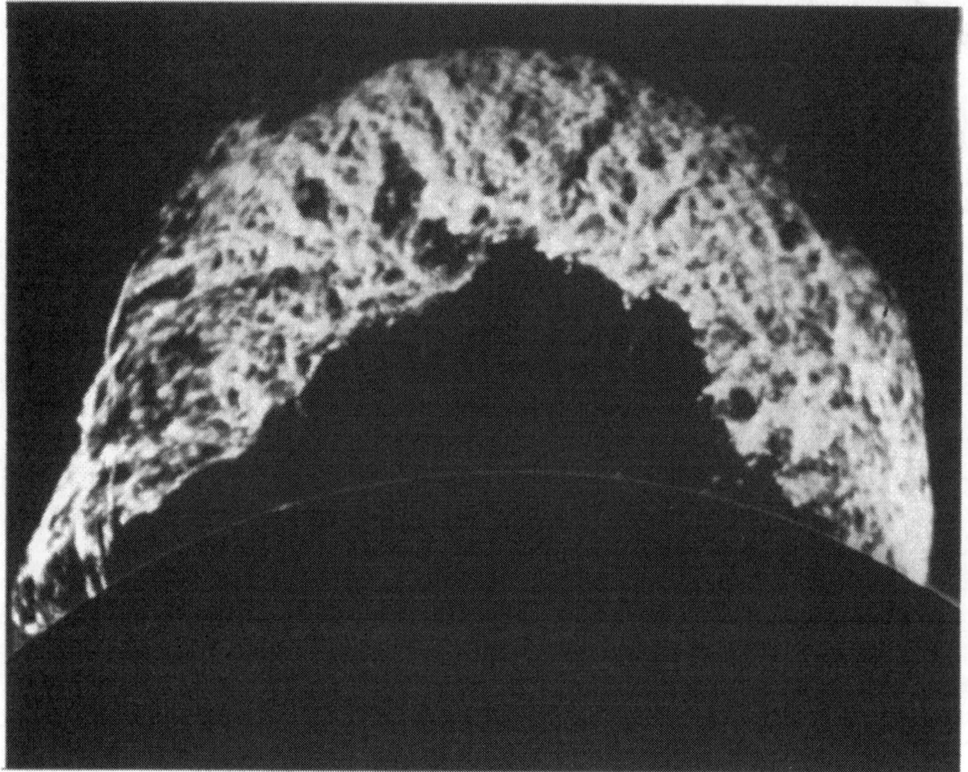

Fig. 12. Famous eruptive prominence of 4 June 1946, photographed by W. O. Roberts at Climax Observatory.

Monochromatic movies of the corona in the green line (λ530.3 nm) show changes at infrequent intervals (Bruzek and Demastus, 1970; Dunn, 1971). Expanding arches are the changes seen most often. Occasionally the whiplike opening of an arch is seen when one of the footpoints appears to disconnect from the photosphere and rise into the corona. Of course, magnetic field lines are solenoidal and cannot be disconnected. What we are seeing, therefore, is an ascending electric current or perhaps current ring, with only one part of the system emitting at the temperature of the coronal green line.

In the radio spectrum, I think a good case can be made that the moving type IV synchrotron-emitting sources contain twisted magnetic fields and therefore electric currents (Riddle, 1970; Smerd and Dulk, 1971; Dulk and Altschuler, 1971).

Well, I have tried to argue the case that X-ray emitting filaments, the flare loop-prominence system, surges, green-line whips, erupting quiescent prominences, and moving type IV radio bursts are associated with the eruption of electric currents from the chromosphere or the photosphere. Certainly these phenomena are not simple perturbations of the potential magnetic field configuration of the corona.

6. Impulsive Coronal Disturbances and Coronal Magnetic Fields

Let me say a few words about the rapid particle acceleration processes indicated by impulsive X-ray and type III radio bursts. In the tenuous coronal plasma, the Hall Effect becomes important for scale sizes of about 100 m. This means that the magnetic field becomes frozen to the electron plasma component rather than to the fluid as a whole (Pikelner, 1966). Since electrons have little inertia, it is likely that bunches of electrons on the 100 m scale can be accelerated in the corona whenever strong non-potential magnetic fields are present (Altschuler *et al.*, 1973). Undoubtedly, X-ray filaments and flare loops provide such fields.

7. Speculations and Conclusion

Now I will enter further into the realm of pure speculation. Spicules have been called mini-surges. They occur at the boundaries of supergranule cells where the magnetic field is enhanced. Often spiral motions can be seen in spicules (see Öhman, 1968) indicating the presence of twisted magnetic fields or electric currents. Spicules apparently play a key role in the heating and mass balance of the corona. On the basis of present observations (Beckers, 1968), it is not necessary to assume hydromagnetic forces to explain the eruption of spicules (Kuperus and Athay, 1967). Nevertheless, there would be a remarkably simple conceptual scheme if such were the case. We could then say that the entire solar corona is merely a manifestation of electric currents of different sizes and shapes being ejected continually from the chromosphere and photosphere. The 'quiet' corona would then be merely a composite of small coronal disturbances in the form of unresolved filaments of electric current. Perhaps confirming this wild idea is the evidence that coronal holes seem to lie above unipolar photospheric regions, away from the neutral lines and the small magnetic arcades which suggest photospheric electric currents (Altschuler *et al.*, 1972). It would be interesting if the spicule density or intensity were smaller under coronal holes.

Let me now carry this speculation to its dire conclusion. If electric currents are being continually ejected from the photosphere, where are they created? We have already mentioned that low magnetic arcades in the strong potential field maps indicate large photospheric electric currents occur near the neutral line, and that a network of concentrated electric currents may well meander all over the photosphere. The recent observation of Howard (1971) that the velocity field of the photosphere near the neutral line is predominantly downward suggests that neutral lines occur at the boundaries of very large convective cells perhaps 300 Mm in diameter

(Bumba, 1967; Simon and Weiss, 1968; Yoshimura, 1971; Wilson, 1972). If so, then these boundaries would contain gradients of pressure, density, and temperature, not necessarily parallel, thus nonpotential forces which could generate electric currents. This was suggested by Kopecky and Kuklin (1971) at the Paris meeting.

The grand vision that I think may soon evolve from the observations is the following. There are large-scale convective cells on the Sun, say 300 Mm in diameter. At the boundaries of these cells, electric currents are being generated by non-potential forces. Smaller convective cells, the supergranules, might also generate electric current at their boundaries. As all of these electric currents build up in the photosphere, they are continually being ejected in the form of spicules. If large electric currents build up faster than they can be ejected by spicules, then they are ejected as eruptive prominences, surges, or large flares. Surprisingly, the energy of a large flare (10^{31}–10^{32} erg) exceeds the kinetic energy of a photospheric convective cell with a density of 10^{17} cm^{-3}, a diameter of 300 Mm, a depth of 10 Mm, and a mean flow speed of 0.1 km s^{-1}. Thus large flares might disrupt or alter the large-scale circulation of the photosphere at the cell boundaries where electric current is generated. The solar cycle then becomes similar to the conflicting two-population (fox and rabbit) problem of Volterra (Davis, 1962). First the number of current-generating regions begins to increase in the photosphere. Then the number of solar active regions begins to grow, eventually disrupting the current-generating regions. Then the Sun goes quiet until new large-scale cells start generating electric current again.

I began this talk by emphasizing the incredible complexity of the coronal plasma, and have ended with the oversimplified picture that the entire mass of the corona derives from fountains and geysers that eject electric current from the photosphere and chromosphere. But actually, until we solve the appropriate system of differential equations, we cannot determine to what extent the ejected electric currents carry entrapped matter and/or produce waves and shocks. Nevertheless, even a simplified picture, if it is correct, can be useful in both theory and observation. The interesting physics of course lies somewhere between the over-simplified and the hopelessly complex.

Acknowledgements

I am deeply indebted to many of my colleagues, in particular to R. G. Athay, R. Dunn, B. Durney, J. C. Henoux, R. Howard, A. Krieger, Y. Nakagawa, W. Neupert, F. Q. Orrall, L. Oster, C. Querfeld, I. Roxburgh, D. F. Smith, E. Tandberg-Hanssen, Y. Uchida, and S. T. Wu, for criticisms, suggestions, references, unpublished manuscripts, and long and lively discussions. The figures for this paper were assembled with the help of D. E. Trotter.

Notes added in proof. (1) After studying very high resolution KPNO magnetographs, J. Harvey makes the point that the large 'unipolar' magnetic region is actually a statistical result of almost-equal numbers of small-scale magnetic elements of opposite polarity. Thus, the unipolar regions, neutral lines, and global magnetic flux connections discussed in this paper should be considered large-area averages.

(2) R. G. Athay points out that S. B. Pikelner (1969) provides convincing arguments against a spicule mechanism based on (1) heat flux from the corona and (2) a passive or static magnetic field. The consensus now seems to be that some form of dynamical magnetic field is essential to the spicule mechanism.

References

Altschuler, M. D.: 1971, *Sky Telesc.* **41**, 146.
Altschuler, M. D., Lilliequist, C. G., and Nakagawa, Y.: 1968, *Solar Phys.* **5**, 366.
Altschuler, M. D. and Newkirk, G., Jr.: 1969, *Solar Phys.* **9**, 131.
Altschuler, M. D. and Perry, R. M.: 1972, *Solar Phys.* **23**, 410.
Altschuler, M. D., Smith, D. F., Swarztrauber, P. N., and Priest, E. R.: 1973, *Solar Phys.* **32**, 153.
Altschuler, M. D., Trotter, D. E., and Orrall, F. Q.: 1972, *Solar Phys.* **26**, 354.
Anderson, G. E.: 1966, Ph.D. Thesis, Univ. of Colorado.
Anzer, U.: 1968, *Solar Phys.* **3**, 298.
Anzer, U. and Tandberg-Hanssen, E.: 1970, *Solar Phys.* **11**, 61.
Beckers, J. M.: 1968, *Solar Phys.* **3**, 367.
Beckers, J. M.: 1971, in R. Howard (ed.), 'Solar Magnetic Fields', *IAU Symp.* **43**, p. 3.
Beckers, J. M. and Wagner, W. J.: 1971, *Solar Phys.* **21**, 439.
Billings, D. E.: 1966, *A Guide to the Solar Corona, Academic Press*, N.Y.
Brown, A.: 1958, *Astrophys. J.* **128**, 646.
Bruzek, A.: 1964, *Astrophys. J.* **140**, 746.
Bruzek, A. and Demastus, H. L.: 1970, *Solar Phys.* **12**, 447.
Bumba, V.: 1967, in P. A. Sturrock (ed.), *Plasma Astrophysics*, Academic Press, p. 77.
Charvin, P.: 1965, *Ann. Astrophys.* **28**, 877.
Charvin, P.: 1971, in R. Howard (ed.), 'Solar Magnetic Fields', *IAU Symp.* **43**, p. 580.
Cheng, C. C., Phillips, K. J. H., and Wilson, A. M.: 1973, *Solar Phys.* **29**, 383.
Coppi, B. and Friedland, A. B.: 1971, *Astrophys. J.* **169**, 379.
Daigne, G., Lantos-Jarry, M. F., and Pick, M.: 1971, in R. Howard (ed.), 'Solar Magnetic Fields', *IAU Symp.* **43**, p. 609.
Davis, H. T.: 1962, *Introduction to Nonlinear Differential and Integral Equations*, Dover Publ.
Dodson, H. W., Hedeman, E. R., and Rovira de Miceli, M.: 1972, *Solar Phys.* **23**, 360.
Dulk, G. A. and Altschuler, M. D.: 1971, *Solar Phys.* **20**, 438.
Dulk, G. A., Altschuler, M. D., and Smerd, S. F.: 1971, *Astrophys. Letters* **8**, 235.
Dunn, R.: 1960, Ph.D. Thesis, Harvard Univ.
Dunn, R. B.: 1971, in C. J. Macris (ed.), *Physics of the Solar Corona*, D. Reidel Publ., p. 114.
Eddy, J. A., Lee, R. H., and Emerson, J. P.: 1973, *Solar Phys.* **30**, 351.
Eddy, J. A. and Malville, J. M.: 1967, *Astrophys. J.* **150**, 289.
Engvold, O.: 1972, *Solar Phys.* **23**, 346.
Foukal, H. and Zirin, H.: 1972, *Solar Phys.* **26**, 148.
Frazier, E. N.: 1972a, *Solar Phys.* **24**, 98.
Frazier, E. N.: 1972b, *Solar Phys.* **26**, 142.
Frazier, E. N. and Stenflo, J. O.: 1972, *Solar Phys.* **27**, 330.
Gold, T.: 1964, in W. N. Hess (ed.), *The Physics of Solar Flares*, NASA SP-50, p. 389.
Gold, T.: 1968, in Y. Öhman (ed.), 'Mass Motions in Solar Flares and Related Phenomena', *Nobel Symp.* **9**, p. 205.
Harvey, J. W.: 1969, Ph.D. Thesis, Univ. of Colorado.
Harvey, J. W.; 1972, in P. McIntosh and M. Dryer (eds.), *Solar Activity Observations and Predictions*, MIT Press, p. 51.
Harvey, K. L., Livingston, W. C., Harvey, J. W., and Slaughter, C. D.: 1971, in R. Howard (ed.), 'Solar Magnetic Fields', *IAU Symp.* **43**, p. 422.
Hildner, E.: 1971, Ph.D. Thesis, Univ. of Colorado.
House, L. L.: 1972, *Solar Phys.* **23**, 103.
Howard, R.: 1971, *Solar Phys.* **16**, 21.

Howard, R., Bumba, V., and Smith, S. F.: 1967, *Atlas of Solar Magnetic Fields*, Carnegie Inst. of Washington, Publication 626.
Howard, R. and Harvey, J. W.: 1964, *Astrophys. J.* **139**, 1328.
Howard, R. and Stenflo, J. O.: 1972, *Solar Phys.* **22**, 402.
Hyder, C. L.: 1965, *Astrophys. J.* **141**, 1382.
Hyder, C. L.: 1966, in *Atti del Convegno sui Campi Magnetici Solari*, Rome Obs., p. 110.
Hyder, C. L.: 1967, *Solar Phys.* **2**, 49.
Hyder, C. L., Mauter, H. A., and Shutt, R. L.: 1968, *Astrophys. J.* **154**, 1039.
Jefferies, J. T. and Orrall, F. Q.: 1964, in W. N. Hess (ed.), *The Physics of Solar Flares*, NASA SP-50, p. 71.
Kippenhahn, R. and Schluter, A.: 1957, *Z. Astrophys.* **43**, 36.
Kleczek, J.: 1964, in W. N. Hess (ed.), *The Physics of Solar Flares*, NASA SP-50, p. 77.
Kopecky, M. and Kuklin, G. V.: 1971, in R. Howard (ed.), 'Solar Magnetic Fields', *IAU Symp.* **43**, p. 534.
Krieger, A. S., Vaiana, G. S., and Van Speybroeck, L. P.: 1971, in R. Howard (ed.), 'Physics of the Solar Corona', *IAU Symp.* **43**, p. 397.
Kuiper, T. B. H.: 1973, *Solar Phys.* **33**, 461.
Kundu, M. R.: 1971, in R. Howard (ed.), 'Solar Magnetic Fields', *IAU Symp.* **43**, p. 642.
Kuperus, M. and Athay, R. G.: 1967, *Solar Phys.* **1**, 361.
Levine, R. and Nakagawa, Y.: 1974, *Astrophys. J.*, in press.
Macris, C. J.: 1971, in C. J. Macris (ed.), *Physics of the Solar Corona*, D. Reidel Publ., p. 168.
McLean, D. J.: 1967, *Proc. Astron. Soc. Australia* **1**, 47.
Meyer, F.: 1968, in K. O. Kiepenheuer (ed.), 'Structure and Development of Solar Active Regions', *IAU Symp.* **35**, p. 485.
Meyer, F. and Schmidt, H. U.: 1968, *Mitt. Astron. Ges.* **25**, 194.
Moreton, G. E. and Ramsey, H. E.: 1960, *Publ. Astron. Soc. Pacific* **72**, 357.
Nakagawa, Y. and Malville, J. M.: 1969, *Solar Phys.* **9**, 102.
Nakagawa, Y. and Raadu, M. A.: 1972, *Solar Phys.* **25**, 127.
Neupert, W. M.: 1971, in C. J. Macris (ed.), *Physics of the Solar Corona*, D. Reidel Publ. Co., Dordrecht, p. 237.
Neupert, W. M., Thomas, R. J., and Chapman, R. D.: 1974, *Solar Phys.* **34**, 349.
Newkirk, G., Jr.: 1971, in C. J. Macris (ed.), *Physics of the Solar Corona*, D. Reidel Publ. Co., Dordrecht, p. 66.
Newkirk, G., Jr. and Altschuler, M. D.: 1970, *Solar Phys.* **13**, 131.
Newkirk, G., Jr., Altschuler, M. D., and Harvey, J.: 1968, in K. O. Kiepenheuer (ed.), 'Structure and Development of Solar Active Regions', *IAU Symp.* **35**, p. 379.
Newkirk, G., Jr., Trotter, D. E., Altschuler, M. D., and Howard, R.: 1972, *A Microfilm Atlas of Magnetic Fields in the Solar Corona*, NCAR-TN/STR-85.
Öhman, Y. (ed.): 1968, 'Mass Motions in Solar Flares and Related Phenomena', *Nobel Symp.* **9**.
Orrall, F. Q.: 1971, in R. Howard (ed.), 'Solar Magnetic Fields', *IAU Symp.* **43**, p. 30.
Parker, E. N.: 1955, *Astrophys. J.* **121**, 491.
Perry, R. M. and Altschuler, M. D.: 1973, *Solar Phys.* **28**, 435.
Petschek, H. E.: 1964, in W. N. Hess (ed.), *The Physics of Solar Flares*, NASA SP-50, p. 425.
Piddington, J. H.: 1972, *Solar Phys.* **27**, 402.
Pikelner, S. B.: 1966, *Fundamentals of Cosmic Electrodynamics* (2nd ed.), Nauka, Moscow.
Pikelner, S. B.: 1969, *Astron. Zh.* **46**, 328 (*Soviet Astron.* **13**, 259).
Pneuman, G. W.: 1967, *Solar Phys.* **2**, 462.
Pneuman, G. W. and Kopp, R. A.: 1971, *Solar Phys.* **18**, 258.
Querfeld, C. W.: 1974 in T. Gehrels (ed.), *Planets, Stars, and Nebulae Studied with Photopolarimetry*, Univ. of Ariz. Press, p. 264.
Raadu, M. A. and Kuperus, M.: 1973, *Solar Phys.* **28**, 77.
Rayrole, J. and Semel, M.: 1968, in K. O. Kiepenheuer (ed.), 'Structure and Development of Solar Active Regions', *IAU Symp.* **35**, p. 134.
Riddle, A. C.: 1970, *Solar Phys.* **13**, 448.
Roy, J. R.: 1972, *Solar Phys.* **26**, 418.
Roy, J. R.: 1973, *Solar Phys.* **28**, 95.
Rust, D. M.: 1966, Ph.D. Thesis, Univ. of Colorado.
Rust, D. M.: 1967, *Astrophys. J.* **150**, 313.
Rust, D. M.: 1968, in K. O. Kiepenheuer (ed.), 'Structure and Development of Solar Active Regions', *IAU Symp.* **35**, p. 77.

Rust, D. M.: 1970, *Astrophys. J.* **160**, 315.
Rust, D. M. and Roy, J. R.: 1971, in R. Howard (ed.), 'Physics of the Solar Corona', *IAU Symp.* **43**, p. 569.
Schatten, K. H.: 1970, *Solar Phys.* **12**, 484.
Schatten, K. H.: 1971a, *Cosmic Electrodyn.* **2**, 232.
Schatten, K. H.: 1971b, in R. Howard (ed.), 'Solar Magnetic Fields', *IAU Symp.* **43**, p. 595.
Schatten, K. H., Wilcox, J. M., and Ness, N. F.: 1969, *Solar Phys.* **6**, 442.
Scherrer, P. H., Wilcox, J. M., and Howard, R.: 1972, *Solar Phys.* **22**, 418.
Schmidt, H. U.: 1964, in W. N. Hess (ed.), *The Physics of Solar Flares*, NASA SP-50, p. 107.
Severny, A. B.: 1969, in C. de Jager and Z. Švestka (eds.), 'Solar Flares and Space Research', *COSPAR Symp.*, p. 38.
Simon, G. W. and Weiss, N. O.: 1968, in K. O. Kiepenheuer (ed.), 'Structure and Development of Solar Active Regions', *IAU Symp.* **35**, p. 108.
Smerd, S. F. and Dulk, G. A.: 1971, in R. Howard (ed.), 'Solar Magnetic Fields', *IAU Symp.* **43**, p. 616.
Smith, S. F. and Harvey, K. L.: 1971, in C. J. Macris (ed.), *Physics of the Solar Corona*, D. Reidel Publ., p. 156.
Smith, H. and Smith, E. v. P.: 1963, *Solar Flares*, Macmillan Co.
Stenflo, J. O.: 1969, *Solar Phys.* **8**, 115.
Stenflo, J. O.: 1971, in R. Howard (ed.), 'Solar Magnetic Fields', *IAU Symp.* **43**, p. 101.
Sturrock, P. A.: 1968, in K. O. Kiepenheuer (ed.), 'Structure and Development of Solar Active Regions', *IAU Symp.* **35**, p. 471.
Sturrock, P. A. and Coppi, B.: 1966, *Astrophys. J.* **143**, 3.
Švestka, Z.: 1969, in C. de Jager and Z. Švestka (eds.), 'Solar Flares and Space Research', *COSPAR Symp.*, 16.
Sweet, P. A.: 1958, in B. Lehnert, 'Electromagnetic Phenomena in Cosmical Physics', *IAU Symp.* **6**, p. 123.
Tandberg-Hanssen, E.: 1971, in R. Howard (ed.), 'Solar Magnetic Fields', *IAU Symp.* **43**, p. 192.
Tandberg-Hanssen, E. (ed.): 1974, *Solar Prominences*, D. Reidel Publ. Co., Dordrecht-Holland.
Thomas, R. J. and Neupert, W. M.: 1971, *Bull. Am. Astron. Soc.* **3**, p. 7.
Uchida, Y.: 1968, *Solar Phys.* **4**, 30.
Uchida, Y.: 1970, *Publ. Astron. Soc. Japan* **22**, 341.
Uchida, Y., Altschuler, M. D., and Newkirk, G., Jr.: 1973, *Solar Phys.* **28**, 495.
Vaiana, G. S. and Giacconi, R.: 1969, in D. G. Wentzel and D. A. Tidman (eds.), *Plasma Instabilities in Astrophysics*, Gordon and Breach, Publ., p. 91.
Valdez, J. and Altschuler, M. D.: 1970, *Solar Phys.* **15**, 446.
Valnicek, B.: 1968, in K. O. Kiepenheuer (ed.), 'Structure and Development of Solar Active Regions', *IAU Symp.* **35**, p. 282.
Veeder, G. J. and Zirin, H.: 1970, *Solar Phys.* **12**, 391.
Wilcox, J. M. and Ness, N. F.: 1965, *J. Geophys. Res.* **70**, 5793.
Wild, J. P.: 1970, *Proc. Astron. Soc. Australia* **1**, 365.
Wilson, P. R.: 1972, *Proc. Astron. Soc. Australia* **2**, 144.
Yoshimura, H.: 1971, *Solar Phys.* **18**, 417.
Zirin, H.: 1972, *Solar Phys.* **26**, 145.
Zirin, H. and Tanaka, K.: 1973, *Solar Phys.* **32**, 173.

DISCUSSION

Sturrock: The magnetic field representation is simpler for the study of coronal disturbances because of the 'frozen-in' plasma condition. The same does not apply to the current representation.

Altschuler: Correct; I used the electric-current representation here to illustrate *where* in the magnetic-field structure the disturbances occur.

Sturrock: Impulsive disturbances need coronal currents over small length scales (~ 1 km); what is the mechanism?

Altschuler: Small scale is certainly important (see Section 6).

Stewart: K-corona transients occur at times of flare loops; do these occur as two distinct disturbances?

Altschuler: The most difficult problem is to explain how a large mass of solar plasma is carried high into the corona. It is here suggested that this is done by many small, hot filaments (currents). Once up, fragmentation may occur in many ways resulting in many different plasma processes.

MAGNETIC STRUCTURE RESPONSIBLE FOR CORONAL DISTURBANCES: THEORY

G. W. PNEUMAN

High Altitude Observatory, National Center for Atmospheric Research, Boulder, Colo., U.S.A.*

Abstract. Due to the strong coupling between the coronal material and magnetic fields, magnetic structures are probably intimately involved, either actively or passively, in virtually all coronal disturbances. The purpose of this paper is to put into perspective the various roles magnetic fields can play in both exciting and guiding these observed transient phenomena. A discussion of our present theoretical concepts relevant to this subject can be conveniently divided into four categories.

We begin by discussing our present understanding of how the gross magnetic structure of the corona is determined. Important considerations here are the tendency for the coronal field to seek its lowest energy state, the effect of convection on the field, and the influence of the solar wind. Secondly, we investigate magnetic structures which reside in elevated energy states (higher than the energy of an equivalent potential field) as well as disturbances which appear to be related to changes in these configurations. Thirdly, the role of the field in guiding coronal disturbances is considered. This is evident for bulk motions (sprays, surges, flare loops, green line events, etc.), wave motions (flare associated waves, Alfvén waves), as well as for individual particle phenomena. Lastly, a special class of magnetic structures which seem to be constantly associated with coronal activity are discussed. These are the magnetic discontinuities such as neutral sheets and current sheets. In this context, the magnetic neutral point and associated reconnection phenomena are considered.

1. Introduction

Certainly the most illuminating observation of the solar corona is a white light photograph taken during a total eclipse (see Figure 1). Despite the difficulties in attempting to visually deconvolve line-of-sight effects, one can learn much about the gross magnetic structure and even its effect upon the distribution of coronal material. For example, in the lower corona below about 2 $R_\odot$, the density structure reveals closed magnetic field lines evidently bottling up the coronal gas which, in the absence of the field would rapidly escape into interplanetary space. The magnetic structure in other places appears to be open permitting outward expansion. This drain of material and energy can result in a significant decrease in density there, causing those regions to appear less bright than the closed regions.

In the lowest part of the corona, the magnetic field energy density is at least comparable to the thermal energy density and, in some regions, probably greater. This fact, however, does not insure any interaction between the field and the coronal gas. It is the high electrical conductivity resulting from the high coronal temperature, so effectively coupling the material with the field, which is primary responsible for all coronal hydromagnetic effects. The magnetic Reynolds number appropriate for coronal conditions is of the order of 10^{10}–10^{14} indicating that there can be virtually no motion of material across field lines.** This conclusion is observationally verified

* The National Center for Atmospheric Research is sponsored by the National Science Foundation.

** A magnetic Reynolds number equal to one corresponds to a condition where diffusion across field lines and the 'freezing-in' effect are of roughly equal importance.

Gordon Newkirk, Jr. (ed.), Coronal Disturbances, 35–68.

by the apparently close correspondence between density structure and field structure.

Another important consequence of its high temperature is the high thermal conductivity of the corona. As pointed out by Chapman (1957), thermal conduction is so effective that, if the corona were static, the resulting temperature distribution would be so flat that the coronal temperature would fall by only a factor of about 5 between

Fig. 1. The solar corona as observed during the 12 November 1966 eclipse. The bright points at the bases of the large helmet streamers are quiescent prominences (courtesy G. Newkirk Jr.).

the Sun and the Earth. A temperature profile of this type can, in fact, be shown to be incompatible with a static corona and, consequently, expansion must take place. The result is a solar wind, a now observationally well documented condition of the interplanetary medium. Another interesting aspect of the thermal conductivity is that it is essentially zero for heat conduction across field lines. This results in non-uniform heating effects which, along with the solar wind, stresses the magnetic fields – giving rise to the possibility of impulsive phenomena.

One unfortunate property of an eclipse photograph is that it is at best only a snapshot. Consequently, little information about coronal transients can be obtained from these pictures. Other evidence, however, suggests that the corona is continually in a disturbed state. For example, type III bursts seem to be constantly taking place and observations of the solar wind at 1 AU reveal temporal variations in interplanetary properties on a time-scale of hours. On the other hand, transients observed in the green line, $\lambda 5303$, are quite rare and type II and IV bursts seem to associate only with the largest flares.

Magnetic fields sometime play a passive role in impulsive events such as, for example, the fields overlying the sites of flares and eruptive prominences. Usually, however, the field configuration and its changes are either instrumental in producing an event or play a role in guiding bulk motions and individual particles involved in the disturbance. The multitude of disturbances which respond to changes in magnetic configuration all seem to be most readily understood in terms of a very few basic physical mechanisms. For example, convection and other motions in the deeper layers transport field lines in such a way that potential field configurations are not always attainable. Consequently, the field can reside in higher energy states which often are either metastable or unstable. Also, again due to the high electrical conductivity of the corona, magnetic discontinuities such as neutral sheets, current sheets and neutral points are present. Reconnection of field lines at these sites is becoming an increasingly more attractive explanation for a great variety of impulsive solar phenomena.

2. Tendency for the Coronal Magnetic Field to Reside in Its Lowest Energy State

If the Sun had no corona or a cool non-conducting corona, the coronal magnetic field would, of course, just be a potential field ($\nabla \times \mathbf{B} = \nabla \cdot \mathbf{B} = 0$) and, if the net flux through the Sun were zero, all field lines would be closed. But the convection zone produces a hot corona which, in turn, couples strongly with the field. It is this coupling and the resultant tug-of-war between fluid and magnetic forces that gives the corona its characteristic appearance. All deviations of the coronal field from its corresponding potential configuration can be ultimately traced to convection. Convective motions not only tangle the field lines setting up strong electric currents but also produce the solar wind which completely dominates the field beyond about $2.5\,R_{\odot}$. The appearances of the polar plumes (Bugoslavskaya, 1950; Van de Hulst, 1950), the over-all flattening of the corona toward the equator during solar minimum, and the observational results of Babcock and Babcock (Babcock and Babcock, 1955; H. D. Babcock, 1959; H. W. Babcock, 1961) has, in the past, suggested to solar astronomers that the Sun's magnetic field, when viewed on the largest scale, resembles that of a dipole with its axis oriented roughly along the axis of rotation. Although more recent higher resolution eclipse photographs and magnetograms have shown that the magnetic fields are generally much more complicated than a simple dipole, there is some evidence that the large-scale fields in the inner corona (as inferred from the brightness structure) do appear to resemble those appropriate for a current-free atmosphere. Support for this comes from a comparison of a computed potential field model using Mt. Wilson magnetograph data covering the period during the November, 1966 eclipse* (Altschuler and Newkirk, 1969). The correspondence be-

* Large-scale twisted force-free fields, for example, are not generally observed.

tween the calculated field lines and the general appearance of the corona was quite good. A similar comparison with the corona during the eclipse of March, 1970, however, was not so satisfactory (Newkirk, 1971), however, this could have been due to a higher level of solar activity during this period violating the steady state assumption inherent in the theory. Also, since the corona is generally believed to be hotter during the maximum phase of the solar cycle, the field may have been more non-potential due to the increased importance of gas pressure forces.

The first model to compare magnetic fields calculated by potential theory with observations was constructed by Schmidt (1964). This model employed rectangular coordinates and, consequently, was valid only for structures whose general dimensions were small as compared to a solar radius. Subsequently, more accurate potential field models were developed by Newkirk *et al.* (1968), Schatten *et al.* (1969), and Altschuler and Newkirk (1969). More recently Schatten (1971) has developed a model in which the field current-free everywhere except at discontinuous surfaces (neutral sheets) over which the polarity of the field changes abruptly. Since in Schatten's model the quantity $B^2/8\pi$ is always continuous across these surfaces it is still valid only in the limit of vanishing gas pressure (since, in general, $P+B^2/8\pi$ should be the conserved quantity). However, comparisons of both Newkirk and Altschuler's model (Newkirk, 1972a) and Schatten's sheet current model (Schatten, 1971) with an appropriate MHD solution using the same boundary conditions (Pneuman and Kopp, 1971) are favorable at low levels for the case when the gas pressure at the coronal base is assumed to be independent of latitude.*

That some caution must be excercised when applying potential theory to actual coronal conditions is borne out by the following consideration. Since, in general, $\nabla\times\mathbf{B}=4\pi\,\mathbf{j}$, the approximation $\nabla\times\mathbf{B}\approx 0$ requires that

$$|\nabla\times\mathbf{B}|\ll|\mathbf{B}|/L_b, \tag{1}$$

where L_b is the scale over which $\mathbf{B}$ spatially varies. If pressure and magnetic forces roughly balance in the low corona, then we also have

$$|\nabla\times\mathbf{B}|\approx\frac{4\pi P}{|B|\,L_p}, \tag{2}$$

L_p being the scale for horizontal pressure variations. Combining Equations (1) and (2), the requirement for validity of potential theory is,

$$P\ll\frac{B^2}{4\pi}\,\frac{L_p}{L_b}\,. \tag{3}$$

Hence Equation (3) could be invalid in places where the gas pressure varies rapidly from place to place over the coronal base (eclipse observations do suggest filamentary structure in the lower corona). Even in the absence of this consideration one notes

* If this were not the case, horizontal pressure gradients might produce significant deviations from potential theory.

that, for $N_e = 2 \times 10^8$ and $T = 1.5 \times 10^6$, the ratio of $B^2/4\pi P$ is about one for a field strength of one gauss. Thus, it is not entirely clear that pressure forces really are negligible in the lower corona.

In addition to the above, two other mechanisms negate the possibility for the coronal magnetic fields to be completely potential. These are the effect of convection on the field and the influence of the overall coronal expansion or solar wind. Both these influences introduce stresses into the field configuration and, hence, are of importance in understanding the mechanisms which could underly coronal disturbances.

3. Magnetic Structures in Elevated Energy States

Beneath about the middle chromosphere, the fluid forces dominate the magnetic field and, as a consequence, field lines can be twisted, tangled, compressed, and, in general, transported in quite an arbitrary manner depending on the fluid motion. Since the coronal field remains tied to this subphotospheric field, it is also transported. The material in the corona is tied intimately to the field however and each field line retains an identity. As a consequence, direct transformation from one topology to another (such as to that of the potential configuration) cannot always occur. If then the magnetic energy density in the corona is large as compared to that in the gas and a potential configuration is topologically unattainable, the field must be either force-free or, as an alternative, potential but with certain regions where sheet currents are present. An example of the latter is shown in Figure 2. Suppose, in a medium of infinite conductivity, a simple bipolar region, labeled (1), exists. Now, further suppose that a second bipolar region (2) emerges beneath (1). If reconnection is not allowed, the resultant configuration will be that of Figure 2a. Configurations (1) and (2) can be potential-like individually but a sheet current will exist between them. Figure 2b

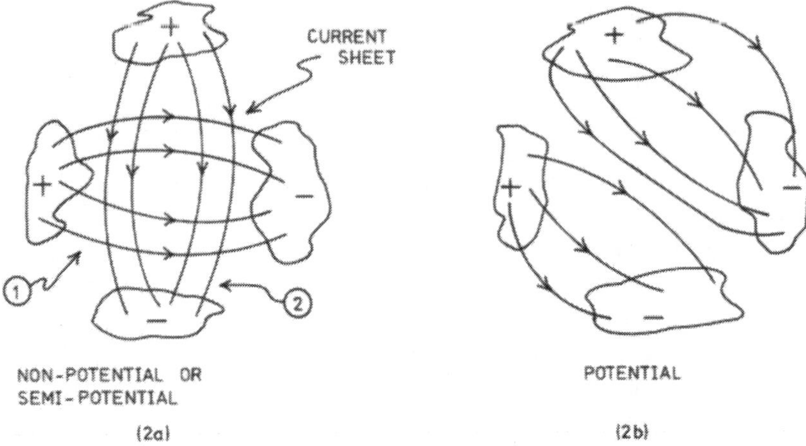

Fig. 2. Two possible resultant configurations when one bipolar magnetic region ((2)) emerges beneath another ((1)). In (2a), reconnection is not permitted and the two flux ropes remain intact with a sheet current separating them. Figure (2b) shows the resulting potential configuration if reconnection occurs easily.

shows the completely potential configuration which cannot be reached without reconnection.

The fact that the chromospheric fields inferred from Hα (Veeder and Zirin, 1970; Zirin, 1972) appear so complicated and the coronal fields relatively much simpler (see Figure 1) suggest that as magnetic flux is transported outward from the convection zone a large amount of simplification has taken place. This implies an energy release mechanism capable of motivating coronal disturbances or even heating the corona on a continuous basis.

3.1. Force-free fields

The importance of force-free fields as a possible exciter of coronal disturbances is obvious. Firstly, this type of structure is expected in at least some regions of the corona and, secondly, the energy density in a force-free field is always higher than that of an equivalent potential field, providing available energy for disturbances. Barnes and Sturrock (1972), for example, have suggested that the transformation from a closed force-free configuration to an open configuration could provide the necessary energy for a solar flare.

A force-free field configuration satisfies the equation

$$\nabla \times \mathbf{B} = \alpha \mathbf{B} \tag{4}$$

$$\nabla \alpha \cdot \mathbf{B} = 0, \tag{5}$$

where α is an arbitrary scalar. Solution of these equations under various assumptions have been discussed by many authors (see Boström (1972) for a comprehensive review). The physical interpretation of the results of these analyses has been rather speculative. More recently, however, comparison of fields calculated from force-free theory with observations near sunspots (Nakagawa *et al.*, 1971) and with chromospheric fibrils and filaments as observed in Hα (Raadu and Nakagawa, 1971; Nakagawa and Raadu, 1972) have been encouraging. In essentially all these models, however, the assumption $\alpha = \text{const}$ is employed, an assumption which is physically difficult to justify. Further investigation for the cases $\alpha \neq \text{const}$ would certainly be fruitful. For example, Low (1973) has recently computed solutions for nonconstant α including resistive diffusion. He fiinds that the field may evolve slowly for an extended period of time, then abruptly develop steep gradients and pass into an explosive phase. A process which may have application to solar flares and eruptive prominences.

In contrast to potential fields, the energy in a force-free configuration can be steadily increased by motion of the foot points. This has been demonstrated by Raadu (1972) for the simple example of differential rotation acting upon a quadrapole field. Assuming the field remained force-free as the footpoints were sheared, Raadu found that the field energy was increased by 25% in only one solar rotation. In addition, the field configuration expanded outward during this period. Although the motion considered here is particularly simple, shearing motions are always occurring in the convective zone. Hence, these results could have significant implications for coronal disturbances such as expanding magnetic arches. bottles, etc.

3.2. Prominences

Historically, the name prominence has been applied to almost any bright protuberance observed in Hα on the limb, whether transient or steady state. In order to understand the physics of these structures and their changes, however, we must clearly differentiate between two types – those associated with active regions and flares such as surges, loop prominences, flare loops, etc. and the so-called quiescent prominences residing usually towards the poles away from active regions under helmet streamers (note the bright points at the base of the helmet streamers in Figure 1). It is questionable whether these two classes of configurations have the same origins or explanations and, consequently, they should be considered separately.

Loop prominences and surges do not seem to be due to alterations in local magnetic structure but rather the result of ejected material from an underlying disturbance such as a flare. The main difference between the two phenomena can be interpreted in terms of the field geometry previously overlying the disturbance. Surges occurring where the field lines are either open or extend over large distances and loop prominences where closed loops directly overlie the disturbance. Surges travel upward rather fast, about 300 km s^{-1} and generally return along more-or-less the same trajectory (Tandberg-Hanssen, 1967). Loop prominences show a great variation in their motion. Some appear to expand extremely slowly – perhaps at 10 km s^{-1} (Bruzek, 1964) but exhibit broad wings in Hα indicating energetic internal motions. Other expand explosively with velocities exceeding 100 km s^{-1} thus separating loop prominences into two distinct classes (Bruzek and Demastus, 1970). Perhaps the cases of slow expansion, since actual deceleration of material is observed (Bruzek and Demastus, 1970), give evidence for magnetic inhibition of the outward motion. Another possibility is that these apparently slow moving prominences do not expand at all but merely reflect an excitation process repeated through successfully higher levels in the solar atmosphere (Goldsmith, 1971) or, according to Schmidt (1969), they could be the result of a slow reconnection of lines of force torn apart by the flare. Perhaps the coronal and interplanetary manifestations of the rapidly expanding loop prominence systems are to be seen in the moving type IV radio bursts and the 'magnetic bottles' observed at 1 AU. Both these phenomena reflect closed field geometries and seem to be always associated with flare activity although all flares certainly do not produce loop prominences or type IV emission.

The mechanism producing type IV radio bursts is now generally accepted to be synchrotron emission (Boishot and Denisse, 1957), however, the magnetic structures associated with these large-scale disturbances are still not fully understood. Observations suggest that they can be of several varieties each of which suggests different physical mechanisms. Some appear to be associated with an advancing shock front (Kai, 1970a; Stewart and Sheridan, 1972, Stewart *et al.*, 1970) occurring generally after a type II burst. Others can be interpreted as an expanding magnetic arch containing trapped electrons while wtill others look like ejected plasma blobs moving radially outward to great heights at uniform velocity (Riddle, 1970). These 'blobs' have been suggested to have a self-contained field structure and move out along the open field lines of the large-scale coronal field (Smerd and Dulk, 1971; Dulk and Altschuler, 1971).

The possibility of magnetic bottles in the interplanetary medium was first suggested by Gold (1963) to explain the Forbusch decrease in galactic cosmic rays observed at the Earth following a large flare. Since then Schatten *et al.* (1968) reported an observation of a magnetic tongue at 1 AU which occurred in connection with a new active region and the birth of a new interplanetary sector. Also, Schatten (1970), from the Faraday rotation of a radio source, inferred a bottle traveling out to 10 $R_{\odot}$ at a velocity of about 200 km s^{-1} (see Figure 3). Another interesting argument for the existence of magnetic bottles as far out as the orbit of Earth is based upon the anomalously low temperatures in the solar wind following interplanetary shock waves (Montgomery *et al.*, 1972; Gosling *et al.*, 1973). These low temperatures could result if the driver gas behind the shock is inclosed in a magnetic geometry which prohibits thermal conduction from the inner corona. The closed field lines of an isolated bottle would provide such a geometry. It seems as though another possible explanation for this phenomena, however, is that the flare ejecta volume could expand outward more rapidly than the r^2 increase of the normal solar wind. If this were the case, simple adiabatic cooling due to overexpansion would produce lower temperatures in these regions. In summary then, there seems to be a great deal of evidence, from observations both in the inner corona and in the interplanetary medium, for closed magnetic loops being expelled from the Sun. One final question is whether these loops remain tied to the solar surface and, perhaps, even return after the disturbance or whether reconnection occurs producing isolated bubbles which can then move freely outward without reverse magnetic forces. As we shall see later, reconnection rates appropriate for these conditions are extremely uncertain so that this question, for the moment, must remain unanswered.

Quiescent prominences are located away from active regions under helmet streamers and appear in Hα on the disc as long thin east-west oriented filaments. They also seem to undergo a general poleward migration on the time-scale of a solar cycle (Lockyer, 1931; d'Azambuja and d'Azambuja, 1948; Hyder, 1965). Contrary to their name, these interesting structures do become involved in coronal disturbances. They differ from the active region prominences, not only in their location but that they seem to undergo some fundamental change in their own structure rather than passively react to the forces of another disturbance. This indicates that energy conversion takes place within the prominence itself.

Most older theories of quiescent prominences are similar to the Kippenhahn and Schluter (1957) and Dungey (1958) concepts of a gas supported against gravity by the sagging lines of force of a magnetic field (see Figure 4).* Observations of the photo-

* See Tandberg-Hanssen (1974) for a comprehensive review of older theories of quiescent prominences.

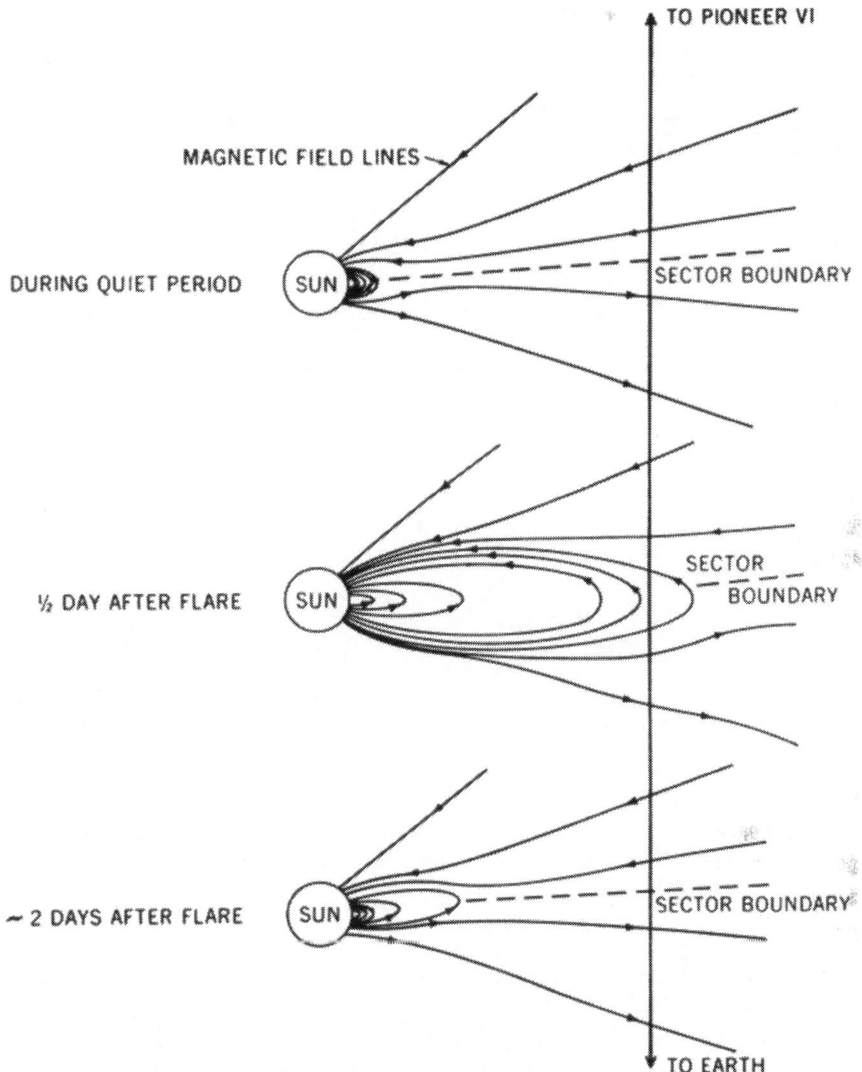

Fig. 3. View from the north of a proposed magnetic bottle observed by Pioneer VI (from Schatten, 1970).

spheric field pattern beneath these prominences by Rust (1970) tend to confirm this hypothesis. Anzer and Tandberg-Hanssen (1970) suppose that quiescent prominences have a helical structure produced by two current systems, one in the photosphere and one along the cylindrical prominence.* Other recent models (Anzer, 1972; Kuperus and Tandberg-Hanssen, 1967; Raadu and Kuperus, 1973) also employ sheet currents of various configurations.

* The classic prominence eruption of June 4, 1946 does give the appearance of untwisting helical lines of force.

Most theories of quiescent prominences assume the prominence is formed by condensation from the corona through a thermal instability (Kleczek, 1957, 1958; Lust and Zirin, 1960; Uchida, 1963; Kuperus and Tandberg-Hanssen, 1967; Raju, 1968, Nakagawa, 1970; Hildner, 1971). Another possibility is that they are the natural consequence of the energy balance requirements at the base of hlmet streamers, i.e.,

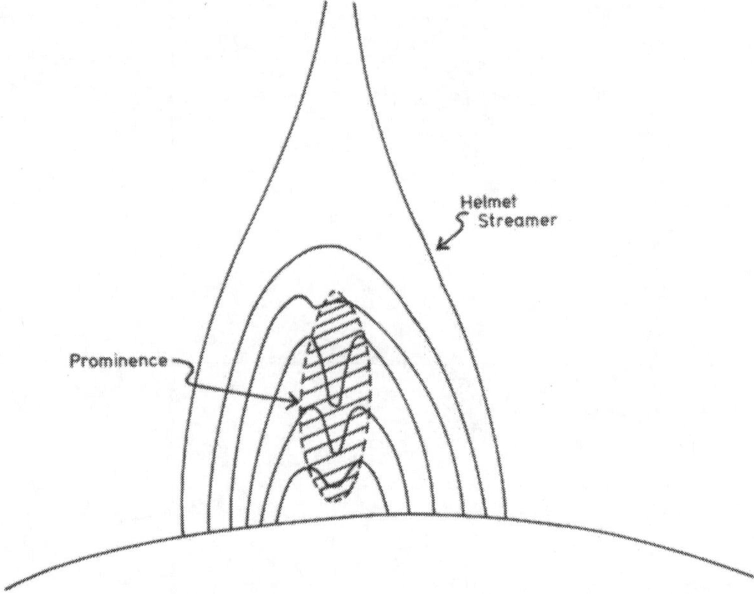

Fig. 4. Kieppenhahn-Schluter model of a quiescent prominence. Here, the dense material is supported against gravity by the sagging lines of force shown in the center portion of the helmet.

that they are necessary in order to balance radiative losses with mechanical heating from below (Pneuman, 1972).

Since quiescent prominences can persist over many solar rotations, we can tentatively conclude that they are *not* unstable. However, they do occasionally disrupt or explosively disappear one or more times either for no apparent reason or due to a triggering wave from a distant flare site (Dodson and Hedeman, 1964) or developing sunspot region (Bruzek, 1952). These sudden disappearances, called *Disparition Brusques* (d'Azambuja and d'Azambuja, 1948) are usually followed by a reappearance of the prominence some time later.*

This type of behavior suggests the quiescent prominence is a *metastable* configuaration – stable to infinitessimal perturbation but not those of sufficient magnitude. Two quite different explanations which could account for this come to mind. One is that some type of change takes place in the underlying magnetic field pattern which excites a reconnection process. Rust (1970), for example, suppose that a flux

* Garcia *et al.* (1971) has indicated that this might not always be the case. If so, the implication toward the theories of these structures are important.

loop of opposite polarity to the field in the prominence emerges from below (see Figure 5). This creates an unstable neutral point configuration and explosive reconnection begins to take place. If, on the other hand, the very existence of the prominence depends upon energy balance considerations then one might speculate that a significant violation of these requirements could also disrupt the system. If, for some reason, the radiative losses from the prominence were temporarily inhibited leaving the incoming mechanical energy flux unaltered, then the closed field lines above the prominence would bottle-up the mechanical energy flux in a manner similar to that suggested by Pneuman (1967) as a possible flare mechanism. If this were the case, the

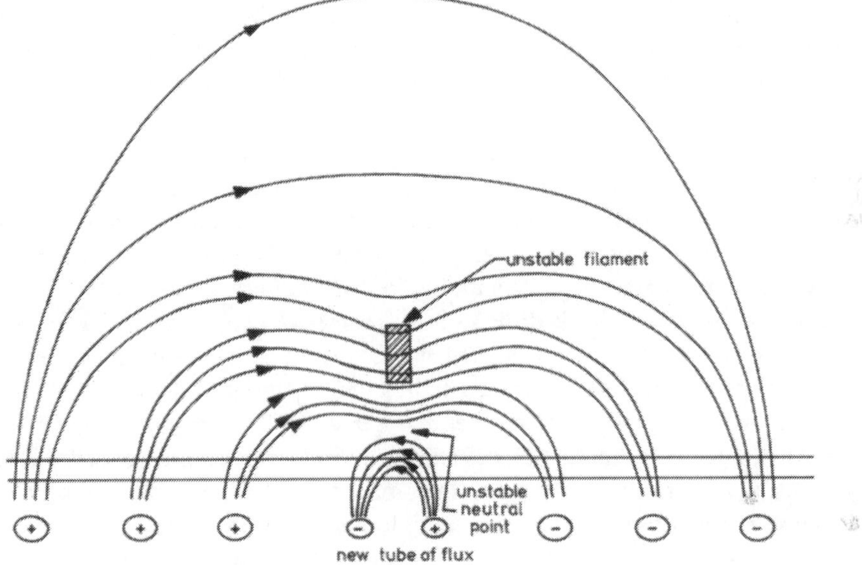

Fig. 5. Unstable neutral point created by the emergence of a new bipolar flux loop beneath a quiescent prominence (from Rust, 1970).

energy content in the prominence would increase until it became equal to that of the restraining fields at which point expansion would begin. This condition can be mathematically expressed as

$$qA_0t = \frac{B^2}{8\pi} A_0L,$$

where q is the incoming mechanical flux, A_0 the cross-sectional area at the base of the prominence, B the field strength, L the length of the prominence, and t the time. Hence,

$$t = \frac{B^2L}{8\pi q}.$$

Taking $1\ \mathrm{G} < B < 50\ \mathrm{G}$, $L = 0.1\ R_{\odot}$, and $q = 4 \times 10^5$ erg cm^{-2} s^{-1} we find that t ranges from about 10 min to several days. Hence, energy balance considerations could conceivably be important.

3.3. Interaction of the solar wind with coronal magnetic fields

Although the existence of a solar wind has been known for only about 15 years, an especially astute eclipse observer could have predicted it many years ago merely from the general appearance of the large-scale coronal features. For example, the form of the helmet streamers and the overall radial configuration of the corona beyond about 2 $R_{\odot}$ can be consistent only with a general expansion. We will not dwell upon the many interesting observational as well as theoretical aspects of solar wind physics here (see Hundhausen (1972) for a review of the subject) but will concentrate upon its relevance in shaping the coronal magnetic field configurations. The solar wind introduces stresses in the large-scale fields, the relief of which could provide energy for coronal disturbances.

The most important influence of the solar wind upon the coronal magnetic field is that it divides the corona into magnetically closed and open regions. Each of these regions is expected to have entirely different physical characteristics due chiefly to influence of the geometry upon the energy balance mechanisms (Pneuman, 1973).

Closed magnetic regions cannot suffer an outward conductive heat loss since heat conduction across field lines is difficult in the corona. Also, as opposed to the open regions, they do not lose energy carried by expansion. As a result, the temperature and density are significantly elevated in the closed regions. The so-called 'coronal holes' (Withbroe *et al.*, 1972; Altschuler and Perry, 1972; Altschuler *et al.*, 1972) are probably just open field lines being constantly drained of their energy content by the solar wind and thermal conduction (Pneuman 1973).* The recurrent high speed streams commonly observed at 1 AU seem to correlate well with these low density open field line regions (Wilcox, 1968; Hundhausen, 1972; Pneuman, 1973; Noci, 1973; Krieger *et al.*, 1973). These streams are sometimes associated with recurrent geomagnetic activity (e.g. Chapman and Bartels, 1940) and provide evidence for the disturbing influence of the solar wind upon the Earth's magnetic field.

When viewed on the large scale, the solar wind appears to be stable. This contention is supported by theory (Parker, 1965, 1966; Carovillano and King, 1966; Jockers, 1968) as well as observation. Small-scale instabilities and disturbances of various kinds, however, are constantly present. Alfvén waves are observed at 1 AU (Coleman, 1967;Unti and Neugebauer, 1968; Belcher and Davis, 1971) and have been invoked to provide momentum to the solar wind (Hollweg, 1971a, b, 1973; Belcher, 1971; Alazraki and Couturier, 1971). These waves could have been produced by the supergranulation network (Hollweg, 1972a, b) and traveled outward along open field lines with very little dissipation. Small-scale disturbances, however, do not appear to influence the corona as a whole. In order to gain insight into the potential large scale changes in the inner corona resulting from the solar wind, we must examine the helmet

* A sample calculation shows that, even for uniform base conditions, a density enhancement of a factor of ten between the closed and open regions can be produced at 2.5 $R_{\odot}$ through these mechanisms.

streamers where most of the energy and mass of the inner corona resides. Those configurations clearly are stressed by the coronal expansion and, thus, reside in an elevated magnetic energy state relative to that of a potential field. Such elevated slates are always suspect when searching for the origins of coronal distrubances.

Streamers appear both over active regions and, at higher latitudes, over quiescent prominences, the latter being both larger and more stable. They are, of course, associated with bipolar magnetic regions on the Sun and, depending upon how high the coronal temperature is, can be either completely open (Parker, 1964a; Pneuman, 1969) or contain regions of closed loops at their base (Pneuman, 1968). In both open streamers and helmet streamers the density is enhanced over the background by a factor of 2–10 (Schmidt, 1953; Michard, 1954; Hepburn, 1955; Saito, 1959; Saito and Billings, 1964; Saito and Owaki, 1967; Leblanc, 1970; Newkirk *et al.*, 1970; Koutchmy, 1971) with the enhancement increasing outward.* This density enhancement is a result of greater energy losses adjacent to the streamer. These losses produce a relatively lower temperature there than in the streamer and, through the scale height effect, a much larger difference in density (Pneuman and Kopp, 1970; Pneuman, 1973).

Streamers containing closed loops, called helmet streamers, appear to be more commonly observed in the corona than open streamers and are probably more relevant to the subject of coronal disturbances. The chief interest in the helmet streamer, shown schematically in Figure 6, lies in the cusp-type neutral point at the

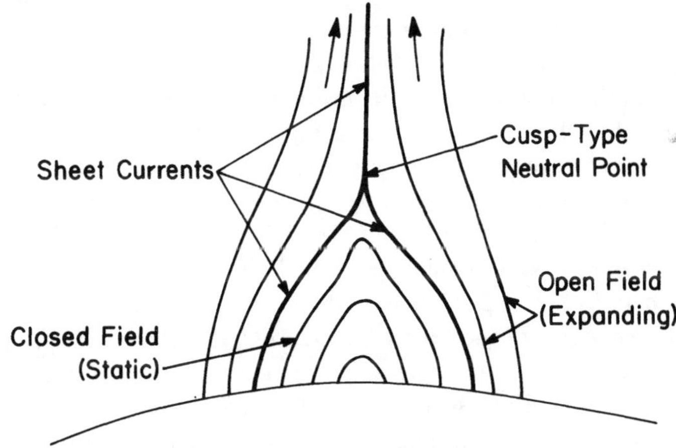

Fig. 6. Schematic of a typical helmet streamer (from Pneuman and Kopp, 1971). Note the 'cusp-type' neutral point and the sheet currents both above and below.

top of the closed loops, the neutral sheet above, and in the sheet currents below the neutral point between the open and closed regions (Pneuman and Kopp, 1971).

On the whole, the helmet seems to be a reasonably stable configuration persisting usually for several rotations (Hansen *et al.*, 1969, 1972; Bohlin, 1970a, b, 1971). Evidently they evolve in accordance with the changes in their underlying field

* See Newkirk (1967) for a comprehensive review of coronal observations.

structure. This picture is consistent with the apparent stability of the sector structure (Wilcox and Ness, 1965, 1967; Wilcox, 1968) assuming sector boundaries are the interplanetary extension of the actual sheets lying above helmet streamers (Wilcox, 1968; Newkirk, 1972a; Hundhausen, 1972). However, sudden disappearances or displacements do occur and the physical reasons underlying these disruptions are not understood. Five possibilities immediately suggest themselves. Firstly, the streamer could be blown out by an eruption of the underlying prominence. Secondly, some basic reorientation of the photospheric and chromospheric field may take place (such as reconnection or the relaxation of a force-free configuration) requiring a major readjustment of the coronal configuration. Thirdly, the streamer itself may be in a *metastable* state and undergo an explosive change due to an outside disturbance such as from a distant flare site.* A fourth mechanism may be that physical conditions in the corona may change such as an increase in temperature requiring a transformation from a closed to open configuration. This can occur for a relatively small change in temperature (Pneuman, 1968) and, once the streamer has opened, the density will rapidly decline giving the appearance of a void in the corona where a streamer once existed. Finally, since outward expansion and thermal conduction are prohibited in the helmet, all the energy dissipated by waves there must be radiated away. If this can't be accomplished, the helmet must expand.

Other transient phenomena associated with streamers perhaps related to the above are the apparent day-to-day changes in streamer locations and orientations observed in white light (Tousey, 1972). Although these changes have been attributed to basic physical changes in the corona, this phenomena could be one of perspective. For example, we expect the neutral sheets above streamers to be quite thin (Pneuman, 1972) due to the high conductivity of coronal material. Suppose the corona consists of a network of thin sheets in the vicinity of which the high density material is located. Since the brightness reflects an integral of density along the line of sight, the material will then be most visible when the plane of the sheet lies along the line of sight and essentially invisible when perpendicular to it. If these surfaces are curved, then the planes which appear bright could either shift about considerably or even flicker in and out of view on a rather short time scale. For example, if t_r is the time scale for solar rotation (≈ 27 days), and l the sheet thickness, then the time scale for this phenomena could be as short as $(l/R_\odot)\,t_r$. For a sheet 1000 km thick, this time is only about an hour.

The overall stability of the helmet streamer configuration has not been investigated. However, the stability of the neutral sheet overlying the streamer has been studied and will be discussed in Section 5. The tearing mode instability is relevant here (Kuperus and Tandberg-Hanssen, 1967; Raadu and Kuperus, 1973), the inception of which may be responsible for the acceleration of electrons responsible for type III bursts (Sturrock, 1966, 1968, 1972). Another interesting type of instability, suggested recently

* Recently, Hansen (1973) has reported three separate observations of a coronal 'twitch' in which the axis of a large helmet streamer was displaced (rotated about its base) by about 5° as the result of a distant flare. Since this flare was accompanied by a type II and moving type IV burst, it is likely that this displacement could be produced by a traveling hydromagnetic shock.

by Cowling (1973), involves the discontinuity in pressure between the open and closed field lines with the larger pressure existing in the closed region. Cowling sees a possible discrepancy in the fact that the large gas pressure in the closed tubes is unable to break open the field lines for the material to escape whereas the smaller pressure outside is able to keep them from closing. He suggests there may be an instability near the top of the arches of closed flux tubes with the tubes becoming filled with plasma, bursting open, and then reclosing after losing their mass and heat content.

4. Role of Magnetic Fields in Guiding Coronal Disturbances

In some regions of the corona the condition $B^2/8\pi > P$ is satisfied and it is in these regions where we expect coronal fields to be effective in constraining particles and waves to move along field lines. We arbitrarily divide these types of disturbances into 3 classes – bulk motions, or those which exhibit continuum properties, wave motions, and individual particles events.

4.1. Bulk motions

The role of the magnetic field in surges is obvious. The surge is evidently produced by some underlying disturbance. Ejected material merely flows outward along the lines of force and generally falls back when its kinetic energy is spent. Loop prom--nences also show motion along the field. It is perplexing that this motion is always downward with the estimated mass falling on the cromosphere much greater than what would be supplied from the corona. Chromospheric brightenings are associated with this falling material motion which have been likened to the flare mechanism (Hyder, 1967a, b). Chromospheric spicules resemble miniature surges in many ways also seem to be a field channeled phenomena. They appear at the boundaries of super-granulation cells where the flow from the cells converges. It is easy to imagine this flow compressing the field lines and setting up strong vertical motions. This process has been evoked in many spicule theories (Ferraro and Plumpton, 1958; Weymann and Howard, 1958; Osterbrock, 1961; Parker, 1964b). Since, neglecting dissipation, ϱv^3 (ϱ being the density and v the velocity) will tend to be constant with height, we find for a vertical field (B=const) that $V \approx \varrho^{-1/3}$. Since the density decreases exponentially, large velocities can result high in the chromosphere.

Among the less obvious mechanisms for producing bulk motions along magnetic field lines is one suggested by Meyer and Schmidt (1968) applicable to closed loops. Although they have used this mechanism only to explain the Evershed motion near sunspots, it may very well be applicable to a great variety of field channeled flow phenomena observed in the corona. Consider a flux tube with both feet rooted in the deeper layers (see Figure 7a). If the temperature of the tube is uniform, it can easily be shown that the gas can be in hydrostatic equilibrium only if the pressure at the base of the tube are equal ($P_1 = P_2$). If not, a flow will be initiated from the higher pressure footpoint to the lower. Although for modest differences in pressure, the total mass flow may not be large, the velocities can be appreciable.

Considering a symmetric flux tube, Figure 7b shows the various possible solutions of the momentum equation for the steady state velocity profile. The only solution which yields $P = P_1$ at the left footpoint and P_2 at the right is one which begins a curve C and terminates on curve D. As can be seen however, there is no continuous solution linking these two curves, the reason being that the flow cannot pass continuously

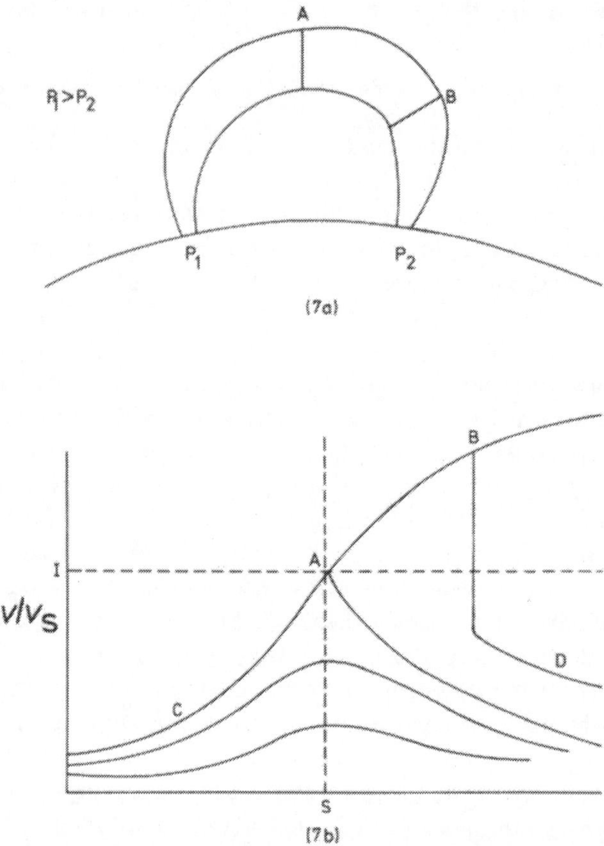

Fig. 7. A possible guiding role of coronal magnetic fields proposed by Meyer and Schmidt (1968). Figure (7a) shows a flux loop with two different gas pressures at its footpoints. As shown in (7b), the resultant flow becomes supersonic at the top of the loop (A) then undergoes a shock transition to subsonic flow at point B.

from a supersonic velocity to a subsonic velocity. The only solution yielding $P = P_1$ at the left and $P = P_2$ at the right is one where the velocity begins on curve C, passes through the sound speed at the top of the loop (point A), then undergoes a shock transition to curve D at point B. In this mechanism, the shocked gas is of higher density and *always* downward moving. It should be more visible and may explain why loop prominence material is only observed when moving downward. The increase in density be the shock can, of course, be only up to a factor of about four – probably not enough to explain the observations.* It should be kept in mind that this theory is a *steady-state* theory. It is more likely that the pressures at the base of closed loops are continuously fluctuating and the resulting mass motions may be more modest.

Another interesting type of coronal event which seems to be field-guided in a certain sense are the so-called 'coronal whips'. In this case mass doesn't appear to be flowing along field lines but, instead, is pulled rapidly through the corona in a whipping fashion by a readjustment of the whole field structure, the entire process beginning gradually and accelerating to velocities of about 100 km s^{-1} (Evans, 1957; Kleczek, 1963; Bruzek and Demastus, 1970; Dunn, 1970). This is clearly produced by a rapid large-scale change in the coronal field structure rather than by the coronal material and probably reveals the after effects of a reconnection process. For example, consider the configuration shown in Figure 8. At the top a loop structure in shown moving toward open field lines of opposite polarity. Eventually, reconnection will

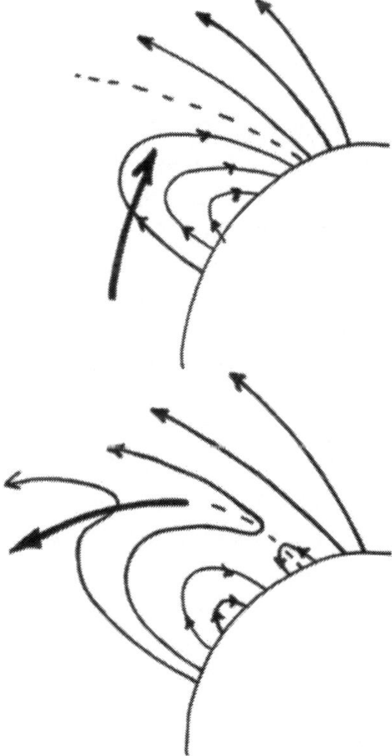

Fig. 8. A possible situation which could result in a coronal 'whip'. The top figure shows a bipolar region encroaching upon an open field region of opposite polarity. Reconnection at the base between the two regions then opens field lines previously closed. The field lines now freed of their photospheric connection can 'whip' out into the corona at approximately the Alfvén speed.

* On the other hand, further compression of the material after being shocked is also certainly possible.

begin to occur freeing the photospheric connection of one side of the loops. Once, released, these noe open field lines will whip out into the corona at approximately the Alfvén speed carrying their load of coronal material along.

4.2. Wave Motions

The contention that the solar corona is heated by waves emanating from the convective zone must at present be considered without observational support. Although attempts have been made to observe coherent disturbances which propagate and dissipate in the low corona, results so far have been inconclusive. Indirect evidence, however, is available from observations of Alfvén waves at 1 AU (Belcher and Davis, 1971) and recently observed coherent oscillations in plage regions (Bhatnager and Tanaka, 1972) which may reflect resonant Alfvén waves trapped in closed magnetic field lines (Pneuman, 1968).

Certainly the most spectacular wave disturbances on the Sun are associated with flares. These disturbances, seen in Hα, are observed to move outward in the chromosphere from the flare site at velocities of the order of 1000 km s^{-1} (Moreton, 1960) and have been of great interest to solar observers (Atahy and Moreton, 1961; Anderson, 1966, Dodson and Hedeman, 1968). The good time correlation between these 'Moreton waves' in the chromosphere and type II radio bursts observed in the corona strongly suggest that these two events are caused by the same shock wave leaving the flare site (Moreton, 1964; Ramsey and Smith, 1966; Wild, 1969).

Considering sonic and Alfvénic velocities appropriate for the chromosphere, Meyer (1968) has argued that a shock traveling there with a velocity of $\approx 10^3$ km s^{-1} would have a mach number of more than 10 and would thus exhibit large wave amplitude and strong dissipation. These difficulties disappear if the Moreton wave is considered to be a coronal phenomenon, the sound and Alfvén velocities being much higher there.

This concept has been amplified by Uchida (1968) in which the disturbance is considered to be a fast-mode MHD wavefront propagating from the flare region into the corona. The chromospheric manifestation is likened to that of a 'sweeping-skirt' of the coronal disturbance. Using a general ray tracing technique (eikonal equation), Uchida *et al.* (1973) have calculated the development of the fast-mode MHD wavefront in the corona in which the Alfvén velocity is computed from potential magnetic field theory (Altschuler and Newkirk, 1969) and deconvoluted density profiles obtained from *K*-coronameter data (Altschuler and Perry, 1972). In general, the wave focuses toward regions of low Alfvén speed so that most of the energy tends toward places where either the magnetic field is weak or the density is high. This brings up an interesting point regarding the neutral sheets above helmet streamers. There, the Alfvén speed goes to zero (or very nearly so) suggesting that these locations may be very effective in concentrating the energy in this type of coronal disturbance.

It is important to note that fast mode MHD waves *do not*, in general, follow field lines. The direction of propagation of these disturbances is sensitive to the distribution of Alfvén speed, not the field direction. Other investigations, on the other hand, have claimed a channeling of the type II shock along open more-or-less radial field lines (Kai, 1969; Dulk *et al.*, 1972).

One shortcoming in all these analyses is that the energy density involved with these type II and IV events is probably so high that the pre-existing field configuration is completely disrupted. Hence, any magnetic field model based upon the pre-existing photospheric magnetic fields could be extremely unrealistic.

4.3. Magnetic fields and individual particles

Perhaps the most common solar event which is essentially an individual particle phenomenon is the type III radio burst. These occur more commonly than flares and are apparently produced by a stream of energetic electrons travelling outward through the corona at a significant fraction of the speed of light exciting plasma oscillations at progressively higher levels. They have recently been observed far from the Sun at distances approaching 1 AU (Hartz, 1964, 1969; Slysh, 1967a, b; Alexander *et al.*, 1969; Haddock and Graedel, 1970; Fainberg and Stone, 1970a, b, 1971).

The coronal densities inferred from type III burst analyses have been consistently high (Newkirk, 1967; Fainberg and Stone, 1971). This, in addition to their apparent association with filaments seen in Hα (McLean, 1969, 1970), has led to the suggestion that the netral sheets associated with coronal streamers are the location of these outward travelling electron streams (Kai, 1970b; Weiss and Wild, 1964). This idea is strengthened by the apparent low degree of circular polarization of the bursts (Kai, 1970b) indicating a weak magnetic field such as could occur very near the neutral sheet of a streamer. Also, Sturrock (1968) has proposed that reconnection at the neutral point at the top of a helmet could be the acceleration site of these particles. One drawback of the neutral sheet hypothesis, however, is that these coronal structures probably contain transverse magnetic fields (Pneuman, 1972). If so, electrons can neither escape along the axis of the sheet nor pull the field lines outward with them (Smith and Pneuman, 1972) They could presumably travel just outside the sheet however.

A great variety of radio emission seems to be associated with closed loop structures in the lower corona. For example, 'U' bursts are evidently the direct counterpart of type III bursts. In this case, the electrons travel along closed field lines rather than open evidenced by the reverse frquency drift and their limited extent in height, their maximum observed height corresponding roughly to the highest observed closed loops in the white light corona. Type V emission also is caused by electrons spiraling back and forth along closed field lines producing broad band continuum radiation (Weiss and Stewart, 1965). They follow type III bursts (about 10% of them) and the radiation is believed to be due to synchrotron emission. Figure 9 is a schematic of a possible field configuration which could be responsible for all three of these phenomena.

The source characteristic of type I noise storms and stationary type IV bursts appear almost indistinbuishable (Kai, 1970b; Wild *et al.*, 1963), both consisting of a narrow band burst component superimposed on a broad band continuum. In the type I storm, the narrow band component is most prominent, while in the stationary type IV the continuum dominates. Type I storms are associated with sunspots and active regions, seem to occur in bipolar fields (Lartos-Jarry, 1970), and are quite persistent. The association of these radio regions with the interplanetary sector structure has been investigated by Sakurai and Stone (1971) (see Figure 10). They found, for the period 13 March to 21 August, 1968, that the number of type I centers was the same as the

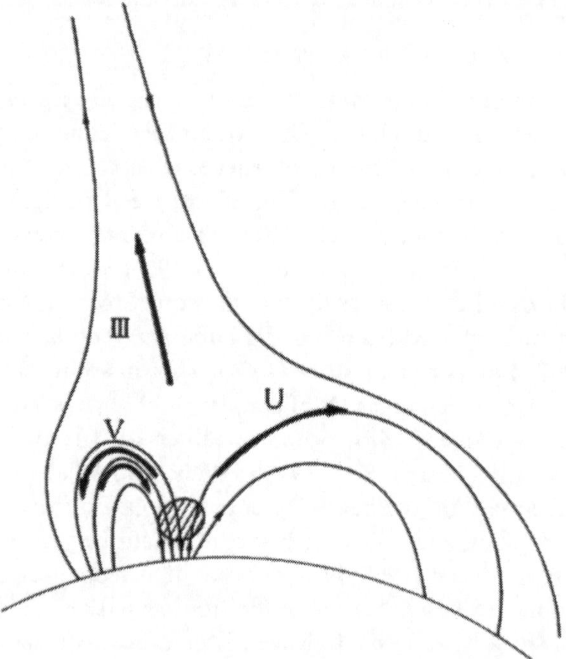

Fig. 9. Possible magnetic field configuration which is consistent with type III, V, and U bursts (from Wild and Smerd, 1972).

number of sector boundaries with the passage of a sector boundary delayed by about 5 days after the central meridian passage of the type I regions.

The overall problem of particle storage in magnetic fields is still quite unsettled. That containment in closed field regions does occur is suggested by X-ray photographs of the lower corona (see Figure 11). The bright arches are attributed to thermal bremsstrahlung and gyro-synchrotron emission of trapped electrons (Benz and Gold, 1971). One peculiar aspect of this subject is the apparent emission of energetic particles (≈ 10 MeV protons) from active regions for days and even weeks following a major flare (Fan *et al.*, 1968; Lin *et al.*, 1968; Krimigis, 1969; Krimigis and Verzariu, 1971). Since stable closed loops probably exist to a maximum height of about 2–2.5 $R_{\odot}$, the density must always be high in the storage regions. It is therefore difficult to understand why these particles don't quickly lose their energy by coulomb collisions with the ambient coronal gas (Ahluwalia, 1972). On the basis of considerations such as these, a continuously operating acceleration mechanism may be indicated (Newkirk, 1972a).

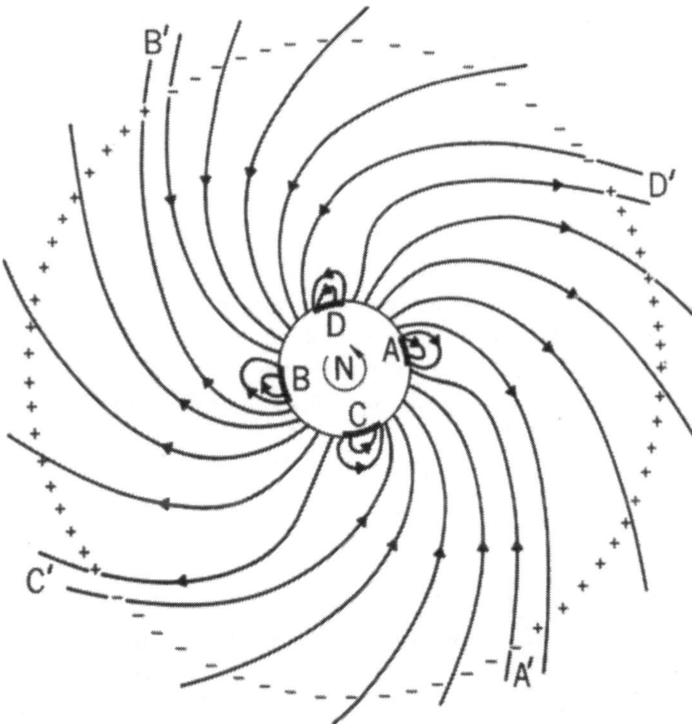

Fig. 10. Relationship between bipolar magnetic regions on the Sun, as inferred from type I storm centers, and the interplanetary sector structure (from Sakurai and Stone, 1971).

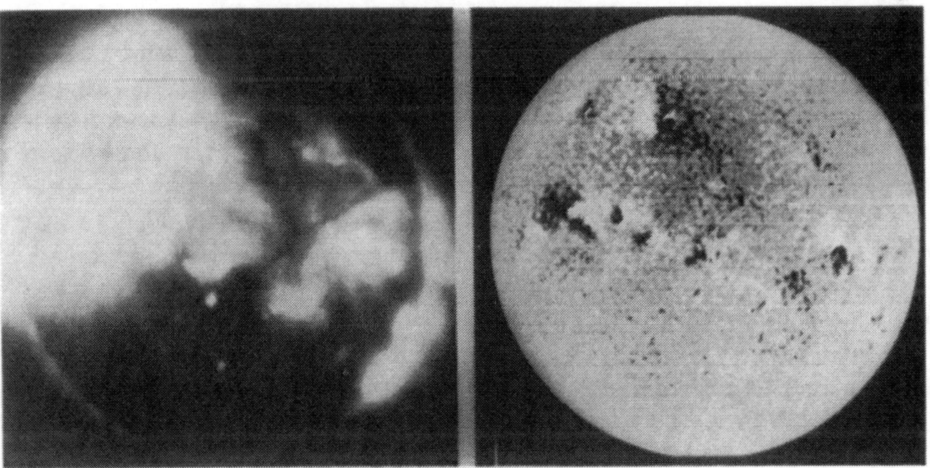

Fig. 11. Concurrent photographs of the Sun for the period covering the 1970 March 7 eclipse. On the left is shown the coronal X-ray emission and, on the right, the polarity pattern of the photospheric magnetic field (taken from Krieger *et al.*, 1970). Note the one-to-one correspondence between bright X-ray sources and bipolar magnetic regions.

5. Magnetic Discontinuities and Reconnection

In the crudest sense, we can estimate that a typical time scale for a magnetic disturbance in a conducting medium of density ϱ where the magnetic field strength is B will be

$$t \propto \frac{l\sqrt{\varrho}}{B},$$

where l is the typical length scale of the region where the disturbance originates. This is just the time required for an Alfvén wave to traverse the distance l. Since our main interest here is in fact coronal transitory phenomena, we search for physical situations where this time is comparatively short. Hence, regions of strong magnetic field, low density, or small characteristic lengths are relevant here. Our attention in this section is turned to those regions where l is small – such as the neutral sheets and current sheets. In spite of the lack of direct observations of these structures in the corona, they are fully expected because of the extremely high electrical conductivity of the medium.

5.1. Current sheets and neutral sheets

Magnetohydrodynamic discontinuities (other than shocks) are commonly observed in the solar wind at 1 AU (Burlaga, 1968, 1969; Siscoe *et al.*, 1968; Burlaga and Ness, 1968, 1969). Direct observations on the Sun or in the inner corona, however, are scant.* Due to their extreme thinness this is to be expected and, hopefully, more observations will be forthcoming from the ATM coronagraph experiments. Neutral sheets are expected above the helmet structures in coronal streamers (Sturrock and Smith, 1968; Pneuman and Kopp, 1971; Endler, 1971) and between adjacent coronal loop systems of opposite polarity such as shown in Figure 12. Current sheets, where the magnetic field is discontinuous but does not reverse polarity, can be expected almost anywhere but especially on the boundary between closed loops and expanding solar wind regions (Pneuman and Kopp, 1971; Endler, 1971).** Quiescent prominences also could possibily be the manifestation of a current sheet consisting of a kink in the magnetic field (Anzer, 1972).

Neutral sheets are formed by the action of the solar wind distending outward the closed magnetic fields of the Sun. As these loops are pulled outward the regions of opposite polarity approach each other and, in the limit of zero resistivity, an infinitely thin sheet develops across which the field polarity reverses. In the actual case however (non-zero resistivity), this sheet contains distended stationary magnetic tongues across which plasma flow takes place (Pneuman, 1972). The equilibrium of the tongues is determined by a balance between the pressure gradient tending to pull

* From a study of old eclipse plates, Eddy (1973) has recently reported an observation of a coronal neutral sheet (1922 eclipse) originating between two coronal helmets.

** Thermal conduction and expansion produce a gas pressure in the open region which is much lower than in the adjacent closed region. Since $P+B^2/8\pi$ must be the same on both sides of the interface, a corresponding discontinuity in B must be present requiring a current sheet.

the loop outward and the reverse $\mathbf{J} \times \mathbf{B}$ force (see Figure 13). Defining a magnetic Reynolds number $R_m = (4\pi/c^2)\,\sigma r_0 v_s$, σ being the electrical conductivity, r_0 the solar radius, and v_s the sound speed, the sheet thickness, transverse magnetic field, and expansion velocity in the sheet are proportional to $R_m^{-1/3}$ whereas the electric current density in the sheet varies as $R_m^{1/3}$. Assuming the sheet cross-section were radial and

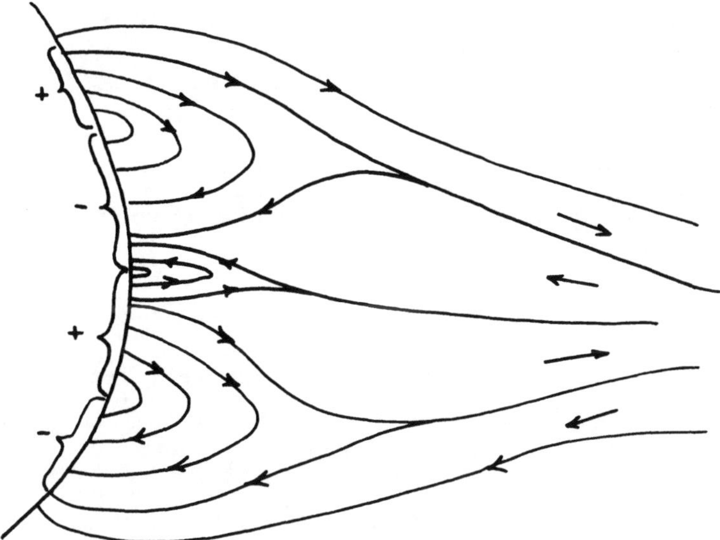

Fig. 12. Resultant field configuration over two bipolar regions of opposing polarity. Here, three current sheets are formed by the solar wind. Because of magnetic forces, these sheets may merge at greater heights. A configuration such as this could produce the neutral sheet seen in the 1922 eclipse (Eddy, 1973) and is the logical result of the action of the solar wind on the magnetic field geometry proposed by Sweet (see Figure 15).

evaluating σ for a current carried by ions perpendicular to the field, thicknesses as small as 500 km were estimated. This, however, is expected to be a lower limit since additional theoretical considerations lead to the expectations that the sheet is not radial but should broaden as one proceeds downward in the corona.* An extension of this concept, incorporating more sophisticated mathematical assumptions, has been carried out by Priest and Smith (1972). In considering these simplified physical models, it should of course be kept in mind that differences in solar wind speed and direction from one side of the sheet to the other could produce a highly sheared and distorted configuration differing drastically from the simple symmetric case.

Although the stability of current sheets in the corona has not been studied, neutral sheets have received some attention. The particular instability suspected in these structures is the tearing mode first suggested by Furth *et al.* (1963). In this instability, reconnection of field lines takes place along the sheet due to finite conductivity effects. Such an instability has been proposed as a mechanism for the formation of quiescent

* The results of Eddy's (Eddy, 1973) observation seems to bear this out.

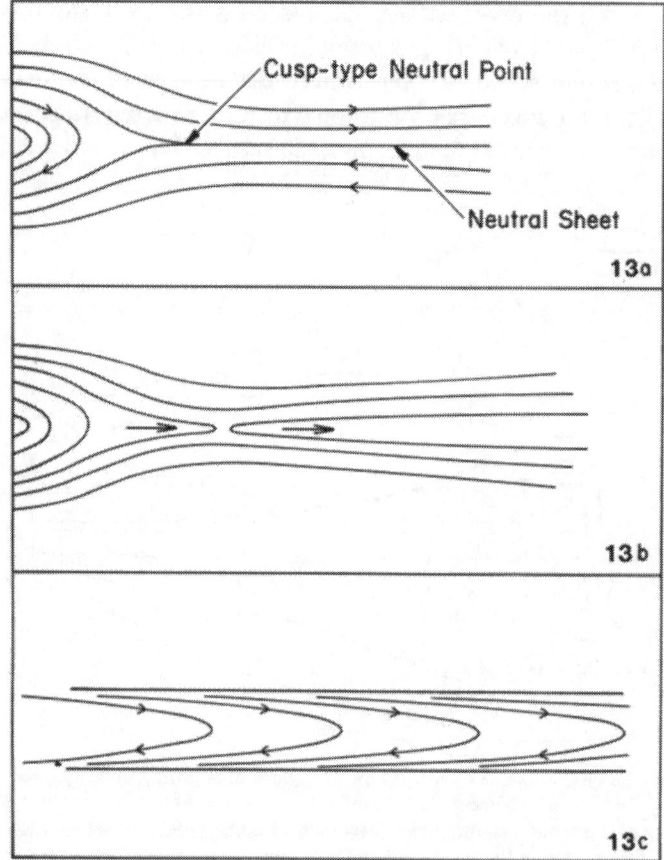

Fig. 13. Schematic of a typical helmet streamer showing how the neutral sheet topology could be developed by finite conductivity effects. (a) Streamer field line topology for the case $\sigma = \infty$ showing the neutral point and current sheet of zero thickness. (b) When finite conductivity is introduced, reconnection of open field lines can occur at the neutral point forming the above shown oppositely directed loops. Arrows denote the subsequent motion of these field lines due to pressure forces. The outer loop is expelled to infinity whereas the inner one expands to a stationary position determined by the reverse $\mathbf{J} \times \mathbf{B}$ force. (c) Resulting topology of field lines in the neutral plane is one of enormously distended magnetic tongues with plasma diffusion taking place across the field (from Pneuman, 1972).

prominences (Kuperus and Tandberg-Hanssen, 1967; Raadu and Kuperus, 1973). In these models, the neutral sheet is assumed thermally unstable giving rise to compression and cooling of material to form the prominences by condensation from the corona. Recent observations however suggest that a coronal origin cannot account for the observed amount of material in prominences (Saito and Tandberg-Hanssen, 1973). If this is so, such a mechanism must be reconsidered.

Although theoretical analyses of the tearing mode instability have shown mixed conclusions (Northrup and Birmingham, 1970; Smith and Raadu, 1972; Coppi and Friedland, 1971; Biskamp and Schlindler, 1971), two considerations lead us to suspect that neutral sheets in the solar corona are stable, at least to gross destruction. Firstly, the long lifetimes of helmet streamers, the stability of the sector structure at 1 AU and of the geomagnetic tail observationally attest to their persistence. Secondly, a neutral sheet such as that pictured in Figure 13 cannot be topologically affected by an infinitesimal displacement as opposed to a sheet that does not contain a transverse magnetic field. This suggests that a disturbance of finite amplitude is necessary to disrupt the sheet – i.e., the configuration could be *metastable*.

An important aspect of neutral sheets to the subject of coronal disturbances is their possible role in guiding both particles and waves. The relevance of these structures to the type III radio burst phenomena and in focusing flare produced MHD waves has already been pointed out. In addition, Bumba and Obridko (1969) have suggested that proton flare activity associated with Bartel's active longitudes occurs in the neighborhood of the sector boundaries of the interplanetary magnetic field. Hence, neutral sheets could be the 'tracks' along which energetic particles travel from the Sun to the Earth. In addition to their guiding role, neutral sheets have been considered in theories of solar flares (Sweet, 1958; Severny, 1958, 1961, 1962a, b; Jaggi, 1964; Sturrock, 1966), prominence formation (Kuperus and Tandberg-Hanssen, 1967; Raadu and Kuperus, 1973) and chromospheric spicules (Uchida, 1969; Pikel'ner, 1969).

5.2. Magnetic Neutral Points and Reconnection Phenomena

Magnetic reconnection at neutral points is believed to be at the root of much solar transient phenomena. Because this process seems to be the most efficient way of converting magnetic energy to other forms, it is often invoked in solar flare theories (Sweet, 1958; Severny, 1958, 1961, 1962a, b; Parker, 1963; Carmichael, 1964; Petschek, 1964; Sturrock, 1966, 1968, 1972; Krivsky, 1968) as well as those involving chromospheric spicules (Pikel'ner, 1969; Uchida, 1969) and prominence disruption (Rust, 1970). It is likely that the so-called 'coronal whips' and other observed transients in $\lambda 5303$ are due to readjustment following magnetic reconnection. According to most theories of the solar cycle, reconnection must take place to explain the observed changes in polarity of the polar field.

Overall coronal evolution is also critically dependent upon how fast field lines can reconnect in the corona – a question that cannot be conclusively answered at the present time. For example, if reconnection rates are rapid as compared to the time scales for evolution of the surface fields, then a *unique* coronal configuration exists for given boundary conditions at the coronal base. If, on the other hand, reconnection is difficult, then the state of the coronal magnetic fields and gas will depend upon the complete time history of the surface changes. This consideration is critical to the overall problem of solar-interplanetary modeling.

Reconnection can occur only at neutral points. Therefore some discussion of their geometry is appropriate here. The 'X-type' neutral point (Dungey, 1953, 1958) (shown schematically in Figure 14) has received the most theoretical interest because of its relevance to flare theories. This configuration can exist between two sunspot pairs as depicted in Figure 15 (Sweet, 1958) or wherever fields of opposite polarity approach each other preferentially at one location.

Three other types of neutral points are possible where field lines undergo a transition from a closed to open configuration such as at the top of coronal helmets. These are the 'Y-type', 'T-type' and 'γ' or 'cusp-type' (Sturrock and Smith, 1968). These are

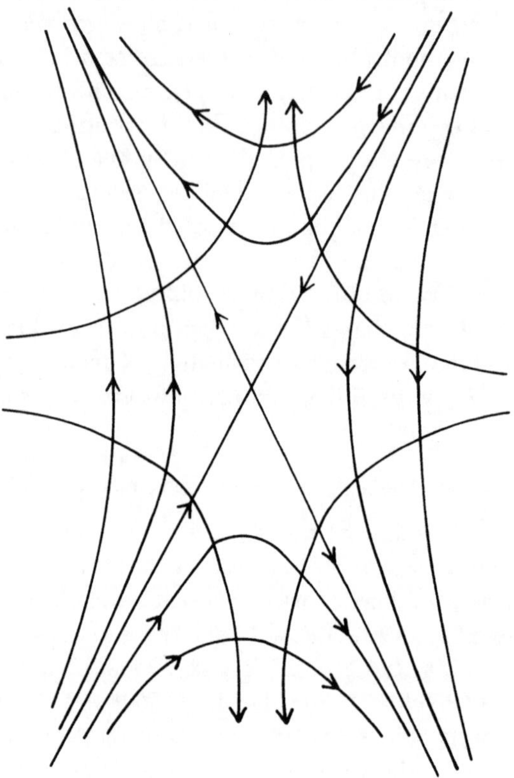

Fig. 14. Schematic of X-type neutral point. The field lines are brought toward the neutral point from the left and right by the flow. After reconnection, they are expelled to the top and bottom.

shown in Figure 16 and each results from a different type of pressure jump between the open and closed regions. If, as seems to be the case, the gas pressure outside the helmet is lower than inside, both the 'Y' and 'T' neutral points can probably be ruled out in the solar corona (Pneuman and Kopp, 1971). The 'Y-type' cannot occur since the field vanishes when approaching the neutral points from all directions and there is consequently no way to balance the pressure jump (since $P+B^2/8\pi$ must be continuous). In the 'T-type', the field vanishes in the open region but not in the closed and the jump in B is of the wrong sign. The only type consistent with the expected mechanical forces is the 'cusp type' in which the field goes to zero approaching the neutral point from the inside but does not vanish on open field lines.

Having reduced the possible neutral point configuration in the corona to two

('X-type' and 'cusp-type') let us now examine the reconnection process. Although reconnection at 'cusp-type' neutral points has received some attention (Kuperus and Tandberg-Hanssen, 1967; Sturrock, 1966, 1968, 1972; Raadu and Kuperus, 1973) the 'X-type' configuration is the usual one considered in theoretical analyses. Moreover, the reconnection processes is essentially the same for both since time dependent reconnection at a cusp would result locally in an X-type geometry.

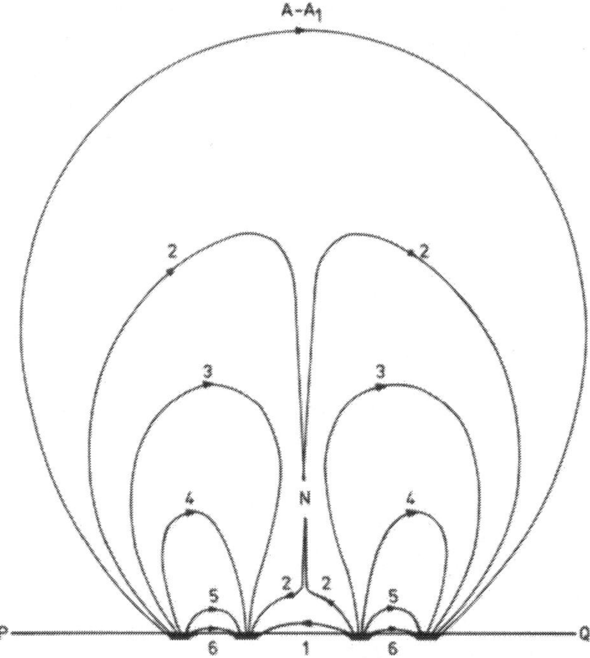

Fig. 15. Neutral point configuration proposed by Sweet (1958) as a model for solar flares.

Early analyses of magnetic reconnection at an 'X-type' neutral point were based upon ordinary dissipation due to the finite conductivity of the plasma. An externally imposed flow approaches the neutral point and, after reconnection took place the fluid was ejected along the axis at essentially the sound speed.

For pressure equilibrium, the sound speed is taken to be equal to the Alfvén speed outside the region. The rate of annihilation is controlled by two factors – the diffusion time and the time required to expel the fluid. This leads to a characteristic reconnection velocity

$$v_d = \left(\frac{B_0}{\sqrt{4\pi\varrho}} \frac{c^2}{4\pi\sigma L} \right)^{1/2},$$

where B_0 is the ambient field outside the boundary, ϱ the density inside and L the characteristic length of the reconnection region. Using reasonable values for these quantities, one can easily verify that the characteristic times obtained through this mechanism are much too long to account for explosive phenomena in the chromosphere and corona. Attempts to alleviate this difficulty have involved corregated neutral surfaces (Parker, 1963) in which the characteristic dimension be reduced and one-dimensional time dependent models (Dungey, 1958; Severny, 1958) in which a nonstationary collapse toward the neutral surface occurs.

In an entirely different mechanism, introduced by Petschek (1964), the magnetic field is annihilated by the propagation of Alfvén waves. The magnetic energy here is directly converted into the kinetic energy of the upward moving plasma. In Petschek's

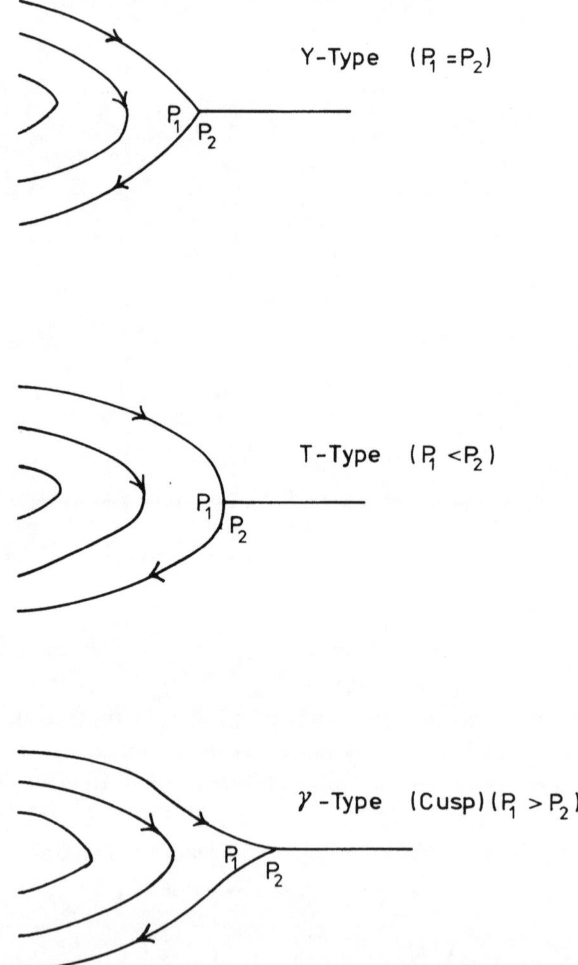

Fig. 16. Three possible neutral point configurations which could exist above a bipolar magnetic region (Sturrock and Smith, 1968). Due to the expected mechanical forces, however, the first two can probably be ruled out.

mechanism, the characteristic velocity is proportional to $(ln\ R_m)^{-1}$ (R_m being the effective magnetic Reynolds number) rather than $R_m^{-1/2}$. In spite of some criticism of the model (Green and Sweet, 1967; Petschek and Thorne, 1967; Priest, 1972), this offers a significant reduction in annihilation time which may be able to account for the observed transient phenomena. Even more in the direction of reduced dissipation times, Yeh and Axford (1970) and Sonnerup (1970) have argued that reconnection is completely independent of the electrical conductivity. Once the input velocity is specified by external conditions, the conductivity just determines the size of the region of annihilation. Hence, reconnection can occur with any speed up to the Alfvén speed.* Instabilities have also been evoked to reduce the reconnection time. For example, Parker (1973) contends that the fluid can escape rapidly from the annihilation region via the interchange and kink instabilities and also finds reconnection speeds of the order of the Alfvén velocity.

In summary, then, the search through the years seems to be toward faster reconnection rates – rates that can now be said to be fully consistent with the observed time scales of coronal disturbances. Processes based upon reconnection alone, however, appear to require very special circumstances to work. If the reconnection rate is too slow, short time scales are not obtained. If, on the other hand, reconnection is too rapid, the event will not be explosive. It is not entirely clear, therefore, that large-scale changes in field topology on a short time scale can be explained by reconnection alone – especially if the region of dissipation is small. For this type of process, the rapid relaxation of a stressed field configuration may be a more attractive explanation.

Acknowledgement

The author wishes to express his thanks to A. Hundhausen for reviewing the manuscript.

References

Ahluwalia, H. S.: 1972, in *Proc. of 12th International Conf. on Cosmic Rays*, Hobart, Tasmania.
Alazraki, G. and Couturier, P.: 1971, *Astron. Astrophys.* **13**, 380.
Alexander, J. K., Malitson, H. H., and Stone, R. G.: 1969, *Solar Phys.* **8**, 388.
Altschuler, M. D. and Newkirk, G., Jr.: 1969, *Solar Phys.* **9**, 131.
Altschuler, M. D. and Perry, R. M.: 1972, *Solar Phys.* **23**, 410.
Altschuler, M. D., Trotter, D. E., and Orrall, F. Q.: 1972, *Solar Phys.* **26**, 354.
Anderson, G. F.: 1966, Ph.D. Thesis, Univ. of Colorado.
Anzer, U.: 1972, *Solar Phys.* **24**, 324.
Anzer, U. and Tandberg-Hanssen, E.: 1970, *Solar Phys.* **11**, 61.
Athay, R. G. and Moreton, G. E.: 1961, *Astrophys. J.* **133**, 935.
Babcock, H. D.: 1959, *Astrophys. J.* **130**, 364.
Babcock, H. W.: 1961, *Astrophys. J.* **133**, 572.
Babcock, H. W. and Babcock, H. D.: 1955, *Astrophys. J.* **121**, 394.
Barnes, C. W. and Sturrock, P. A.: 1972, *Astrophys. J.* **174**, 659.
Belcher, J. W.: 1971, *Astrophys. J.* **168**, 509.
Belcher, J. W. and Davis, L., Jr.: 1971, *J. Geophys. Res.* **76**, 3534.

* Recently Priest (1973) has re-examined this mechanism and concludes that the fastest reconnection rate is about an order of magnitude smaller.

Benz, A. D. and Gold, T.: 1971, *Solar Phys.* **21**, 157.
Bhatnager, A. and Tanaka, K.: 1972, *Solar Phys.* **24**, 87.
Biskamp, D. and Schindler, K.: 1971, *Plasma Phys.* **13**, 1013.
Bohlin, J. D.: 1970a, *Solar Phys.* **12**, 240.
Bohlin, J. D.: 1970b, *Solar Phys.* **13**, 153.
Bohlin, J. D.: 1971, *Solar Phys.* **18**, 450.
Boischot, A. and Denisse, J. F.: 1957, *Compt. Rend. Acad. Sci. Paris* **245**, 2149.
Bostrom, R.: 1972, *Models of Force-Free Magnetic Fields in Resistive Media*, Royal Institute of Technology of Stockholm Rept. TRITA-EPP-72-24.
Bruzek, A.: 1952, *Z. Astrophys.* **31**, 99.
Bruzek, A.: 1964, *Astrophys. J.* **140**, 746.
Bruzek, A. and Demastus, H.: 1970, *Solar Phys.* **12**, 447.
Bugoslavskaya, E. J.: 1950, *Publ. Sternberg Inst.* **19**.
Bumba, V. and Obridko, V. N.: 1969, *Solar Phys.* **6**, 104.
Burlaga, L. F.: 1968, *Solar Phys.* **4**, 67.
Burlaga, L. F.: 1969, *Solar Phys.* **7**, 57.
Carmichael, H.: 1964, in W. Hess (ed.), *AAS-NASA Symposium on Physics of Solar Flares*, NASA SP-50, p. 451.
Carovillano, R. L. and King, J. H.: 1966, *Astrophys. J.* **145**, 426.
Chapman, S.: 1957, *Smithsonian Contrib. Astrophys.* **2**, 1.
Chapman, S. and Bartels, J.: 1940, *Geomagnetism*, Clarendon Press, Oxford.
Coleman, P. J., Jr.: 1967, *Planetary Space Sci.* **15**, 953.
Coppi, B. and Friedland, A. B.: 1971, *Astrophys. J.* **169**, 379.
Cowling, T. G.: 1973, private communication.
d'Azambuja, M. and d'Azambuja, L.: 1948, *Ann. Obs. Paris, Meudon* **6**, fasc. 7.
Dodson, H. W. and Hedeman, E. R.: 1964, in W. Hess (ed.), *AAS-NASA Symposium on Physics of Solar Flares*, NASA SP-50, p. 15.
Dodson, H. W. and Hedeman, E. R.: 1968, in Y. Öhman (ed.), 'Mass Motions in Solar Flares and Related Phenomena', *Nobel Symp.* **9**, 37.
Dulk, G. A. and Altschuler, M. A.: 1971, *Solar Phys.* **20**, 438.
Dulk, G. A., Altschuler, M. D., and Smerd, S. F.: 1972, *Astrophys. Letters* **8**, 235.
Dungey, J. W.: 1953, *Phil. Mag.* **44**, 725.
Dungey, J. W.: 1958, *Cosmic Electrodynamics*, Cambridge Univ. Press, p. 98.
Dunn, R.: 1970, *AIAA Bull.* **7**, 564.
Eddy, J.: 1973, *Solar Phys.* **30**, 385.
Endler, F.: 1971, Ph.D. Thesis, Göttingen Univ.
Evans, J. W.: 1957, *Publ. Astron. Soc. Pacific* **69**, 421.
Fainberg, J. and Stone, R. G.: 1970a, *Solar Phys.* **15**, 222.
Fainberg, J. and Stone, R. G.: 1970b, *Solar Phys.* **15**, 433.
Fainberg, J. and Stone, R. G.: 1971, *Solar Phys.* **17**, 392.
Fan, C. Y., Pick, M., Pyle, R., Simpson, J. A., and Smith, D. R.: 1968, *J. Geophys. Res.* **73**, 1555.
Ferraro, V. C. A. and Plumpton, C.: 1958, *Astrophys. J.* **127**, 495.
Furth, H. P., Killeen, J., and Rosenbluth, M. N.: 1963, *Phys. Fluids* **6**, 459.
Garcia, C., Hansen, R. T., Hull, H., Lacey, L., and Lee, R.: 1971, *Bull. Am. Astron. Soc.* **3**, 261.
Gold, T.: 1963, *Proc. Pontificial Acad. of Sciences, Vatican* **25**, 431.
Goldsmith, D. W.: 1971, *Solar Phys.* **19**, 86.
Gosling, J. T., Pizzo, V., and Bame, S. J.: 1973, *J. Geophys. Res.* **78**, 2001.
Green, R. M. and Sweet, P. A.: 1967, *Astrophys. J.* **147**, 1153.
Haddock, F. T. and Graedel, T. E.: 1970, *Astrophys. J.* **160**, 293.
Hansen, R. T.: 1973, private communication.
Hansen, R. T., Garcia, C. J., Hansen, S. F., and Loomis, H. G.: 1969, *Solar Phys.* **7**, 417.
Hansen, S. F., Hansen, R. T., and Garcia, C. J.: 1972, *Solar Phys.* **26**, 202.
Hartz, T. R.: 1964, *Ann. Astrophys.* **27**, 823.
Hartz, T. R.: 1969, *Planetary Space Sci.* **17**, 267.
Hepburn, N.: 1955, *Astrophys. J.* **122**, 445.
Hildner, E.: 1971, Ph.D. Thesis, Univ. of Colorado.
Hollweg, J. V.: 1971a, *J. Geophys. Res.* **76**, 5155.

Hollweg, J. V.: 1971b, *Phys. Rev. Letters* **27**, 1349.
Hollweg, J. V.: 1972a, *Cosmic Electrodyn.* **2**, 423.
Hollweg, J. V.: 1972b, *Astrophys. J.* **177**, 255.
Hollweg, J. V.: 1973, *Astrophys. J.* **181**, 547.
Hulst, H. C. van de: 1950, *Bull. Astron. Inst. Neth.* **11**, 150.
Hundhausen, A. J.: 1972, *Coronal Expansion and Solar Wind*, Springer-Verlag, New York.
Hyder, C. L.: 1967a, *Solar Phys.* **2**, 49.
Hyder, C. L.: 1967b, *Solar Phys.* **2**, 267.
Jockers, K.: 1968, *Solar Phys.* **3**, 603.
Kai, K.: 1969, *Solar Phys.* **10**, 460.
Kai, K.: 1970a, *Solar Phys.* **11**, 310.
Kai, K.: 1970b, *Solar Phys.* **11**, 456.
Kippenhahn, R. and Schluter, A.: 1957, *Z. Astrophys.* **43**, 36.
Kleczek, J.: 1957, *Bull. Astron. Inst. Czech.* **8**, 120.
Kleczek, J.: 1958, *Bull. Astron. Inst. Czech.* **9**, 115.
Kleczek, J.: 1963, *Publ. Astron. Soc. Pacific* **75**, 9.
Koutchmy, S.: 1971, *Astron. Astrophys.* **13**, 79.
Krieger, A. S., Timothy, A. F., and Roelof, E. C.: 1973, *Solar Phys.* **29**, 505.
Krieger, A. S., Vaiana, G. S., and Van Speybroeck, L. P.: 1971, in R. Howard (ed.), 'Solar Magnetic Fields', *IAU Symp.* **43**, 397.
Krimingis, S. M.: 1969, in *Proc. 11th International Conf. on Cosmic Rays* **2**, 125.
Krimingis, S. M. and Verzariu, P.: 1971, *J. Geophys. Res.* **76**, 792.
Krivsky, L.: 1968, in K. O. Kiepenheuer (ed.), 'Structure and Development of Solar Active Regions', *IAU Symp.* **35**, 465.
Kuperus, M. and Tandberg-Hanssen, E.: 1967, *Solar Phys.* **2**, 39.
Lantos, J.: 1970, *Solar Phys.* **15**, 40.
Leblanc, Y.: 1970, *Astron. Astrophys.* **4**, 315.
Lin, R. P., Kahler, S. W., and Roelof, E. C.: 1968, *Solar Phys.* **4**, 338.
Lockyer, N.: 1931, *Monthly Notices Roy. Astron. Soc.* **91**, 797.
Low, B. C.: 1973, *Astrophys. J.* **181**, 209.
Lust, R. and Zirin, H.: 1960, *Z. Astrophys.* **49**, 8.
McLean, D. J.: 1969, *Proc. Astron. Soc. Australia* **1**, 188.
McLean, D. J.: 1970, *Proc. Astron. Soc. Australia* **1**, 315.
Meyer, F.: 1968, in K. O. Kiepenheuer (ed.), 'Structure and Development of Solar Active Regions', *IAU Symp.* **35**, 485.
Meyer, F. and Schmidt, H. U.: 1968, *Angew. Math. Mech.* **48**, 218.
Michard, R.: 1954, *Ann. Astrophys.* **17**, 429.
Montgomery, M. D., Asbridge, J. R., Bame, S. J., and Feldman, W. C.: 1972, *EOS, Trans. Am. Geophys. Union* **53**, 503.
Moreton, G. E.: 1960, *Astron. J.* **65**, 494.
Moreton, G. E.: 1964, *Astron. J.* **69**, 145.
Nakagawa, Y.: 1970, *Solar Phys.* **12**, 419.
Nakagawa, Y. and Raadu, M. A.: 1972, *Solar Phys.* **25**, 127.
Nakagawa, Y., Raadu, M. A., Billings, D. E., and McNamara, D.: 1971, *Solar Phys.* **19**, 72.
Newkirk, G., Jr.: 1967, *Ann. Rev. Astron. Astrophys.* **5**, 213.
Newkirk, G., Jr.: 1971, in R. Howard (ed.), 'Solar Magnetic Fields', *IAU Symp.* **43**, 547.
Newkirk, G., Jr.: 1972a, in C. P. Sonett, P. J. Coleman, and J. M. Wilcox (eds.), *Proc. Asilomar Conf. on the Solar Wind*, NASA, Washington, p. 11.
Newkirk, G., Jr.: 1972b, in R. Ramaty and R. G. Stone (eds.), *Proc. Symposium on High Energy Phenomena on the Sun*, Greenbelt, Maryland, 1972, NASA/GSFC Preprint X-693-73-193, p. 453.
Newkirk, G., Jr., Altschuler, M. D., and Harvey, J. W.: 1968, in K. O. Kiepenheuer (ed.), 'Structure and Development of Solar Active Regions', *IAU Symp.* **35**, 379
Newkirk, G., Jr., Dupree, R. G., and Schmahl, E. J.: 1970, *Solar Phys.* **15**, 15.
Noci, G.: 1973, *Solar Phys.* **28**, 403.
Northrup, T. G. and Birmingham, T. J.: 1970, *Solar Phys.* **14**, 226.
Osterbrock, D. E.: 1961, *Astrophys. J.* **134**, 347.
Parker, E. N.: 1963, *Astrophys. J. Suppl.* **77**, 8.

Parker, E. N.: 1964a, *Astrophys. J.* **139**, 690.
Parker, E. N.: 1964b, *Astrophys. J.* **140**, 1170.
Parker, E. N.: 1965, *Space Sci. Rev.* **4**, 666.
Parker, E. N.: 1966, *Astrophys. J.* **143**, 32.
Parker, E. N.: 1973, *Astrophys. J.* **180**, 247.
Petschek, H. E.: 1964, in W. Hess (ed.), *AAS-NASA Symposium on Physics of Solar Flares*, NASA SP-50, p. 425.
Petschek, H. E. and Thorne, R. M.: 1967, *Astrophys. J.* **147**, 1157.
Pikel'ner, S. B.: 1969, *Soviet Astron.* **13**, 259.
Pneuman, G. W.: 1967, *Solar Phys.* **2**, 462.
Pneuman, G. W.: 1968, *Solar Phys.* **3**, 578.
Pneuman, G. W.: 1969, *Solar Phys.* **6**, 255.
Pneuman, G. W.: 1972, *Astrophys. J.* **177**, 793.
Pneuman, G. W.: 1972, *Solar Phys.* **23**, 223.
Pneuman, G. W.: 1973, *Solar Phys.* **28**, 247.
Pneuman, G. W. and Kopp, R. A.: 1970, *Solar Phys.* **13**, 176.
Pneuman, G. W. and Kopp, R. A.: 1971, *Solar Phys.* **18**, 258.
Priest, E. R.: 1972, *Monthly Notices Roy. Astron. Soc.* **159**, 389.
Priest, E. R.: 1973, *Astrophys. J.* **181**, 227.
Raadu, M. A.: 1972, *Solar Phys.* **22**, 443.
Raadu, M. A. and Kuperus, M.: 1973, *Solar Phys.* **28**, 77.
Raadu, M. A. and Nakagawa, Y.: 1971, *Solar Phys.* **20**, 64.
Raju, P. K.: 1968, *Monthly Notices Roy. Astron. Soc.* **138**, 479.
Ramsey, H. E. and Smith, S. F.: 1966, *Astron. J.* **71**, 197.
Riddle, A. C.: 1970, *Solar Phys.* **13**, 448.
Rust, D. M.: 1970, *Astrophys. J.* **160**, 315.
Saito, K.: 1959, *Publ. Astron. Soc. Japan* **11**, 234.
Saitko, K. and Billings, D. E.: 1964, *Astrophys. J.* **140**, 760.
Saito, K. and Owaki, N.: 1967, *Publ. Astron. Soc. Japan* **19**, 535.
Saito, K. and Tandberg-Hanssen, E.: 1973, *Solar Phys.* **31**, 105.
Sakurai, K. and Stone, R. G.: 1971, *Solar Phys.* **19**, 247.
Schatten, K. H.: 1970, *Solar Phys.* **12**, 484.
Schatten, K. H.: 1971, *Cosmic Electrodyn.* **2**, 232.
Schatten, K. H., Ness, N. F., and Wilcox, J. M.: 1968, *Solar Phys.* **5**, 240.
Schatten, K. H., Wilcox, J. M., and Ness, N. F.: 1969, *Solar Phys.* **6**, 442.
Schmidt, M.: 1953, *Bull. Astron. Inst. Neth.* **12**, 59.
Schmidt, H. U.: 1964, in W. Hess (ed.), *AAS-NASA Symposium on Physics of Solar Flares*, NASA SP-50, p. 107.
Schmidt, H. U.: 1969, in K. O. Kiepenheuer (ed.), 'Structure and Development of Solar Active Regions', *IAU Symp.* **35**, 95.
Severney, A. B.: 1958, *Astron. Zh.* **35**, 335.
Severney, A. B.: 1961, *Astron. Zh.* **38**, 402.
Severney, A. B.: 1962a, *Astron. Zh.* **39**, 990.
Severney, A. B.: 1962b, *Izv. Krymk. Astrofiz. Obs.* **27**, 71.
Siscoe, G. L., Davis, L., Jr., Coleman, P. J., Jr., Smith, E. J., and Jones, D. E.: 1968, *J. Geophys. Res.* **73**, 61.
Slysh, V. I.: 1967a, *Astron. Zh.* **44**, 94.
Smerd, S. F. and Dulk, G. A.: 1971, in R. Howard (ed.), 'Solar Magnetic Fields', *IAU Symp.* **43**, 616.
Smith, D. F. and Pneuman, G. W.: 1972, *Solar Phys.* **25**, 461.
Smith, D. F. and Raadu, M. A.: 1972, *Cosmic Electrodyn.* **3**, 285.
Sonnerup, B. V. O.: 1970, *J. Plasma Phys.* **4**, 161.
Stewart, R. T. and Sheridan, K. V.: 1970, *Solar Phys.* **12**, 229.
Stewart, R. T., Sheridan, K. V., and Kai, K.: 1970, *Proc. Astron. Soc. Australia* **1**, 313.
Sturrock, P. A.: 1966, *Nature* **211**, 695.
Sturrock, P. A.: 1968, in K. O. Kiepenheuer (ed.), 'Structure and Development of Solar Active Regions', *IAU Symp.* **35**, 471.
Sturrock, P. A.: 1972, *Solar Phys.* **23**, 438.
Sturrock, P. A. and Smith, S. M.: 1968, *Solar Phys.* **5**, 87.

Sweet, P. A.: 1958, in B. Lehnert (ed.), 'Electromagnetic Phenomena in Cosmical Physics', *IAU Symp.* **6**, 123.
Tandberg-Hanssen, E.: 1967, *Solar Activity*, Blaisdell Publ. Co., Waltham, Mass.
Tandberg-Hanssen, E. (ed.): 1974, *Solar Prominences*, Reidel Publ. Co., Dordrecht, Holland.
Tousey, R.: 1972, presented at *Conference on Flare-Produced Shock Waves in the Corona and Interplanetary Space*, Boulder, Colorado.
Uchida, Y.: 1963, *Publ. Astron. Soc. Japan* **15**, 65.
Uchida, Y.: 1968, *Solar Phys.* **4**, 30.
Uchida, Y.: 1969, *Publ. Astron. Soc. Japan* **21**, 128.
Uchida, Y., Altschuler, M. D., and Newkirk, G., Jr.: 1973, *Solar Phys.* **28**, 495.
Unti, T. W. J. and Neugebauer, M.: 1968, *Phys. Fluids* **11**, 563.
Veeder, G. J. and Zirin, H.: 1970, *Solar Phys.* **12**, 391.
Weiss, A. A. and Stewart, R. T.: 1965, *Australian J. Phys.* **18**, 143.
Weiss, A. A. and Wild, J. P.: 1964, *Australian J. Phys.* **17**, 282.
Weymann, R. and Howard, R.: 1958, *Astrophys. J.* **128**, 142.
Wilcox, J. M.: 1968, *Space Sci. Rev.* **8**, 258.
Wilcox, J. M. and Ness, N. F.: 1965, *J. Geophys. Res.* **70**, 5793.
Wilcox, J. M. and Ness, N. F.: 1967, *Solar Phys.* **1**, 437.
Wild, J. P.: 1969, *Proc. Astron. Soc. Australia* **1**, 181.
Wild, J. P. and Smerd, S. F.: 1972, *Ann. Rev. Astron. Astrophys.* **10**, 159.
Wild, J. P., Smerd, S. F., and Weiss, A. A.: 1963, *Ann. Rev. Astron. Astrophys.* **1**, 291.
Withbroe, G. L., Dupree, A. K., Goldberg, L., Huber, M. C. E., Noyes, R. W., Parkinson, W. H., and Reeves, E. M.: 1972, *Solar Phys.* **21**, 272.
Yeh, T. and Axford, W. I.: 1970, *J. Plasma Phys.* **4**, 207.
Zirin, H.: 1972, *Solar Phys.* **22**, 34.

DISCUSSION

Dryer: The interpretation that led to the suggestion that a closed 'magnetic bottle' had been observed may not be correct; the observations can also be explained without involving closed fields. Also, enhancements at 1 AU have been observed without the corresponding enhancement in B_r which would occur in a 'bottle'. The white light observations certainly suggest a 'bottle' but we are not yet certain that closed magnetic fields are involved.

Pneuman: Another explanation could be that the ejected plasma flows out from the Sun in a much wider angle than the r^2 expansion of the solar wind. This could result in an anomalously low temperature by adiabatic cooling due to overexpansion.

Schutten: What characterizes 'coronal holes'?

Pneuman: In my opinion they are regions of open field lines where the solar wind flows resulting in a drain in energy content due to thermal conduction and expansion. Calculations, using this hypothesis, show that density differences of a factor of ten and temperature differences of a factor of about two can be expected between closed and open regions.

Schmidt: In defense of 'magnetic bottles', I should note that the cool electrons seem to flow in from both directions as would be expected.

Mullaly: Decimetre (21, 43 cm) bright region auto-correlations show persistence of stable magnetic fields in the corona for ~year. The possibility remains that the persistent regions may be 'holes' rather than 'bright' regions.

Newkirk: Similar persistence of some coronal features has been known for some time.

Smith: As correctly pointed out by Parker, none of the analyses of reconnection to date can tell what determines the rate of reconnection. Thus, all analyses which have increased the reconnection rate have done so by guessing, and you must accept these analyses pretty much on faith.

Dryer: A recent paper (*J. Plasma Phys.*) by Fukao and Tsuda provided this kind of computation for the diffusion region with finite electrical conductivity. They found the Yeh and Axford suggestion, that (for situations where the proton gyro radius is small relative to the diffusion region's scale size) the reconnection rate is independent of the conditions within the diffusion region and depends only on the external boundary conditions, is valid.

Smith: Tsuda obtained the result which he did because he set the problem up that way. His analysis was also a steady-state analysis and thus gives us no new information about what determines the rate of reconnection.

Dryer: I believe you are referring to their earlier paper. They have recently published a time-dependent solution.

Zirin: Filaments lie *along* field lines; this does not seem compatible with Kippenhahn-Schlüter model.

Pneuman: How certain is it that magnetic-field is directed along the filament?

Zirin: I am sure it is.

Meyer: Kippenhahn and Schlüter showed that any stable prominence supporting field configuration must *already* contain a trough even before the filament material has settled into this trough. The filament material will bring its own magnetic field with it. The natural configuration which that process will produce is then one in which the magnetic field of the heavy filament gas lies along the trough. Thus, the magnetic field of the filament gas is along the filament, the supporting field is, however, across.

INVERSION LINES OF PHOTOSPHERIC MAGNETIC FIELDS AND SOLAR CORONA

F. AXISA

Service d'Electronique Physique, Centre d'Études Nucléaires de Saclay, France

and

M.-J. MARTRES and C. MERCIER*

Solar Dept. Meudon Observatory, 92190 Meudon, France

Abstract (*Solar Phys.*). Two results were obtained recently at Meudon concerning the metric type III production by flares which both emphasize the importance of the local magnetic configuration involved.

The first point explores the possibility that the particular location of Flare Productive Sites in an active region is closely connected to the production of type III radio bursts (Axisa, 1974). It is shown that there is a significant difference between the type III productivity of flares which occur in different Flare Productive Sites, even in the same active region. Qualitative informations about the photospheric magnetic pattern support the point that the type III production is stimulated in the case the flares are located near the periphery of the facular pattern where the strength of the magnetic field is rather low at the photospheric level.

For example, Figure 1 shows the photospheric magnetic field pattern of the active center McMath region 8905 observed on July 27, 1967.

North polarity is delineated by full lines, south polarity by dotted lines, hatchings indicate the location of the flare producing sites (FPS).

– FPS 1 is located inside strong magnetic fields, on penumbra of a complex spot and was very poor in producing type III bursts.

– FPS 2 is located at the periphery of the facular region near the preceding spot: it produced many type III bursts.

From the study of the FPS's for several active regions, it was found that for FPS in such a situation as FPS 1, among 67 flares, 14% only was accompanied by type III bursts; for FPS in such a situation as FPS 2, 60 flares appeared and 60% of them were associated with groups of type III bursts. Incidently it is to note that FPS 2 was also responsible for the electron events observed and discussed by Lin.

So it was found that the magnetic structure giving rise to a flare associated with type III bursts is very often characterised by the existence of an inversion line of longitudinal magnetic field between a spot (preceding or following) and its 'parasitic polarity'. *This line appears in region of high gradient of longitudinal magnetic fields and vanishes very near weak fields outside the facular region.*

The second point concerns the association between filaments and type III bursts (Mercier, 1973).

* Equipe de Recherche Associée du C.N.R.S. No. 306.

Gordon Newkirk, Jr. (ed.), Coronal Disturbances, 69–72. All Rights Reserved.

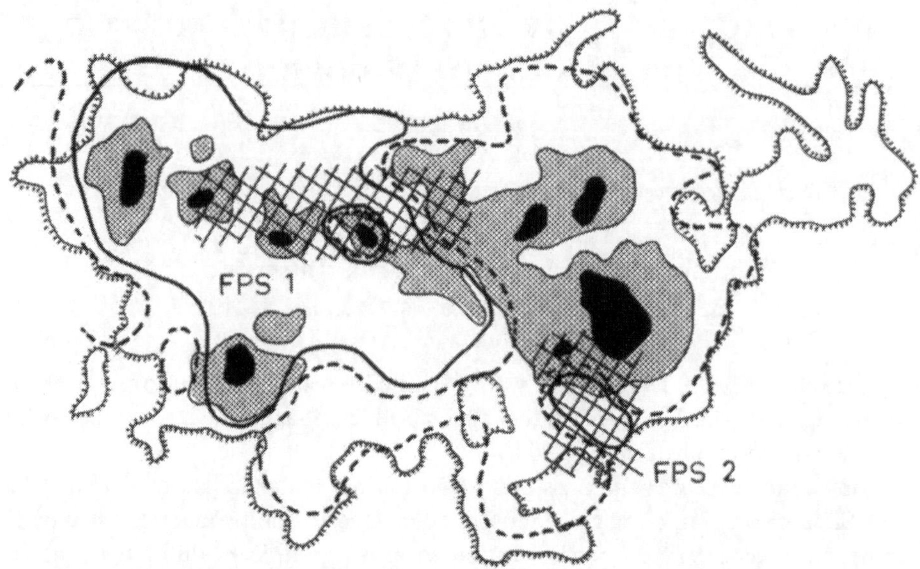

Fig. 1. Spot group McMath 8905. Observed on 1967, July 27. The spots are presented as visible on spectroheliogram K_{1v}: black areas for umbrae, and dotted for penumbrae. The facula is outlined by a 'comb'. The longitudinal magnetic polarities are indicated by a full line (north polarity) or dashed line (south polarity). Hatched regions shows the flare producing sites (FPS) of this day (from Axisa, 1973).

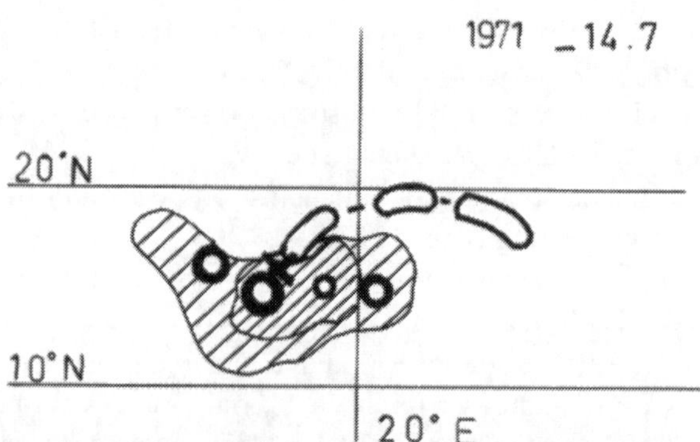

Fig. 2. Active region as seen on Synoptic Maps of the chromosphere. The facula and the spots are schematically drawn. Differences in hatching mark differences in brightness in the facula. The cross indicates the approximate position of the flare. The filament concerned appears as closed double parallel lines (from Mercier, 1973).

It is often observed that a filament (called plagefilament) appears along such an inversion line. As it is schematically shown in Figure 2, this filament is often of small size and does not last more than 3 or 4 days. Occasionally it may be directed toward a greater system of quiescent filaments.

It is worth emphasizing that the plage filament is not always present at the time of type III occurences. The delay between the type III burst and the appearance of the filament very rarely exceeds one day. The histogram of distribution of this delay in the observed cases is presented in Figure 3. Negative delays correspond to cases where filaments appeared after the type III burst occurence.

In addition, it is observed, in one case, that the filament becomes more visible one day after the first observation, suggesting the time scale of the formation of such a

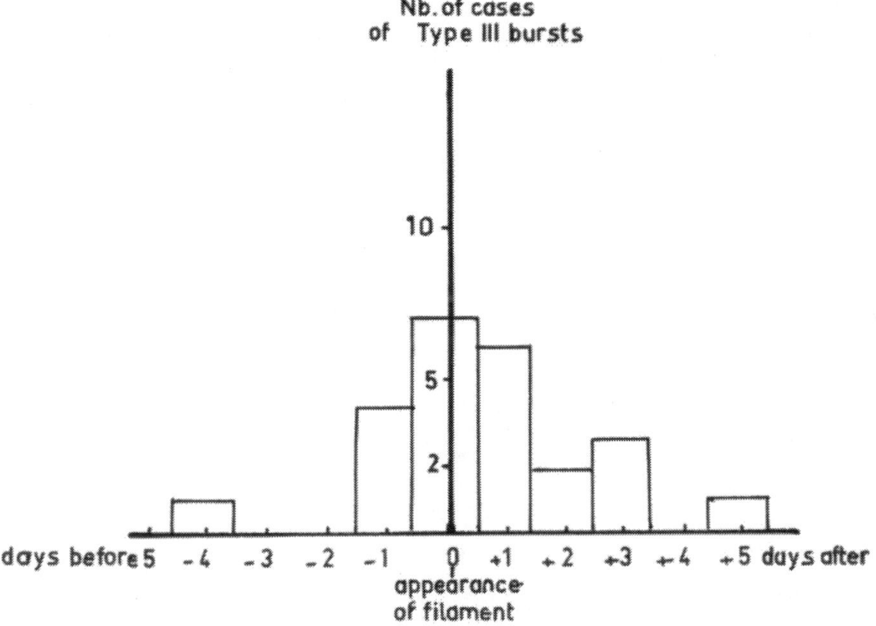

Fig. 3. Histogram of the distribution of delays between the occurrence of type III bursts and the appearance of filaments. Negative delay indicates that the type III bursts occurred before this appearance (from Mercier, 1973).

filament in the order of one day. This fact lead to informations about the condensation process of the prominence.

If it is accepted that type III bursts are generated in coronal neutral sheets, these results would suggest that filaments are condensed from preexisting neutral sheets with a time scale of one day or less. This interpretation is consistent with the theoretical picture given by Kuperus and Raadu in which filaments result on thermal instability in neutral magnetic sheets.

Inside such coronal structures, the electronic density may be estimated by observational radio positions, as is to be discussed by Rosenberg in the present Symposium. It is difficult to reach such information by ground coronometer observations (eclipses apart) due to the extreme location, the fast evolution, the size and the integration along the line of sight.

The observational radio results reveal, with good conditions, the exact root of this structure in the low solar atmosphere.

References

Axisa, F.: 1974, *Solar Phys.* **35**, 207–224.
Mercier, C.: 1973, *Solar Phys.* **33**, 177.

DISCUSSION

Leblanc: What is the position of type III bursts relative to quiescent filaments?

Rosenberg: We studied the III-filament relation from one-dimensional Nancay observations and concluded that filaments certainly do influence the location of type III bursts. This is only a partial answer to your question.

LARGE-SCALE MAGNETIC STRUCTURES RESPONSIBLE FOR CORONAL DISTURBANCES

V. BUMBA

Astronomical Institute of the Czechoslovak Academy of Sciences, Ondřejov Observatory, Czechoslovakia

and

J. SÝKORA

Astronomical Institute of the Slovak Academy of Sciences, Skalnaté Pleso Observatory, Czechoslovakia

Abstract. In the first part of our communication, in a short summary of our recent results, it is demonstrated that over the last 15 years (the time interval for which the magnetic synoptic charts are available) the largest solar activity, usually connected with proton-flare occurrence, has been very closely related to a characteristic large-scale pattern in the magnetic field distribution with a life-time of the order of 8–10 solar rotations. These regular features are seen in the negative as well as in the positive polarity, although they can be seen better in the negative polarity where their forms are more pronounced, regardless of the activity cycle. The form alternates with the location in the northern or the southern solar hemispheres due to the mutual relations of both polarities, individual active regions, the influence of the differential rotations, the shift in longitude, etc. This pattern could be observed up to the large August 1972 proton-flare activity.

The large-scale magnetic patterns described are accompanied by characteristic large-scale features in the green (λ 5303) coronal emission, presented in the form of isophotes on the synoptic charts (reduced to a unified photometric scale).

In the second part of the presentation, preliminary results, concerning the correlation of the longitudinal distribution of the green coronal emission with the negative and the positive polarity fields for two time intervals (August 1960–September 1961 and January 1969–December 1969) are described. The existence of two 'coronal active longitudes' in both intervals, as well as the close relation of these longitudinal emission maxima to certain parts of the large-scale characteristic bodies of negative polarity, is discussed. Also, the existence of one heliographic longitude, connected with 'coronal holes' (minimal green corona emission), and its relation to the positive polarity large-scale pattern are proved.

1. Introduction

The greatest coronal disturbances are connected with large solar flares, above all with particle-emitting flares. The active regions in which these flares occur, have not only a specific magnetic field configuration, but this magnetic situation seems to develope within the frame of a slowly changing large-scale distribution of the background magnetic field patterns in the solar atmosphere. Using the solar magnetic field synoptic charts, constructed from the Mt. Wilson Observatory daily disk maps, we have tried several times to study the regularities and recurrences in this distribution and the dynamics of its development, especially in relation to the distribution of solar activity, and of the green corona and large solar flares (see, for example, Bumba and Howard, 1969; Ambrož *et al.*, 1971; Bumba, 1972a, b; Bumba and Sýkora, 1973). In the present note we would, first of all, like to give a short summary of our results as concerns the close relationship of particle emitting flares and a certain regular type of large-scale magnetic field distribution, which is reflected in the distribution of green corona emission, over the last fifteen years. Using the com-

Gordon Newkirk, Jr. (ed.), Coronal Disturbances, 73–83.

parison of the longitudinal distribution of the green coronal emission with that of the magnetic fields, we are trying to find some physical meaning for the striking morphological regularities.

2. Particle-Emitting Flares and the Regular Patterns of the Large-Scale Distribution of Solar Magnetic Fields and Green Corona

Two years ago, it was demonstrated that during the declining part of the last 19th cycle of solar activity, the flares which are followed by a particle emission, observed as cosmic ray or PCA events, tend to correlate with a specific pattern of the background magnetic field (Bumba, 1972a). Of 46 flares, accompanied by type IV radiobursts, observed between August 1959 and September 1963 in 22 active regions, about 32, or possibly 40 flares in all or possibly 17 active regions occurred in this characteristic magnetic field configuration. (4 flares in 3 active regions were not correlated because of the poor quality of the synoptic maps.) Although the number of such flares is also relatively large during the present 20th cycle, this problem has not yet been treated statistically, because it will still take years to complete the list of flares. But several of the largest events have been examined with regard to this behaviour, and the same relation to practically the same form of magnetic features has been found, including the August 1972 large active region (Bumba *et al.*, 1972; Bumba and Sýkora, 1972a, b; Bumba, 1973).

The best example of the complex situation in the background magnetic field distribution connected, with a proton-flare region, may be seen displayed in both polarities in colour and in the purest form on the synoptic chart for rotation No. 1428 in our Atlas (Howard *et al.*, 1967), where the mean heliographic position of the centre is about 130° in longitude and about 12° N in latitude. The magnetic field pattern may be described even better on synoptic charts where the polarities are drawn separately (Figures 1, 2 and 3). The events in negative polarity are related to a drop-shaped (we call it 'supergiant') structure with its tail in the higher latitudes of the northern hemisphere. The drop with its head to the west, formed from older activity regions, more stable and regular, is usually about 90°–100° long. The centre of gravity of the major solar activity and of the particle-emitting flare occurrence is located at the eastern part of the feature, where the activity changes more rapidly, just below the root of the tail. The area of positive polarity forms a 'mirror' image to that of the negative polarity with its tail stretched out to the southern latitudes and with its body fitting into the drop of the negative polarity feature.

The development of the described magnetic patterns is a very complex process, because morphologically both magnetic polarities studied separately on a large scale do not develop simultaneously and in phase with regard to time and heliographic longitude. The life-time of a 'supergiant' structure is of the order of one year although it can best be observed only for a few rotations.

Correlating the synoptic charts of the green (λ5303) coronal emission distribution, drawn in the form of isophotes from coronal data which were reduced to a unified

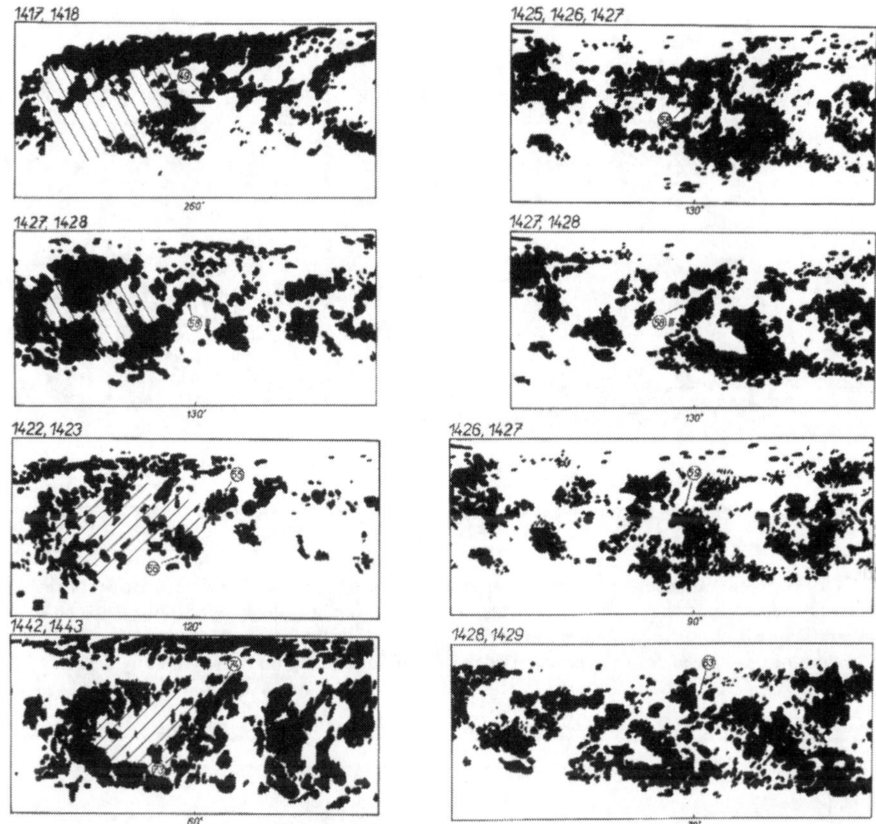

Fig. 1. Examples of the drop-shaped ('supergiant') structures to which the occurrence of particle-emitting flares is related in negative (to the left) and positive (to the right) polarity large-scale magnetic fields. The numbers of rotations of the magnetic charts, two of which were overlapped for integration, are shown, as well as the heliographic longitude of the centre of the magnetic map in the neighbourhood of which the flare centre is located (indicated by an arrow and a number in a circle). The tails, located at higher latitudes of the northern hemisphere for the negative polarity and of the southern hemisphere for the positive polarity bodies, can be seen.

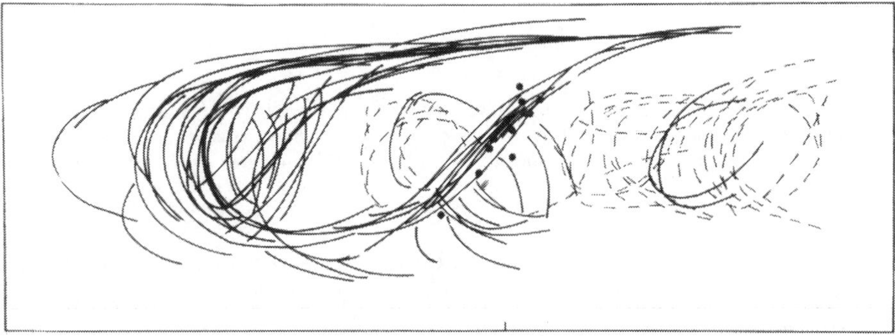

Fig. 2. Diagram of overlapping of best visible drop-shaped ('supergiant') negative polarity structures to which the particle-emitting flares with their positions, indicated by full circles, are related.

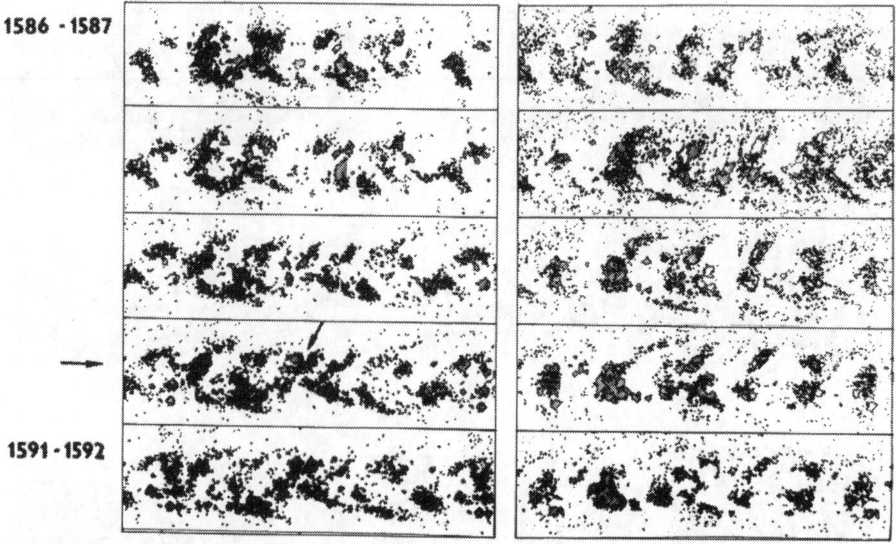

Fig. 3. Development of discussed magnetic patterns, demonstrated in separated polarities for the August 1972 active region. Negative polarity large-scale fields are to the left, positive polarity fields to the right. For integration two consecutive charts, one of which is repeated, are overlapped. The numbers of the rotations are given, and the position of the August active region is indicated by arrows.

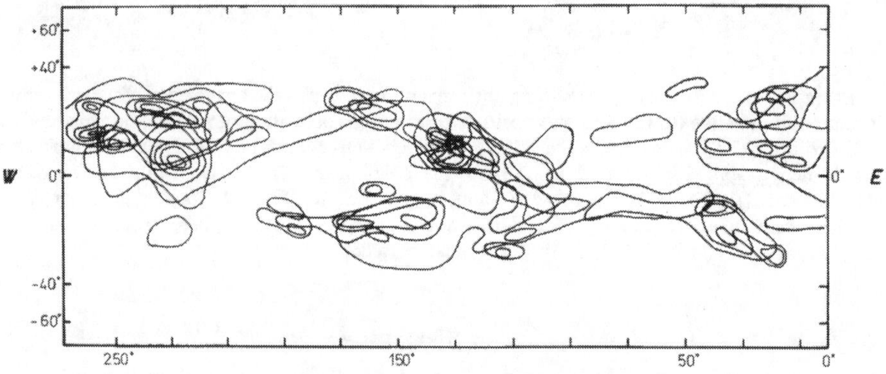

Fig. 4. Example of the 'supergiant' structure to which the occurrence of particle-emitting flares is related (positions of flares are indicated by full circles) in the distribution of the green coronal emission (only the maximal isophotes were drawn). Three synoptic charts (Nos. 1425, 1426 and 1427) were overlapped.

photometric scale (Sýkora, 1971, 1972) with the same large flare active regions, we may see that big elliptical features, formed from an enhanced coronal emission, conform practically with the 'supergiant' body of the magnetic field. The emission is concentrated to the periphery of this body and is usually higher above the proton-flare region (Figures 4 and 5).

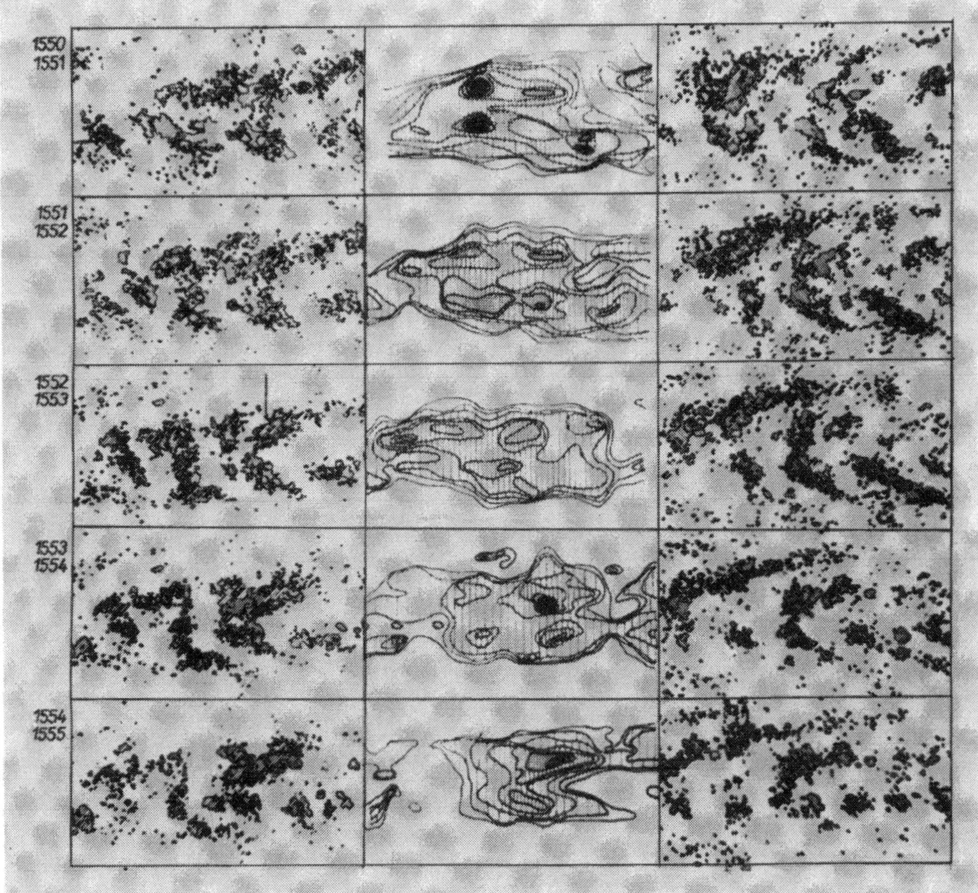

Fig. 5. Parts of synoptic charts (Nos. 1530–1555) of the distribution of the green coronal emission (in the middle part of the figure) and of the same magnetic synoptic charts, drawn in separated polarities (positive on the left side, negative on the right side of the picture) for the proton-flare event of November 2, 1969. For integration two consecutive magnetic maps, one of which is repeated, are overlapped. The development of the large complex magnetic pattern, connected with the evolution of green coronal characteristic feature, may be observed. The position of the particle-emitting flare is visualized by an arrow.

3. Longitudinal Distribution of the Green Coronal Emission and its Relation to the Background Magnetic Fields

Several attempts have already been made to show how the 'supergiant' regular structures of the background magnetic fields develope from regularities in the longitudinal distribution of these fields (see, for example, Ambrož *et al.*, 1971; Bumba, 1972a). It is a very complicated problem, connected with the question of the reality of the so-called 'active longitudes' and their reasons.

It seemed to us, therefore, that it may be of some interest to study the longitudinal distribution of the green coronal emission and its relation to the same distribution

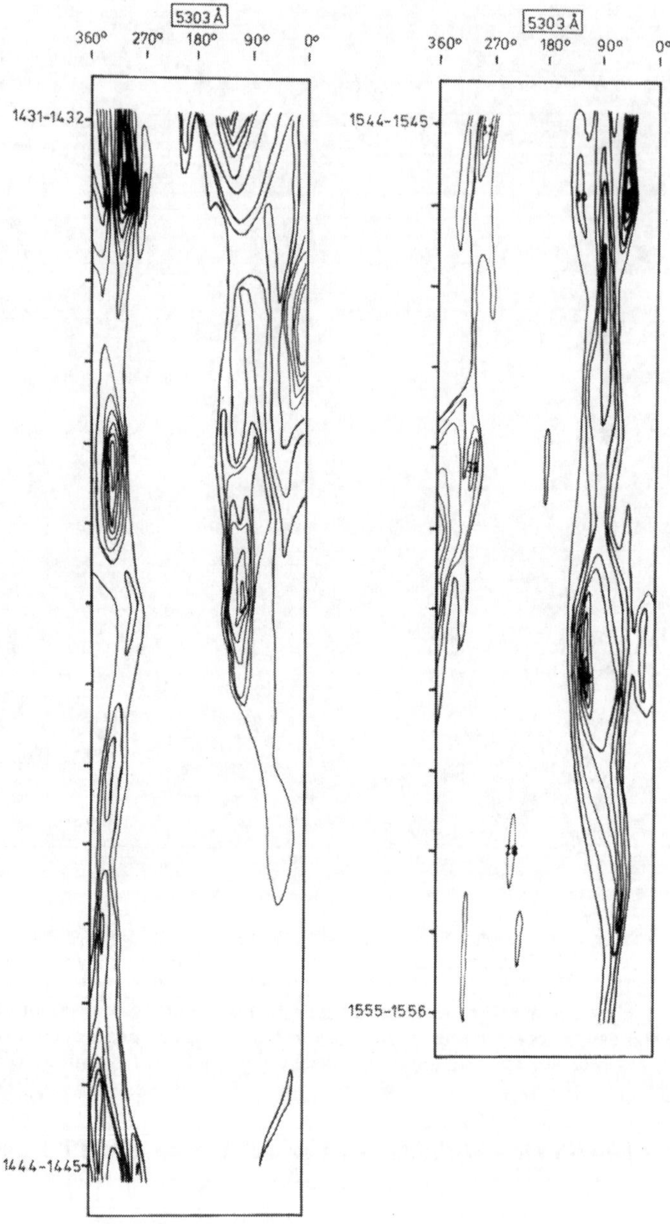

Fig. 6. The longitudinal distribution of the green (λ5303) coronal emission, demonstrated by the isophotes in hundreds of absolute coronal units for two investigated time intervals: August 1960–September 1961 (Rotations Nos. 1431–1444) to the left and to the right January 1969–December 1969 (Rotations Nos. 1544–1556). For integration two consecutive rotations are overlapped. The heliographic longitudes are indicated at the top of each picture.

of the opposite polarities of the background magnetic fields. We hoped that it could help us to understand at least some of the mutual relations between the magnetic fields and the corona in the formation of the regular patterns described, certain regions of which are responsible for large coronal disturbances.

Because it is still time consuming, only two time intervals were chosen for them; the synoptic charts of the green corona emission were integrated for each rotation and then drawn in the form of isophotes in hundreds of absolute coronal units in the same manner and on the same scale as the magnetic synoptic charts: August 1960–September 1961 (Rotations Nos. 1431–1444) and January 1969–December 1969 (Rotations Nos. 1544–1556). Figure 6, displaying this green corona longitudinal dis-

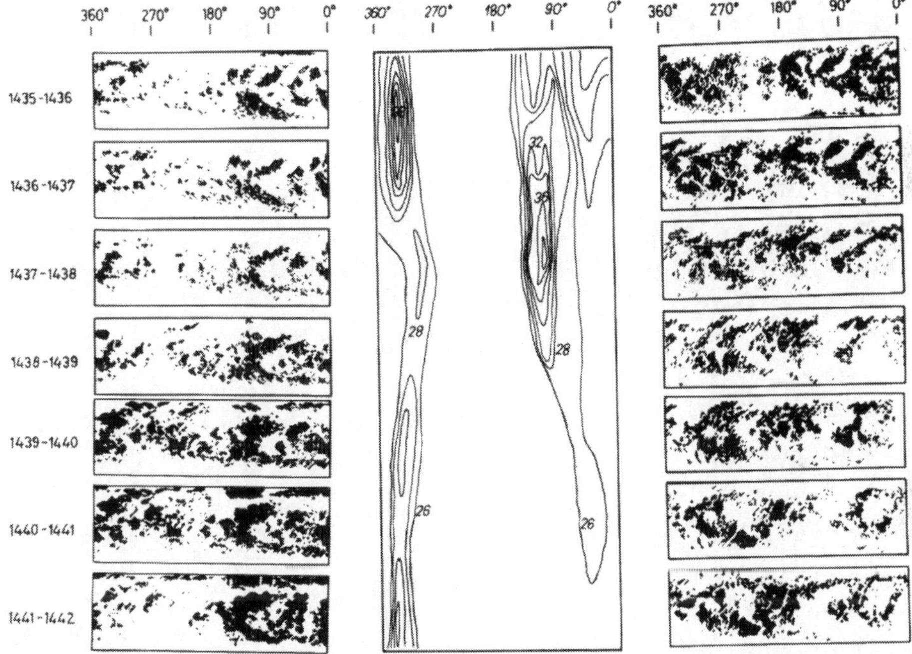

Fig. 7. The longitudinal distribution of green (λ5303) coronal emission for rotations Nos. 1435–1442 demonstrated by isophotes in hundreds of absolute coronal units drawn in the same manner as the magnetic synoptic maps shown on each side of the coronal data. Magnetic synoptic charts are constructed in separated polarities (negative to the left, positive to the right). For integration two consecutive maps, one of which is repeated, are overlapped. The heliographic longitudes are indicated at the top of each picture.

tribution, demonstrates that during both periods the green coronal emission had two maxima in the heliographic longitude: one around 90°–120° and the second one around 320°–340°. This means that they are again not symmetrically spaced in heliographic longitude like many other solar phenomena. The sudden shifts in position of this maximal coronal emission, amounting to several tens of degrees in the longitude seemed to be striking. The internal structure of the individual emission maxima – their changes in time – not only seem to be connected with the evolution of activity,

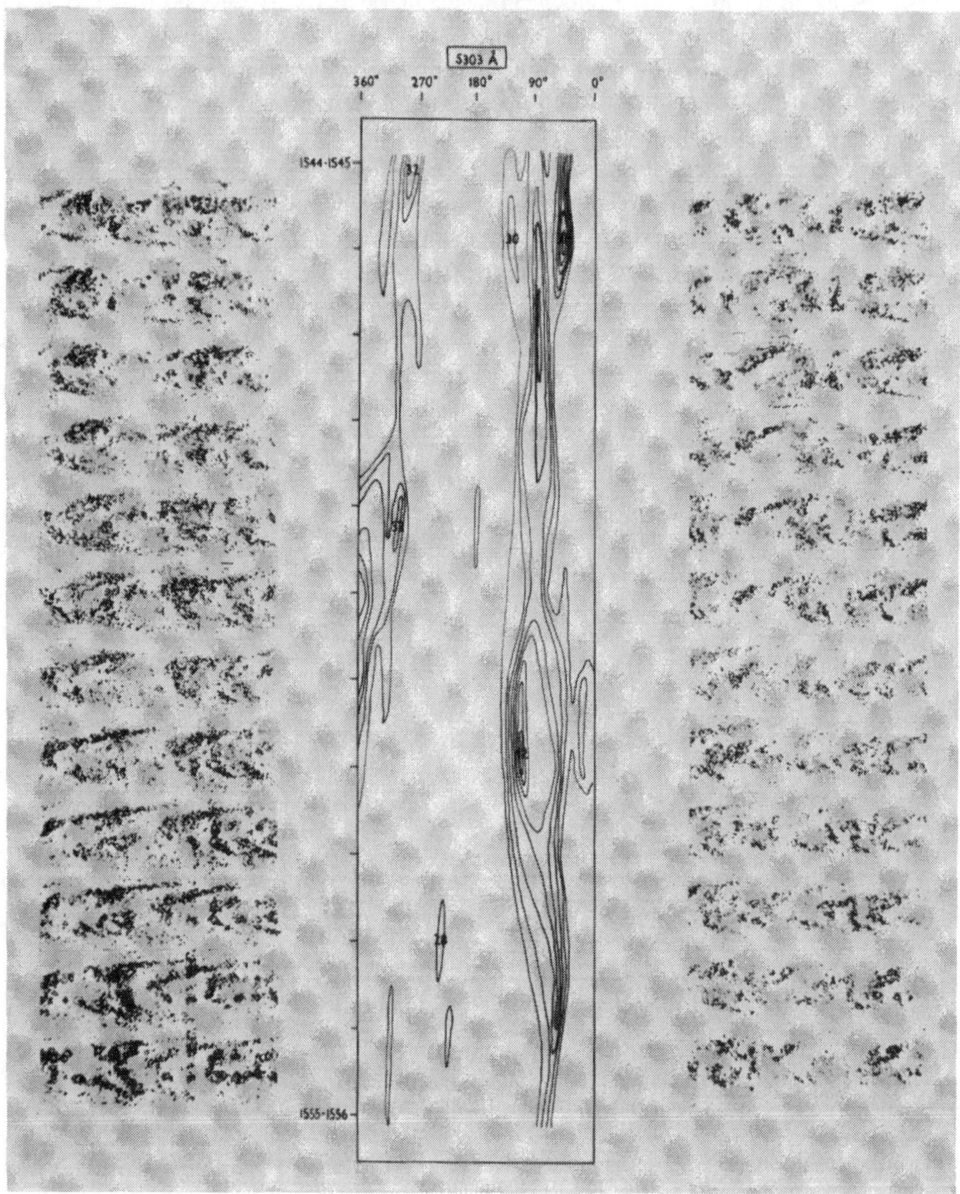

Fig. 8. The longitudinal distribution of the green (λ 5303) coronal emission for rotations Nos. 1544–1556 demonstrated by isophotes in hundreds of absolute coronal units drawn in the same manner as the magnetic synoptic maps shown on each side of the coronal data. Magnetic synoptic charts are constructed in separated polarities (negative to the left, positive to the right). For integration two consecutive maps, one of which is repeated, are overlapped. The heliographic longitudes are indicated at the top of the middle picture.

but also with different rotational periods of the active regions in various latitudes, etc. This problem certainly has to be studied separately for the whole time interval, for which the basic observational material for the corona (Sýkora, 1972) is available.

It has already been demonstrated (Ambrož *et al.*, 1971; Bumba, 1972a, b; Bumba and Sýkora, 1973) that the large-scale magnetic features of negative polarity tend to be closely related, at least during the declining phase of the recent cycle and during the increasing phase of the present cycle of solar activity, to the regions with developing solar activity and the positive polarity features, formed from older fields, to the regions which seem to be the source of the quiet solar wind. The same facts seem to be reflected in the correlation of the green coronal emission maxima with the background magnetic field patterns. During the first time interval, the maxima of the coronal emission coincide well in position with the negative polarity features (Figure 7). When a 'supergiant' structure is fully developed, the position of the coronal maximum agrees with the eastern part of this structure, i.e. with the region the large flares tend to be connected with. Even the longitudinal shift of the coronal maximum

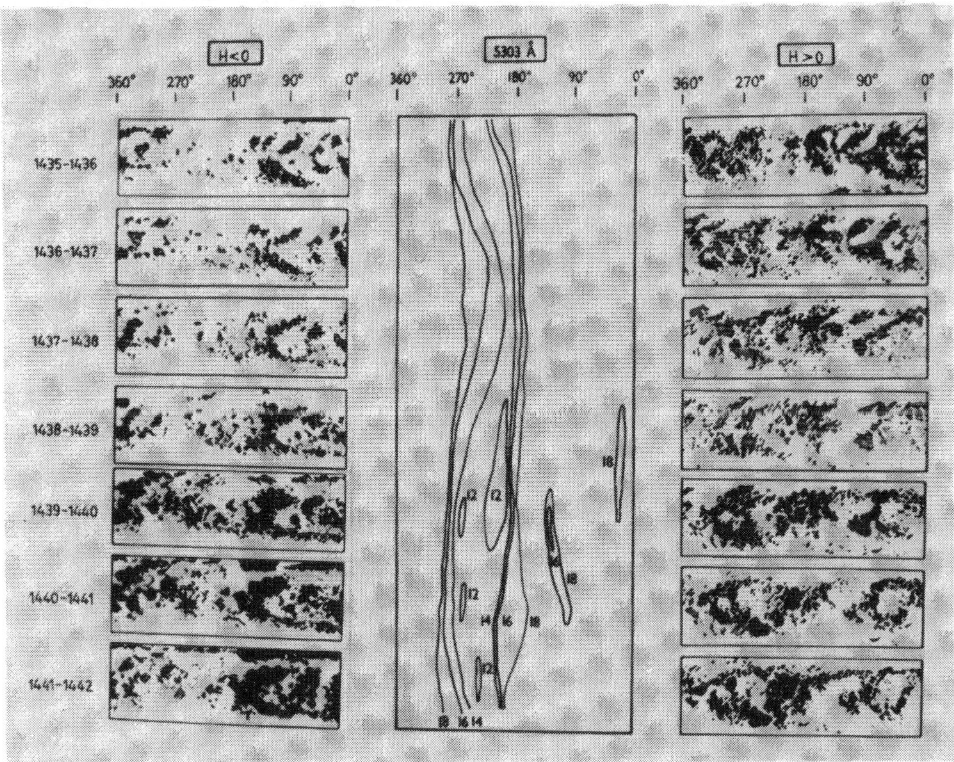

Fig. 9. The isophotes of the minimal green coronal emission in hundreds of absolute coronal units for rotations Nos. 1435–1442, drawn in the same manner as the magnetic synoptic charts, shown on each side of the coronal data: negative polarity maps to the left and positive polarity charts to the right. Again two consecutive rotations are overlapped for integration and the heliographic longitudes are indicated at the top of each picture.

demonstrates the shift of solar activity from the head of such a structure to the root of its tail. As concerns the correlation with positive polarity patterns, it again agrees with what was said above. But we have to underline that the 'mirror positive polarity inclusion' of the negative 'supergiant' body does not agree with the large positive regular 'supergiant' structure, which developed, for example, during the studied time interval in the position ($\sim 200°$) opposite to that of the negative polarity 'supergiant' structure. The same seems to be true of the second time interval (Figure 8), although the magnetic situation there is more complicated. During both time intervals proton-flare regions were observed in agreement with our results.

Just recently 'holes' in the corona have been discussed frequently. We succeeded in constructing the isophotes of the minimum green coronal emission only for our first studied time interval. Comparing the position of the minimum coronal emission, in Figure 9, in heliographic longitude with the synoptic charts of magnetic field, drawn in separate polarities, we may see that it coincides well with the areas minimally occupied by negative polarity fields, on the one hand, but, on the other, it agrees very well with the position of the positive 'supergiant body' which seems to be, as demonstrated (Ambrož *et al.*, 1971), very closely related to the period of enhanced geomagnetic activity.

4. Conclusion

We did not plan to discuss the physical reasons for the observed regularities in the solar magnetic field distribution, so closely related not only to the coronal disturbances, but also to the distribution of the green coronal emission in the solar atmosphere, in our note. We do not understand many problems connected with the morphology of this phenomenon yet. We would only like to draw your attention to this question, we consider interesting, which may prove very important not only for the solar physicists, especially because it promises to lead to long-term forecasts of situations, responsible not only for coronal disturbances, but also other phenomena.

References

Ambrož, P., Bumba, V., Howard, R., and Sýkora, J.: 1971, in R. Howard (ed.), 'Solar Magnetic Fields', *IAU Symp.* **43**, 696.

Bumba, V.: 1972a, in C. P. Sonett, P. J. Coleman, and J. M. Wilcox (eds.), *Asimolar Conf. on the Solar Wind*, NASA, Washington, p. 31.

Bumba, V.: 1972b, in C. P. Sonett, P. J. Coleman, and J. M. Wilcox (eds.), *Asimolar Conf. on the Solar Wind*, NASA, Washington, p. 151.

Bumba, V.: 1973, World Data Center A Report UAG-28, Part I, 77.

Bumba, V. and Howard, R.: 1969, *Solar Phys.* **7**, 28.

Bumba, V., Křivský, L., and Sýkora, J.: 1972, *Bull. Astron. Inst. Czech.* **23**, 85.

Bumba, V. and Sýkora, J.: 1972a, World Data Center A Report UAG-24, Part I, 43.

Bumba, V. and Sýkora, J.: 1972b, World Data Center A Report UAG-24, Part II, 311.

Bumba, V. and Sýkora, J.: 1973, *Space Research XIII*, Akademie-Verlag, Berlin, d 23-1.

Howard, R., Bumba, V., and Smith, S.: 1967, Carnegie Inst. of Washington Publ. No. 626, Washington, D.C.

Sýkora, J.: 1971, *Bull. Astron. Inst. Czech.* **22**, 12.

Sýkora, J.: 1972, *Contr. Astron. Obs. Skalnaté Pleso* **5**, in press.

DISCUSSION

Mullaly: I would like to draw attention to a 'gap' in the positions of microwave bursts (1958–61) as reported by Krishnan and Mullaly.

Boischot: Why does the differential rotation not show in your synoptic maps?

Bumba: Differential rotation appears with individual active regions but the large scale magnetic pattern indicates rigid rotation.

Stix: The hydromagnetic equations governing the solar dynamo have, besides the axisymmetric solutions describing the 22 year cycle, also ϕ-dependent solutions. Such solutions may provide a theoretical explanation of the supergiant structures mentioned by Dr Bumba. They have the interesting property that their sector boundaries do *not* change their shape as they slowly drift in longitude.

DISTRIBUTION OF CIRCULARLY POLARIZED EMISSION ACROSS THE SOLAR DISK AT $\lambda=4.3$ cm*

PIERRE KAUFMANN and E. SCALISE, JR.

Centro de Rádio-Astronomia e Astrofísica, Universidade Mackenzie, São Paulo, Brazil

Abstract (*Solar Phys.*). The emission of circularly polarized radiation of reversed senses from the respective solar hemispheres at $\lambda=4.3$ cm is discussed, referring to consistent eclipse results obtained in November 1966 (Figure 1), March 1970, and January 1973 (Figure 2). Further qualitative information were obtained by performing a series of 18 daily solar maps (Figure 3) at the same wavelength, with moderate angular resolution (i.e., 12′). We were not able to resolve individual centers' polarities, but have confirmed that reversed senses of circular polarization occur, tending to be right-handed at southern areas of the solar disk, and left-handed at northern areas. These senses correspond to preceding polarities at the hemispheres for the 20th solar cycle, and a 'neutral line' divide the two polarized hemispheres, showing a contour that depends strongly on the presence of active centres and on their displacement with solar rotation. This suggest that the polarized features are strongly controlled, if not completely, by the presence of photospheric underlying magnetic areas.

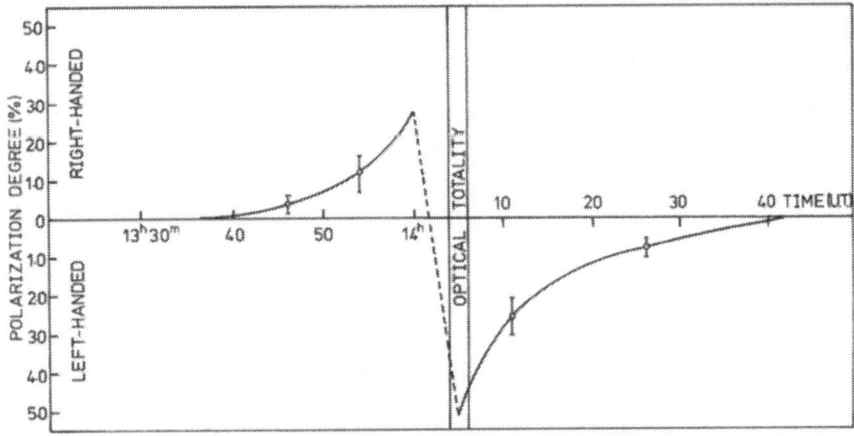

Fig. 1. The variation of the polarization degree (in percent) at $\lambda=4.3$ cm with time, during 12 November 1966 total solar eclipse. Areas seen in excess from the southern hemisphere before totality, and from the northern hemisphere after totality, had provided right-handed and left-handed senses of circular polarization, respectively.

* Work supported by brazilian agencies FAPESP, BNDE-FUNTEC and CNPq, and by U.S. AFCRL(LIR).

Gordon Newkirk, Jr. (ed.), Coronal Disturbances, 85–87.

Polarized radiation was attributed to net magnetic fluxes existent on the areas observed in the case of eclipses, or on the areas subtended by the antenna beam in the case of the maps. Gross structures of coronal magnetic fields, as proposed by Altschuler and Newkirk (*Solar Phys.* **9**, 131, 1969), can well account for the existence of the microwave polarized radiation. Over active centres, the so-called 'missing'

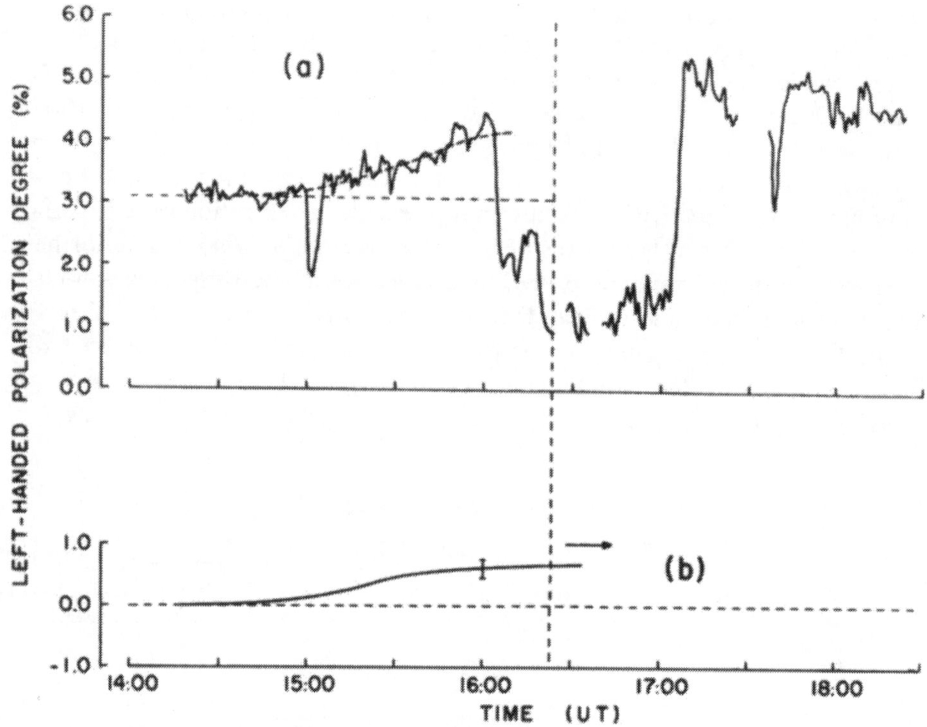

Fig. 2. Result obtained during the 4 January 1973 partial solar eclipse at $\lambda = 4.3$ cm. The polarization degree (in percent) has shown a gradual left-handed increase as the northern hemisphere was observed in excess (or as the southern hemisphere has been progressively eclipsed). Curve (a) shows the original result, with the presence of a small flare, and the eclipsing of two active centers. Curve (b) is the same data reduced for the presence of the sources. The data for the second half of the eclipse was spoiled due to thunderstorm and power failures at the observing site (Itapetinga Radio Observatory, Atibaia, SP, Brazil). At the elipse's maximum no active radio source was present in the uncovered area of the solar disk.

magnetic fluxes could be invoked to explain the effect. Solar polarization maps of moderate resolution could then provide useful information on morphology and temporal variations of large scale magnetic fields on the solar corona.

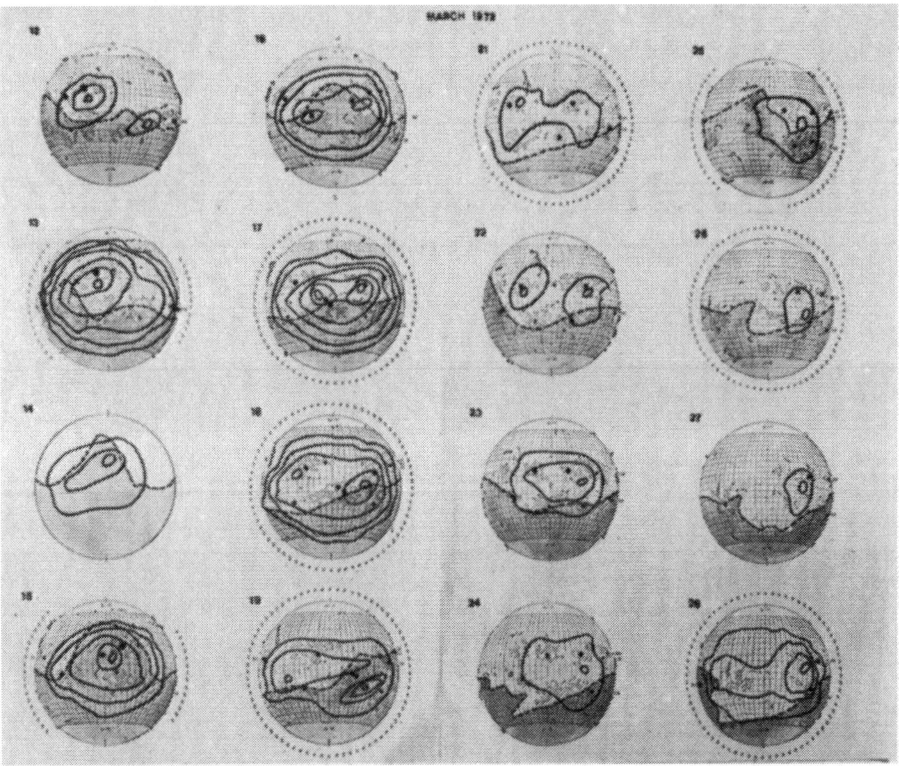

Fig. 3. Polarization maps from the Sun taken in 12–29 March 1972 at Itapetinga Radio Observatory with a 45 ft radio telescope, providing 12′ resolution at $\lambda = 4.3$ cm. The data was plotted over Fraunhofer Institut maps and North is at the top. Days 20 and 29 were omitted due to incomplete data. Dark areas are for right-handed polarization senses and clear areas are for left-handed polarization. Unrestored intensity contour lines obtained at the same wavelength are also shown, in relative but uncalibrated antenna temperatures, being the outer contour for 6000 K, and the inner contours increasing by steps of 500 K. It is clearly shown the strong dependence of the 'neutral line' on the presence of net magnetic fluxes from active centres, and on their displacement with solar rotation.

DISCUSSION

Kundu: Are you measuring quiet Sun polarization or active region polarization. If it is active region polarization, your discussion in terms of one sense of polarization for the northern hemisphere and the opposite sense for the southern hemisphere is misleading. If it is the quiet Sun polarization, how do you calibrate your instrument for zero circular polarization?

Kaufmann: Several of the large-scale magnetic structures due to active centers may well spread into uneclipsed areas even if no apparent spot or radio active center was present. The eclipse data are then not representative of quiet sun conditions. The polarization maps, the strong dependence of polarized areas on the active centers, and their evolution with solar rotation, suggests that nearly all net magnetic fluxes necessary to explain the polarization are due to the centers.

BOUNDARY FITTING PROBLEMS ASSOCIATED WITH CORONAL MAGNETIC MODELS

KENNETH H. SCHATTEN
Victoria University, Wellington, New Zealand

Abstract. The calculation of coronal magnetic fields was first suggested by Gold (1958). Altschuler and Newkirk (1969) and Newkirk *et al.* (1968) used a Legendre polynomial fit to the photospheric observations of magnetic fields whereas Schatten (1968) with Wilcox and Ness (Schatten *et al.*, 1969) use a magnetic monopole fit, first incorporated by Schmidt (1964).

Altschuler and Newkirk berate the monopole fit as not being 'mathematically valid'. Although physical reasons are not provided in their paper, the only physical reason which may be difficult for some to understand is the following. In the monopole fit one uses $\nabla \cdot B = 4\pi\varrho_m$ where ϱ_m is a magnetic monopole charge density necessary to fit the photospheric observations. The non-zero divergence of the magnetic field *is* mathematically valid, but *not* physically valid, if one were to use the magnetic field solution in the region where the monopoles were located. As the monopoles are only embedded beneath the photosphere (where the solution is not calculated) the solution obtained by those authors who use this technique is valid in the regions of space where the field has been calculated (the corona). The remaining question is which technique fits boundary conditions (in the photosphere) more accurately. As no one has obtained a method of objective comparison with observations, the only test one has available, is to see which technique best fits the observations.

As a test to measure the 'quality of fit' of these two procedures, I have calculated the mean square difference between a calculated magnetic field (at 1.05 solar radii – a small distance above the photosphere) and the raw photospheric observations. A value of 0.0 would mean a perfect fit was obtained. The values were normalized so that a quality of fit equal to 1.0 would indicate the difference between the raw data and a constant value of zero. Using the Legendre polynomial technique with $N=9$ (the same value used by Altschuler and Newkirk (1969)) a quality of fit equal to 0.77 was obtained for the November 12, 1966 solar eclipse photospheric data and a value of 0.36 was obtained using a monopole fit with a comparable grid spacing. Similar values were obtained with data at other eclipse periods.

The closer fit obtained by the monopole technique appears, at first, surprising. However when one looks at the behaviour of the photospheric field, it is not. The photospheric field is 'clumpy', there are weak fields and strong fields. The monopole technique can fit combinations of these fields equally well. The Legendre polynomial technique attempts the fit of a 'smooth' function to this clumpy data. Hence a poor fit can result with this technique. Furthermore there is no Legendre index N, however large, which will adequately describe a fit to a point source. If the photospheric

Gordon Newkirk, Jr. (ed.), Coronal Disturbances, 89–91.

field were 'smoother' the Legendre technique would be better, however it is not. The problem of field lines 'going through the Sun' is solved through the use of a Green's function solution in my work.

It is interesting to note that the same problem occurs in gravitational anomalies associated with mass concentrations on the moon (mascons). The problem is resolved in an identical fashion. It is found that these mascons can only be described (gravitationally) by the inclusion of point sources in the gravitational field. Thus the Sun appears in a magnetic sense, similar to the Moon in a gravitational sense – the effect of active regions being identical to that of mascons.

Acknowledgements

I am grateful to Dave Howell, who helped with the calculations and for the benefit of working with R. Howard, J. Wilcox and N. F. Ness. I thank R. Howard of the Hale Observatories and Carnegie Institute for the use of his photospheric field data obtained in a program in part by the Office of Naval Research under contract NR 013-023, N000 14-66-C-0239.

References

Altschuler, M. D. and Newkirk, G., Jr.: 1969, *Solar Phys.* **9**, 131.
Gold, T.: 1958, in B. Lehnert (ed.), 'Electromagnetic Phenomena in Cosmical Physics', *IAU Symp.* **6**, 275.
Newkirk, G., Jr., Altschuler, M. D., and Harvey, T. W.: 1968, in K. O. Kiepenheuer (ed.), 'Structure and Development of Solar Active Regions', *IAU Symp.* **35**, 379.
Schatten, K. H.: 1968, Ph.D. Thesis, Univ. of Calif. Berkeley.
Schatten, K. H., Wilcox, J. M., and Ness, N. F.: 1969, *Solar Phys.* **6**, 442.
Schmidt, H. O.: 1964, in W. Hess (ed.), *AAS-NASA Symposium on Physics of Solar Flares*, NASA SP-50, p. 107.

DISCUSSION

Schmidt: I proposed the monopole method only for fitting the surface field on a plane such as might be applied to active regions. Loop prominences represent the only case where the force-free field method has given a demonstrably correct picture. This can be understood because during the flare the energy has been released and the field relaxed to the potential configuration.

Schatten: You say all large-scale current-free models are *not* appropriate. Thus you have the same criticism of the Legendre technique as you have with monopole technique. Both techniques make the same assumptions outside the Sun.

Altschuler: The problem is that we cannot use the line-of-sight magnetic measurements to choose values for monopoles in a spherical surface. On a spherical surface, any monopole produces a contribution on the other side of the Sun. Consequently, the total distribution of monopoles you finally get is not consistent with the originally given boundary condition.

Schatten's statistical test is also misleading; when he compares the calculated field values at $r = 1.05\ R_{\odot}$ with the raw data, he is only showing that the field values he chooses at the surface can be regurgitated immediately above the surface. The problem, however, is to obtain the *large-scale* pattern of magnetic field directions. In the Legendre polynomial method we have limited resolution at the surface with principal index $N = 9$, but higher in the corona only the lowest harmonics survive anyway, so the crucial matter is not high resolution but the proper magnetic field directions on the large-scale. There is no difficulty in applying the Legendre polynomial method with a much larger principal index than $N = 9$, but we have not done so because, as Dr. Schmidt points out, there are only limited applications for the potential field calculation.

Schatten: I used only magnetic monopole sources from regions less than about 30–40 deg removed from the point of field computation. This, then, prevents magnetic field from going through the Sun. Effectively, I used a Green's function $\bar{G} \approx (\hat{r} - \hat{r}_m)/r_m^2$ for $\theta < 30$–$40°$ and $\bar{G} \approx 0$ for $\theta > 30$–$40°$, where θ is the angle between the source and the point of field computation. Thus, the magnetic field from a point on one side of the Sun does *not* produce a significant flux on the opposite side of the Sun as you suggest. I think the way to go in future analyses is in the direction of Pneuman and Kopp, that is, one should consider the interaction of the solar wind with the magnetic field.

ANALYSIS OF EUV OBSERVATIONS OF A CORONAL ACTIVE REGION MADE DURING THE 7 MARCH 1970 ECLIPSE

A. H. GABRIEL and C. JORDAN

SRC Astrophysics Research Division of the Appleton Laboratory, Culham Laboratory, Abingdon, Berkshire, England

Abstract (*Monthly Notices Roy. Astron. Soc.*). Rocket observations of the EUV solar spectrum obtained during the total eclipse of 7 March 1970 showed the presence of a large coronal condensation on the NE limb (Gabriel *et al.*, 1971). The condensation shows the existence of loop structure, which defines the local magnetic fields, and is apparent in forbidden lines including those of Si VIII, Si IX, S XI, Fe IX, Fe XI and Fe XII which lie between 1200 Å and 2050 Å (Jordan, 1971). These lines are formed in the temperature range 9.3×10^5 K to 2.0×10^6 K. The photographic observations of the active region show clearly that the spatial distribution of material varies considerably with its temperature.

The populations of the excited levels emitting the forbidden lines were calculated as a function of density and temperature using currently available atomic data. The $^1S_0$–$^3P_1$ line of Fe XI and the $^2P_{1/2}$–$^4S_{3/2}$ line of Fe XII have intensities which are proportional to N_e^2. The decrease in the intensity of these lines as a function of increasing height along the apparent loop structures closely follows that expected in hydrostatic equilibrium at the temperatures where these ions are predominantly formed. Therefore the change with height in the line of sight path-length, L, must be small for these emission lines.

An initial analysis using calculated level populations showed inconsistences with the apparent temperature structure. Therefore the following normalisation process was used to determine the level populations.

A region of the quiet corona was chosen from the eclipse data and the variation with height of the density dependent line ratios in Fe XII, $^2P_{1/2}$–$^4S_{3/2}$/$^2D_{5/2}$–$^4S_{3/2}$ was used to show that the electron density was constant apart from hydrostatic variations. Similarly the temperature dependent line ratios Si VIII/Si IX and Fe XI/Fe XII were used to show that one temperature could be attributed to the region. The temperature giving the best fit to all the observed intensities was found to be $T_e = 1.45 \times 10^6$ K. The scale height at this temperature was used to determine the line of sight distance in a spherically symmetric atmosphere. The electron density was calculated from the absolute intensity of the Si VIII line and was found to be 3.3×10^8 cm^{-3}. Thus through knowing N_e, T_e and L the excited level population could be found independently of the atomic data. These populations are up to an order of magnitude larger than those calculated using available atomic data, and the populations depending on the smallest excitation rates need the largest correction.

Gordon Newkirk, Jr. (ed.), Coronal Disturbances, 93–95.

The abundances which give intensities consistent with the data are $N(\text{Si})/N(\text{H}) = 4 \times 10^{-5}$; $N(\text{S})/N(\text{H}) = 1 \times 10^{-5}$; $N(\text{Fe})/N(\text{H}) = 8 \times 10^{-5}$.

The normalised populations were then used in the analysis of the active region data. In the active region it is found that each emission line is formed predominantly at a temperature close to the peak of the ionization equilibrium distribution, indicating a range of temperatures present in a given line of sight. The active region contains material both hotter and cooler than the average quiet corona. The temperature at which most of the material exists varies with height and position in the active region; the higher, outer regions contain more cool material, and the lower, central regions more hot material. At a chosen point at a height of $\simeq$20000 km in the left side of the loop (see Figure 1) the electron density increases with temperature

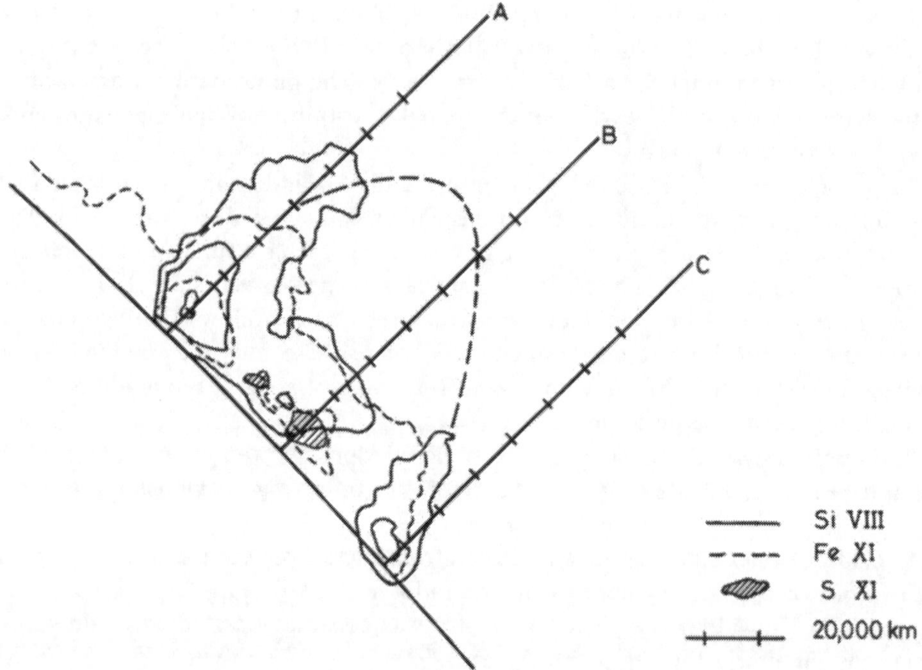

Fig. 1. A diagram of the active region seen at the limb during the 7 March 1970 eclipse. For Si VIII and Fe XI the isophotes for maximum and half-maximum intensity are shown. Only the maximum intensity isophote is shaded for S XI.

from 1.8×10^9 cm^{-3} at 9.3×10^5 K to 4.2×10^9 cm^{-3} at 1.7×10^6 K. The line of sight path-length decreases by a factor of 9 over the same temperature range. The minimum magnetic field strength needed to balance the gas pressure varies between 4 G and 18 G. These values are typical of the entire region at 20000 km, but the structure as a function of position will be described in more detail later.

References

Gabriel, A. H., Garton, W. R. S., Goldberg, L., Jones, T. J. L., Jordan, C., Morgan, F. J., Nicholls, R. W., Parkinson, W. H., Paxton, H. J. B., Reeves, E. M., Shenton, D. B., Speer, R. J., and Wilson, R.: 1971, *Astrophys. J.* **169**, 595.

Jordan, C.: 1971, *Solar Phys.* **21**, 381.

DISCUSSION

Aller: I agree that population calculations could be in error by a factor ≈ 2.

Pneuman: I believe Werner Neupert, using X-ray data, has observed cases in loops where the temperature actually increases with height in the corona, rather than being isothermal. If this is the general case, the implications for coronal heating mechanisms are interesting. Do you find any such indications from your observations?

Jordan: We find that high loops are cooler than low loops.

Stewart: Calcium observations give higher temperature at the tops of loops.

Jordan: This is probably a short-lived flare loop.

Jefferies: High-density, high-temperature concentrations in the center of loops were previously found by Orrall and Lyot (1952).

Jordan: We use the loop geometry for our condensation model while he used cylindrical geometry. Also, we find the core of the loop to be both denser and hotter than the outside.

Gabriel: The present model is very much in line with that proposed by J. Parkinson from analysis of X-ray measurements from OSO-5. By measuring the change in emission as active regions pass over the limb, he derives a model in which the hot dense material is low in the structure.

CORONAL MAGNETIC FIELD STRUCTURE DERIVED FROM TWO-FREQUENCY RADIOHELIOGRAPH OBSERVATIONS

K. KAI* and K. V. SHERIDAN
Division of Radiophysics, CSIRO, Sydney, Australia

Abstract (*Solar Phys.*). An exceptional variety of positions and polarizations was found for two type I storms and numerous sporadic bursts observed during 15 consecutive days with the Culgoora radio-heliograph at 80 and 160 MHz.

The main observational points for the type I storms are as follows:

(A1) The type I storm centres were not situated radially above the associated sunspots but were displaced systematically northwards.

(A2) They did not always show a uniform daily motion across the disk. The second storm, in particular, shows anomalous movements between 1972 October 27 and 30.

(A3) The sense of circular polarization of the storms was left-handed (L.H.) and remained so as the sources rotated across the disk, except on one day, October 30, when the L.H. polarized source was accompanied by a right-handed (R.H.) polarized companion at both 80 and 160 MHz.

In addition to the persistent storm centres a large number of sporadic bursts occurred. These were mainly of types III and V, with some showing inverted-*U* structure. The sporadic bursts generally showed a variety of temporal and spatial distributions, particularly at 80 MHz. One sequence of such bursts observed at 80 MHz between 00^h37^m and 01^h05^m UT on 1972 October 29 is illustrated in Figure 1. Some remarkable features shown by the group of the sporadic bursts are the following:

(B1) All sources except No. 2 appeared to lie on an arc extending around the active region in which the associated Hα-flare occurred.

(B2) Irrespective of spectral type all the sources were weakly L.H. circularly polarized.

(B3) No sequence was evident in the positions of successive bursts.

(B4) Sources showing different spectral features usually appeared at different places.

The two-frequency radioheliograph data are combined with optical data to derive a model of the coronal magnetic field structure for this complex of active regions. In attempting this we assume that

(a) type I storms occur in the stronger magnetic fields of active regions, probably in magnetic fields forming closed loops;

(b) sporadic bursts are associated with weaker magnetic fields; when bursts have inverted-*U* structure their exciting agencies (electrons) are guided around closed

* On leave from Tokyo Astronomical Observatory, University of Tokyo.

Gordon Newkirk, Jr. (ed.), Coronal Disturbances, 97–101. *All Rights Reserved.*

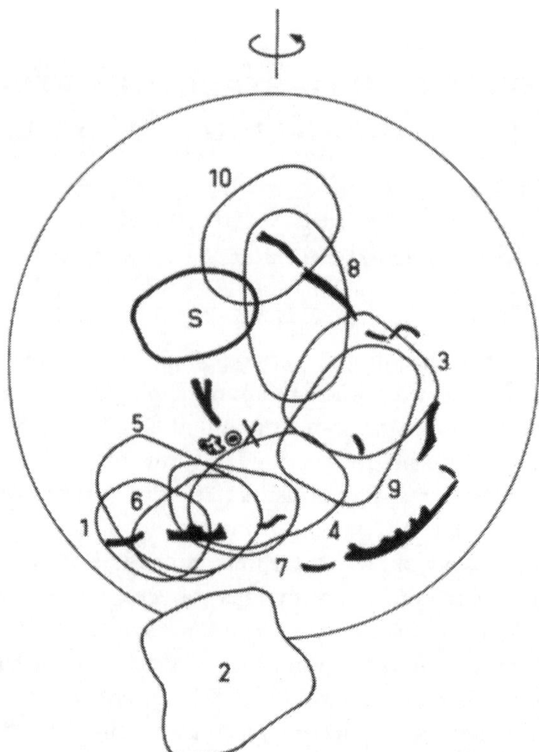

Fig. 1. The distribution of the 80 MHz sources of the sporadic bursts (numbered 1 to 10 for reference) and some prominent quiescent filaments on 1972 October 29. The associated Hα flare ($-N$, 00^h37^m to 00^h39^m UT) indicated by the cross occurred in the region north-west of the associated leading spot. The heavy full contour labelled S is the persistent type I storm (~80% L.H. circular polarization). The sporadic sources are: (1) III or V: $00^h37^m38^s$, ~6% L.H.; (2) III or V: $00^h38^m16^s$, ~0; (3) III or V: $00^h38^m18^s$, ~10% L.H.; (4) III: $00^h39^m24^s$, ~30% L.H.; (5) III: $00^h39^m26^s$, ~5% L.H.; (6) V: $00^h40^m08^s$, ~5% L.H.; (7) III: $00^h40^m42^s$, ~10% L.H.; (8) III?: $01^h00^m11^s$, ~15% L.H.; (9) U or III: $01^h03^m06^s$, ~10% L.H.; (10) ?: $00^h35^m58^s$, ~30% L.H.

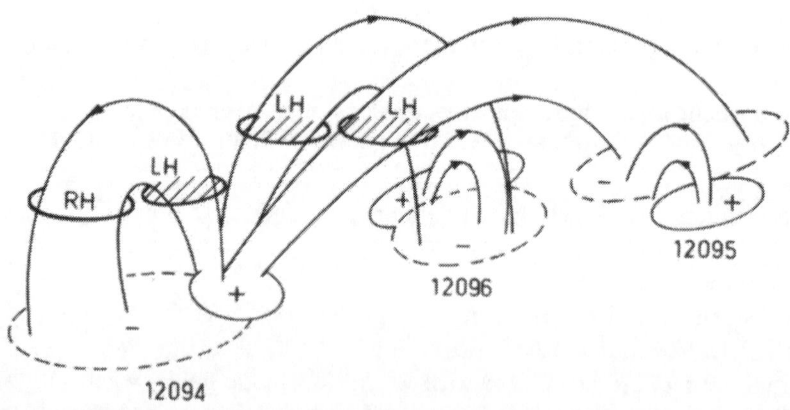

Fig. 2. Model proposed to account for the observed positions of type I storm centres relative to the associated optical features (the numbers refer to McMath plage regions). The polarized radio sources labelled L.H. and R.H. are drawn on magnetic loops; the direction of the field is indicated by arrows and corresponds to the observed polarities of the underlying plage regions (marked + and −).

magnetic loops whereas in normal type III bursts the electrons have access to 'open' magnetic field lines; type V bursts are due to radiation from electrons trapped and reflected in closed magnetic loops.

Figure 2 depicts a model of the magnetic fields associated with the second series of the type I storms. It is suggested that there are three possible sites where type I source regions can exist, and that the apparently anomalous positions (A1) and position shifts (A2) are simply due to the decay of the source at one site and its reappearance at one of the previously inactive sites. It is also suggested that the bipolar structure (A3) is associated with the magnetic field which emerges from the leading spots and closes into the trailing spots (or the region surrounding them) of the active region 12094. This is based on the observation that the leading, L.H. polarized and the trailing, R.H. polarized radio sources of October 30 seemed to overlie respectively the leading positive and the trailing negative magnetic fields of that active centre. These combined heliograph and magnetograph results are probably the clearest evidence yet in support of the ordinary-mode hypothesis for type I storm radiation.

The large quiescent filaments (see Figure 1) indicate boundaries dividing regions of opposite polarity of the weak magnetic fields measured at the photospheric level. Inspection of Figure 1 suggests that the arc joining the 80 MHz sporadic sources is related to the configuration of these filaments. Open weak magnetic fields must di-

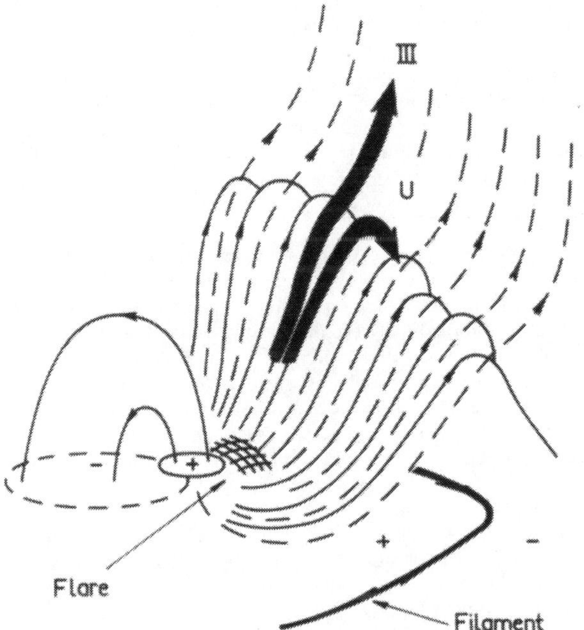

Fig. 3. A suggested model of magnetic fields (light lines with light arrow heads) associated with the widely-spread sporadic sources observed at 80 MHz. The heavy arrows indicate the flow of electron streams which are presumably accelerated in the flare region (shown hatched) and which are responsible for exciting the sources of the aradio bursts of type III (and U). The filament is shown under the field lines.

verge from the flare region to guide the electron streams to the widely-spread radio source sites. The field lines are generally directed toward the filaments, but spread into a wide angle (~180°). A suggested model for these fields is shown in Figure 3. Some of the magnetic fields emerging from the active region (12094) cross over the filaments and return into regions of opposite polarity on the other side. Electron streams guided along these field lines give rise to inverted-*U* bursts and trapped electrons 'mirroring' on these lines produce the type V sources. Other field lines emerging from both sides of the filaments form a cusp-like entrance to the open 'streamer' configuration above the filaments; electrons guided from the active centre along these lines produce the normal type III bursts (B4). The sporadic bursts lying on the arc are partially L.H. polarized, consistent with ordinary mode emission in the field of the model (B2). The sporadic bursts were scattered widely in position and randomly in time (B3). This can be readily explained by the model, since electrons accelerated in the flare region have ready and varied access to many parts of the weak magnetic field bridging the filaments.

DISCUSSION

Smith: Is there any evidence that moving IV did not open the field lines?

Sheridan: No.

Rosenberg: Type III bursts are not always associated with flares but many appear with rising filaments, dark features, etc. But this may be due to an unobserved flare's exciting the filament. Also, there are flares without type III bursts. Acceleration may not occur exclusively in flares – there is abundent evidence for particle acceleration in the corona.

Pick: The right hand polarized bipolar region was observed only on one day – have you any explanation? Was there observed any modification of the magnetic field pattern?

Kai: We have no direct evidence for the intensification of magnetic fields associated with the trailing spot. But the region surrounding the trailing spot had become active one day before.

Pick: It seems that, according to the aspect of the chain of filaments, another possibility would be a direct injection of electrons from the flaring site into the current sheet (see paper Mercier: 1973, *Solar Phys.* **33**, 177).

Sheridan: Successive (within ~2 s) type III bursts have occurred in widely-separated regions so the idea is quite plausible.

Zirin: 'Decent' type III bursts (other than type III storm bursts) do *not* occur without flares.

McKenna-Lawlor: I have observed that certain localized parts of active regions seen in Hα repeatedly show brightenings which are time associated within ±1 min with the production of type III bursts.

Wild: I haven't heard Zirin press so strongly for a type III-flare association before. Maybe, it is a question of 'decent' vs 'indecent' type III's. There are many 'indecent' type III's which have only low frequency components which appear to be initiated high in the corona.

Rosenberg: There is ample evidence for acceleration high in the corona from decametric type III bursts.

Fainberg: This carries on to kilometric III's as shown by satellite.

Hartz: Confirm Fainberg's observations.

Erickson: At decametric wavelengths it is possible to associate some Hα feature with each type III, however, the association is not with obvious flares. Many type III's are associated with small twitchs and wiggles in Hα, while rather similar twitchs produce no III's. I agree with Zirin that major type III's can practically always be associated with flares.

Leblanc: I would like to confirm that there are two kinds of type III bursts in decameter wavelengths. The first kind are storms of type III burst which appear only at high levels in the corona ($>0.5\ R_\odot$). The second kind are type III bursts which come from the very low corona and can be seen at all wavelengths. The storms of type III bursts are not associated with flares, but the other type III bursts which come from the low corona are associated with flares.

Steinberg: What makes you believe that the size of type I's is proportional to the cross section of a magnetic tube of force at the critical surface at f_p, your observing frequency. I understand that the model accounts qualitatively for the observations, but we know that the apparent diameter of type I's vary from one event to the next and sometimes within an event.

Uchida: Was there any indication in the event shown in your last slide that the curved path of the region of appearance of the type III-moving IV events had something to do with dark filaments?

Sheridan: Yes. However, the filament position was not indicated in the slide since the figure becomes too crowded if everything is shown.

PART II

THE FLASH PHASE OF SOLAR FLARES

IMPULSIVE (FLASH) PHASE OF SOLAR FLARES: HARD X-RAY, MICROWAVE, EUV AND OPTICAL OBSERVATIONS

S. R. KANE

Space Sciences Laboratory, University of California, Berkeley, Calif. 94720, U.S.A.

Abstract. Recent observations of impulsive hard X-ray, microwave, EUV and optical emissions during solar flares are briefly reviewed in order to deduce the characteristics of the impulsive (flash) phase phenomenon in small solar flares particularly from the point of view of the acceleration of electrons and their role in producing the various impulsive phase emissions. Observed and deduced characteristics of the various electromagnetic emission sources are summarized (Table II). The deduced characteristics of the electron acceleration process (Table III) indicate a process with high acceleration efficiency. The observations are found to be consistent with a model in which electrons are accelerated in a series of short pulses each lasting for $\lesssim 1$ s and the accelerated electrons provide the energy necessary for all the observed electromagnetic emissions produced during the flash phase of small solar flares. Models of the impulsive phase emissions in which energetic electrons play a prominant role are examined and crucial tests to check the accuracy of these models are indicated (Table IV).

1. Introduction

It is well known that the apparent characteristics of a solar flare depend strongly on the frequency of the observed radiation as well as the spacial and time resolution of the instrument used. In addition there are large variations in these characteristics from one flare to another. However, certain broad characteristics of solar flares can still be deduced from the observed time variation of the electromagnetic radiation. A flare can, in general, be roughly divided into three phases: Precursor, Impulsive Phase and Gradual Phase. These phases are schematically illustrated in Figure 1. The *precursor* is a relatively small slow increase in the radiation indicating a possible ocurrence of the impulsive and/or gradual phase within the following 10 min or so. The *impulsive phase*, which has a duration of ~ 100 s, is characterized by the most rapid increase and decrease in the radiation flux during the lifetime of the flare. The radiation flux may have one or more maxima which may occur quasiperiodically. In the *gradual phase* the radiation flux continues to increase, reaches a maximum after the impulsive phase maximum, and then decreases to the pre-flare level. The total duration of the gradual phase is $\gtrsim 10$ min. It should be emphasized here that all the three phases do not necessarily occur in all flares and in a given flare a particular phase may not be detectable at certain radiation frequencies. However, it can be stated as a general rule that if impulsive phase occurs in a given flare, the gradual phase will almost certainly occur in that flare.

In this paper we will be primarily concerned with the impulsive phase. The principal characteristics of this phase is the non-thermal radiation, such as impulsive hard X-ray, microwave and type III radio emission, which indicates acceleration of particles, particularly electrons from 10 keV to a few hundreds of keV. In the past

Gordon Newkirk, Jr. (ed.), Coronal Disturbances, 105–141.

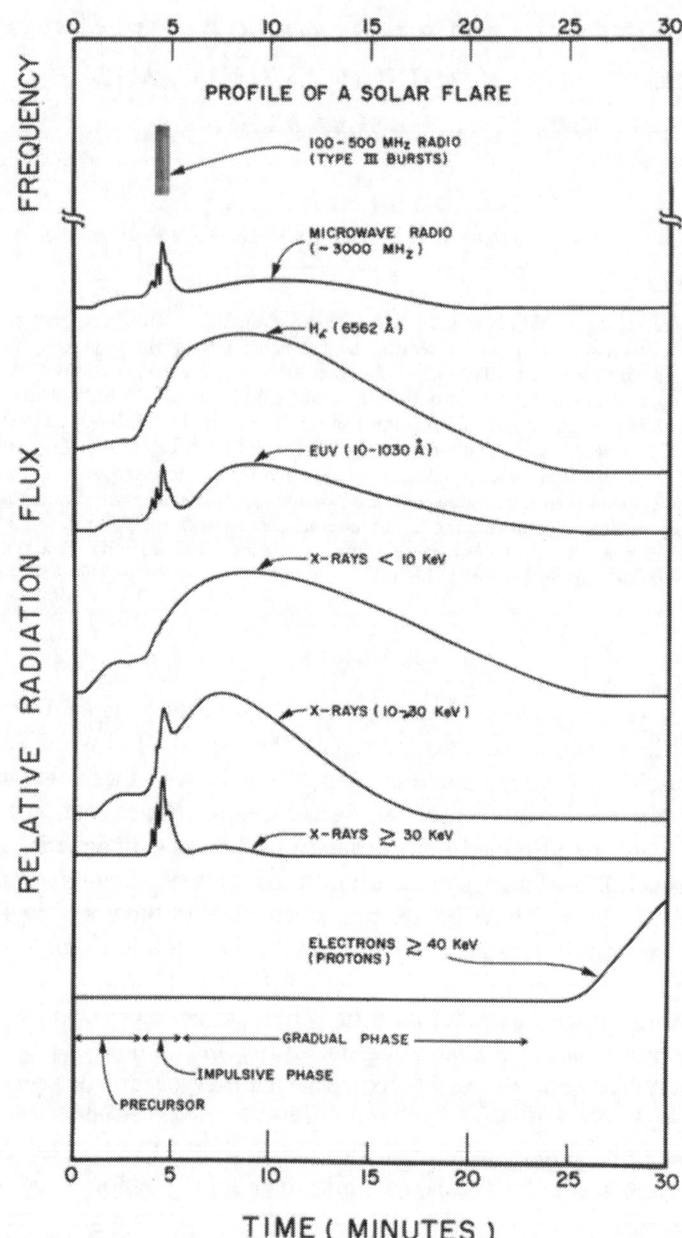

Fig. 1. A schematic representation of the different phases of a solar flare as observed in the electromagnetic and particle radiation.

this phase of the flare has been discussed in terms of two different phases observed in Hα emission: Flash Phase (Ellison, 1949) and Explosive Phase (Moreton, 1964). As we will find later in this paper, these two phases are probably a subgroup of the more general impulsive phase defined above.

As a result of the work of Anderson and Winckler (1962), de Jager (1964, 1969), Kundu (1965), Moreton (1964), Peterson and Winckler (1959), Takakura and Kai (1966), Wild *et al.* (1963), and others, a conceptual picture of the flash phase or explosive phase had emerged in the early 1960's. It was recognized, for example, that acceleration of electrons was the most characteristic feature of the flash phase. Concepts regarding the propagation of electrons in the solar atmosphere and the associated hard X-ray, microwave and type III radio emission were also developed. What was lacking at that time was a quantitative picture. This was due to two reasons: (1) the number of well correlated observations, especially in the X-ray region, were small; (2) observations which could give quantities such as electron spectrum were not available.

With the launch of a simple hard X-ray detector aboard the OGO-1 satellite in 1964 (Arnoldy *et al.*, 1967, 1968a, b, c) the number of correlated observations dramatically increased and have been steadily increasing since then. New phenomena such as EUV flares have been discovered and optical and radio observations with higher spacial and time resolution have been made. In this paper we briefly review these new observations and discuss their impact on the earlier concept of the flash phase. Since most of the new observations are related to flares of Hα-importance $\lesssim 1$, the conclusions will be primarily applicable to small flares.

2. Objectives of the Electromagnetic Radiation Measurements

There are two principal objectives of the impulsive emission measurements from flares:

(1) To study the time variation of the emission from different parts of the flare region;

(2) To determine the location of the flare with respect to the magnetic field structure of the associated active region.

Both of these studies have important bearing on the physical mechanism of solar flares. It is not always possible to achieve both of the above objectives. For example, in the case of the hard X-ray emission, the spacial resolution of the detectors used so far was very poor and so only a part of the objective (1) could be fulfilled, viz. the study of the time variation of the X-ray emission from the entire flare region. On the other hand, this difficulty is much less severe for the optical observations where spacial resolution of $\sim 0.5''$ is available with most of the recent instruments. In fact, the principal limitation of the optical measurements made so far has been the relatively poor time resolution ($\gtrsim 10$ s).

In general, the quantities of interest are the spectrum, directivity, polarization, spacial extent and altitude of the emission and the variation of these quantities with time. Since the impulsive phase is relatively short lived, measurements with highest available time resolution are to be emphasized. The characteristic time constants for the increase and decrease of the radiation flux at a given frequency are of particular interest because they indicate the rate of deposition and loss of non-thermal energy in that radiation source. This is of primary importance in determining the nature of the radiation process and the agency by which energy is transported from the point of primary energy release to the radiation source.

Since most of the recent observations are related to the flares of Hα-importance $\lesssim 1$, the conclusions will be directly applicable only to relatively small flares as compared to the previous studies (Bruzek, 1967; de Jager, 1969) which were mainly related to large flares (Hα-importance $\gtrsim 3$). There are certain advantages in studying small flares. They occur more frequently and often have a relatively simple emission structure. Therefore the interpretation of observations is usually more reliable. Whenever possible, the small flare observations can of course be supplemented with the large flare observations so as to obtain a more general picture of the impulsive phase.

It has been apparent for some time that observations of one type of radiation, however extensive they may be, are not likely to reveal the true nature of the impulsive phase. Fortunately we now have a large number of simultaneous observations of impulsive X-ray, EUV, optical and radio emissions made during the past few years. The present problem of the impulsive phase is to construct a self-consistent picture based on the bits and pieces of information deduced from these different types of observations.

3. X-Ray Observations

Spectral observations of solar flare X-rays have been recently discussed by Neupert (1971), de Feiter (1972), Doschek (1972), and Kane (1973a). Although impulsive X-ray emission may occasionally be detectable at X-ray energies of ~ 5 keV (Kahler and Kreplin, 1971; Peterson *et al.*, 1973), in most flares it can be unambiguously measured only at energies > 10 keV. The observations of impulsive line emissions are scarce. In fact calculations of possible Kα line emission from S, Ar, Ca and Fe in the flare region show that the photon flux in these lines is expected to be weak during most impulsive events (Phillips and Neupert, 1973). We will therefore confine our discussion primarily to the impulsive continuum emission at X-ray energies $\gtrsim 10$ keV.

The early hard X-ray measurements were made with balloons (Peterson and Winckler, 1959; Vette and Casal, 1961; Anderson and Winckler, 1962). These experiments suffered from short observation time (~ 24 h per flight) and the effects of the Earth's atmosphere on the incident solar X-ray spectrum. However, such measurements gave rise to the discovery of hard solar X-ray bursts and even now new significant results are being obtained with balloon borne detectors (Parks and Winckler, 1969; Takakura *et al.*, 1971; Kane *et al.*, 1972). The rocket measurements were relatively free from atmospheric effects and hence the X-ray spectrum could be accurately measured down to a few keV (Chubb *et al.*, 1966). However, the problem of the short observation time (a few minutes) was even more severe in this type of measurement. Consequently only a few hard X-ray bursts from rather large solar flares were observed with balloons and rockets during the period 1958–1964.

With the advent of the satellites the difficulties due to the short observation time and effects of the Earth's atmosphere were simultaneously removed. Since 1964 a large number of hard X-ray bursts have been recorded with the ionization chambers aboard the OGO-1 and OGO-3 satellites (Arnoldy *et al.*, 1968a, b, c; Kane and Winckler, 1969a, b) and the scintillation spectrometers aboard the following satellites: OSO-3 and OSO-7 (Hudson *et al.*, 1969; Elcan, 1973; Peterson *et al.*, 1973), OGO-3 (Cline *et al.*, 1968), OGO-5 (Kane and Anderson, 1970; Kane, 1971), and OSO-5 (Frost, 1969; Frost and Dennis, 1971). In addition, polarization of 10–20 keV X-rays has been measured with the Intercosmos-1 and 4 spacecraft (Tindo *et al.*, 1970, 1972a; Mandelstam, 1972).

Figure 2 shows an example of a relatively simple impulsive X-ray burst observed by the OGO-5 satellite (Kane and Anderson, 1970). The associated microwave emission is also shown for comparison. The impulsive X-ray emission, which was most prominent at X-ray energies $\gtrsim 20$ keV, reached its maximum simultaneously with the maximum of the impulsive microwave emission but ~ 1.5 min before the maximum of the Hα emission from the *total* flare area. Sometimes the X-ray bursts consist of several impulsive peaks which may occur quasi-periodically with a period of 10–30 s (Parks and Winckler, 1969, 1971; Frost, 1969).

In X-ray bursts such as the one shown in Figure 2 one can identify two X-ray components: (1) an impulsive component which is more prominent at X-ray energies $\gtrsim 30$ keV and may exhibit quasi-periodic fluctuations; (2) a gradual component with a relatively smooth time-intensity profile and which reaches its maximum several minutes after the maximum of the impulsive component. The existence of two components in the X-ray emission $\gtrsim 10$ keV was suggested earlier by several workers (Acton, 1968; Takakura, 1969; de Jager, 1969). The two component structure was first observed in the OGO-5 data (Kane, 1969; Kane and Anderson, 1970) had has since then been found to be consistent with the OSO-3, OSO-5 and OSO-7 measurements.

The X-ray spectrum is related to the spectrum of the energetic electrons in the X-ray source. For example, a power law X-ray spectrum $\sim E^{-\gamma}$ photons cm^{-2} s^{-1} keV^{-1} indicates a power law electron spectrum $\sim E_e^{-\alpha}$ electrons keV^{-1} where $\gamma - \alpha \approx 1$ (Kane and Anderson, 1970; Brown, 1971). It is therefore important to have accurate high time resolution measurements of the X-ray spectrum so that the characteristics of the electron spectrum can be deduced. Figure 3 shows the X-ray spectrum in 10–100 keV range measured at the maxima of three impulsive X-ray bursts observed by the OGO-5 satellite (Kane, 1973a). A spectrum measurement of 10–200 keV X-rays made with the OSO-5 satellite has been reported by Frost (1969) and Frost and Dennis (1971) (Figure 9). For X-rays in the energy range $E_{C1} - E_{C2}$, the observed spectrum is consistent with the form

$$\frac{\mathrm{d}J}{\mathrm{d}E} = KE^{-\gamma} \text{ photons cm}^{-2}\,\text{s}^{-1}\,\text{keV}^{-1}. \tag{1}$$

Above some critical energy E_{C2} the spectrum steepens rapidly (larger γ). The value of E_{C2} varies from one flare to another but usually lies in the range 60–100 keV. A large

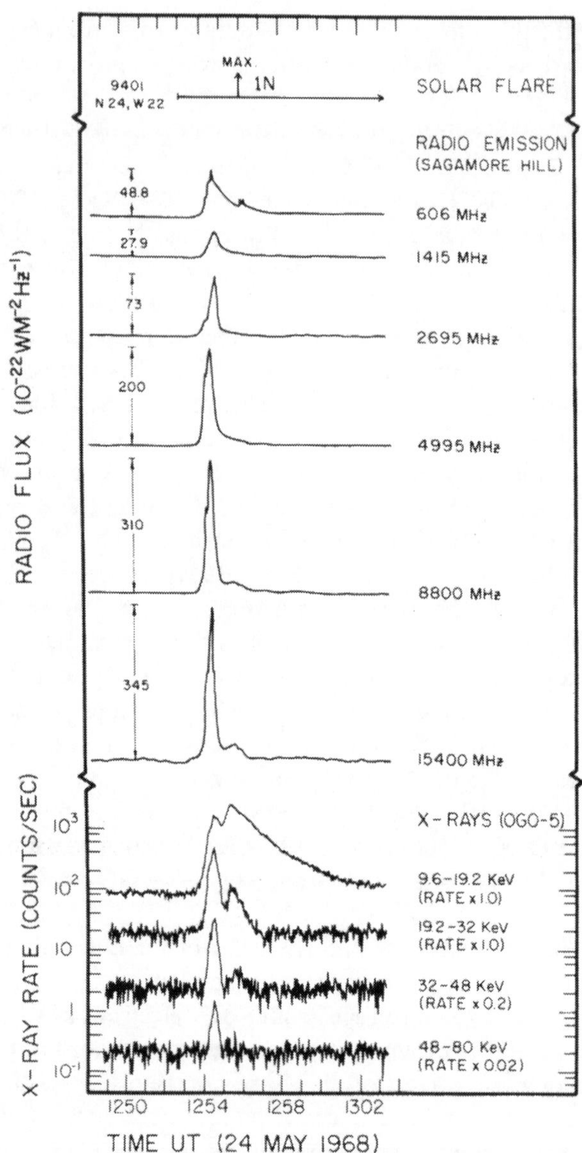

Fig. 2. An example of a relatively simple impulsive X-ray and microwave burst (Kane and Anderson, 1970).

flare in which E_{C2} might have been ~500 keV has been reported by Gruber *et al.* (1973). Below some critical energy E_{C1} the gradual X-ray component dominates and the impulsive X-ray spectrum cannot be determined un-ambiguously. Usually E_{C1} is ~20 keV. However, occasionally it may be as low as ~5 keV as was observed by Kahler and Kreplin (1971), and Peterson *et al.* (1973) (Figure 4).

The spectral exponent γ also varies from one flare to another. Figure 5 shows the

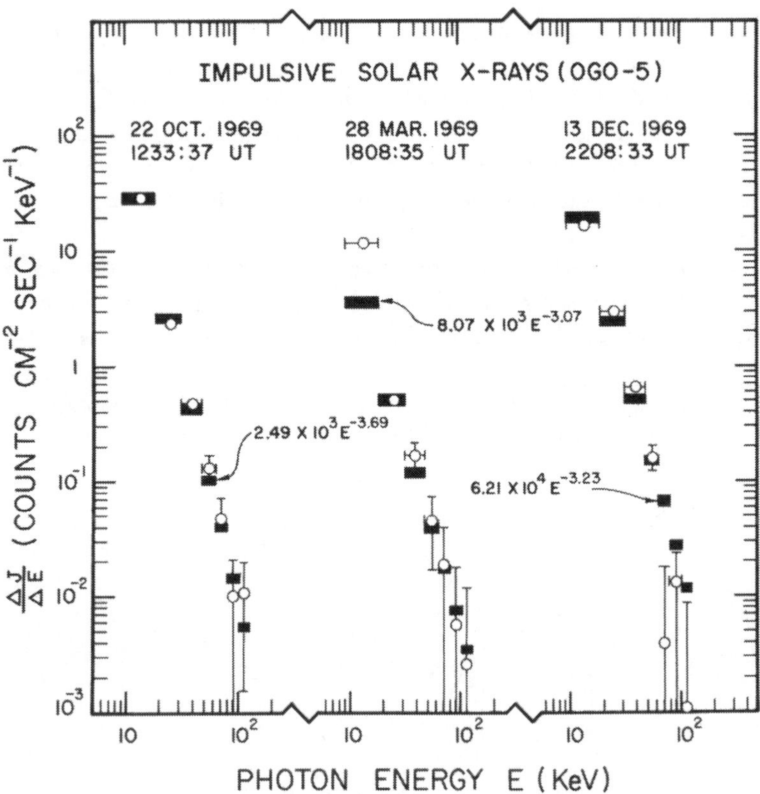

Fig. 3. The X-ray spectra at the maxima of three impulsive bursts. Open circles are observations and dark rectangles are computed response of the OGO-5 detector to a power law X-ray spectrum (Kane, 1973a).

frequency of occurrence of different values of γ in small solar flares (Kane, 1973a). For $2.5 \lesssim \gamma \lesssim 5.0$ the probability of occurrence of a given γ in a flare increases systematically with the increase in the value of γ. The probability of occurrence of a flare with $\gamma < 2.5$ is extremely small indicating that there is essentially an upper limit on the hardness of the impulsive X-ray spectrum (and hence the electron spectrum) produced in solar flares (Kane, 1971).

The time variation of the impulsive X-ray spectrum during a single flare is of particular interest because it is expected to give information about the acceleration and energy loss processes for the energetic electrons in the flare region. At present there is no general agreement on the nature of this variation. In several small X-ray bursts Kane and Anderson (1970) find that the X-ray spectrum in 10–60 keV range hardens from the onset of the X-ray burst to its maximum and then softens during the decay (Table I). Similar spectral variation has been observed by Parks and Winckler (1969) and Kodama *et al.* (1971) in a large X-ray burst recorded with balloon born detectors. The softening of the X-ray spectrum during the decay has also been reported by McKenzie *et al.* (1973) and Peterson *et al.* (1973). One case of hardening of the spectrum during the increasing phase of the X-ray burst has been observed by Peterson *et al.* (1973). In some events no systematic variation of the spectrum is observed during the increasing phase (McKenzie *et al.*, 1973; Datlowe and Peterson, 1973). Frost (1969) and Frost and Dennis (1971) have each reported one burst associated with a large flare in which the X-ray spectral exponent was constant throughout the duration of the X-ray burst. On the other hand, for one large flare, Cline *et al.* (1968) found that the X-ray spectrum $\gtrsim$80 keV hardened during the decay phase. A similar effect has been reported by Gruber *et al.* (1973) for the 23 May 1967 flare events. The discrepancy between these different observations could be due to the inherent variations from one flare to another. Another factor could be the fine time structure ($\lesssim$1 s) in the X-ray time-intensity profile and the associated spectral variations. Clearly, high time resolution ($\sim$0.1 s) measurements of a large number of flares are required to resolve this question satisfactorily.

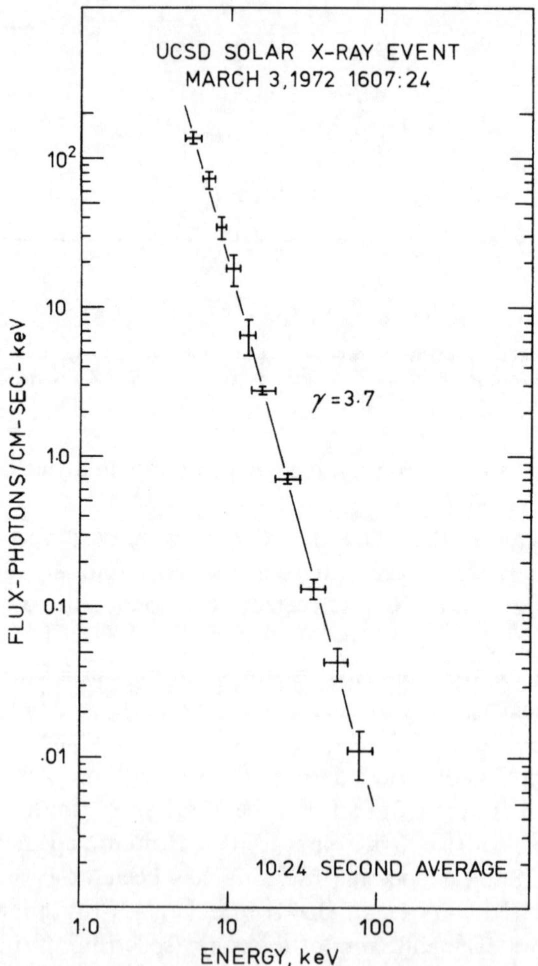

Fig. 4. An example of the power law X-ray spectrum extending down to $\sim$5 keV (Peterson *et al.*, 1973).

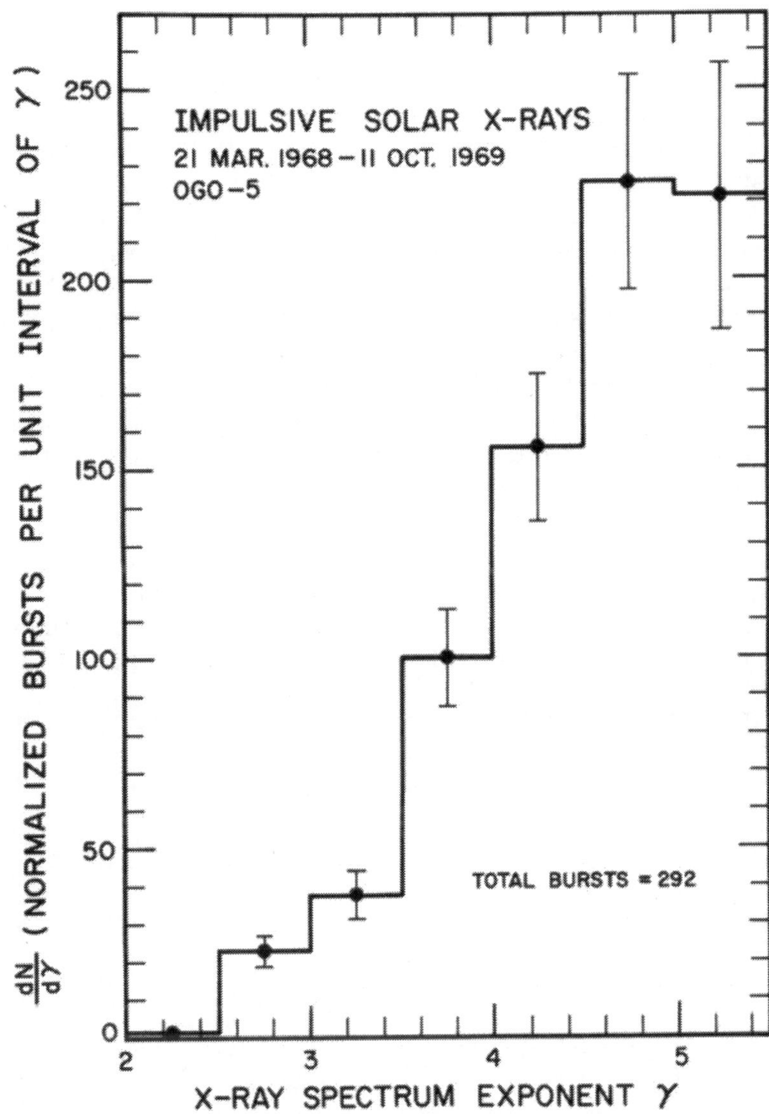

Fig. 5. The relative frequency of occurrence of the different values of the X-ray spectral exponent γ in small solar flares (Kane, 1973a).

TABLE I

Impulsive X-rays ($\gtrsim 10$ keV) from small solar flares

Spectrum at maximum (at 1 AU)	$\sim E^{-\gamma}$ photons cm^{-2} s^{-1} keV^{-1}
	$2 < \gamma \lesssim 6$ for $10 \lesssim E \lesssim 100$ keV
	Much steeper for $E > 100$ keV
Maximum energy flux at 1 AU	10^{-7}–10^{-5} erg cm^{-2} s^{-1} for $10 \leqslant E \leqslant 100$ keV
Rise time	2–5 s for $E \sim 40$ keV less for $E > 40$ keV
Decay time ($\gtrsim$ rise time)	3–10 s for $E \sim 40$ keV less for $E > 40$ keV
Correlated emissions	Impulsive microwave, EUV, Hα type III radio

A question of some interest is the variation of the peak flux and the spectral exponent γ with the location of the associated Hα-flare on the solar disc. Since the absorption of hard X-rays in the solar atmosphere is negligible, any variation in the characteristics or number of the X-ray bursts with the location of the flare may be an indication of the directivity of the X-ray emission. Since large variations in the burst characteristics are observed even for flares having essentially the same location on the disc, it is essential to analyze a large number of events before a statistically significant result can be obtained. Two somewhat different results have been reported so far. From an analysis of 46 hard X-ray burst recorded by the OGO-1 and OGO-3 ionization chambers (Arnoldy *et al.*, 1968c) and the OSO-1 spectrometer (Frost, 1964), Ohki (1969) has concluded that the number of X-ray bursts decreases systematically from the center to the limb of the solar disc indicating a limb-darkening. On the other hand, using similar data Pinter (1969) found that the number distribution of the X-ray bursts has two maxima, one at 20–30° E and the other at 20–30° W heliographic longitude. The difference between these two results is most probably due to the statistical uncertainties caused by the relatively small number of events used in the analysis. Figure 6 shows the number distribution obtained for ~300 impulsive X-ray bursts observed with the OGO-5 satellite. All of these bursts were associated with

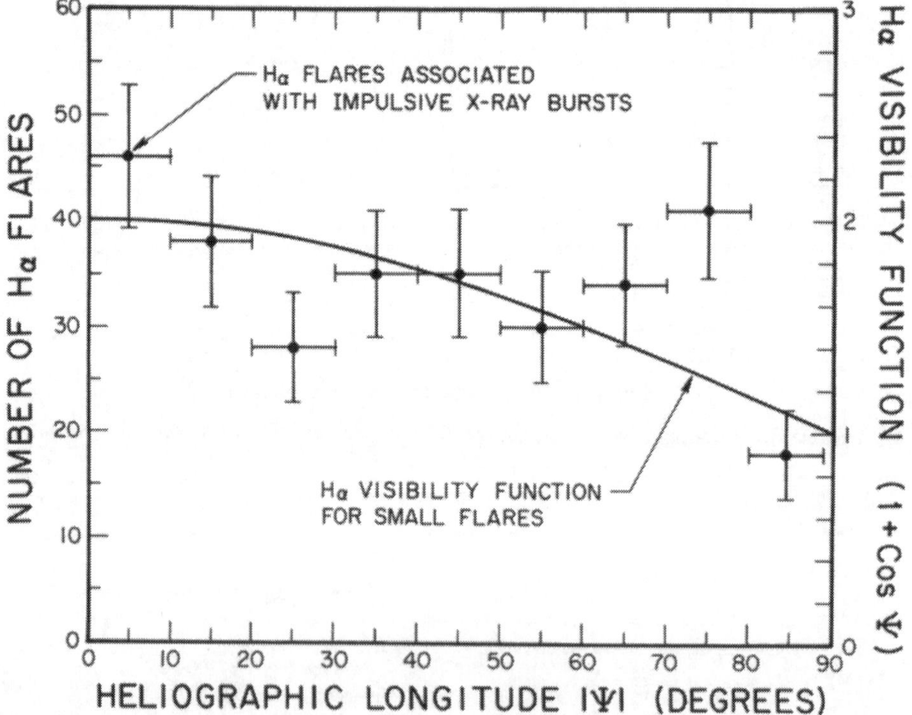

Fig. 6. Center-to-limb variation of the location of the Hα-flares associated with ~300 impulsive X-ray bursts observed by the OGO-5 satellite. The solid line represents the Hα visibility function for small flares given by Smith and Smith (1963).

flares of Hα-importance $\lesssim 1$. Therefore the 'Hα visibility function' for small solar flares (Smith and Smith, 1963) is also shown in Figure 6 for comparison. It can be seen that at least in the case of small flares there is no significant center-to-limb variation in the frequency of occurrence of X-ray bursts.

Another indication of the directivity of the X-ray emission and/or the directivity of the energetic electrons producing the observed X-ray emission is its polarization. The linear polarization of 10–20 keV X-rays in several solar flares has been measured with the Intercosmos-1 and 4 spacecraft (Tindo *et al.*, 1970, 1972a, b; Mandelstam, 1972). An example taken from Tindo *et al.* (1972a) is shown in Figure 7. Here the three curves (a), (b) and (c) respectively represent the intensity of the radiation, degree of polarization and the angle of polarization (angle between the plane of polarization and the 'reference' plane of the polarimeter). The measurements have been *normalized* in such a way that the polarization is zero during the late decay phase of the gradual X-ray burst. The authors have found significant polarization for time intervals lasting from a few up to ten minutes. At the maximum of the impulsive burst the polarization is $\sim 20\%$ and the plane of polarization is so oriented that its projection on the solar disc passes through the associated flare region and the denter of the disc. An increase in polarization is also found at the maximum of the gradual X-ray component.

Direct measurements of the location and size of the impulsive hard X-ray source are difficult because of the high time resolution necessary to obtain meaningful results. One dimensional measurements of one X-ray burst has been made by Takakura *et al.* (1971) with a balloon born modulation collimator. They found the center of the X-ray source to be on a line passing through the center of the associated Hα flare (Figure 8). Although the Hα flare region was $\sim 3'$, the size of the hard X-ray source was $\lesssim 1'$.

At present the only source of information about the altitude of the impulsive hard X-ray source are the behind-the-limb flares. These are invariably relatively large flares and the information is therefore directly applicable only to large flares. Figure 9 shows an impulsive X-ray burst from a flare $\sim 14°$ behind the west limb (Frost and Dennis, 1971). It shows that the source of the observed emission was located at an altitude $> 10^4$ km above the photosphere. McKenzie and Peterson (1973) have recently reported observations of five behind-the-limb flares which are consistent with such a conclusion for small flares.

It should be pointed out here that X-ray emission produced at the flare site is subject to Compton scattering in the solar atmosphere. The X-rays observed at 1 AU therefore consist of the relatively unaffected direct beam of photons as well as the Compton back-scattered photons. Calculations show that the relative flux of the direct and scattered photons depends on the inial energy of the photons and the location of the X-ray source on the sun (Tomblin, 1972; Santangelo *et al.*, 1973). At present it appears that the corrections due to this 'albedo effect' are of the same order of magnitude as the uncertainties in the calculations themselves and the uncertainties in the measured parameters, such as spectral index and polarization, caused by statistical fluctuations and the limited spectral, temporal and spatial

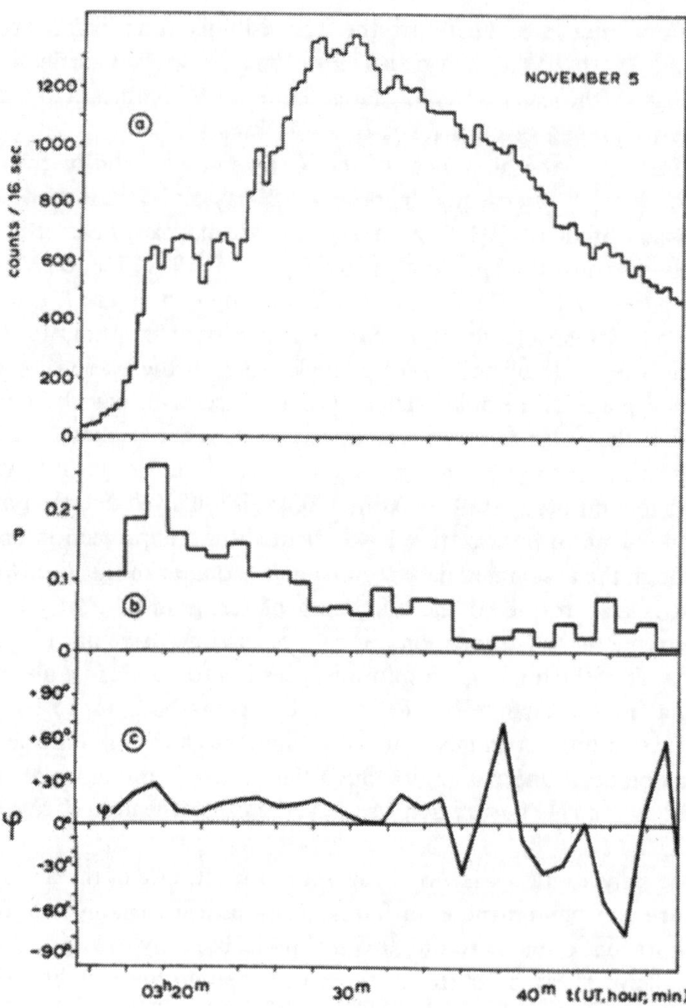

Fig. 7. Linear polarization of 10–20 keV X-rays measured with the Intercosmos 4 spacecraft. The three curves are: (a) intensity of the radiation, (b) degree of polarization, and (c) angle of polarization (Tindo *et al.*, 1972a).

resolution of the instruments. Therefore not much is likely to be gained by applying corrections for the albedo effect to the presently available observations. However as the spatial, temporal and spectral resolution of the instruments improves in the future, the albedo effect must be taken into account.

4. Radio Observations

The observations of flare associated radio emission have been widely discussed in the past (Kundu, 1965; Maxwell, 1965; Takakura, 1969; Boischot, 1972). The radio bursts most intimately associated with the impulsive phase of flares are the impulsive

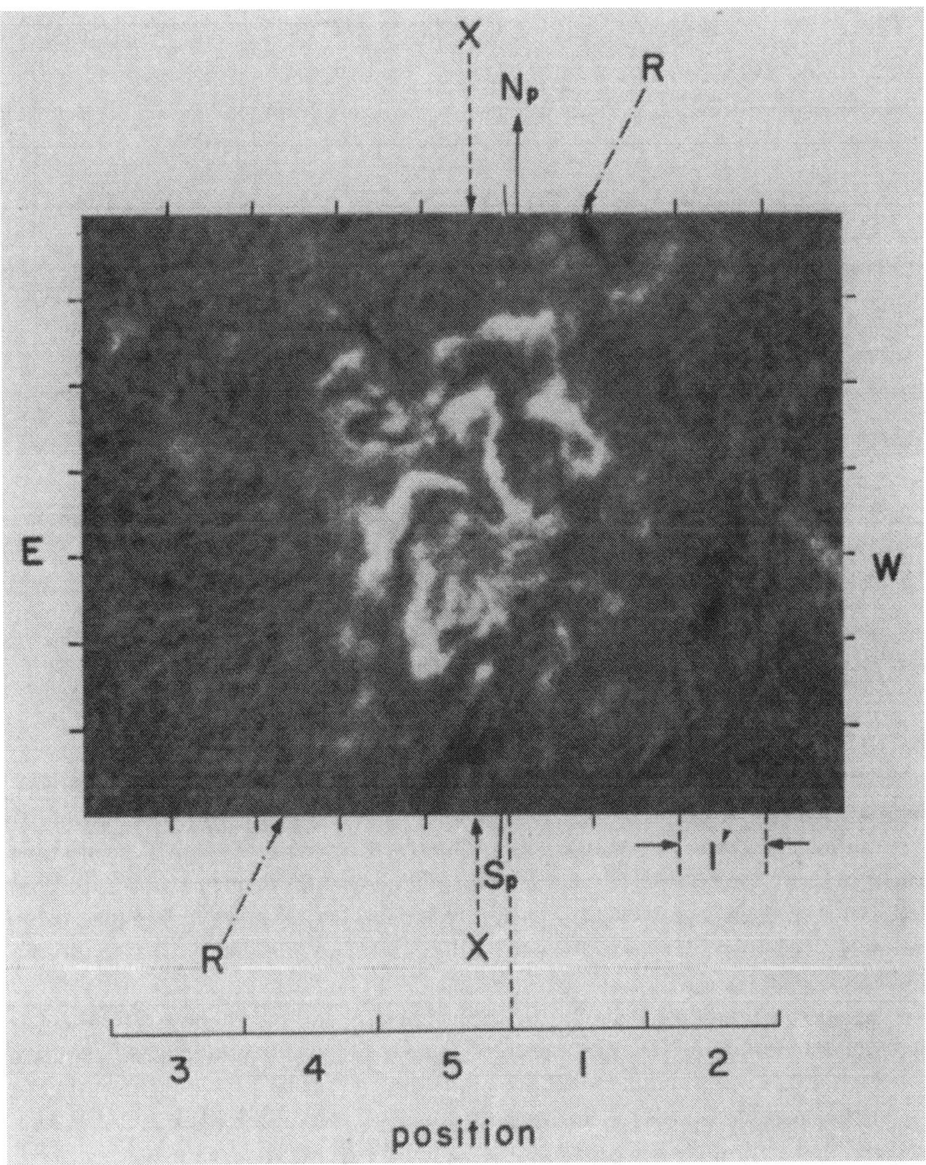

Fig. 8. Location of the centers of the hard X-ray source (line X) and 3750 MHz source (line R) with respect to the associated Hα-flare region (Takakura *et al.*, 1971).

microwave bursts ($\lambda \lesssim 10$ cm, $f \gtrsim 3$ GHz) and the type III radio bursts at decimeter and meter wavelengths. In this section we will be mostly concerned with the impulsive microwave bursts. The ground based and satellite observations of the type III bursts are discussed elsewhere in the proceedings of this symposium (Stewart, 1973; Fainberg, 1973). Therefore we will only mention here the recent correlation studies related to the type III bursts, particularly those extending into the decimeter range.

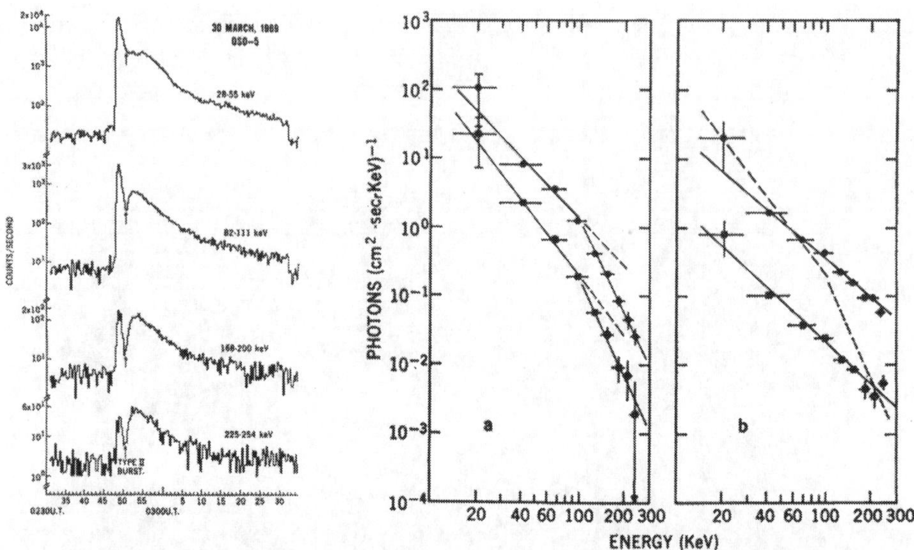

Fig. 9. An impulsive X-ray burst followed by a slower long lasting burst observed by the OSO-5 satellite. The event was associated with a flare located behind the west limb of the Sun. (a) and (b) show the X-ray spectra at two different times during the impulsive and slow bursts respectively. The impulsive burst is believed to be caused by acceleration of electrons during the impulsive (flash) phase. The slow burst is considered evidence for a second stage of acceleration (Frost and Dennis, 1971).

As in the case of X-ray measurements the quantities of interest in the microwave measurements are the spectrum, polarization, spatial extent and altitude of the emission source above the photosphere and the variation of these quantities with time. Some of these characteristics, as deduced from the measurements made before 1964, have been described in detail by Kundu (1965, Chapters 7 and 13). We give below some of the results of the more recent studies and also summarize the results obtained in the past.

An example of an impulsive microwave burst observed at single frequencies between 0.606 and 15.4 GHz is shown in Figure 2. The emission generally consists of a broad band continuum extending from $\sim$0.5 to $\sim$100 GHz. In the centimeter range the maximum flux at 1 AU varies from 1 to 5×10^4 sfu (1 sfu$\equiv10^{-22}$ WM^{-2} Hz^{-1}), the lower limit being primarily set by the instrument sensitivity.

The spectra of the impulsive microwave bursts have been recently compiled by Guidice and Castelli (1973) and Croom (1973). The spectrum most often associated with small and medium flares (Hα-importance $\lesssim$2) is what Guidice and Castelli (1973) have called type C. It is schematically shown in Figure 10. This is a relatively simple spectrum in which the flux reaches a single maximum at a frequency f_{max} in the centimeter range and decreases on both sides of this maximum. The frequency of maximum emission f_{max} is usually $\sim$5 GHz and the emission at mm wavelengths ($f>10$ GHz) is $<50\%$ of the emission at f_{max}. At the low frequency side the spectrum has an effective cut-off at a frequency $f_{cut}\approx0.3\,f_{max}$ (Guidice and Castelli, 1973).

In general, the observed emission does not reach its maximum simultaneously at all frequencies. A statistical analysis made by Kakinuma *et al.* (1969) is shown in Figure 11. It shows that the emission at high frequencies (~9400 MHz) tends to reach its maximum a few seconds before the emission at low frequencies (~1000 MHz).

In case of large flares (Hα-importance $\gtrsim 3$) considerable emission is observed at mm wavelengths and in some events $f_{max} \gtrsim 20$ GHz (Croom, 1973; Kundu, 1973).

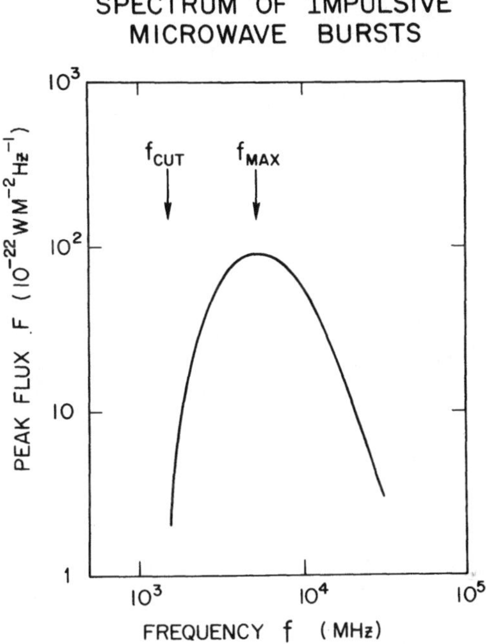

Fig. 10. A schematic representation of the emission spectrum most often observed in impulsive microwave bursts associated with small flares.

Such events tend to be closely associated with flares which produce significant number of energetic protons. In fact a rough correlation exists between the duration of an impulsive burst and the peak flux of protons >10 MeV observed at 1 AU (Croom, 1973).

High resolution measurements at 9.4 GHz by Tanaka *et al.* (1967) show that some impulsive bursts consist of two different emission sources separated by ~2′ and coinciding with the two sunspots of a bipolar group. Both sources are circularly polarized but in opposite directions. Figure 12 shows the time development of the intensity and polarization distribution observed at 9.4 GHz by Enome *et al.* (1969) during the 16 December 1967 flare event. A hard X-ray burst was also observed in this event (Kane and Winckler, 1969b) approximately in time coincidence with the impulsive microwave burst which reached its maximum at 0252 UT. Initially (0246–0249 UT) and after the flux maximum (0255–0258 UT) the right hand polarization (*R*) was dominant in the whole radio source. However, near the burst maximum (0252 UT) the left handed polarization (*L*) dominated in the eastern part of the source. This change in polarization during the increasing phase of the burst, for which no optical counterpart was found in Hα photographs, was observed at 9.4 GHz but not at 3.75 GHz. The authors have interpreted this observation as indirect evidence for a

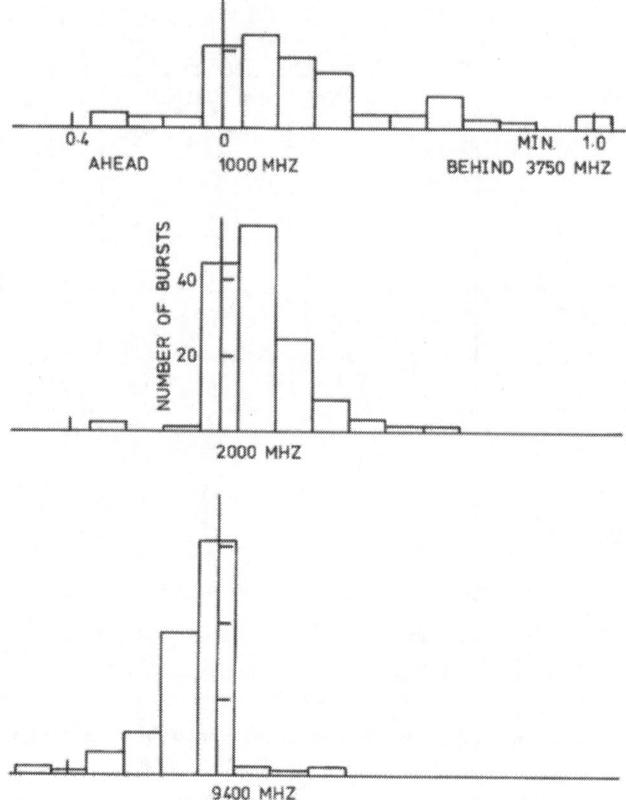

Fig. 11. A statistical analysis of the differences in the times of maximum intensity 1000, 2000 and 9400 MHz with respect to the time of maximum at 3750 MHz. On an average, the emission at 1000 and 2000 MHz tends to reach its maximum later than the emission at 9400 and 3750 MHz (Kakinuma *et al.*, 1969).

change in the magnetic field configuration of the associated active region during the impulsive burst.

The center-to-limb variation of the characteristics of the microwave bursts have been studied by several workers (Dodson *et al.*, 1954; Akabane, 1958; Harvey, 1964; Kundu, 1965; Castelli and Barron, 1969; Kakinuma *et al.*, 1969; Castelli and Guidice, 1972). In some studies (see e.g. Akabane, 1956) no distinction was made between the impulsive and gradual rise and fall (GRF) bursts. This is unfortunate because the GRF bursts show a much larger center-to-limb variation in occurrence frequency than the impulsive bursts. For the impulsive bursts, Kakinuma *et al.* (1969) found

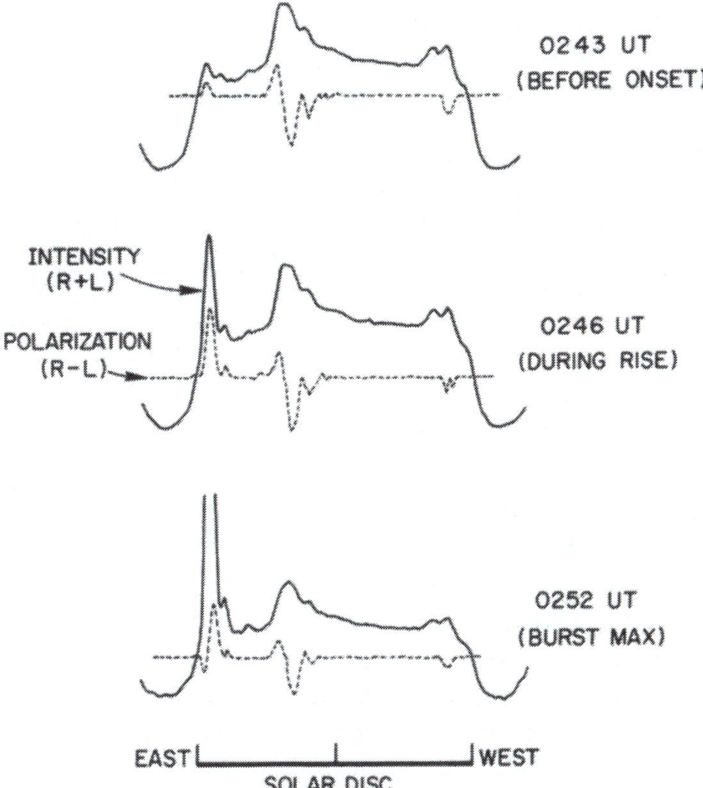

Fig. 12. The time development of the intensity (solid line) and circular polarization (dashed line) at 9.4 GHz during the increasing phase of the impulsive burst on 16 December 1967, Initially the right hand plarization (R) was dominant in the whole radio source (located near the east limb). However near the burst maximum the left handed polarization (L) dominated in the eastern part of the source (Enome *et al.*, 1969).

that, as compared to the bursts associated with flares located near the center of the solar disc, the intensity of bursts for flares located at a central meridian distance (CMD) of $\sim 75°$ is about 0.56, 0.58, 0.65 and 0.51 at 9.4, 3.75, 2.0 and 1.0 GHz respectively. This decrease in the observed emission for flares at large CMD could be due to the unisotropic emission by the source itself and/or due to the attenuation of the emission by an absorbing layer above the source. In the latter case, the observations indicate that the optical depth of the source at CMD$=0$ is ~ 0.2, nearly independent of the emission frequency in 1.0–9.4 GHz range. No significant center-to-limb variation in the degree of polarization of the impulsive microwave bursts has been observed (Kakinuma *et al.*, 1969).

The recent high resolution measurements by Enome *et al.* (1969) show that the size of the microwave source is $\sim 0.5'$. For the microwave burst associated with the hard

X-ray burst (Figure 8) observed by Takakura *et al.* (1971), Tanaka and Enome (1971) found that at 3.75 GHz the width of the radio source was $<1.8'$. The observations for this event are consistent with the location and size of the hard X-ray and microwave sources being essentially the same.

The altitude of the microwave source is believed to be $\gtrsim 2\times10^4$ km above the photosphere (Kundu, 1965). This estimate is based on observations of behind-the-limb flares (Covington and Harvey, 1961a; Bruzek, 1964; Enome and Tanaka, 1971). The small optical depth of the source deduced from the center-to-limb variation of the microwave bursts is also consistent with a high altitude for the microwave source.

The correlation between the impulsive hard X-ray and microwave emissions was noticed earlier by Peterson and Winckler (1959) and Kundu (1961). This correlation has now been confirmed by observations of a large number of flare events (Arnoldy *et al.*, 1967, 1968a, b; Kane and Anderson, 1970; Kane, 1972a; McKenzie, 1972). Although the correlation is most apparent for the times of microwave and X-ray maxima, quantitative relationship between the X-ray and microwave fluxes also exists (Figure 13). The ratio of the peak X-ray ($\gtrsim 20$ keV) and microwave (~ 3 cm) energy fluxes is given by

$$R_{xm} = \frac{\xi(\gtrsim 20\ \text{keV})}{\xi(\sim 3\ \text{cm})} \sim 10^2 . \tag{1}$$

The type III radio bursts represent energetic electrons passing outward through the corona. Since the early work of Malville (1962) the flare associated type III bursts have been recognized as an impulsive phase phenomenon. The fact that electrons in essentially the same energy range (10–100 keV) are responsible for the impulsive hard X-ray and type III radio emission led de Jager (1960, 1962) to predict a close relationship between these two emissions. Although correlated type III and hard X-ray bursts were occasionally observed (Anderson and Winckler, 1962; Winckler *et al.*, 1961), in general no significant relationship was found (Kundu, 1961, 1965). The recent high time resolution (~ 2.3 s) measurements of impulsive hard X-ray bursts indicate that even though the overall correlation between the hard X-ray and type III bursts is only $\sim 30\%$, the relationship between these two emissions is probably more intimate than was believed in the past (Kane, 1972b). This is particularly true about the type III bursts which extend into the decimeter range. Figure 14 shows an example of a type III radio event which seems to be well correlated even in some of the fine structure with an impulsive X-ray burst observed with the OGO-5 satellite. In this particular case, there is a suggestion that individual groups of bursts in the type III emission have their counterparts in the intensity spikes in the associated X-ray burst. This indicates that the true time constant for X-ray intensity variation during the flash phase may in fact be ~ 0.1 s.

5. Extreme Ultraviolet Observations

The impulsive 10–1030 Å EUV emission during solar flares was first deduced in-

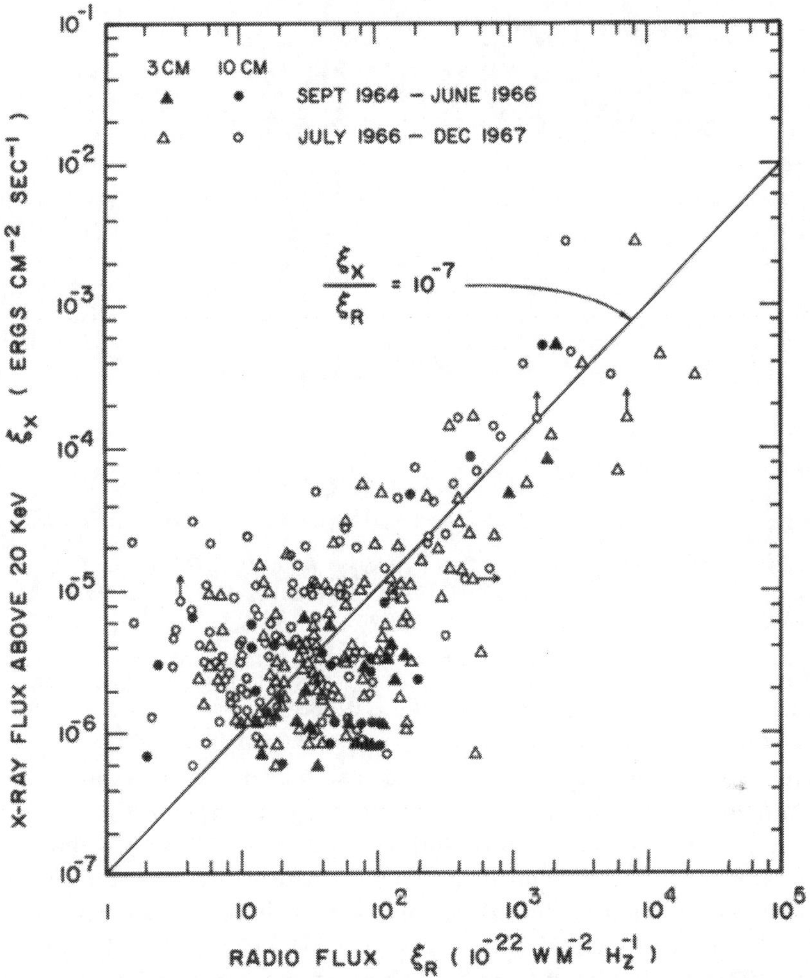

Fig. 13. A regression plot of the peak energy flux of X-rays ≳20 keV measured by the OGO-1 and OGO-3 ionization chambers and the peak energy flux of the microwave emission. Two separate time intervals are considered: September, 1964–June, 1966 (full symbols) and July, 1966–December, 1967 (open symbols). There is no observable difference between the regression lines for these two time intervals (Arnoldy *et al.*, 1968a, b, c; Kane and Winckler, 1969a, b).

directly by Donnelly (1968, 1969, 1971) from the observations of the ionospheric effect 'Sudden Frequency Deviation' (SFD) and a good correlation was found between the impulsive X-ray, EUV and microwave bursts (Kane and Donnelly, 1971). Since then satellite measurements of the line spectrum (Hall and Hinteregger, 1969; Castelli and Richards, 1971; Hall, 1971; Wood *et al.*, 1972; Wood and Noyes, 1972; Donnelly *et al.*, 1973) and the broad band emission (Kelly and Rense, 1973) have con-

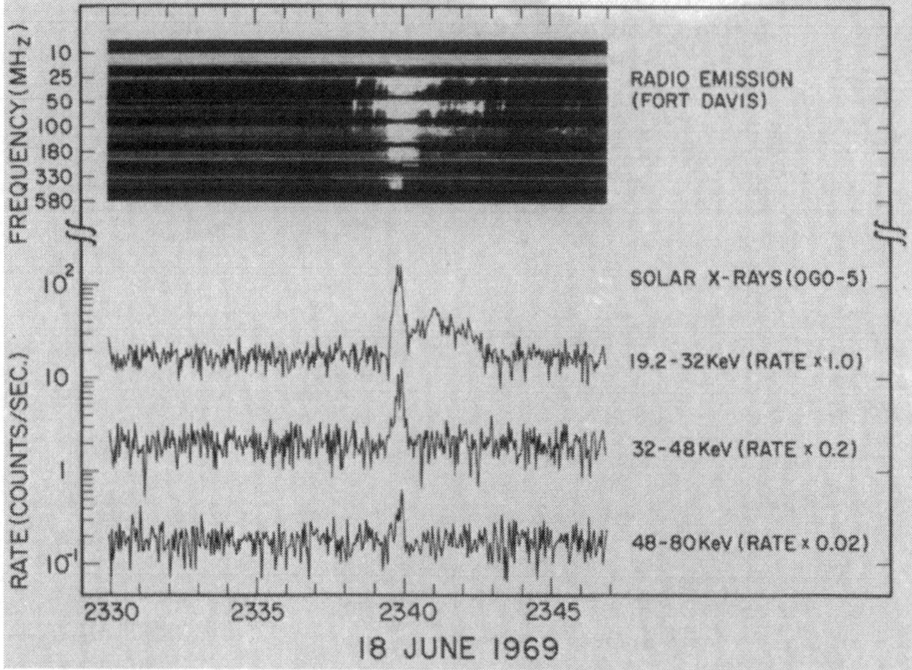

Fig. 14. An example of a type III radio event which seems to be well correlated even in some fine structure with an impulsive X-ray burst observed with the OGO-5 satellite. The individual groups of bursts in the type III emission seem to have their counterparts in the intensity spikes in the X-ray emission.

firmed the existence of the impulsive EUV emission by direct observations and have given some insight into the characteristics of energy transport in the flare region. The ground based and satellite observations of the 10–1500 Å flare emission have been recently reviewed by Donnelly (1973) and Noyes (1973) respectively. Here we will only briefly summarize the characteristics of the impulsive EUV bursts.

Figure 15 shows an impulsive burst at 630 Å (O v) reported by Castelli and Richards (1971). Similar bursts have been observed in other EUV lines such as 304 Å (He II Lα) and 972.5 Å (H Lγ) and broad band EUV emission. A comparison of the rise times of the EUV and X-ray emission is shown in Figure 16. There is an approximate proportionality between the peak EUV and X-ray energy fluxes. This is illustrated in Figure 17 (Kane and Donnelly, 1971).

Since all the desirable features, such as a high time and spatial resolution and observations of a large number of EUV lines and continuum over a wide spectral range, are not simultaneously available for many flare events, the available information is far from complete. Therefore only a very crude picture of the EUV flare can be constructed at this time. We present below what appears to be some general properties of the impulsive EUV emission:

(1) The EUV emission from flares consists of two components, viz. impulsive and gradual. (2) The increase in the impulsive component coincides in time with the in-

crease in other impulsive emissions, such as hard X-ray and microwave emission. (3) The rise time of the broad band 10–1030 Å emission is comparable to that of ~10 keV X-rays. (4) For flares of Hα-importance $\lesssim 3$, the peak energy flux in 10–1030 Å range varies from 0.1–10 ergs cm^{-2} s^{-1} at 1 AU. (5) The ratio of the peak EUV flux to the peak flux of X-rays $\gtrsim 10$ keV, which is $\sim 10^5$ for flares located near

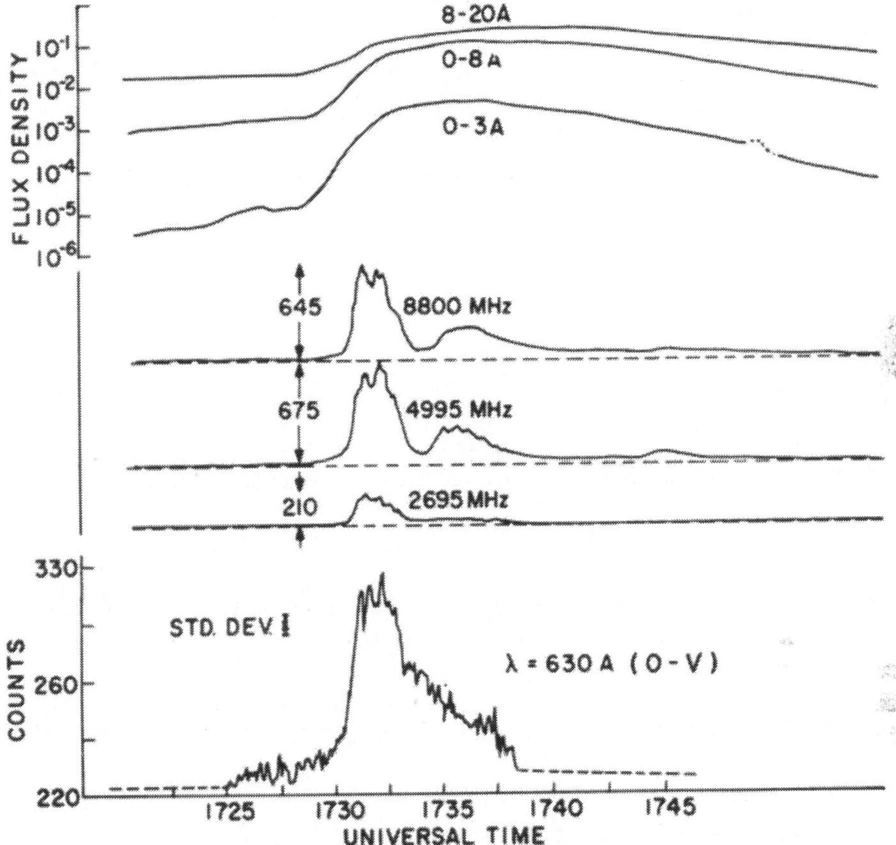

Fig. 15. An impulsive burst at 630 Å (O V) well correlated with the associated microwave emission (Castelli and Richards, 1971).

the central meridian, decreases with the increase in the central meridian distance of the flare. (6) The impulsive EUV source is located at an altitude $< 10^4$ km above the photosphere. (7) The lines enhanced during the impulsive phase represent characteristic temperatures varying from 10^4 K (chromospheric) to 1.5×10^6 K (coronal). (8) The relative enhancement of the various lines is approximately the same at the emission maximum although there is some evidence that the 'transition zone lines' tend to increase more than the coronal lines. (9) At a given time, the increase in the total line emission is of the same order of magnitude as the increase in the total continuum emission.

6. Optical Observations

Since the discovery of solar flares, the optical observations have played a very important role in the classification and understanding of the flare phenomenon. Excellent reviews of solar flare observations made before 1963 have been given by Ellison (1963) and Smith and Smith (1963). The more recent optical observations of flares have been discussed by Švestka (1969, 1970) and Zirin (1973). Since we are primarily

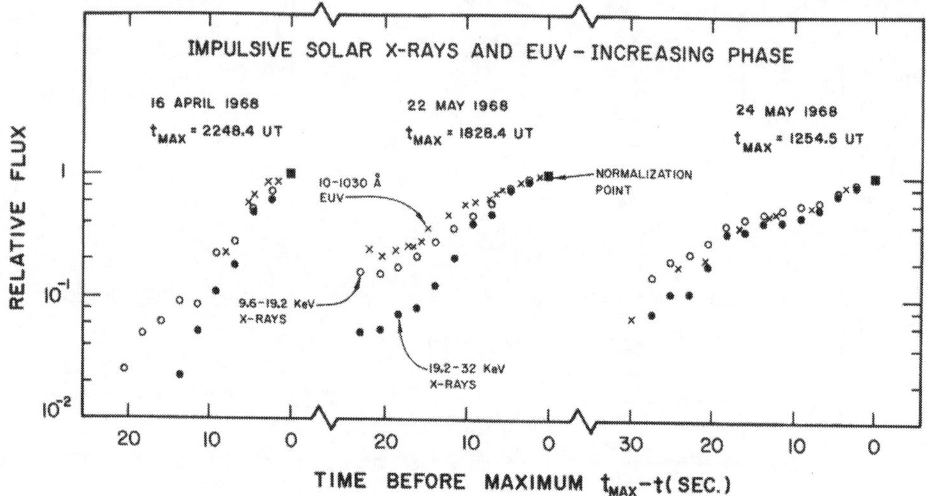

Fig. 16. A comparison of the rise times of the broad band EUV and hard X-ray emission for three impulsive bursts. The rise time of EUV emission is comparable to that of ~10 keV X-rays (Kane and Connelly, 1971).

concerned here with the impulsive phase of flares, the observations most relevant to the present discussion are those made with a high time and spatial resolution.

6.1. 'Monochromatic' observations

Most 'monochromatic' (narrow band) observations of the impulsive optical emission have been made in Hα. Few observations of the K line of calcium have also been made. However the high resolution observations made at the Big Bear Observatory show that there is no significant difference in the appearance of flares in Hα and the K line (Zirin, 1973). Therefore only the Hα observations will be discussed here.

In the past, the observations of the impulsive Hα emission from flares have been discussed in terms of two phases, viz. the flash phase and the explosive phase. Following Ellison (1949, 1963) the flash phase is usually considered as the time of rapid increase in Hα line-width associated with the rapid increase in the Hα emission during the early (pre-maximum) part of a flare. On the other hand, the explosive phase is marked by an abrupt increase in the Hα flux caused primarily by the rapid expansion of the flare borders (Moreton, 1964). A common factor in the two phases is the rapid increase in the observed Hα flux. Whereas there is no specific requirement in the flash phase on the cause of the increase in Hα flux, the explosive phase requires that this increase be primarily due to the expansion of the flare borders.

The flash phase of small flares and their correlation with the type III radio bursts has been studied by Malville (1962). Following the early work of Covington and Harvey (1961b), the explosive phase and its relationship with the impulsive X-ray, EUV and microwave emission has been studied by Moreton (1964), Davies and Donnelly (1966), Angle (1968), and Harvey (1971). The observed good correlation clearly

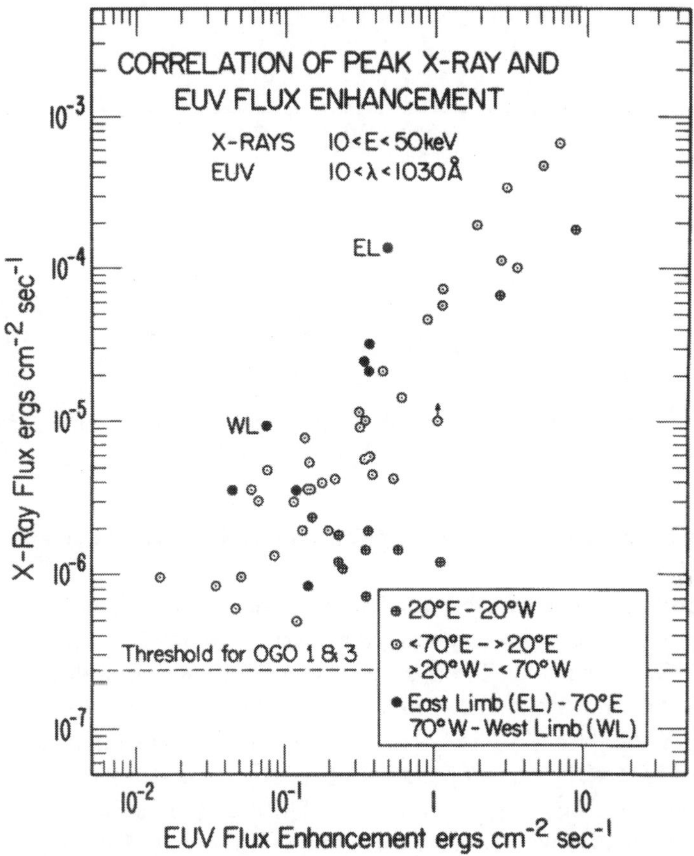

Fig. 17. A regression plot of 10–1030 Å EUV flux and 10–50 keV X-ray flux observed at the maxima of impulsive flare events (Kane and Donnelly, 1971).

indicates that the explosive phase is nearly coincident with the impulsive phase defined here.

It has been known for some time that the impulsive phase may occur only in a small part of an Hα-flare (Dodson *et al.*, 1956; Dodson and Hedeman, 1968; Harvey, 1971; Vorpahl, 1972). An example of the development of a large flare (Hα-importance $\gtrsim 3$) is shown in Figure 18 (Harvey, 1971). Among the A, B and E sections of the flare only the section E of the flare had an explosive phase. A similar effect

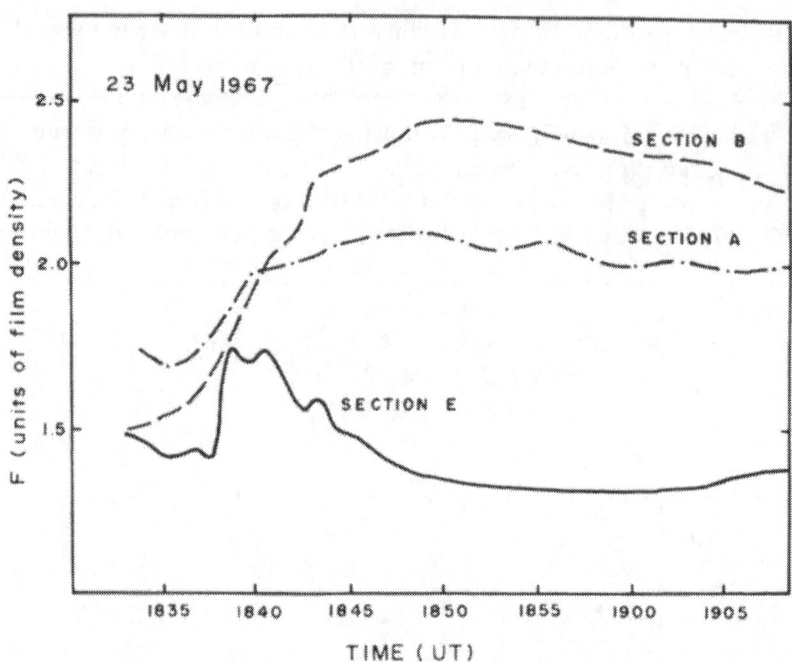

Fig. 18. The development of different regions (A, B, E,) of a large Hα-flare with time. Only the region E had an explosive phase (Harvey, 1971).

can be seen in Figure 19 where the development of a relatively small flare is illustrated (Vorpahl, 1972).

The impulsive phase of more than 55 flare events, most of which were of Hα-importance $\lesssim 1$, has been recently studied by Zirin *et al.* (1971), Vorpahl (1972), and Zirin (1973). Their principal findings may be summarized as follows: (1) The Hα kernels which undergo impulsive development are much more intense than the remaining part of the flare. (2) The Hα intensity of the impulsive kernels begins to increase 20–30 s before the onset of the impulsive X-ray increase and peaks approximately at the time of the X-ray maximum. (3) The increase in intensity of the impulsive kernels is associated with an increase in the Hα line-width. However, this increase in line-width is confined to the Hα emission from the impulsive kernels only; the remaining flare area is relatively unaffected. (4) The total energy radiated in Hα by an impulsive kernel is $\sim 5 \times 10^{25}$ erg in a small flare. (5) The impulsive kernels are located lower in the chromosphere ($\lesssim 5000$ km altitude) than the remaining part of the Hα flare. (6) Although outward motion is sometimes seen simultaneously with, or following, the impulsive phase, an impulsive burst is more likely to occur when the optical flare exhibits little expansion of material or light. (7) The effective diameter of the Hα kernel is 3000–6000 km and normally represents $<\frac{1}{8}$ to $\frac{1}{2}$ of the total flare area at the Hα maximum. (8) The occurrence of the impulsive phase has little to do with the size of the Hα-flare. (9) Bright kernels invariably occur near steep magnetic field gradients. (10) Almost all flares are bipolar. When only one kernel is present, the brightness occurs only on one side of the neutral line. Multiple kernels, when present, are located in areas of opposite polarity and are apparently connected by the same field lines.

6.2. Broad-band Observations

The most common broad band observations are the observations of 'white light flares' (DeMastus and Stover, 1967), which are usually very large and complex flares. For small and medium flares with which we are primarily concerned here, the only

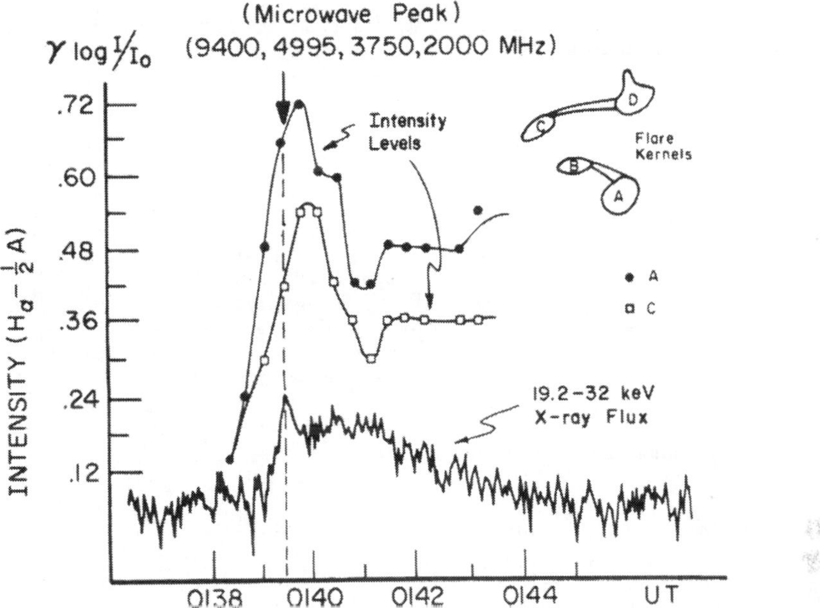

Fig. 19. The development of different regions or kernels (A, B, C, D) of a small Hα-flare with time. Only the kernels A and C had an impulsive phase (Vorpahl, 1972).

broad band measurements reported in the literature are those made by Zirin (1972) in the 15 Å band centered at 3835 Å. They did not find any increase in the emission of 3835 Å even in fairly large flares indicating that in most flares there is no significant heating of the lower chromosphere at a height where the density is $\sim 10^{17}$ particles cm^{-3}. However, in the very large flares of 2 August 1972, Tanaka and Zirin (1973) did find a large increase in the 3835 Å emission in good time coincidence with the hard X-ray and microwave enhancements.

7. Summary and Discussion

Table II summarizes the characteristics of the various electromagnetic emissions during the impulsive phase. The quantities such as the rise time and peak energy flux are based on observations discussed above. Some characteristics of the source region, such as altitude above the photosphere, linear dimension and double-source nature are also based primarily on observations. In some cases very few observations of a given quantity are available. For example, the altitude of the hard X-ray source has been observed only in one well documented behind-the-limb flare. However, in view of the studies involving comparison of the impulsive X-ray and interplanetary electron measurements (Datlowe and Lin, 1973) and the fact that no known observation is inconsistent with an altitude of $\sim 10^4$ km for the hard X-ray source, we have included this altitude in Table II as the best estimate presently available. The quantities such as the ion density n_i, neutral density n_0 and Temperature T for the radiation sources are deduced on the basis of the observed spectral and temporal characteristics of a given radiation and a plausible theoretical model specifically applicable to

TABLE II

Impulsive phase electromagnetic emissions

Characteristic	Radiation				
	Microwave	X-rays ($\gtrsim$20 keV)	Type III decimeter and meter	EUV	Hα
Emission process	Synchrotron	Brems-strahlung	Plasma hypothesis	Bound-bound free-bound and free-free	Bound-bound
Rise time (s)	~ 1	$\lesssim 1$	~ 0.1	~ 10	~ 10
Peak energy flux relative to peak X-ray energy flux	$\sim 10^{-2}$	1	$\sim 10^{-4}$	$\sim 10^5$	$\sim 10^2$
Total energy radiated in a small flare (ergs)	$\sim 10^{21}$	$\sim 10^{23}$	$\gtrsim 5 \times 10^{18}$	$\sim 10^{28}$	$\sim 5 \times 10^{25}$
Source region:					
Altitude h (km)	$\sim 10^4$	$\sim 10^{4}$ [a]	$\gtrsim 5 \times 10^4$	$\lesssim 5 \times 10^3$	$\sim 5 \times 10^3$
Linear dimension L (km)	$\sim 10^4$	?	?	?	$\sim 5 \times 10^3$
Double-source nature	Yes	?	–	?	Yes
Ion density n_i (cm^{-3})	$\sim 10^{10}$	$\sim 10^{10}$ [a]	$\lesssim 10^9$	$\gtrsim 10^{11}$	$\gtrsim 10^{11}$
Neutral density n_0 (cm^{-3})	~ 0	~ 0 [a]	~ 0	$\gtrsim 10^{12}$	$\gtrsim 10^{12}$
Temperature T (K)	$\sim 10^6$	$\sim 10^{6}$ [a]	$\sim 10^6$	$\lesssim 10^4$	$\lesssim 10^4$

[a] Although there is considerable uncertainty in these numbers, they are the best estimates available at the present time.

that radiation source. For example the value of $n_i \sim 10^{10}$ cm^{-3} for the microwave source is based on the theoretical models of Takakura (1968, 1973), Holt and Ramaty (1969), and Ramaty and Petrosian (1972) which are consistent with the observations.

A number of important features of the impulsive phase may be noted from Table II. Most of the radiated energy is emitted in the EUV range. The energy radiated as optical emission is not well known at the present time. The Hα emission carries a relatively small fraction of the energy. The study of white light flares recently made by McIntosh and Donnelly (1972) indicates that the energy radiated as white light

(3500–6500 Å) is probably of the same order of magnitude as the energy in the broad band EUV emission.

Although the energy radiated as microwave, X-ray and type III radio emission is relatively small, these emissions are the evidence for the existence of non-thermal electrons in the flare region. 'Thermal' electron spectra have been invoked from time to time to explain fully or in part the observed hard X-ray spectra (Chubb *et al.*, 1966; Chubb, 1972; Brown, 1973). It was also reported that the emission measure remains constant during an X-ray burst (Hudson *et al.*, 1969). This was, however, found to be an instrumental effect (Kane and Hudson, 1970). The power law X-ray spectrum, its rapid time variations and the close relationship with the microwave and type III radio emissions strongly suggest that the electron spectrum inside the hard X-ray source is non-thermal. The X-ray polarization measurements (Tindo *et al.*, 1970, 1972a, b) are also consistent with this interpretation.

The rapid time variations in the hard X-ray, microwave and type III radio emission indicate an electron acceleration process which is continuous in time or more probably a continuous series of impulses with a time constant $\lesssim 1$ s. Although the parameters of the electron acceleration spectrum deduced from the hard X-ray observations depend on the details of the assumed X-ray source model, some characteristics of the acceleration process can still be determined. Table III shows the properties of the electron acceleration process deduced from the X-ray observations summarized in Table I. The efficiency of acceleration is expressed as the ratio of the total kinetic energy of the accelerated electrons $\gtrsim 20$ keV to the total energy (thermal and non-thermal) of the flare. In constructing Table III a thin-target model for the hard X-ray source with continuous electron injection has been used (Kane and Lin, 1972; Kane, 1973a, b). If, instead, a thick-target model is assumed (Hudson, 1972a, b; Syrovatskii and Shmeleva, 1971), the deduced electron spectrum exponent φ will be larger by ~ 1, but the overall characteristics of the acceleration process indicated in Table III will remain relatively unchanged. Two important facts are revealed by Table III: (1) the acceleration process has a very short time constant, (2) the total kinetic energy of the accelerated electrons $\gtrsim 20$ keV is comparable to the total energy radiated during the impulsive phase. Thus the observations are consistent with a model of the impulsive

TABLE III

Acceleration of electrons in small solar flares

Nature of the process	Continuous (or continuous series of impulses)		
Characteristic time	$\lesssim 1$ s		
Total duration	~ 100 s		
Electron spectrum (averaged over ~ 2 s)	$\sim E_e^{-\varphi}$ electrons s^{-1} keV^{-1} ($10 \lesssim E_e \lesssim 100$ keV)		
Limits on exponent φ	$\varphi > 1$		
Time variation of φ	$\varphi(>\varphi_0)$ onset	φ_0 X-ray max	$\varphi(>\varphi_0)$ end
Probability of occurrence of a given φ_0 in a flare	Increases with increase in φ_0 for $1 < \varphi_0 \lesssim 3$		
Total kinetic energy of electrons $\gtrsim 20$ keV	$\sim$ total energy radiated during the impulsive phase		
Efficiency of acceleration	$\sim 10\%$		
Probable location of the acceleration region	$n_i < 10^9$ ions cm^{-3} (lower corona)		

phase in which energetic electrons play a central role and provide most of the energy required for the various electromagnetic emissions during the impulsive phase.

At present there seems to be no clear evidence for the unisotropic emission of either the hard X-ray or the microwave emission. Models of hard X-ray source with highly directional beams of energetic electrons, similar to those suggested earlier by de Jager and Kundu (1963), predict a relatively high degree of energy dependent polarization and unisotropy in the hard X-ray emission (Korchak, 1967; Elwert and Haug, 1970; Petrosian, 1973; Phillips, 1973). Polarization measurements reported by Tindo *et al.* (1972a, b) are consistent with beams of electrons moving radially on the Sun. However, the absence of significant center-to-limb variation in the observed location of the X-ray producing Hα-flares indicates that the unisotropy in the X-ray emission, if present, is relatively small. Hard X-ray observations with two or more spacecraft separated widely in heliographic longitude will be required before this question can be resolved satisfactorily.

The microwave as well as Hα emission shows a double-source structure. Thus far it has not been possible to resolve a similar structure in the EUV and X-ray sources. The double-source structure is sometimes considered as evidence for a magnetic field structure where the two radiation sources are connected by closed field lines (Vorpahl, 1972). However, it is also possible that the two sources are located on open field lines which pass through the region of primary energy release. Thus, the observations do not necessarily require an acceleration region located in the closed field region.

The observations are in general agreement with models of solar flares, such as those proposed by Alfvén and Carlqvist (1967), Sturrock (1968, 1973), Syrovatskii (1969), and Takakura (1971), in which induced or quasi-static electric fields accelerate particles during the impulsive phase. Except for a model recently discussed by Smith (1973), most flare models do not predict the spectrum of the accelerated particles and hence direct comparison with the observations is not possible. A similar difficulty arises with respect to the basic time constant for the acceleration process. The observations require that the basic time constant for acceleration be $\lesssim 1$ s and the rate of energy release in a small flare be $\sim 10^{27}$ erg s^{-1}.

8. Models of the Flash Phase

Although the role of energetic (>1 MeV) protons in the impulsive phase emissions has been emphasized from time to time (Švestka, 1971; Najita and Orall, 1970; Biswas and Radhakrishnan, 1973), the principal characteristic of the impulsive phase seems to be the acceleration of electrons to energies up to a few hundred keV. This was earlier recognized by de Jager (1960, 1962) and Wild *et al.* (1963) and is still fully consistent with the observations described here. As suggested by Wild *et al.* (1963) and de Jager (1969), the acceleration of protons to energies greater than a few hundred keV probably takes place during a second phase of acceleration which occurs soon after the impulsive phase. New evidence to support such a two step acceleration process has been recently reported by Frost and Dennis (1971). We will therefore consider here electrons from a few keV to a few hundreds of keV as the principal component of the particles accelerated during the impulsive phase, keeping in mind that some protons in a similar energy range may also be accelerated.

There is a growing opinion that the accelerated electrons probably play a central role in the development of the impulsive phase (Anderson *et al.*, 1970; Kahler *et al.*, 1970; Lin and Hudson, 1971; Syrovatskii and Shmeleva, 1971; Zirin *et al.*, 1971; Hudson, 1972a, b; Brown, 1971, 1973; Kane, 1973a, b; Lin, 1970, 1973, 1974a, b; Tanaka and Zirin, 1973). This inference is primarily based on the interpretation of the hard X-ray measurements.

The impulsive X-ray emission is generally believed to be bremsstrahlung from energetic electrons interacting with the ambient solar gas. If the effect of Compton back scattering is neglected, it is relatively simple to deduce the instantaneous electron spectrum inside the X-ray source from the observed X-ray spectrum. The procedure has been described by Kane and Anderson (1970), Brown (1971) and others. In such a deduction the only unknown parameter is the ion density n_i in the X-ray source. However, the problem gets complicated and highly model dependent when we try to deduce the electron spectrum produced at the acceleration region. Several important assumptions are involved:

(1) Location of the acceleration region

(2) Structure of the magnetic field connecting the acceleration region to the hard X-ray source

(3) Initial pitch angle distribution of the accelerated electrons

(4) Scattering of the electrons during their passage from the acceleration region to the X-ray source

(5) Average ion density n_i in the X-ray source

(6) Energy loss processes for the electrons

Since most of these quantities are not known observationally, one is forced to take a 'black box' approach, i.e., substitute quantities such as deflection time and scale height for the detailed physical processes.

The models of the impulsive phase proposed so far can be classified into three types: Their basic properties are presented in Figure 20 and Table IV. The three models have many common features. For example, the emission processes are basically the same in all the three types of models. The hard X-ray and microwave emissions are produced as bremsstrahlung and synchrotron radiation respectively from the energetic electrons accelerated during the flash phase. The EUV and optical radiations are produced by bound-bound, free-bound and free-free emission processes. In all models the electron acceleration region is located in the lower corona and the EUV and optical emission sources are located in the lower chromosphere. The three models differ primarily in the details of the electron and/or energy transport from the acceleration region to the various emission sources.

In Model 1, which is similar to the models proposed by Hudson (1972a, b) and Syrovatskii and Shmeleva (1971), most of the accelerated electrons moving initially downward reach the lower chromosphere, deposit all of their energy there and produce localized heating. The hard-X-rays are produced by thick-target bremsstrahlung in the lower chromosphere. The location of the microwave source has not been specified. The motion of the electrons from the acceleration region towards the hard X-ray – EUV source may be in the form of a well directed beam or as a diffusion process depending on the magnetic field structure in the region and the amount of

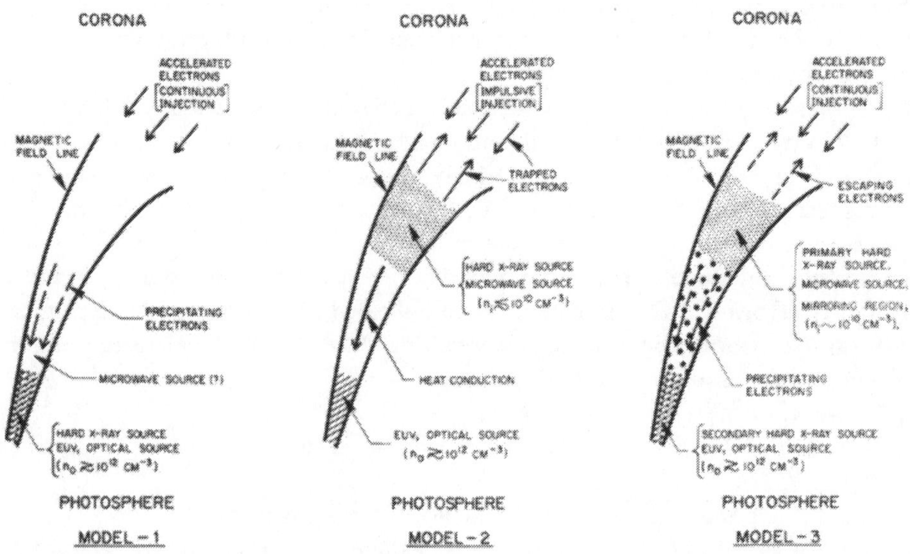

Fig. 20. Models of impulsive (flash) phase in which electrons play a central role. Model 1 is similar to the models proposed by Syrovatskii and Shmeleva (1971) and Hudson (1972). Model 2 is similar to the model proposed by Takakura (1973). Model 3 has been proposed by Kane (1973b). In Model 3 the relative strength of the primary and secondary hard X-ray sources may vary from event to event.

scattering the electrons have to undergo. In models where electron beams are considered, an implicit assumption is that the pitch angles α_0 of the electrons leaving the acceleration region are relatively small and that the electrons do not undergo significant scattering. If B_0 is the magnetic field strength near the acceleration region and B_x the field strength in the hard X-ray – EUV source, the required pitch angles are given by $\alpha_0 \lesssim \sin^{-1}(B_0/B_x)^{1/2}$. Electrons with larger pitch angles will mirrore before they reach the hard X-ray – EUV source. Inside the EUV source we expect $B_x \sim 10^3$ G. If the acceleration region is located in the lower corona where $B_0 \sim 10$ G, we find $\alpha_0 < 6°$. Since some scattering is almost certain to occur, which will further reduce the required values of α_0, we find that very small initial pitch angles for the accelerated electrons are required if the electrons are to move as a well directed beam.

At the other extreme is Model 2, which is similar to the one proposed by Takakura (1973). Here most of the accelerated electrons have large initial pitch angles and are trapped in a magnetic arch at a level where the ion density n_i is $\lesssim 10^{10}$ cm^{-3}. This is also the location of the hard X-ray and microwave sources. The hard X-ray emis-

TABLE IV

Models of flash phase in which electrons play a prominent role

Characteristic feature	Model 1	Model 2	Model 3
Location of acceleration region	Lower corona	Lower corona	Lower corona
Time dependence of acceleration process	Continuous	Impulsive	Continuous
Hard X-rays (bremsstrahlung)	Thick-target	Thin-target	Thin and thick-target Thin-target dominant
Electron escape into corona from hard X-ray source	Negligible	Negligible	Substantial
Hydrogen (atom/ion) density in emission sources (cm^{-3})			
Microwave	?	10^9–10^{10}	$\sim 10^{10}$
Hard X-ray	$\gtrsim 10^{12}$	$\sim 10^{10}$	Extended source $\sim 10^{10}$ (major part) $\sim 10^{12}$ (minor part)
EUV and optical	$\gtrsim 10^{12}$	$\gtrsim 10^{12}$	$\gtrsim 10^{12}$
Transport from acceleration region to EUV and optical sources			
Electron propagation	Complete	Negligible	Partial
Energy transport (primary mode)	Electrons $\gtrsim 5$ keV	Heat conduction	Electrons $\lesssim 10$ keV
Crucial test for the model	Low altitude of hard X-ray source	Short time constant (< 10 s) for heat conduction	Presence of escaped electrons in lower corona
Reference	Hudson (1972) Syrovatskii and Shmeleva (1972)	Takakura (1973)	Kane (1973b)

sion is produced by thin-target bremstrahlung. The energy for the EUV and optical emissions is transported from the acceleration region to the lower chromosphere by means of heat conduction.

Model 3, proposed by Kane (1973b) takes into account two possibilities:

(a) The electrons can mirror in the magnetic field before they reach the chromosphere and then escape into lower corona.

(b) Some of the mirroring electrons, particularly the lower energy ones ($\lesssim 10$ keV), can scatter into the loss cone and precipitate into the lower chromosphere.

In this model, the hard X-ray source extends from the lower chromosphere to the lower corona. The most intense part of the hard X-ray source is nearly coincident with the microwave source and has an ion density $n_i \sim 10^{10}$ cm^{-3}. The X-ray emission in this part of the source is thin-target bremsstrahlung. The energy transport from the acceleration region to the lower chromosphere is through electrons with energy $\lesssim 10$ keV. The value of n_i inside the mirroring region may vary from one flare to another. When n_i is $\ll 10^{10}$ cm^{-3}, the hard X-ray, microwave, EUV and optical emission may not be detectable although the type III radio burst produced by the electrons escaping through the corona may be easily observed. This will correspond to the observations of type III bursts without optical flares (Malville, 1962; Teske *et al.*, 1971; Axisa *et al.*, 1973).

The most attractive feature of Models 1 and 2 is their relative simplicity. However, there are several difficulties with both these models. Model 1 predicts very low altitude of the hard X-ray (and microwave?) sources, which is not consistent with the available behind-the-limb flare observations. Another difficulty is the dominant thick-target X-ray emission which is not favored by some observations involving hard X-rays and interplanetary solar electrons (Datlowe and Lin, 1973). In addition, Model 1 does not offer a natural explanation of the observations of type III bursts without impulsive hard X-ray, microwave, EUV and optical emission. Model 2 overcomes some of these difficulties. However, in order to explain the observed energy dependence of the decay of X-ray bursts, it requires a specialized ion density structure and pitch angle distribution of the electrons inside the X-ray source such as the one considered by Brown (1972). Moreover, it does not explain the similarity in the time variation of ~ 10 keV X-rays and broad band EUV emission unless the time constant for heat conduction from the acceleration region to the lower chromosphere happens to be $\lesssim 10$ s. Model 3 removes these difficulties and seems to be consistent with most of the available observations. However, it is considerably more complex than Models 1 and 2. A possible difficulty with Model 3 is the escape of a significant fraction of the energetic (> 20 keV) electrons into the lower corona where presumably their energy is slowly dissipated.

On the basis of the available observations above it is not possible at the present time to completely rule out any one of the three models. Observations with higher spatial and time resolution will have to be made before any one of these models can be fully proved or disproved.

9. Future Work

Although the problem of the impulsive phase is far from being fully solved, we have arrived at a stage where self-consistent models of emission sources can be constructed in some detail. Several such models have been published in the past two years and hopefully more theoretical work in this direction will be done in the years to come.

Since the problem of the impulsive phase is now fairly well defined, new observations to study specific aspects of the problem are required. For example, the basic time constant of the electron acceleration process can be determined by high time resolution (~ 0.1 s) measurements of hard X-ray spectrum. Such measurements with two spacecraft widely separated in heliocentric longitude will help to determine the directivity of hard X-ray emission and the altitude of the hard X-ray source above the photosphere. The question of escape of energetic electrons into the lower corona and their subsequent trapping in the coronal magnetic field can be resolved by a hard X-ray heliograph flown on a suitable satellite.

The measurements of EUV emission with satellites have just begun. Better measurements with higher time resolution ($\lesssim 1$ s) and spacial resolution are required before the energy transport during the impulsive phase can be understood.

The high resolution Hα measurements now available have helped to remove many uncertainties regarding the relationship between the Hα emission and other impulsive emissions. However the broad band optical measurements are still behind the times, partly due to instrumental difficulties and perhaps partly due to lack of interest. Hopefully this will be rectified soon.

The high resolution microwave measurements have now shown the existence of structure in the microwave source. Hopefully comparable hard X-ray and EUV measurements will be available in the future so that relative locations of the impulsive emission sources can be determined.

It is very satisfying to see measurements in the infra-red (in progress) and milimeter region completing the observed range of electromagnetic emission spectrum from flares. Like white light emission, the detectable infra-red and mm radiation are apparently associated with relatively large flares. However, these measurements are important for determining the energy balance of a flare.

Acknowledgements

The author wishes to acknowledge discussions with Drs R. P. Lin, K. A. Anderson, R. F. Donelly, D. L. McKenzie, D. Datlowe, H. S. Hudson, T. Takakura, S. Enome and J. C. Brown. This research was supported in part by the National Aeronautics and Space Administration under grant NGL05-003-017 and NGR-05-003-510. The author's travel to Surfer's Paradise, Australia was partially funded through a travel grant from the National Academy of Sciences.

References

Acton, L. W.: 1968, *Astrophys. J.* **152**, 305.

Akalane, K.: 1956, *Publ. Astron. Soc. Japan* **8**, 173.

Alfvén, H. and Carlqvist, P.: 1967, *Solar Phys.* **1**, 220.

Anderson, K. A., Kane, S. R., and Lin, R. P.: 1970, 'Fast Electrons from Solar Flares', presented at the *International Seminar on the Problem of Cosmic Ray Generation on the Sun*, Leningrad, U.S.S.R., December.

Anderson, K. A. and Winckler, J. R.: 1962, *J. Geophys. Res.* **67**, 4103.

Angle, K. L.: 1968, *Astron. J.* **73**, 553.

Arnoldy, R. L., Kane, S. R., and Winckler, J. R.: 1967, *Solar Phys.* **2**, 171.

Arnoldy, R. L., Kane, S. R., and Winckler, J. R.: 1968a, *Astrophys. J.* **151**, 711.

Arnoldy, R. L., Kane, S. R., and Winckler, J. R.: 1968b, in K. O. Kiepenheuer (ed.), 'Structure and Development of Solar Active Regions', *IAU Symp.* **35**, 490.

Arnoldy, R. L., Kane, S. R., and Winckler, J. R.: 1968c, Tech. Rep. CR-108, Cosmic Ray Group, School of Physics and Astronomy, Univ. of Minnesota, Minneapolis.

Axisa, F., Martres, M. J., Pick, M., and Soru-Escant, I.: 1973, *Solar Phys.* **29**, 163.

Biswas, S. and Radhakrishman, B.: 1973, *Solar Phys.* **28**, 211.

Boischot, A.: 1972, in E. R. Dyer (ed.), *Solar Terrestrial Physics/1970*, D. Reidel Publ. Co., Dordrecht-Holland, p. 87.

Brown, J. C.: 1971, *Solar Phys.* **18**, 489.

Brown, J. C.: 1972, *Solar Phys.* **25**, 158.

Brown, J. C.: 1973, this volume, p. 395.

Bruzek, A.: 1964, in W. N. Hess (ed.), *AAS-NASA Symposium on the Physics of Solar Flares*, NASA SP-50, p. 301.

Bruzek, A.: 1967, in J. N. Xanthakis (ed.), *Solar Physics*, Interscience Publ., New York, p. 399.
Castelli, J. P. and Barron, W. R.: 1969, *Bull. Am. Astron. Soc.* **1**, 274.
Castelli, J. P. and Richards, D. W.: 1971, *J. Geophys. Res.* **76**, 8409.
Chubb, T. A.: 1972, in E. R. Dyer (ed.), *Solar Terrestrial Physics/1970*, Part I, D. Reidel Publ. Co., Dordrecht-Holland, p. 99.
Chubb, T. A., Kreplin, R. W., and Friedman, H.: 1966, *J. Geophys. Res.* **71**, 3611.
Cline, T. L., Holt, S. S., and Hones, E. W.: 1968, *J. Geophys. Res.* **73**, 434.
Covington, A. E. and Harvey, G. A.: 1961a, *Phys. Rev. Letters* **6**, 51.
Covington, A. E. and Harvey, G. A.: 1961b, *Nature* **192**, 152.
Croom, D. L.: 1973, in R. Ramaty and R. G. Stone (eds.), *Proc. Symposium on High Energy Phenomena on the Sun*, held at Goddard Space Flight Center, Greenbelt, Maryland, September 28–30 1972, NASA/GSFC Preprint X-693-73-193, p. 114.
Datlowe, D. and Lin, R. P.: 1973, *Solar Phys.* **32**, 459.
Datlowe, D. and Peterson, L. E.: 1973, 'OSO-7 Observations of Solar X-Ray Bursts from 28 July to 9 August, 1972', preprint.
Davies, K. and Donnelly, R. F.: 1966, *J. Geophys. Res.* **71**, 2843.
De Feiter, L. D.: 1972, *Space Sci. Rev.* **13**, 827.
De Mastus, H. L. and Stover, R.: 1967, *Publ. Astron. Soc. Pacific* **79**, 615.
Dodson, H. W., Hedeman, E. R., and McMath, R. R.: 1956, *Astrophys. J. Suppl.* **2**, 241.
Dodson, H. W. and Hedeman, E. R.: 1968, *Solar Phys.* **4**, 229.
Donnelly, R. F.: 1968, *Solar Phys.* **5**, 123.
Donnelly, R. F.: 1969, *Astrophys. J. Letters* **158**, L165.
Donnelly, R. F.: 1971, *Solar Phys.* **20**, 188.
Donnelly, R. F.: 1973, in R. Ramaty and R. G. Stone (eds.), *Proc. Symposium on High Energy Phenomena on the Sun*, held at Goddard Space Flight Center, Greenbelt, Maryland, September 28–30, 1972, NASA/GSFC Preprint X-693-73-193, p. 242.
Donnelly, R. F., Wood, A. T., Jr., and Noyes, R. W.: 1973, *Solar Phys.* **29**, 107.
Doschek, G. A.: 1972, *Space Sci. Rev.* **13**, 765.
Elcan, M. J.: 1973, *Bull. Am. Astron. Soc.* **5**, 340.
Ellison, M. A.: 1947, *Monthly Notices Roy. Astron. Soc.* **109**, 3.
Ellison, M. A.: 1963, *Planetary Space Sci.* **11**, 597.
Elwert, G. and Haug, E.: 1970, *Solar Phys.* **15**, 234.
Enome, S., Kakinuma, T., and Tanaka, H.: 1969, *Solar Phys.* **6**, 428.
Enome, S. and Tanaka, H.: 1971, in R. Howard (ed.), 'Solar Magnetic Fields', *IAU Symp.* **43**, 413.
Fainberg, J.: 1973, this volume, p. 183.
Frost, K. J.: 1964, in W. N. Hess (ed.), *AAS-NASA Symposium on the Physics of Solar Flares*, NASA SP-50, 139.
Frost, K. J.: 1969, *Astrophys. J. Letters* **158**, L159.
Frost, K. J., and Dennis, B. R. : 1971, *Astrophys. J.* **165**, 655.
Gruber, D. E., Peterson, L. E., and Vette, J. I.: 1973, in R. Ramaty and R. G. Stone (eds.), *Proc. Symposium on High Energy Phenomena on the Sun*, held at Goddard Space Flight Center, Greenbelt, Maryland, September 28–30 1972, NASA/GSFC Preprint X-693-73-193, p. 147.
Guidice, D. A. and Castelli, J. P.: 1973, in R. Ramaty and R. G. Stone (eds.), *Proc. Symposium on High Energy Phenomena on the Sun*, held at Goddard Space Flight Center, Greenbelt, Maryland, September 28–30 1972, NASA/GSFC Preprint X-693-73-193, p. 87.
Hall, L. A.: 1971, *Solar Phys.* **21**, 167.
Hall, L. A., and Hintuegger, H. E., 1969, in C. de Jager and Z. Švestka (eds.), 'Solar Flares and Space Research', *COSPAR Symp.*, 81.
Harvey, G. A.: 1964, *Astrophys. J.* **139**, 16.
Harvey, K. L.: 1971, *Solar Phys.* **16**, 423.
Holt, S. S. and Ramaty, R.: 1969, *Solar Phys.* **8**, 119.
Hudson, H. S.: 1972a, *Solar Phys.* **24**, 414.
Hudson, H. S.: 1972b, paper presented at the *Seminar on the Generality of Particle Acceleration of Various Scales in Cosmos*, Leningrad, August 16–18, 1972.
Hudson, H. S.: 1973, in R. Ramaty and R. G. Stone (eds.), *Proc. Symposium on the High Energy Phenomena on the Sun*, Goddard Space Flight Center, Greenbelt, Maryland, September 28–30 1972, NASA/GSFC Preprint X-693-73-193, p. 207.

Hudson, H. S. and Ohki, K.: 1972, *Solar Phys.* **23**, 155.
Hudston, H. S., Peterson, L. E., and Schwartz, D. A.: 1969, *Astrophys. J.*, **157**, 389.
Jager, C. de: 1960, in H. K. Bijl (ed.), *Space Research I*, North Holl. Publ. Co., Amsterdam, p. 628.
Jager, C. de: 1962, *Space Sci. Rev.* **1**, 487.
Jager, C. de: 1964, in H. Odishaw (ed.), *Research in Geophysics*, Vol. 1, M.I.T. Press, Cambridge, Mass., Chapter 1.
Jager, C. de: 1969, in C. de Jager and Z. Švestka (eds.), 'Solar Flares and Space Research', *COSPAR Symp.* 1.
Jager, C. de and Kundu, M. R.: 1963, in W. Priestu (ed.), *Space Research III*, North Holl. Publ. Co., Amsterdam, p. 836.
Kahler, S. W. and Kreplin, R. W.: 1971, *Astrophys. J.* **168**, 531.
Kahler, S. W., Meekins, J. F., Kreplin, R. W., and Boyer, L. S.: 1970, *Astrophys. J.* **162**, 293.
Kakinuma, T., Yamashita, T., and Enome, S.: 1969, *Proc. Res. Inst. Atmospherics, Nagoya Univ. Japan* **16**, 127.
Kane, S. R.: 1969, *Astrophys. J. Letters* **157**, L139.
Kane, S. R.: 1971, *Astrophys. J.* **170**, 587.
Kane, S. R.: 1972a, *Space Sci. Rev.* **13**, 822.
Kane, S. R.: 1972b, *Solar Phys.* **27**, 174.
Kane, S. R.: 1973a, in R. Ramaty and R. G. Stone (eds.), *Proc. Symposium on High Energy Phenomena on the Sun*, Goddard Space Flight Center, Greenbelt, Maryland, September 28–30, 1972, NASA/GSFC Preprint X-693-73-193, p. 55.
Kane, S. R.: 1973b, paper presented at the *3rd Meeting of the Solar Physics Division of AAS*, Las Cruces, New Mexico, Jan. 4–6.
Kane, S. R. and Anderson, K. A.: 1970, *Astrophys. J.* **162**, 1003.
Kane, S. R. and Donnelly, R. F.: 1971, *Astrophys. J.* **164**, 151.
Kane, S. R. and Hudson, H. S.: 1970, *Solar Phys.* **14**, 414.
Kane, S. R., Kahler, S. W., and Kurfess, J. D.: 1972, *Solar Phys.* **25**, 418.
Kane, S. R. and Lin, R. P.: 1972, *Solar Phys.* **23**, 457.
Kane, S. R. and Winckler, J. R.: 1969a, Tech. Rep. CR-134, Cosmic Ray Group, School of Physics and Astronomy, Univ. of Minnesota, Minneapolis.
Kane, S. R. and Winckler, J. R.: 1969b, *Solar Phys.* **6**, 304.
Kelly, P. T. and Rense, W. A.: 1972, *Solar Phys.* **26**, 431.
Kodama, M., Kusunosi, M., and Ogura, K.: 1971, *Rep. Ionosphere Space Res. Japan* **25**, 285.
Korchak, A. A.: 1967, *Soviet Astron. AJ*, **11**, 258.
Kundu, M. R.: 1961, *J. Geophys. Res.* **66**, 4308.
Kundu, M. R.: 1965, *Solar Radio Astronomy*, Chaper 13, Interscience Publ., New York.
Kundu, M. R.: 1973, in R. Ramaty and R. G. Stone (eds.), *Proc. Symposium on High Energy Phenomena on the Sun*, held at Goddard Space Flight Center, Greenbelt, Maryland, September 28–30 1972, NASA/GSFC Preprint X-693-73-193, p. 104.
Lin, R. P.: 1970, 'Acceleration of 10–100 keV Electrons in Solar Flares', presented at *Seminar on the Acceleration of Particles in Near-Earth and Interplanetary Space, Galaxy and Metagalaxy*, Leningrad, U.S.S.R., July 13–15.
Lin, R. P.: 1973, in R. Ramaty and R. G. Stone (eds.), *Proc. Symposium on High Energy Phenomena on the Sun*, held at Goddard Space Flight Center, Greenbelt, Maryland, September 28–30 1972, NASA/GSFC Preprint X-693-73-193, p. 242.
Lin, R. P.: 1974a, *Space Sci. Rev.* **16**, 189.
Lin, R. P.: 1974b, this volume, p. 201.
Lin, R. P. and Hudson, H. S.: 1971, *Solar Phys.* **17**, 412.
Malville, J. M.: 1962, *Astrophys. J.* **135**, 834.
Mandelstam, S. L.: 1972, *Acta Physica Academiae Scientiarum Hungaricae* **32**, 307.
Maxwell, A.: 1965, in C. de Jager (ed.), *The Solar Spectrum*, D. Reidel Publ. Co., Dordrecht, Holland, p. 342.
McIntosh, P. S. and Donnelly, R. F.: 1972, *Solar Phys.* **23**, 444.
McKenzie, D. L.: 1972, *Astrophys. J.* **175**, 481.
McKenzie, D. L., Datlowe, D., and Peterson, L. E.: 1973, *Solar Phys.* **28**, 175.
McKenzie, D. L. and Peterson, L. E.: 1973, *Bull. Am. Astron. Soc.* **3**, 340.
Moreton, G. E.: 1964, in W. N. Hess (ed.), *AAS-NASA Symposium on the Physics of Solar Flares*, NASA-SP 50, p. 209.

Najita, K. and Orrall, F. Q.: 1970, *Solar Phys.* **15**, 176.
Neupert, W. M.: 1971, *Phil. Trans. Roy. Soc. London* **270**, 143.
Noyes, R. W.: 1973, in R. Ramaty and R. G. Stone (eds.), *Proc. Symposium on High Energy Phenomena on the Sun*, held at Goddard Space Flight Center, Greenbelt, Maryland, September 28–30, 1972, NASA/GSFC Preprint X-693-73-193, p. 231.
Ohki, K. I.: 1969, *Solar Phys.* **7**, 260.
Parks, G. K. and Winckler, J. R.: 1969, *Astrophys. J. Letters* **155**, L117.
Parks, G. K. and Winckler, J. R.: 1971, *Solar Phys.* **16**, 186.
Peterson, L. E. and Winckler, J. R.: 1959, *J. Geophys. Res.* **64**, 697.
Peterson, L. E., Datlowe, D. W., and McKenzie, D. L.: 1973, in R. Ramaty and R. G. Stone (eds.), *Proc. Symposium on the High Energy Phenomena on the Sun*, Goddard Space Flight Center, Greenbelt, Maryland, September 28–30 1972, NASA/GSFC Preprint X-693-73-193, p. 132.
Petrosian, V.: 1973, *Astrophys. J.* **186**, 291.
Phillips, K. J. H.: 1973, *Observatory* **93**, 17.
Phillips, K. J. H. and Neupert, W. M.: 1973, *Solar Phys.* **32**, 209.
Pinter, S.: 1969, *Solar Phys.* **8**, 142.
Ramaty, R. and Petrosian, V.: 1972, *Astrophys. J.* **178**, 241.
Santangelo, N., Horstman, H., and Horstman-Moretti, E.: 1973, *Solar Phys.* **29**, 143.
Smith, D. F.: 1973, this volume, p.
Smith, H. J. and Smith, E. v. P.: 1963, *Solar Flares*, The McMillan Co., New York.
Stewart, R. T.: 1973, this volume, p.
Sturrock, P. A.: 1968, in K. O. Kiepenheuer (ed.), 'Structure and Development of Solar Active Regions', *IAU Symp.* **35**, 471.
Sturrock, P. A.: 1973, this volume, p.
Švestka, Z.: 1969, in C. de Jager and Z. Švestka (eds.), 'Solar Flares and Space Research', *COSPAR Symp.*, 16.
Švestka, Z.: 1970, *Solar Phys.* **13**, 471.
Syrovatskii, S. I. 1969, in C. de Jager and Z. Švestka (eds.), 'Solar Flares and Space Research', *COSPAR Symp.*, 346.
Syrovatskii, S. I. and Shmeleva, O. P.: 1971, Lebedev Institute Preprint 158.
Takakura, T.: 1968, *Solar Phys.* **6**, 133.
Takakura, T.: 1969, in C. de Jager and Z. Švestka (eds.), 'Solar Flares and Space Research', *COSPAR Symp.*, 165.
Takakura, T.: 1971, *Solar Phys.* **19**, 186.
Takakura, T.: 1972, *Solar Phys.* **26**, 151.
Takakura, T.: 1973, in R. Ramaty and R. G. Stone (eds.), *Proc. Symposium on the High Energy Phenomena on the Sun*, Goddard Space Flight Center, Greenbelt, Maryland, September 28–30, 1972, NASA/GSFC Preprint X-693-73-193, p. 179.
Takakura, T. and Kai, K.: 1966, *Publ. Astron. Soc. Japan* **18**, 57.
Takakura, T., Ohki, K., Shibuya, N., Fujii, M., Matsuoka, M., Miyamoto, S. Nishimura, J., Oda, M., Ogawara, Y., and Ota, S.: 1971, *Solar Phys.* **16**, 454.
Tanaka, H., Kakinuma, T., and Enome, S.: 1967, *Proc. Res. Inst. Atmospherics, Nagoya Univ. Japan* **14**, 23.
Tanaka, H. and Enome, S.: 1971, *Solar Phys.* **17**, 408.
Tanaka, H. and Zirin, H.: 1973, in R. Ramaty and R. G. Stone (eds.), *Proc. Symposium on the High Energy Phenomena on the Sun*, Goddard Space Flight Center, Greenbelt, Maryland, September 28–30, 1972, NASA/GSFC Preprint X-693-73-193, p. 26.
Teske, R. G., Soyumer, T., and Hudson, H. S.: 1971, *Astrophys. J.* **165**, 615.
Tindo, I. P., Ivanov, V. D., Mandelstam, S. L., and Shuryghin, A. I.: 1970, *Solar Phys.* **14**, 204.
Tindo, I. P., Ivanov, V. D., Mandelstam, S. L., and Shyryghin, A. I.: 1972a, *Solar Phys.* **24**, 429.
Tindo, I. P., Ivanov, V. D., Valnicek, B., and Livshitz, M. A.: 1972b, *Solar Phys.* **27**, 426.
Tomblin, F. F.: 1972, *Astrophys.* **171**, 377.
Vette, J. I. and Casal, F. G.: 1961, *Phys. Rev. Letters* **6**, 334.
Vorpahl, J. A.: 1972, *Solar Phys.* **26**, 397.
Wild, J. P., Smerd, S. F., and Weiss, A. A.: 1963, *Ann. Rev. Astron. Astrophys.* **1**, 291.
Winckler, J. R., May, T. C., and Masley, A. J.: 1961, *J. Geophys. Res.* **66**, 316.
Wood, A. T., Jr. and Noyes, R. W.: 1972, *Solar Phys.* **24**, 180.

Wood, A. T., Jr., Noyes, R. W., Dupree, A. K., Huber, M. C. E., Parkinson, W. H., Reeves, E. M., and Withbroe, G. L.: 1972, *Solar Phys.* **24**, 169,
Zirin, H.: 1972, *Solar Phys.* **26**, 393.
Zirin, H.: 1973, 'Solar Flares', review paper for *Vistas in Astronomy* **16**, 1.
Zirin, H., Pruss, G., and Vorpahl, J.: 1971, *Solar Phys.* **19**, 463.

DISCUSSION

Brown: I would like to disassociate my name from your third schematic flare model. There seems to be a misunderstanding since, although I have studied trapped electron models for the hard X-ray source, I have always advocated a thick target geometry if electrons are to heat the optical and EUV flare. Secondly, it is worth adding that Jerry Drake has shown that the only statistically significant amount of data on hard X-ray burst longitude distribution is in the 1–6 keV band. He found this to be isotropic – which it should be since it is mostly thermal.

Kane: Higher energy bursts should be used for such studies. Petrosian has predicted that the X-ray spectra should be harder at the limb. This arises because of the directivity of the high energy bremsstrahlung.

Zirin: There is an empirical way of determining the height of an X-ray burst based on Hα data and a cut-off in the microwave spectrum at 3000 MHz. These data indicate a low height.

Kane: This may happen in some events but the lack of center-to-limb variation in the microwave radiation indicates that the source is high.

Zirin: Those results may not be trustworthy.

Smith: There are several mechanisms for the microwave cut-off. With our present knowledge of a flare, selection of one particular michanism over the others, namely absorption below the plasma frequency, is not necessarily valid.

Zirin: I have looked at free-free absorption, Razin effect and others and they all imply a low source height.

Takakura: Self absorption could produce a cut-off for a high source if the source were small and the number of nonthermal electrons were high, say 10^3–10^4 cm^{-3} above 100 keV.

Zirin: Wouldn't this produce visible absorption?

Brueckner: How do you obtain the number 10^{28} erg in the UV for a major flare?

Kane: From SFD data which has now been calibrated by spacecraft observations in the EUV.

Kundu: Have you any data available on the X-ray spectrum of bursts associated with gradual rise and fall centimeter burst?

Kane: Soft X-rays – we don't have much data but should expect the gradual rise and fall to have a thermal spectrum.

Kundu: That is precisely the reason I am asking. Recent high resolution interferometric observations with resolution of seconds of arc of a gradual rise and fall type cm burst indicates that a significant fraction of the burst energy is emitted within a volume of 2″ diam. The brightness temperature that we get is about 10^9 K. Consequently, this type of cm burst, which was previously thought to be thermal, is certainly non-thermal. I am just wondering if you have any nonthermal component of X-rays associated with such bursts.

Kane: The X-ray component of the gradual rise and fall may be too weak to have been measured yet.

Sturrock: Could you comment further on the emission process in the flash phase; is it thin or thick target?

Kane: Since the ion density in (or altitude of) the X-ray source is not known, it is not possible at present to rule out either the thick- or thin-target models. In the few events where simultaneous X-ray and interplanetary electron observations have been compared, the thin target model fits the observations better. But the question is not resolved.

Brown: Your comparison of interplanetary electron spectra and those inferred from hard X-rays unfortunately does not settle the thick-thin target controversy. The problem is that the interplanetary electrons comprise only 0.1% (Lin and Datlowe, 1973) of those in the X-ray source and are therefore probably most uncharacteristic of the latter. Thus, if more energetic electrons escaped preferentially, the spectral evidence would favor a thick rather than thin target.

SPECTRAL ASSOCIATION OF THE 7 AUGUST 1972 SOLAR RADIO BURST WITH PARTICLE ACCELERATION

J. P. CASTELLI and A. L. CARRIGAN

Air Force Cambridge Research Laboratories, Bedford, Mass., U.S.A.

and

H. C. KO

Ohio State University, Columbus, Ohio, U.S.A.

Abstract (*Solar Phys.*). Microwave burst data from the August 7, 1972 event recorded on nine discrete frequencies between 245 and 35 000 MHz at the Sagamore Hill Radio Observatory (Figure 1) provide a basis for correlation studies (especially timing in-

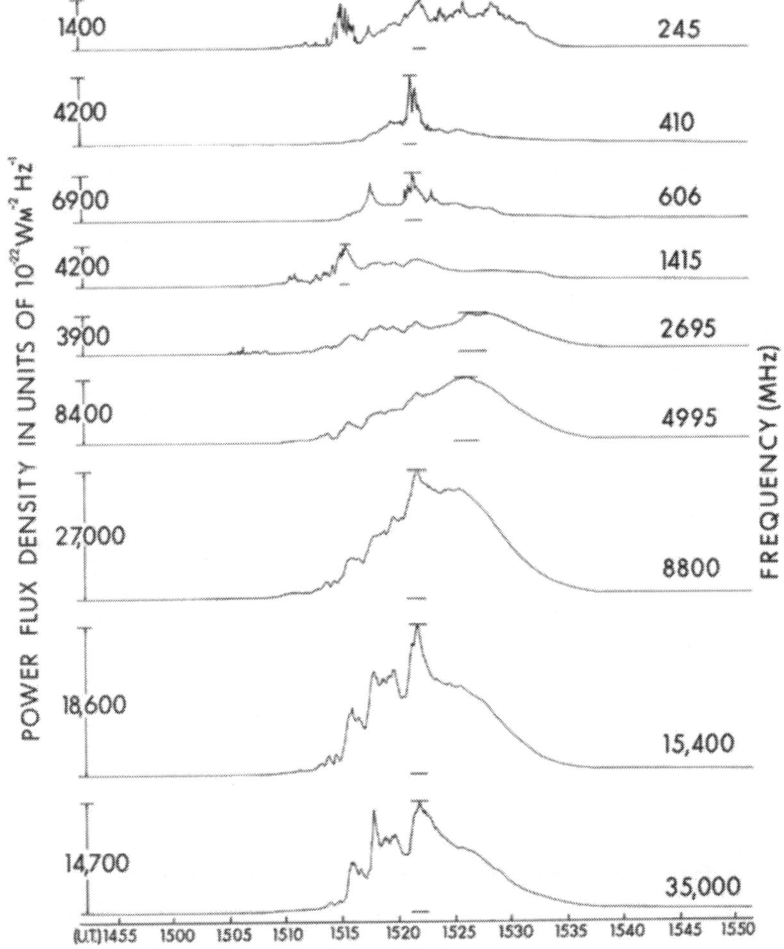

Fig. 1. Great burst observed 7 August, 1972 at Sagamore Hill Radio Observatory, Hamilton, Mass.

Gordon Newkirk, Jr. (ed.), Coronal Disturbances, 143–146. *All Rights Reserved.*

formation) of the associated white light flare, the high energy particle emission, type II bursts, and many other phenomena. This is perhaps the first time that sufficient radio coverage (i.e., data above 10000 MHz) was available to obtain the spectral slope information (related to electron-energy distribution) which is inherent in this part of the spectrum. Heretofóre, timing was related to burst flux density profile variations. Improved correlations resulted from shorter centimeter wavelength data which supplied more accurate timing information than that derived from Hα observations. The shape and intensity of the burst peak flux density spectrum has also been used for qualitative analysis of energetic particle and white light events. The ultimate good may possibly come from spectral analysis of the minute by minute variation of the burst microwave radiation spectral slope α (above f_{max}) in the area between 15000 and 35000 MHz. This may be used alone or in relation to the position of f_{max}, where f_{max} is the frequency of burst maximum emission at a given time. This is basically our present investigation.

It is well known that the synchrotron radiation from an ensemble of highly relativistic electrons with the power-law energy distribution $N(E) \propto E^{-\gamma}$ has the radiated

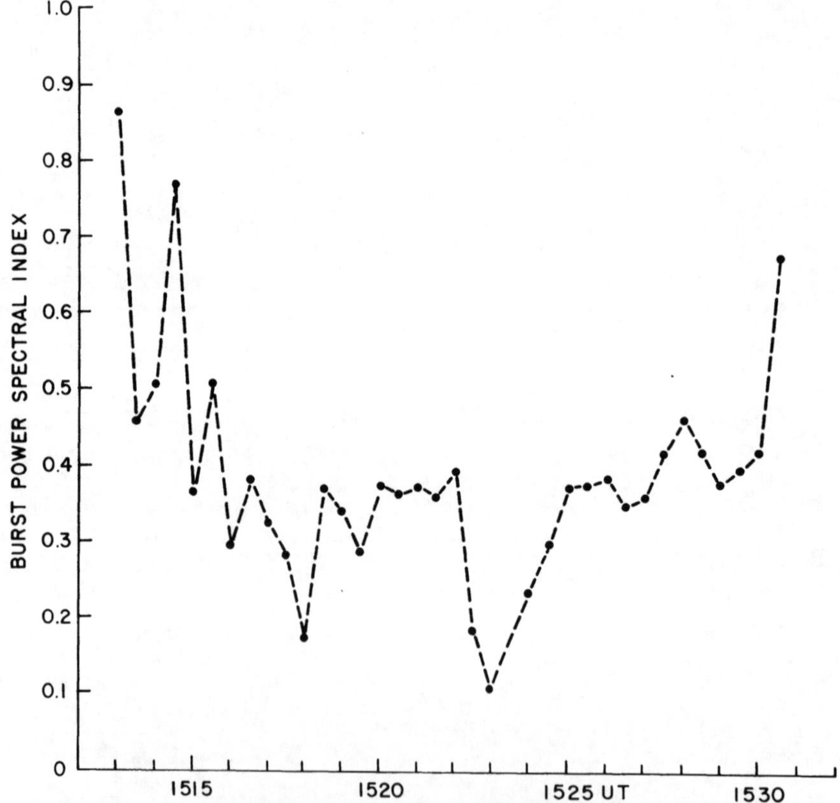

Fig. 2. Burst power spectral index variation during principal part of event of 7 August 1972, based on measured data at 15000 and 35000 MHz.

power spectrum given by $f^{-\alpha}$ where $\alpha=(\gamma-1)/2$ (γ and α are the electron-energy spectral index and the microwave radiation power spectral index, respectively). Unfortunately, this simple relation is not directly applicable to microwave solar bursts since the low and medium energy electrons play dominant roles (gyro-synchrotron emission) instead of ultrarelativistic electrons. Still, there is a strong tendency with very energetic events for this $\alpha=(\gamma-1)/2$ relation to apply, especially as we go further into the millimeter range. This is evidenced by the increasing flatness (hardness) of the radiation slopes above f_{max}. To evaluate the functional relation between α and γ, one has to integrate the general synchrotron radiation formulas with respect to the powerlaw energy distribution of electrons. By means of a computer program, we have calculated the volume emissivity as a function of frequency for various values of γ and θ, where the latter is the angle of observation measured from the direction of the magnetic field. The spectral index is found to vary with frequency. From these numerical data we have obtained a set of curves relating α and f/f_H (f_H is the gyrofrequency) for various γ and θ. From these curves, the electron-energy spectral index γ can be estimated from an assumption of θ and a given set of α and f/f_H.

For the August 7, 1972 event during the most intense part of the burst, the microwave radiation spectral index varies between $\alpha=0.2$ and 1.0 (see Figure 2). The corresponding electron-energy spectral index is estimated at between $\gamma=1.2$ and 2.0. Thus, the distribution of electrons during this period had an energy spectral index as hard as 1.2 (1.6 on an average). The α variation (0.2 to 1.0) over about 15 min is found to be very much flatter than for non-white lights flares and non-ground level events. In relating the white light flare development on August 7, 1972 to burst spectral hardness, an α of approximately 0.2 is apparent at the white light flare start near 1518 UT. Its maximum near 1523 UT agrees well with the time of burst maximum spectral hardness. The white light flare seemed to persist until about 1530 UT at which time the burst α rose abruptly above 0.4. These burst spectral data are correlated only with timing aspects of the flare phenomena at present. As such, they serve as a signature. This study does not address the problem of whether the white light flare was produced by high or low energy electrons or protons, all of which were abundant during the period.

While the radio burst spectrum was varying in hardness, the frequency of maximum radio emission was also varying; the position of f_{max} reached its highest excursion at about 1518 UT (Figure 3). This tends to suggest that the greatest influx of electrons occurred at this time. If the emission is primarily from non-thermal electrons, then an increase in the average electron-energy (injected into the microwave source) would shift the position of f_{max} upward. An increase in the influx of thermal electrons contributing to ionized-medium suppression would also effectively shift f_{max} upward. In either case, an increase in the magnetic field intensity is not required.

There are several points which are still incomplete. We have not investigated sufficiently the total number of electrons nor the total amount of energy involved. A preliminary calculation of the radio burst energy transformed into radiation in the 600–35000 MHz range yields $\sim 0.5\times10^{26}$ ergs. Since the same electrons may be re-

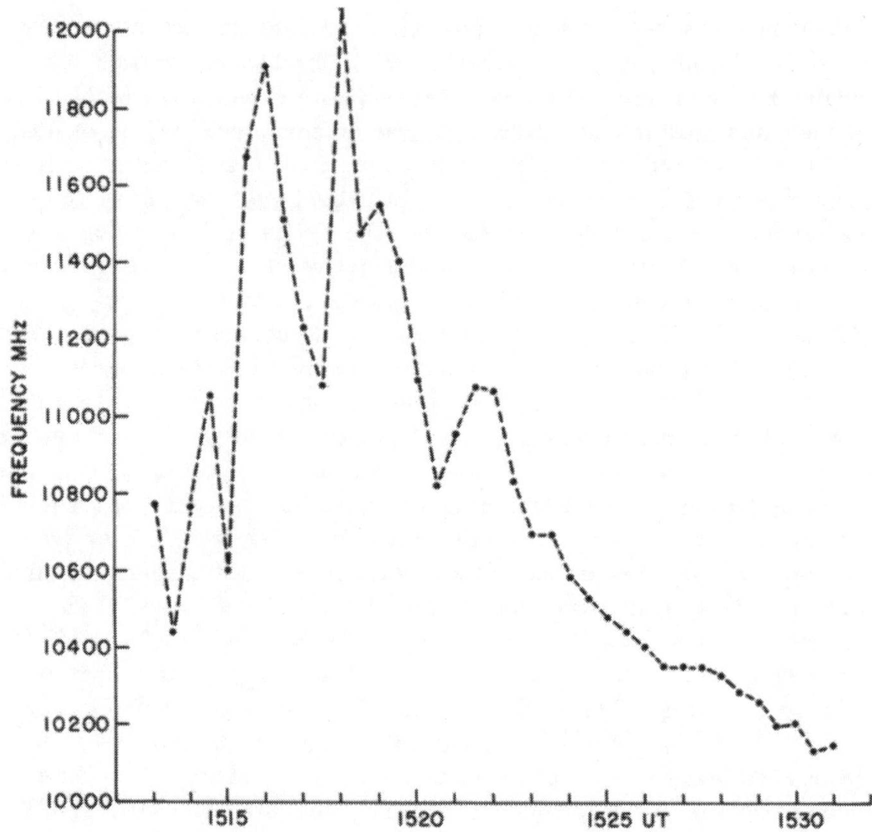

Fig. 3. Variation of centimeter wavelength burst spectral maximum (f_{max}), during principal part of event of 7 August 1972. f_{max} was determined from best fit quadratic to measured flux densities at 4995, 8800 and 15400 MHz.

sponsible for hard X-rays, we will investigate whether the energy spectrum derived in our study is consistent with the hard X-ray spectrum. (For this purpose we have to choose either the thin or thick target interpretation of the X-ray spectrum which will give considerably different results.) Hard X-ray data are now available from which to complete the study.

X-RAY EMISSION IN ABSENCE OF FLARES RELATED TO Hα ACTIVITY AND TYPE III BURST PRODUCTION

S. R. KANE

Space Sciences Laboratories, University of California, Berkeley, Calif. 94720, U.S.A.

R. W. KREPLIN

E. O. Hulburt Center for Space Research, Naval Research Laboratory, Washington, D.C. 20390, U.S.A.

and

M. J. MARTRES, M. PICK, and I. SORU-ESCAUT

Observatoire de Paris, 92 Meudon, France

Abstract (*Solar Phys.*). The relationship between Hα absorption features, type III radio bursts and soft X-ray emission has been examined in order to determine the characteristics of the particle acceleration process operating when a Hα-flare may or may not be detectable. The Hα observations were made by Meudon Observatory with a Hα telescope fitted with a 0.75 Å band pass Lyot filter. During a 10 s period, three pictures were obtained – one at the Hα line center, one at Hα+0.75 Å and one at Hα−0.75 Å. This sequence of three pictures was repeated every one minute. Each picture covered a rectangular area 18×24 mm^2, the diameter of the complete solar image being 38 mm on this scale. In addition, Meudon Hα films of the whole solar disc were also used. The X-ray observations were made with the University of California (Berkeley) experiment aboard the OGO-5 satellite and the NRL experiment aboard Solrad-9. The wavelength range covered was 0.5–20 Å. The type III radio data was obtained from two sources: The 169 MHz radio-heliograph at Nancay which provided east–west position of the radio burst on the Sun with an accuracy of $\sim 1'$ and the radio spectra measured by various ground based observatories. The findings are as follows:

Transient Hα activity observed in the absence of reported flares is associated with production of type III radio and soft X-ray emission. Since such optical phenomena are much more frequent than flares themselves, we conclude that instabilities generating fast particles may be produced in the corona in a quasi-continuous way with coincident perturbations in the lower solar atmosphere.

The soft X-ray component is not necessarily the direct product of fast particles, but is probably associated with some type of heating since both the soft X-ray emission and the Hα features exhibit a comparable evolution. The type III bursts, when they are produced, occur near the maximum of this perturbation.

We identify the transient Hα activity (emission or absorption) with the existence of a metastable situation which may or may not lead to the triggering of a flare.

Gordon Newkirk, Jr. (ed.), Coronal Disturbances, 147–148. *All Rights Reserved.*

DISCUSSION

Newkirk: Aren't your results similar to the observations which led Hyder to formulate his impact infall model of some flares? He suggested that in these flares the X-ray emission emerged from the shock wave produced by the impact of prominence material into the chromosphere.

Pick: I believe that Hyder observed red shifts and we observe both blue and red shifts.

Martin: What is the spatial resolution of the Hα observations that you have used in your study.

Martres: Approximately 2″.

Enome: What is the position of the absorbing feature in Hα in the active region? Is it located near the sunspot or the neutral line?

Pick: Along the inversion line of polarity.

CORONAL DISTURBANCES OBSERVED IN THE OPTICAL EMMISSION LINES

M. WALDMEIER

Swiss Federal Observatory, Zürich, Switzerland

Abstract. Some examples are presented of fast variations of the structure and the intensity of coronal condensations. For the observations the lines $\lambda 5303$ and $\lambda 5694$ have been used. Most of the observed phenomena are flare-connected, but all of them are related to some kind of Hα-activity. Sudden brightening of the emission lines may occur after the disappearance of a filament. The events have a duration of up to several hours and consist of hot clouds of high density that are slowly rising from the upper chromosphere into the lower corona.

Coronal disturbances can optically best be observed in the lines 5303 and 5694 Å. The first is reasonably intense and the second visible only in the disturbance itself.

The monochromatic corona is mostly quiet. In the course of hours generally one observes but gradual variations of the line intensities. Even variations from day to day are small and mostly due to the solar rotation. Most of the features can be followed on several consecutive days. In this respect the corona has some similarity with the stationary filaments. These are formations in the innermost corona and undergo very slow evolutions like the corona.

Under coronal disturbance we understand variations of the intensity or the structure of the corona that take place in a time not longer than a few hours. Practically all these disturbances are in close connection with centers of activity and most of them in connection with a special Hα-activity in that center. If no Hα-activity is seen by observing the coronal disturbance, it may be that the Hα-activity is not at the limb but on the disk or behind the limb. From observations over 35 years, I got the impression that all coronal disturbances are in connection with some kind of Hα-activity. The relation between the importance of a coronal disturbance and the importance of the responsible chromospheric phenomenon is poor. Most of the flares are of importance 1, having little height and not reaching the level that is observed by coronagraphs. These flares in general are not accompanied by coronal disturbances. There are a few exceptions when small flares have produced noticeable coronal disturbances. Larger flares which penetrate higher into the corona always produce disturbances, especially when they are followed by surges or loops. Surges not connected with flares and ascending prominences may or may not affect the corona. Other activities not flare connected, like active prominences streaming down into a center of attraction, may also disturb the corona.

The nature of these disturbances may consist of:

(a) fluctuations of the intensity;

(b) expansions of already existing structures, especially arches;

(c) disruptions of existing structures;

(d) ejections; or

(e) formation of new structures.

Gordon Newkirk, Jr. (ed.), Coronal Disturbances, 149–154.

Most of the coronal disturbances have a sporadic character. In general, when the disturbance is over, the corona looks very much the same as before. As Lyot (1944) has noticed already 30 yr ago, the variations of the structure of the corona are not arising from mass motions, but from brightening of some regions and fading of others. This is certainly true for the slow events. The phenomena we are calling disturbances are fast events, and in these as a rule mass motion is involved. Doppler shifts of the emission lines indicate mass motion in the corona. If the spectrogram shows chromospheric lines too, these lines exhibit Doppler shifts of the same sign as the coronal lines, but in general not of the same amount as the coronal features; even if they appear at the same place, they are spaced from each other along the line of sight.

After these general remarks we will present some examples.

1. Disturbance by Ascending Prominences

A rising prominence was observed at heliographic latitude north 45° on the east limb on April 9, 1956. It was violent ascent with Doppler shifts corresponding to 300 km s^{-1}. The distribution of the intensity of the line $\lambda 5303$ was observed one hour before the ascent began and again half an hour after the disappearance of the prominence. Not the slightest change of the structure or the intensity of the green line was noticed. So it seems that chromospheric material may move through the corona without affecting it. This example is not conclusive as the rising filament was not at the limb. Therefore, the region where the filament crossed the line of sight was rather far above the Sun's limb and did contribute but little to the total line intensity. Even if this region was affected by the filament, the influence upon the line intensity could hardly be detected.

Another example is more conclusive. On March 18, 1961, a formerly stable filament at latitude north 27° on the east limb was rising. The intensity distribution of the line $\lambda 5303$ immediately before the ascent did not show anything unusual. It was almost the same as on the day before, and it was still unchanged after the disappearance of the filament. Two hours after the filament had vanished, the region, occupied before by the stable prominence, became suddenly very bright, much brighter than the maxima over the main zones of activity (Waldmeier, 1961). The red line $\lambda 6374$ also became very bright at the same time. Two arches, not very regular in shape, were formed, the outer reaching a height of 100000 km and the lower one 50000 km. The region of the inner arch showed the line $\lambda 5694$. This is quite unusual, and is the only case where this line has been observed in a spotless region. Two hours later, the arches had disappeared, but the now structureless region was still unusually bright. It was only about six hours after the appearance of the condensation that it had faded away completely. This phenomenon has some similarity with the behaviour of the chromosphere in connection with rising filaments. It was on September 2, 1937 that the author for the first time observed a flare-like brightening about half an hour after the dissolution of a filament (Waldmeier, 1938). It is by now well known that the sudden disappearance of a filament is, not always but often, followed by flare-like Hα emission. Bruzek (1957) has given several examples. The brightening of the corona after a 'disparition brusque' is unique. I have observed this phenomenon only once.

2. Disturbance by a Flare

On September 21, 1955, the whole development of a relatively small but very bright limb flare and of the coronal disturbance connected with it was observed (Figure 1). The line λ5303 began to brighten at the beginning of the flare. The brightening of this line was still going on when, after 15 min, the flare faded away. The maximum brightness occurred about one hour after the outbreak of the flare. The intensity

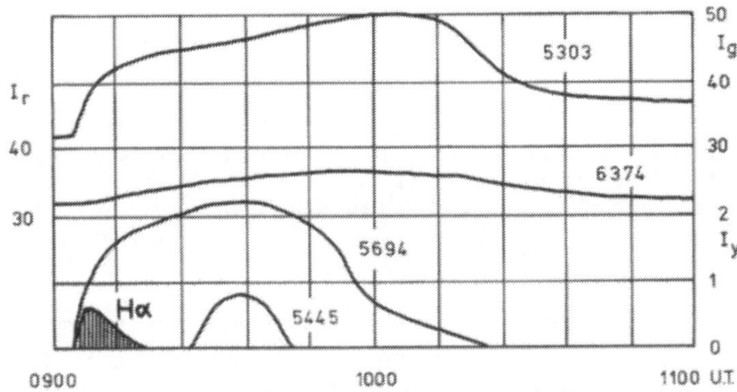

Fig. 1. The variations of the intensities of the green, I_g, the red, I_r, and the two yellow lines, I_y, in connection with a small flare on September 21, 1955 at latitude south 25° on the west limb.

decreased slowly returning to its preflare value only after three hours. The line λ6374 was little affected. It showed a gradual rise and fall, with a maximum 40 min after the beginning of the flare. A strong reaction was observed in the line λ5694. At the beginning of the flare the intensity of this line, that was not visible before, rose sharply, reached its maximum half an hour later and became invisible 70 min later. At its maximum, when the flare had already disappeared, the second line of Ca XV, λ5445, became visible for 15 min. Out of the flare, a cloud of hot gases was emerging at a low speed of a few km per second. Half an hour after the flare it reached its highest temperature at a height of about 25000 km. In the center of the hot cloud, gases condensed into bright Hα knots. The hot cloud persisted for about three hours. This hot cloud, originating from a flare, may be responsible for the postburst increase of the microwave radiation. The microwave brightness temperature of the hot cloud is about one hundred times that of the undisturbed solar disk, but its area is one thousand times smaller than that of the Sun. Therefore, the postburst increase should be of the order of 10% of the preflare intensity.

The formation of hot clouds, actually the hottest regions of the corona, is a characteristic disturbance following a flare. During its formation, the cloud is rising and becoming hotter. The hottest points are not found closest to the limb but at an altitude of roughly 50000 km (Figure 2).

A similar event was observed on August 14, 1970. At the latitude north 9° on

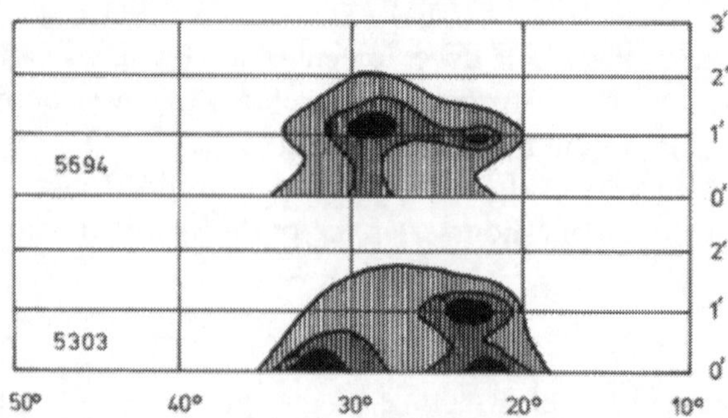

Fig. 2. A hot cloud condensation, observed on April 1, 1957 at 1120 UT at the west limb. Heights are given in arc minutes above the edge of the occulting disk. The knots visible only in the line $\lambda 5303$ are low lying and of relatively low temperature. The knot at latitude north 29°, visible only in the line $\lambda 5694$ at a height of 50000 km, is the hottest feature in the condensation.

the east limb, the appearance of the line $\lambda 5694$ indicated that a strong center of activity was approaching the limb. At that time, a big spot of type H was one day behind the limb. The bright but quiet corona became suddenly active in that place. The region of the $\lambda 5694$ emission was strongly increasing in intensity and rising. Half an hour later, the hot cloud reached its greatest brightness at a level of 40000 km above the limb. At that time, according to observations in Hα and D_3, loops were fully developed. In the line $\lambda 5303$ the structure became very complex. The appearance of the corona in this line agreed only in the major features with that in the line $\lambda 5694$ or with the Hα loops. The whole phenomenon is so characteristic for a postflare event that there must have been a flare behind the limb.

3. What is First, the Hot Cloud or the Flare?

The two examples I have mentioned cannot answer this question. In the first one, the intensities of both the chromospheric and the coronal lines were rising simultaneously, and in the second one, the flare occurred behind the limb. Active centers likely to produce flares are always surrounded by a permanent coronal condensation. If it is large and dense enough, it emits the line $\lambda 5694$; therefore it is hot. Such a hot cloud could become the origin of a flare, or a flare may produce the hot cloud. The material now available indicates clearly that the flare precedes the formation of hot coronal clouds.

On March 18, 1956, at 1015 UT, a small flare appeared at the latitude north 28°

on the east limb (Figure 3). The spectrum was carefully watched from the beginning. The chromospheric lines were extremely bright, but the line λ5303 was unchanged compared to the preflare intensity while λ5694 was invisible. At 1035 UT the flare came to end. It was only at that moment, and not before, that the coronal lines became affected. The line λ5694 was first seen at 1040 UT, reached its highest in-

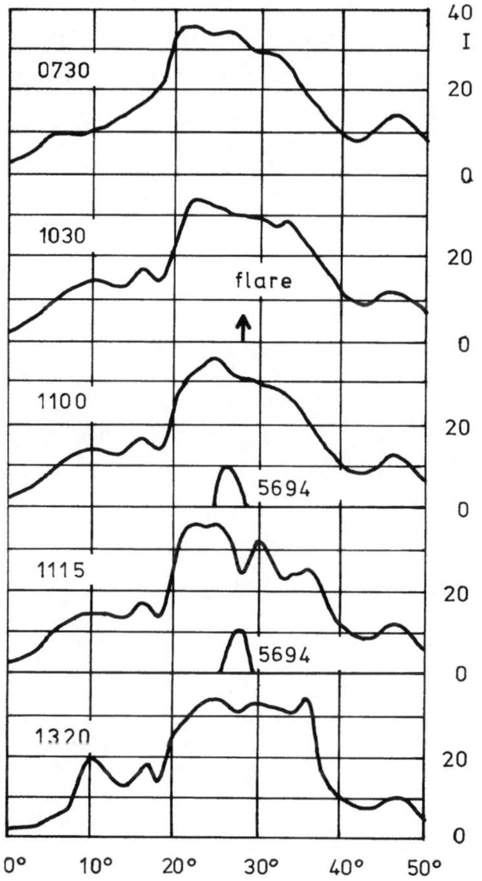

Fig. 3. The variations of the intensities of the lines λ5303 and λ5694 following a flare at latitude north 28° on the east limb on March 18, 1956.

tensity around 1100 UT, and disappeared at 1130 UT. At the time of the visibility of this line, the green line was strongly weakened. Two hours after the flare, the line intensities were the same as before the flare.

This example, one out of many others, shows clearly that the hot coronal cloud comes after the flare, and has a long duration. We do not conclude that the coronal cloud is produced by the optical flare. Probably, both the flare and the cloud have a common primary origin that is not visible, and not yet known.

4. Conclusions

Disturbances of the corona that take place in a time not longer than a few hours, are rare phenomena. The variations in most of these fast events consist of changes both in structure and intensity but mass motion is also involved in many of them. They always are connected with flares or some other Hα activity. The most remarkable optical features are a very large intensity of the line λ5694 and a strong continuoum, indicating high temperature and density. Typically the highest, densest and hottest condensations occur not close to the Sun's limb but at a height of about 40000 km.

References

Bruzek, A.: 1957, *Z. Astrophys.* **42**, 76.
Lyot, B.: 1944, *Ann. Astrophys.* **7**, 31.
Waldmeier, M.: 1938, *Z. Astrophys.* **15**, 310.
Waldmeier, M.: 1961, *Z. Astrophys.* **53**, 206.

DISCUSSION

Altschuler: What mechanism forms the yellow line region?

Waldmeier: I have only presented observational results; but we might assume that they are related to the X-ray regions.

FORBIDDEN LINE EXCITATION DATA FOR CERTAIN CORONAL LINES

S. J. CZYZAK

Astronomy Dept., The Ohio State University, Columbus, Ohio, U.S.A.

L. H. ALLER

Astronomy Dept., University of California, Los Angeles, Calif., U.S.A.

and

R. N. EUWEMA

Aerospce Research Laboratories, Wright Patterson Air Force Base, Ohio, U.S.A.

Abstract. Plasma diagnostics of the active corona require data on transition probabilities and collision cross sections for relevant coronal emission lines. Atomic parameters pertaining to a number of coronal transitions have been derived as follows:

Ion	Con-fig.	λ (Å)	A (s^{-1})	Ω	Ion	Con-fig.	λ (Å)	A (s^{-1})	Ω
S II	$2p$	7536	20.817	0.1178	Ni XVI	$3p$	3600.9	190.80	0.170
Ca XIII	$2p^4$	4086.5	315.7	0.1140	Ni XV	$3p^2$	8024.2	22.121	0.249
K XI	$2p^5$	4256	231.13	0.1093			6701.9	55.820	0.038
Mn XIII	$3p$	6535.4	31.922	0.2270	Cr IX	$3p^4$	4451	4.308	0.112
Fe XIV	$3p$	5302.9	60.056	0.1950	Fe X	$3p^5$	6374.5	69.149	0.286
Co XV	$3p$	4351.4	108.16	0.1851	Ni XII	$3p^5$	4232.3	235.26	0.219

We have calculated and tabulated occupation numbers of the upper excited levels and corresponding emissivities as a function of electron density and temperature following a procedure similar to that of Zirker (*Solar Phys.* **11**, 68, 1970). These numbers almost certainly represent lower limits to the quantities involved since at coronal temperatures one must consider collisional excitation to high permitted levels from which cascade can occur, dielectronic recombination of the pertinent ions, and subsequent decay to the level involved, and also collisional excitation and ionization which serve to remove ions from the level considered. A rigorous calculation would require accurate knowledge of all relevant transition probabilities, collision cross sections, and dielectronic recombination parameters. The effects are more severe at the higher densities and can be estimated only when we know the necessary parameters. The present numbers may prove useful for lower densities and temperatures and higher stages of ionization.

Gordon Newkirk, Jr. (ed.), Coronal Disturbances, 155.

MULTIPLE HARD X-RAY BURSTS AND ASSOCIATED EMISSIONS

J. VORPAHL

Sacramento City College, Sacramento, Calif., U.S.A.

Multiple hard X-radiation, along with associated emission at optical and radio wavelengths, is discussed for three events in particular: December 13, 1970–1832 UT; December 12, 1970–1843 UT; June 28, 1970–2001 UT. Characteristics of these events as well as other multiple hard X-ray bursts observed by the author (e.g., Vorpahl, J.: *Solar Phys.* **29**, 447, 1973) are given at the end. Data originated from hard X-ray experiments on the OGO-5 (Kinsey Anderson) and OSO-5 (Ken Frost) satellites and were compared so that the effect of pulse pile-up in the OGO-5 data could be observed. The two December flares occurred in different active regions separated by about 100000 km yet both produced hard X-ray bursts with similar structure. This suggests that the two regions were joined by extended field lines with the acceleration mechanism somewhere in between. Although nothing was visible in Hα (such as surges or post flare ejections) that would indicate connecting field lines, higher coronal structures could explain the similarity in hard X-rays from two separate regions.

DECEMBER 3, 1970–1832 UT

Hα Characteristics

(1) Hα kernals were very abrupt and intense, exhibiting secondary enhancements at the time of the first two small X-ray spikes in the 20–32 keV range. The film was too saturated by the time of the third spike to register any increase.

(2) Kernals were directly over good size sunspots of opposite polarity. The spots were growing very fast in a new flux region that had appeared near older spots.

(3) A small filament passed directly through the space between the opposite polarity spots and *into* one of them.

(4) Only a small amount of material ejection was visible in Hα concurrent with the radio emission at metric wavelengths.

(5) Although very intense, the flare was small in area and of very short duration, lasting only about 12 min before fading.

Hard X-Ray Burst

Fairly large with at least three individual spikes as high as 111 keV.

Microwave

Medium size burst with 49 flux units at 8800 MHz and 36 flux units at 15400 MHz.

Gordon Newkirk, Jr. (ed.), Coronal Disturbances, 157–159.

Long Radio Wavelengths

Intense type II and III, with weak type IV.

DECEMBER 12, 1970–1843 UT

Hα Characteristics

(1) Multiple X-ray spikes corresponded to an optical brightening initially at one end of a big filament, followed by secondary enhancements at both ends of the same filament simultaneous with the secondary hard X-ray spikes.

(2) As in the 12/13/70 event, the flare originated at the site of a new magnetic region growing near some older spots.

(3) Also similar to the 12/13/70 flare, the large filament involved in the kernal brightenings was rooted in a region of mixed polarities resulting from the newly emerging flux.

Hard X-Ray Burst

Fair size, although both the maximum intensity and the highest energy attained (82 keV) were less than the 12/13/70 flare.

Microwave

Event twice as large as that a day later – 115 flux units at 8800 MHz and 22 flux units at 15400 MHz.

Long Radio Wavelengths

Short duration type III bursts of medium intensity. No type II or type IV emission recorded.

JUNE 28, 1970–2001 UT

Hα Characteristics

(1) Several Hα kernels coinciding with the initial hard X-ray spike, however, the time resolution on the film was 10 s and therefore not good enough in this case to know whether the brightening of the many knots really was simultaneous.

(2) The various Hα kernels were in opposite polarities but due to the complexity of the event, it was impossible to identify which pairs were connected by the same flux tube.

(3) Gas ejections were visible in Hα during the approximate time of type II, III, and IV radio emission, but again the event was too complex to identify the optical and long wavelength radio component unambiguously.

(4) The eruption and apparent untwisting of a small filment was observed concurrent with the radio emissions discussed in #3.

Hard X-Ray Burst

Large burst with energies up to 141 keV observed. This event is an example of an event with type IV-like hard X-ray emission described by Ken Frost in these proceedings.

Microwave

Intense burst with 820 flux units at 8800 MHz and 435 flux units at 15400 MHz; Major spikes in frequencies above 2695 MHz tracked the hard X-ray profile.

Long Radio Wavelengths

Intense type II, III, and IV bursts, with major components in the decimetric type IV mirroring the hard X-rays.

General Characteristics of Multiple X-Ray Bursts

(1) When multiple hard X-ray bursts occur, there are simultaneous enhancements in the Hα kernel if the film has not already been saturated. Problems in over-exposure are important usually after the second or third component in a multiple burst.

(2) The Hα flare need not be large although the kernels are always intense and brighten abruptly, i.e. within about 20 s.

(3) Ejections in Hα are not necessary but are often observed at the time of III and II or IV events. On the other hand, this author feels that it is still not possible to determine whether there is type III (or type II/IV) flux strictly from observing the Hα flare.

(4) There seems to be a tendency for the microwave and hard X-ray bursts to be more intense when the simultaneous type III is weaker, and vice versa. This suggests either that: (1) there is a preferential beaming of non-thermal electrons up or down; or (2) the magnetic field lines near the acceleration site determine whether any electrons escape upwards, to produce the long wavelength radio emissions or downwards to explain the hard X-ray and microwave flux.

GROUND-BASED OBSERVATIONS OF TYPE III BURSTS

R. T. STEWART

Division of Radiophysics, CSIRO, Sydney, Australia

Abstract. Observations over the past 20 yrs or so are reviewed, with emphasis on recent high spatial resolution observations. The results lend support to earlier ideas on the propagation of type III electron streams through coronal regions of weak magnetic field strength but have not as yet settled the question whether the electrons propagate along the axes of coronal streamers. Several important burst properties appear to be significantly affected by ray scattering on small-scale size density irregularities in the corona.

1. Introduction

The type III burst was first recognized as a distinct class of solar radio emission from spectral studies at metre wavelengths (Wild, 1950). During the past two decades the type III burst has been studied in great detail with groundbased spectrographs, interferometers and heliographs, and also with satellite-based spectrographs. The bursts have been observed at frequencies as high as ~600 MHz and as low as 30 kHz, corresponding to source heights near the base of the corona and the Earth's orbit respectively. The radio emission has been shown to be the result of plasma waves excited by a stream of fast electrons ejected from near the region of a chromospheric flare and travelling outwards, sometimes as far as the Earth's orbit. Since the electrons escape along magnetic field lines they act as tracers of the coronal and inter-planetary magnetic fields. The source heights at various frequencies give information about the electron density distribution along the electron path. The exponential decay rates of the burst at different frequencies have been used to estimate the electron temperatures in the corona. These and other aspects of the type III burst will be discussed in the light of recent ground-based observations. For earlier reviews of ground-based observations of type III bursts see Wild *et al.* (1963), Kundu (1965), Wild and Smerd (1972), and Solar Radio Group Utrecht (1974).

2. Burst Characteristics

Type III bursts are a common form of sporadic radio emission, characterized by rapid frequency drift from high to low frequencies and by short duration. They occur as isolated bursts (as members of a connected group), sometimes regularly spaced (Figure 1a), and in very large numbers as members of a persistent storm. Isolated bursts often accompany the flash phase of a chromospheric flare (Loughead *et al.*, 1957; Malville, 1962). Storms usually follow a large flare and can last several days. During the storm, type I bursts occur at higher frequencies (> 100 MHz) than type III bursts, which at lower frequencies may blend into a 'decametric continuum' (de la Noë *et al.*, 1973).

Bursts are rich in variety. At one extreme is the short-duration type III burst with

Gordon Newkirk, Jr. (ed.), Coronal Disturbances, 161–181.

a rapid drift from high to low frequencies (Figure 1a). At the other extreme is the long-duration type III-V burst lasting $\simeq 1$ min at metre wavelengths (Figure 1c). In between are intermediate-duration bursts, which may be a mixture of type III and type V (Figure 1b). The starting frequency varies from burst to burst and can be as high as 600 MHz or as low as 25 MHz. Sometimes bursts have both fundamental and second-harmonic components, which are difficult to distinguish because of the rapid drift rate (Figure 1b). Some bursts bend over at low frequencies and are known as 'inverted-U' bursts (Figure 1d). Another sub-type, the type IIIb burst, will be described

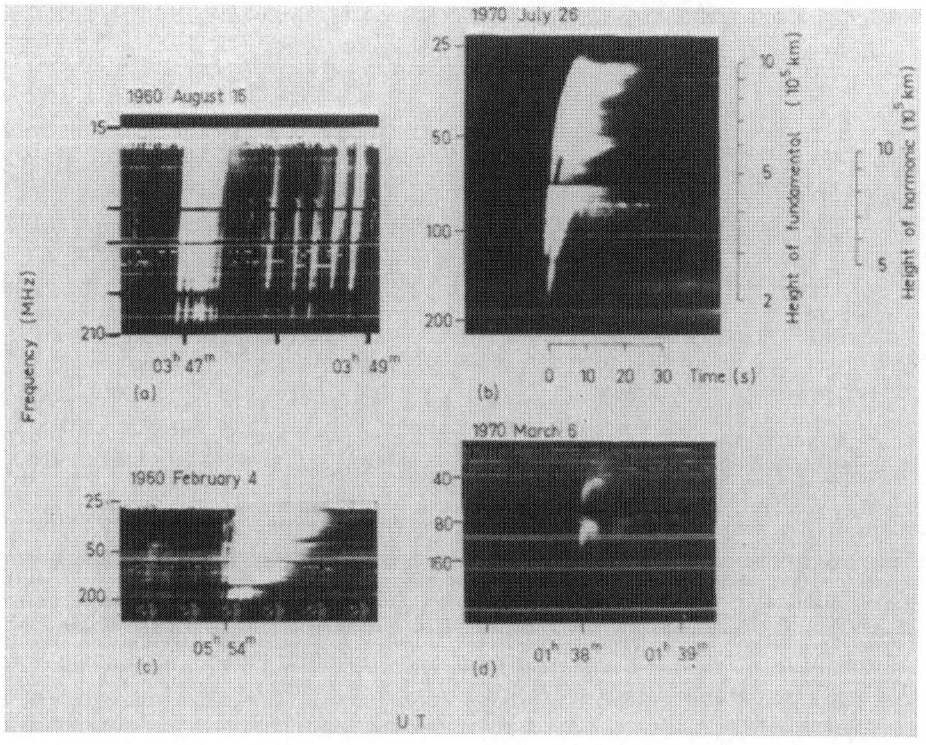

Fig. 1. (a) Groups of regularly spaced type III bursts (Wild *et al.*, 1963). (b) A type III burst showing fundamental and second-harmonic structure followed by brief type V continuum (Wild and Smerd, 1972). (c) An intense type III–V burst (Weiss and Stewart, 1965). (d) An inverted-U burst showing fundamental and second harmonic structure.

in Section 11. From the ground, the lowest observing frequency (determined by the ionospheric cut-off) is $\gtrsim 7$ MHz.

At any frequency the intensity profile of a type III burst shows a rapid rise followed by an exponential decay. The decay rate decreases with decreasing frequency and so the duration increases. Decay rates, if due to collisional damping, give an upper limit to electron temperature in the corona. (See Section 10 for discussion on the effects of scattering of the radiation on coronal irregularities.)

3. Source Height

According to the plasma hypothesis (Wild, 1950) type III emission occurs at the plasma frequency $f_p = 9 \times 10^{-3} N^{1/2}$ MHz, where N cm^{-3} is the electron density. The frequency drift results from the outward movement of the source through regions of decreasing electron density, i.e. decreasing plasma frequency. Hence the frequency drift, which is most clearly defined by the leading edge of the burst, represents the outward velocity of the source; the more rapid the drift, the faster the source velocity.

The plasma hypothesis was confirmed by one-dimensional interferometer measurements of the type III source height at different frequencies which showed successively lower frequencies being emitted from successively greater heights (Wild *et al.*, 1959). The observed heights of type III bursts (if assumed to be fundamentals) give coronal electron densities which are about four times the values derived by white-light observations of the quiet *K*-corona (Figure 2). Hence it has been assumed

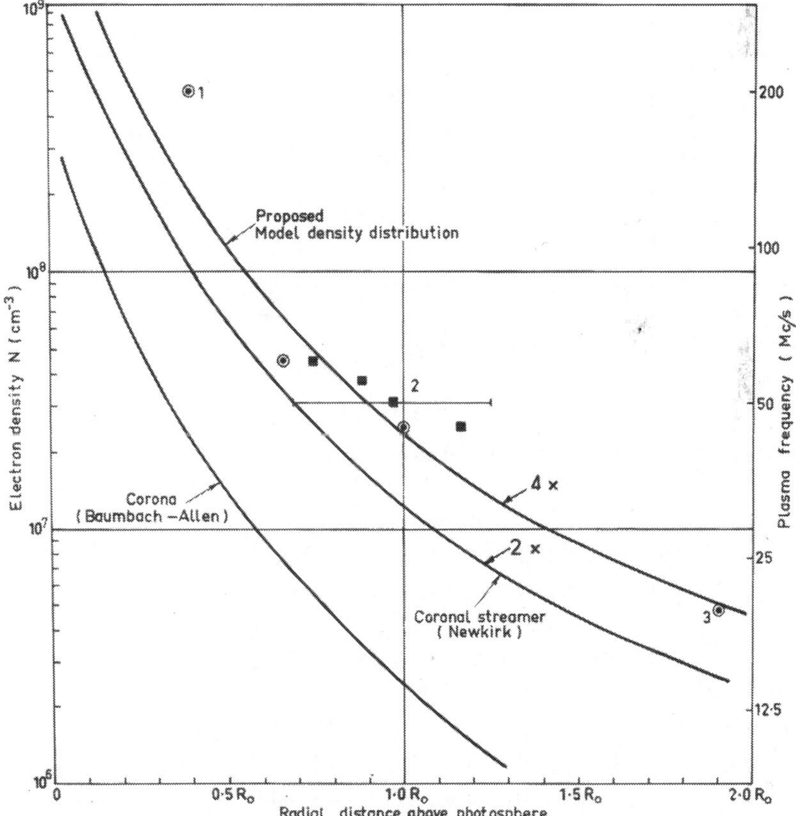

Fig. 2. A comparison of radio and optical determinations of electron density in the low corona. The plotted points give the heights of emission (fundamental) of type II and type III bursts measured by ground-based observers (Weiss, 1963). Note that Newkirk (1961) coronal streamer densities = 2 × Newkirk (1961) quiet Sun densities (Wild *et al.*, 1963).

by many authors that the type III disturbance propagates along the axis of a coronal streamer, where the densities are higher than the surrounding corona. (See discussion on the effects of scattering in Section 8.)

4. Source Velocity

Interferometer observations show that the average radial velocity varies from 0.2 to 0.8 c (Figure 3) between the 60 and 45 MHz plasma levels. Individual type III bursts observed over a frequency range from 200 to 12 MHz, i.e. from 0.15 $R_\odot$ to 2.0 $R_\odot$ above the photosphere, have drift rates which correspond to radial source velocities

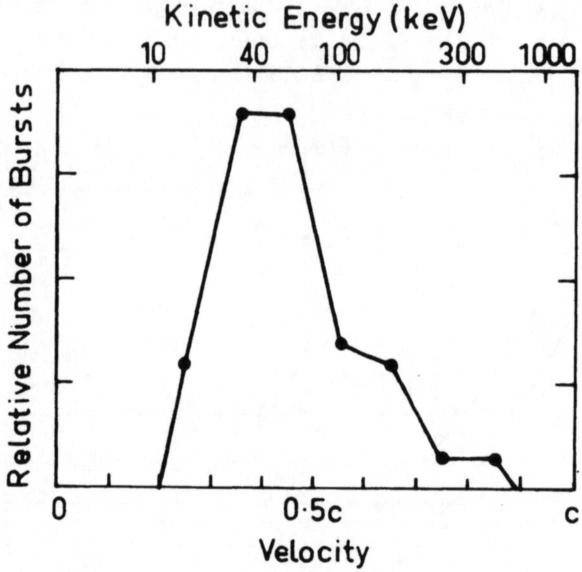

Fig. 3. Distribution of the speed of exciters of type III bursts assumed to be electrons traversing radial path (Wild and Smerd 1972, from velocity data of Wild *et al.*, 1959).

$\sim c/3$. The constancy of the radial component of the source velocity over such great distances suggests that the electrons in the type III disturbance travel without substantial spiralling along magnetic field lines (Stewart, 1965). Confirmation has come from satellite observations at hectometre wavelengths (Slysh, 1967; Fainberg and Stone, 1970; Haddock and Alvarez, 1973) which show that the average radial velocity of type III bursts is more or less constant, $\sim$0.3 to 0.4 c, out to heights $\sim$100 $R_\odot$, although Fainberg *et al.* (1972) claim that some type III bursts do show a systematic deceleration with distance from the Sun. (See also review article by Fainberg (1974) in this issue.) The derived electron densities are in reasonable agreement with the density model based on metre wavelength observations if it is assumed that the observed radiation at low frequencies ($\leqslant$1 MHz) is second harmonic, as the results of Haddock and Alvarez (1973) suggest. Recent work by Alvarez and Haddock (1973) has shown that the observed frequency drift rates, for frequencies between 550 MHz and 75 kHz, can be used to derive electron density models out to 1 AU which are in good agreement with models based on optical and particle observations. (See Figures 3 and 5 of Alvarez and Haddock (1973).) Their theory assumes the exciter particles travel with constant velocity and zero pitch angle along an Archimedes spiral and ignores ray scattering and refraction effects.

5. Electrons or Protons

The question of whether the source particles are electrons or protons has been more or less settled by satellite observations of high fluxes of impulsive bursts of solar electrons at the Earth's orbit within 30 min following a flare and a type III burst (Van Allen and Krimigis, 1965; Anderson and Lin, 1966). Simultaneous observations of solar electrons and hectometric type III bursts from the Imp-6 space craft show that the onset of radio emission located at 1 AU corresponds to the arrival of electrons of ~ 100 keV energy. The emission maximum corresponds to the arrival of slower (~ 10 keV) electrons (Lin *et al.*, 1973). Hence the observed energy range of solar electrons 10 to 100 keV is in good agreement with the speed of type III bursts observed at metre wavelengths (see Figure 3). Lin *et al.* (1973) also point out that the apparent deceleration of the type III drift rate with distance from the Sun may be due to pitch-angle scattering and not to energy loss. (See also review article by Lin (1974) in this issue.) Fluxes of energetic protons have been reported (McDonald and Van Hollebeke, 1973) but these fluxes are very low.

In addition, Kane (1972) has shown that impulsive hard X-ray bursts often occur almost simultaneously with type III bursts which extend into the decimetre band (Figure 4). The X-ray bursts are attributed to bremsstrahlung from non-thermal electrons located in the lower corona ($<0.1\ R_\odot$ above the photosphere). The derived electron energy is in the range 10 to 100 keV, i.e. similar to that of type III bursts (Kane, 1972). (See also review paper by Kane (1974) in this issue.)

6. Origin of Type III Electrons

Statistical studies by various authors (Loughead *et al.*, 1957; Swarup *et al.*, 1960; Malville, 1962) have shown that the correlation of isolated type III bursts (as distinct from storm bursts) with flares is quite high. Recent high-resolutions Hα studies (Kuiper and Pasachoff, 1973; Vorphal and Zirin, 1972) show that the occurrence of type III bursts is often closely related in time with Hα brightenings, which are also associated with X-ray and microwave bursts (see also Kane, 1972). Martres *et al.* (1972) and Axisa *et al.* (1973) find that some type III bursts occur associated closely in time (± 1 min) with the appearance of Hα absorbing features even when there is no detectable flare brightening. These features occur at the border of active regions along a line separating opposite sense of the longitudinal photospheric magnetic field ($H_{\parallel}=0$). Axisa *et al.* (1973) suggest that the transient disturbance which causes

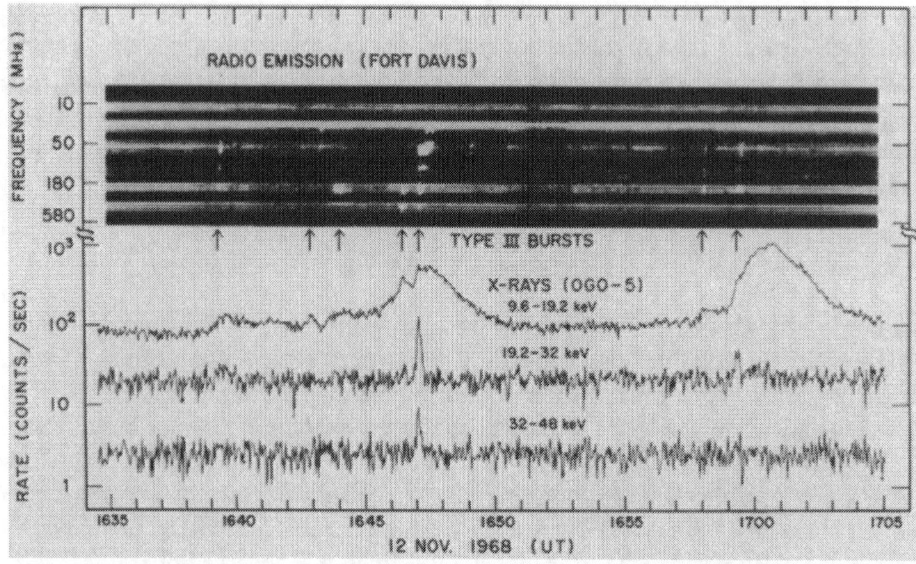

Fig. 4. An example of time-correlation between impulsive solar X-rays $\geqslant 10$ keV and type III solar radio bursts in the 10 to 580 MHz frequency range. Note the large spike at 1647 UT (Kane, 1972).

this chromospheric feature may also at times trigger the type III emission and the Hα flare. Teske *et al.* (1971) report several type III bursts occurring in the absence of flares (perhaps associated with Hα absorbing features (Kane *et al.*, 1974)) but preceded by weak soft X-ray bursts, an observation which suggests that a thermal event precedes the type III instability.

However, the starting frequencies of type III storm bursts and also of some isolated type III bursts can be less than 100 MHz. If the type III electrons are always accelerated in the lower corona, say $\leqslant 0.1\ R_{\odot}$ above the photosphere, one wonders why the starting frequencies of the type III bursts are often so small. Do the electrons travel over considerable distances in the lower corona without emitting radiation, or are the electrons accelerated locally near their starting frequency?

7. Role of the Coronal Magnetic Field

Wild and Smerd (1972) have summarized published observations and ideas on type III bursts in the model of Figure 5.

Electron acceleration and injection are assumed to occur in an unstable region containing opposing magnetic lines of force. Electrons which have access to open field lines around neutral planes (supposed to occur along the axis of the coronal streamer) give rise to type III bursts, while others injected into closed loop structures produce either *U* bursts if the electrons are guided only once around the loop and then dispersed, or type V bursts if the electrons are trapped for periods ~ 1 min.

Supporting evidence for the propagation of electrons in a weak field region has come from observations of a low degree of circular polarization* in type III bursts (Kai, 1970; Chernov *et al.*, 1972). (See Section 9 for a discussion of linear polarization measurements.) Kai (1970) estimates the field strength in the type III source to be <0.14 G while Melrose and Sy (1972) estimate it to be <0.04 G (see also Rosenberg, 1973). Melrose and Sy base their estimates on observations of a smaller degree of

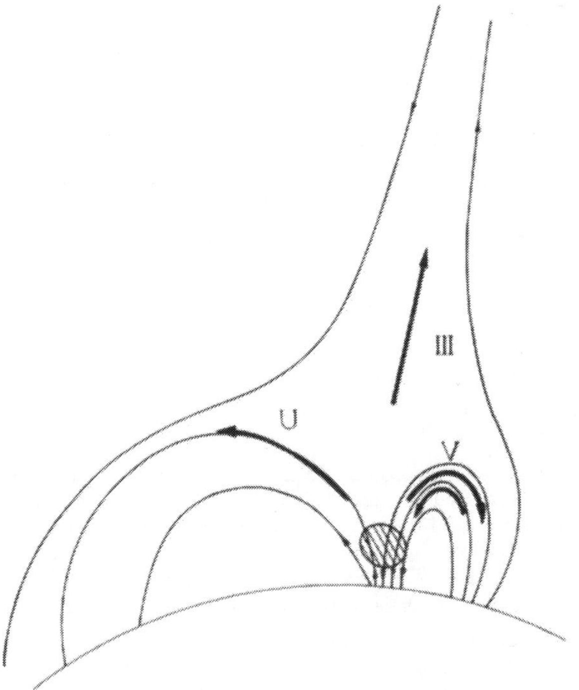

Fig. 5. Schematic magnetic field configuration to illustrate the possible different paths taken by electrons, ejected in the flash-phase explosion (hatched region) in classical type III, type V and inverted-*U* bursts (Wild and Smerd, 1972).

circular polarization in harmonics than in fundamental (McLean, 1971). Kai's observation of bipolar structure in type I bursts at 80 MHz is a strong argument for the source of the type I bursts being in a closed field region above a sunspot group (Figure 6a). His observation of a displacement between associated type I and type III bursts is attributed to the propagation of type III electrons along open field lines away from the sunspot region (Figure 6b). For storms, displacements between type I and type III burst centres have also been observed (Stewart and Labrum, 1972; see also Figure 9).

Several authors (Weiss and Wild, 1964; Gleeson, 1965) have shown that if a group of electrons are injected isotropically into a diverging neutral sheet region then they will be guided outwards along the neutral plane. Evidence for the possible channelling

* An exceptionally high degree of circular polarization has been measured at the beginning of some type III bursts (Slottje, 1974; Rao, 1965).

of type III bursts along a neutral plane has been reported by McLean (1970): at 80 MHz the members of a type III group have been observed to be strung out along a narrow segment of the solar disk (Figure 7). Position shifts between the leading and trailing arms of *U* bursts at 80 MHz have been observed (Labrum and Stewart, 1970). This result is consistent with the model of Figure 5. Position shifts have also been

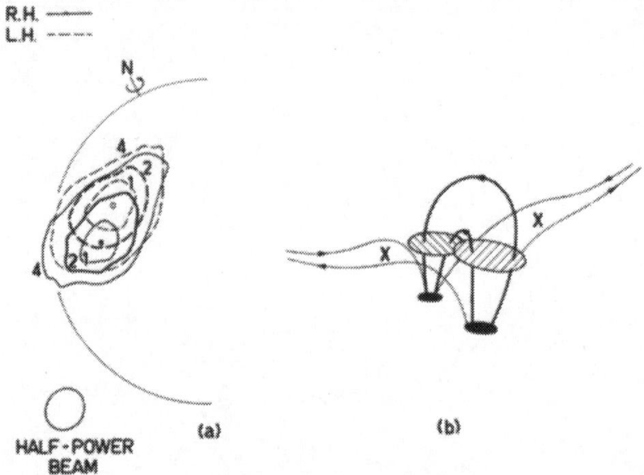

Fig. 6. (a) 80 MHz contours of a type I storm source in the two senses of circular polarization showing bipolar structure. Contours marked 1, 2 and 4 signify intensities $1/\sqrt{2}$, 1/2 and 1/4 of the peak total (RH+LH) value (Wild, 1970, after Kai, 1970). (b) Kai's (1970) model of a bipolar storm centre. Type I emission originates from the shaded regions where strong magnetic fields intersect the plasma level. Type III emission originates from weak-field or neutral regions, such as those marked × (Wild, 1970, after Kai, 1970).

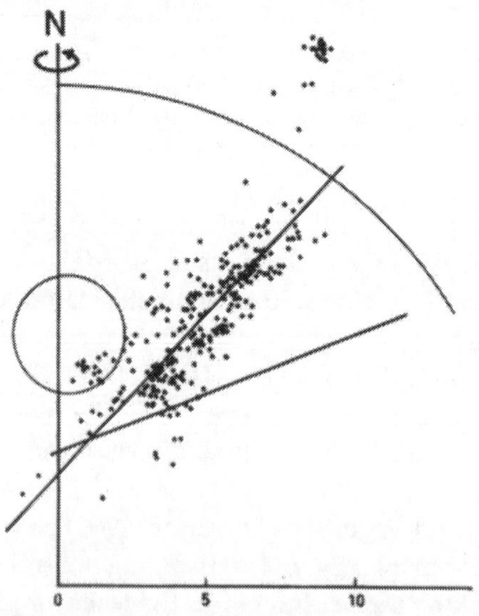

Fig. 7. The 80 MHz radioheliograph centroids of burst sources, for each second of available data from 1968 November 13, $00^h18^m32^s$ to $00^h36^m00^s$. The lower line and the circle show the scatter of points for earlier phases (McLean, 1970).

observed between type V bursts and associated type III bursts (Labrum, unpublished; Kundu *et al.*, 1970), while the type V source is considerably larger than the type III (Figure 8), as one would expect from the model of Weiss and Stewart (1965). Labrum and Duncan (1974) find the position shift between type III and type V sources to be predominantly outward.

A slight variation on the model of Wild and Smerd (1972) has been proposed to

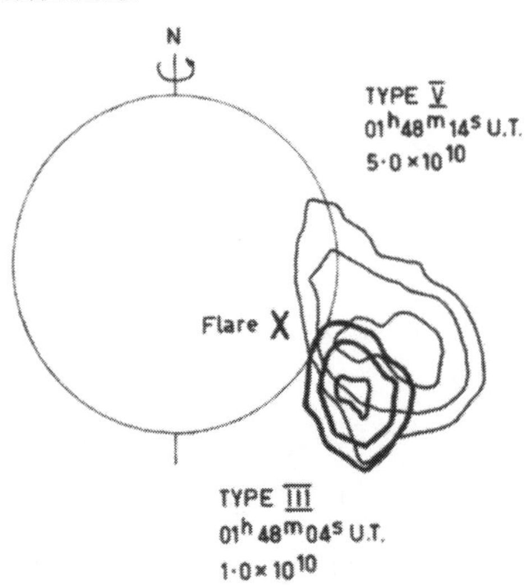

Fig. 8. 80 MHz radioheliograph observations showing the much larger source of a type V burst (light contours) displaced from that of the preceding type III burst (heavy contours). Successive contours are in the ratio 2:1. The numbers indicate peak beam temperature (K) (Labrum, unpublished data; Wild and Smerd, 1972).

explain the observed frequency and spatial displacements between type I and type III storm bursts (Stewart and Labrum, 1972). Unlike isolated type III bursts, the storm bursts tend to occur only at low frequencies $\leqslant 80$ MHz and are always accompanied by type I bursts at higher frequencies (Boischot *et al.*, 1971). If the type III electrons are accelerated near the starting height of the type III burst ($\geqslant 0.6\, R_\odot$ above the photosphere) the acceleration region is considerably displaced, both radially and horizontally, from the type I region (Figure 9). It is proposed that an unstable region (such as the cusp of a helmet streamer (Pneuman, 1968)) is triggered by a m.h.d. disturbance from the flare region and electrons are accelerated locally and escape along the axis of the streamer.

Recently, Smith and Pneuman (1972) have cast doubts on the ability of type III electrons to escape and to be channelled along the axis of a coronal streamer. They argue that if account is taken of the finite conductivity and the azimuthal magnetic field component of a coronal streamer then the field lines do not remain open but close across the neutral sheet. Direct observations of the relative positions of type III sources and coronal streamers should settle this question. Such observations have been made for one type I–type III storm event by the author in collaboration with Shirley Hansen and R. Hansen and for several type III bursts by Leblanc *et al.* (1974). The results, shown in Figures 10a and b, (and also by Leblanc *et al.* (1974)) are not conclusive. There is no definite evidence for a coronal streamer near the type III

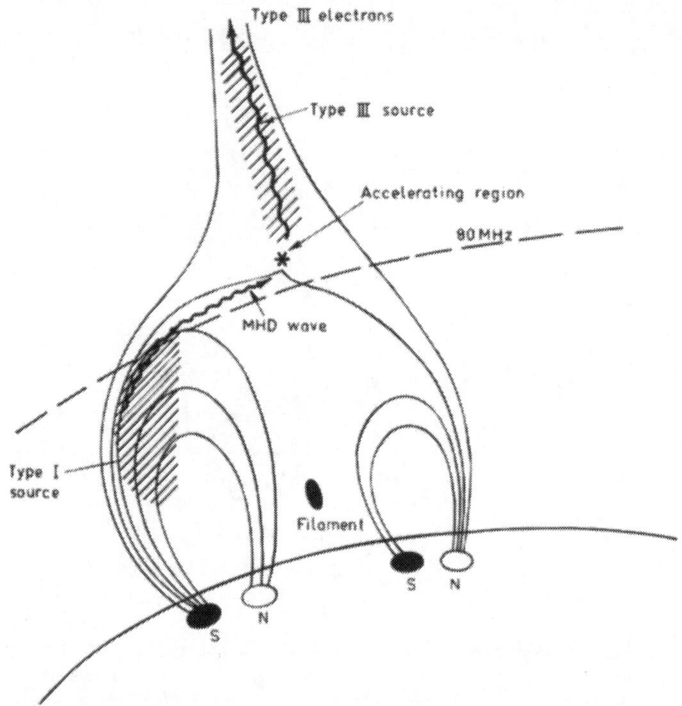

Fig. 9. Proposed model to explain the non-radial displacement between the sources of type I and type III storm bursts, observed during 1968 August 19–21 with the 80 MHz radioheliograph (see text) (Stewart and Labrum, 1972).

position (see A of Figure 10). It may be possible to resolve this question when three-dimensional density models are derived from *K*-corona and ATM-Skylab data.

8. Propagation and Scattering of Radiation

The propagation of type III radiation from the source to the Earth is determined by refraction by large-scale features of the solar corona, such as streamers, and by scattering on relatively small-scale irregularities. Scattering is crucial for the escape of fundamental radiation from the plasma level. Fokker (1965) showed that scattering would cause an increase in the angular size of the source and an outward shift in its apparent position. More complete studies by Steinberg *et al.* (1971) and Riddle (1972a) have combined the effects of refraction in a spherically symmetric corona with scattering on randomly-distributed density inhomogeneities. The results show that scattering can explain why fundamental sources are seen sometimes near the limb –

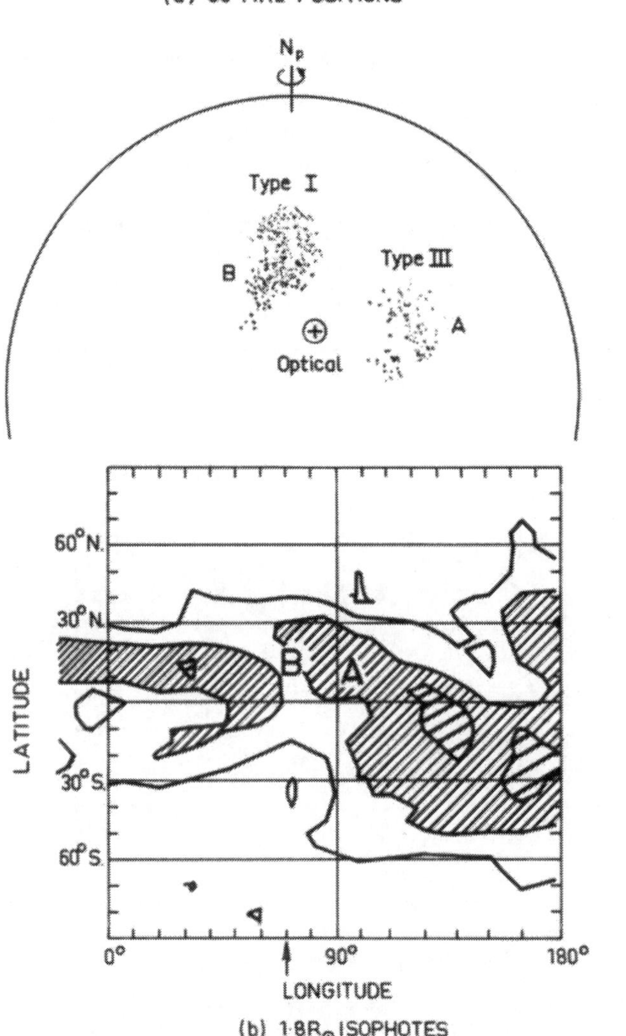

Fig. 10. (a) Scatter plot of centroids of all the 80 MHz radioheliograph sources, A (type III) and B (type I) for 1971 October 4–5, 23^h10^m to 01^h06^m. The cross indicates the centre of the active region McMath 11 537. (b) *K*-corona isophotes (arbitrary units) derived from 1.8 $R_\odot$ west limb scans with the coronal activity monitor on Mauna Loa, Hawaii (courtesy HAO Boulder). CMP for October 4–5 is shown by the arrow.

without scattering, the radiation is strongly beamed into the radial direction by refraction and the fundamental should only be seen near the centre of the solar disk. Riddle (1972b) has also considered a coronal streamer model. His results show that scattering can account for most of the discrepancy between electron densities derived from radio observations (Figure 2) and values obtained from optical observations (Newkirk, 1961). Weiss (1963) derived densities of 2 × Newkirk's values for a coronal streamer by assuming that scattering would cause the apparent position to coincide with the true projected position. According to Riddle this method over-estimates densities by a factor of 1.5 (Figure 11).

Riddle (1972a, b) and Leblanc (1973) have shown that scattering also causes an increase in the angular size of the harmonic source and a position shift inward. The position shift and angular size are largest at large longitudes ($\geqslant 60°$) for radiation beamed into the backward direction but are considerable even for isotropic emission

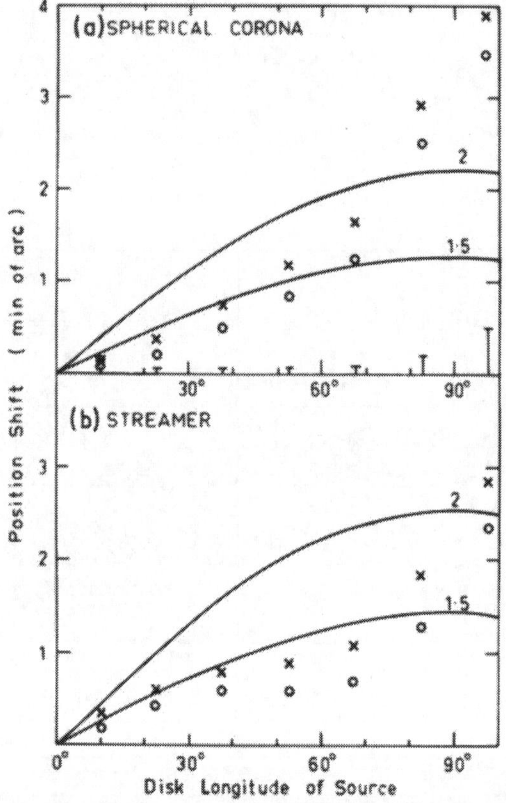

Fig. 11. Computed position shifts due to scattering and refraction of 80 MHz radiation from a point source at the plasma level (a) in a spherically symmetric corona and (b) on the axis of a Newkirk streamer superimposed on the spherically symmetric corona model (a), with coronal temperatures of 10^6 K (crosses) and 2×10^6 K (circles). The curves show that similar displacements can be obtained by increasing the electron density model by 1.5 and 2 times and by neglecting the effects of scattering and refraction (Riddle, 1972a).

(Figure 12b(ii) and (iv)). This result tends to weaken the argument for backward emission (Smerd *et al.*, 1962) based on the observation that fundamental and harmonic type III bursts appear at nearly the same radial distances (Stewart, 1972; McLean, 1971; Bougeret *et al.*, 1970). However, the position shift and angular size depend on the assumed density distributions of the large-scale as well as the small-scale features – e.g. if the source is located on the axis of a coronal streamer the position shift and angular size are less than in a spherically symmetrical corona (see Figure 12b(i) and (ii)).

A study of fundamental and harmonic source sizes observed at 80 MHz (Stewart, unpublished data) shows that: (a) the sources are usually elongated in the transverse direction; (b) the size of the harmonic source is usually equal to but sometimes larger than that of the fundamental (Figure 13); (c) the harmonic source size increases somewhat with time (Figure 14). Current scattering models may account for (c) by showing

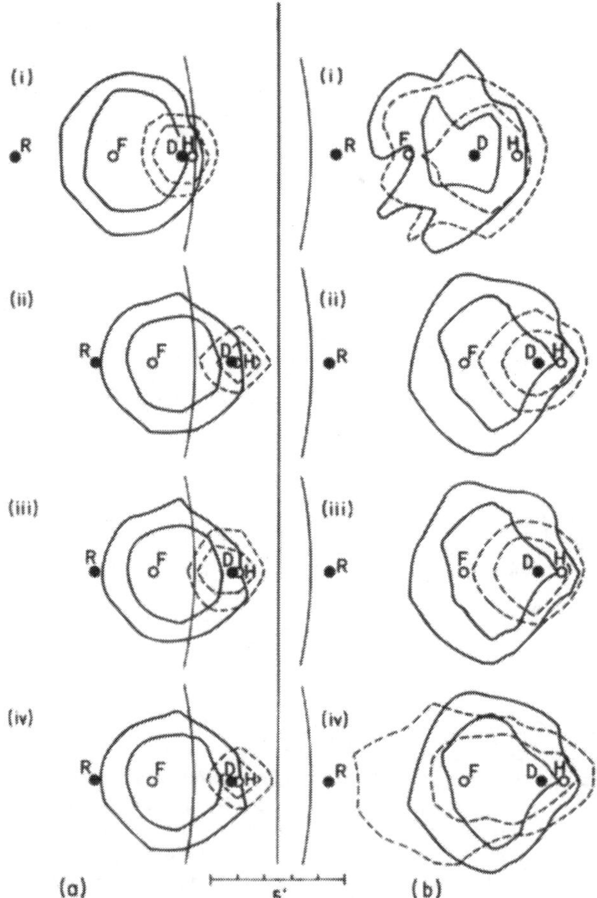

Fig. 12. Diagram showing the half-power and quarter-power contours of 80 MHz radiation as derived from computations which include the effects of scattering and refraction. Full contours refer to radiation from an isotropically emitting 'fundamental' point source, F, located near the 80 MHz level. The broken contours refer to a 'harmonic' point source, H, located at the 40 MHz level. The points D and R show the direction of arrival of the direct and reflected rays of the source H in the absence of scattering. The arc on each diagram indicates a portion of the limb of the visible disk of the Sun. The sources are located at longitude $37\frac{1}{2}°$ (a) and $67\frac{1}{2}°$ (b) in Newkirk (1961) spherically symmetric model for (i) and on the axis of Newkirk (1961) streamer model for (ii), (iii) and (iv). The source H is assumed to emit isotropically in (i) and (ii) and to have a radiation pattern as given by Smerd *et al.* (1962) in (iii) and by Zheleznyakov and Zaitsev (1970) in (iv). The exciter velocity is assumed to be one-third the speed of light (Riddle, 1972b).

that rays scattered at large angles are more delayed than rays scattered at small angles, although the calculations have not been done as yet; they cannot account for (a) by assuming randomly-distributed scattering centres or (b) by assuming isotropic emission. The following interpretations are worth considering: (a) the density irregularities are not randomly distributed but aligned radially like filaments; (b) the

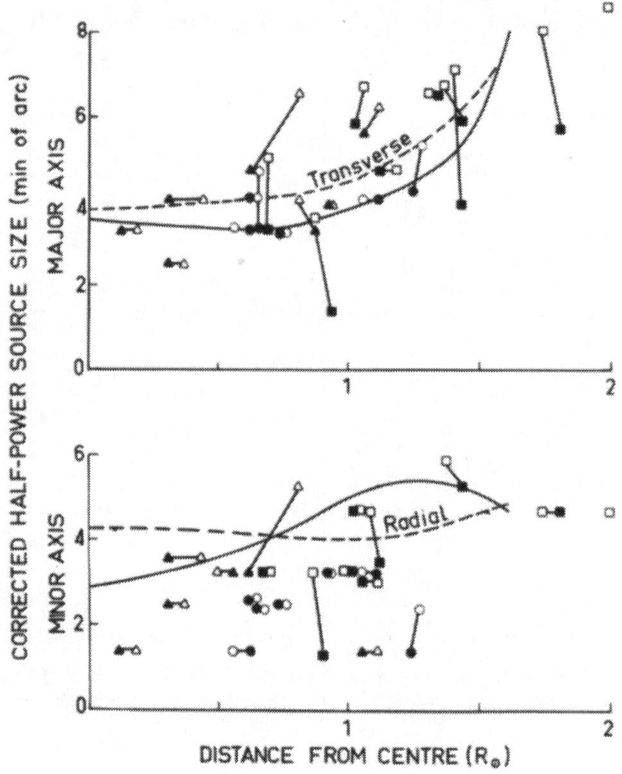

Fig. 13. Observed 80 MHz source sizes (corrected for beam-broadening) of fundamental (filled symbols) and second-harmonic (open symbols) type III bursts. The circle indicates short-duration bursts, the square long-duration bursts, and the triangle inverted-*U* bursts. The full and dashed curves are derived from the data of Riddle (1972a) and show the transverse (upper figure) and radial (lower figure) dimensions of the computed angular sizes for point sources in a spherically symmetric corona with randomly-distributed density irregularities. The dashed line refers to fundamental radiation from the 78 MHz plasma level and the full line refers to harmonic radiation from the 40 MHz plasma level. It is assumed that the sources have Gaussian brightness distributions.

emission of the harmonic is in the backward direction; (c) the intrinsic source size increases owing to stream dispersion.

Recent observations at 169 MHz (Caroubalos and Steinberg, 1974) from the ground and from space ('Stereo-1' experiment) show that the directivity of type III bursts can be higher than the value derived by ray-tracing in a spherically symmetric corona where the scattering power is large enough to account for the observed source size.

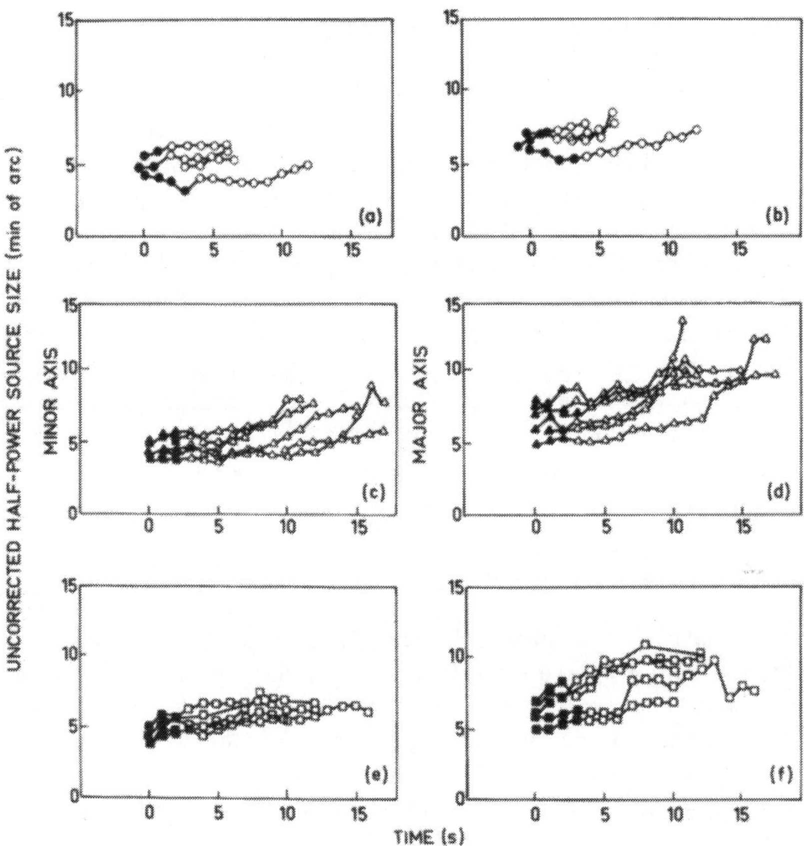

Fig. 14. Observed variation of 80 MHz source size with time. Minor axes are shown at the left, major axes at the right. Data are from four short-duration type III bursts (a and b), five long-duration type III bursts (c and d) and five inverted-*U* bursts (e and f). Filled symbols refer to fundamentals and open symbols to harmonics.

Large deviations from spherical symmetry and a directive harmonic source may be required to satisfy the observations. However, the 'Stereo' results are only preliminary at this stage.

9. Linear Polarization

Several authors (Akabane and Cohen, 1961; Bhonsle and McNarry, 1964; Daene and Voigt, 1964; Chin *et al.*, 1971) have attempted to measure linear polarization in type III bursts by using narrow bandwidths to eliminate depolarization by Faraday rotation between the source and the observer. Their results remain controversial because no account was taken of the linear polarization caused by ground reflections.

Dodge (1972, 1973) has attempted to correct for ground reflections in his observations at 34 MHz. He measured a high degree of linear polarization in bursts but no Faraday rotation across the 3 kHz bandwidth. If the result is not caused by ground reflection, then it most likely implies that the linear polarization is produced by mode-coupling of the radiation in regions of the corona well above the source, where the Faraday rotation is low and the effects of scattering are small.

To avoid the effects of ground reflection Grognard and McLean (1973) used a correlation technique to measure the Faraday group delay between the coherent parts of the two circular components of the radiation at 80 MHz. They found the delay was <10 μs. One would expect a much higher delay from a polarized source in the corona because of differential Faraday rotation near the source.

However, Fokker (1971) has pointed out that any coherence between the two circular components of a linearly polarized signal should be washed out by scattering effects. The analysis of Steinberg *et al.* (1971) shows that scattered rays from a point source in the corona reach the observer with a considerable dispersion of differential time delays. The scattered rays suffer different Faraday rotations and hence even if the coherence between the two circular components were high at the source it would be considerably reduced by the time it reached the observer. Hence scattering can explain Grognard and McLean's (1973) negative result.

10. Decay Rate and Coronal Temperature

Many authors have derived coronal electron temperatures from the observed decay rates of the type III bursts. At a single frequency the intensity rises rapidly to a maximum then decays exponentially, as shown in Figure 15 (Aubier and Boischot, 1972). If it is assumed that the decay is due to the damping of plasma oscillations by electron-ion collisions then

$$\tau = T^{3/2}/CN,$$

where τ is the observed exponential decay constant (s), T is the electron temperature (K) and N is the electron density (cm^{-3}); C is a constant usually taken to be 40. For emission at the fundamental plasma frequency this equation can be used to express the electron temperature in terms of the decay constant and the observing frequency, f (Hz):

$$T = 0.65\ 10^{-4} f^{4/3} \tau^{2/3}\ (\mathrm{K}).$$

For harmonic emission the above equation over-estimates T by a factor of 1.5, because the observed frequency is twice the plasma frequency and the observed decay constant is half that of the plasma oscillations (assuming that the harmonic results from the scattering of two collisionally damped plasma waves).

Ground-based observations ($f > 7$ MHz) of type III bursts give values of T ranging from 1 to 5×10^6 K, if the emission is fundamental (Figure 16) and from 0.6 to 3×10^6 K if it is harmonic. These values are in reasonable agreement with temperatures derived from coronal emission lines (Billings, 1966). At frequencies <7 MHz or so the derived electron temperatures are too low and some other damping mechanisms such as Landau damping by the stream (Zaitsev *et al.*, 1972) or the surrounding plasma (Harvey and Aubier, 1973) may apply.

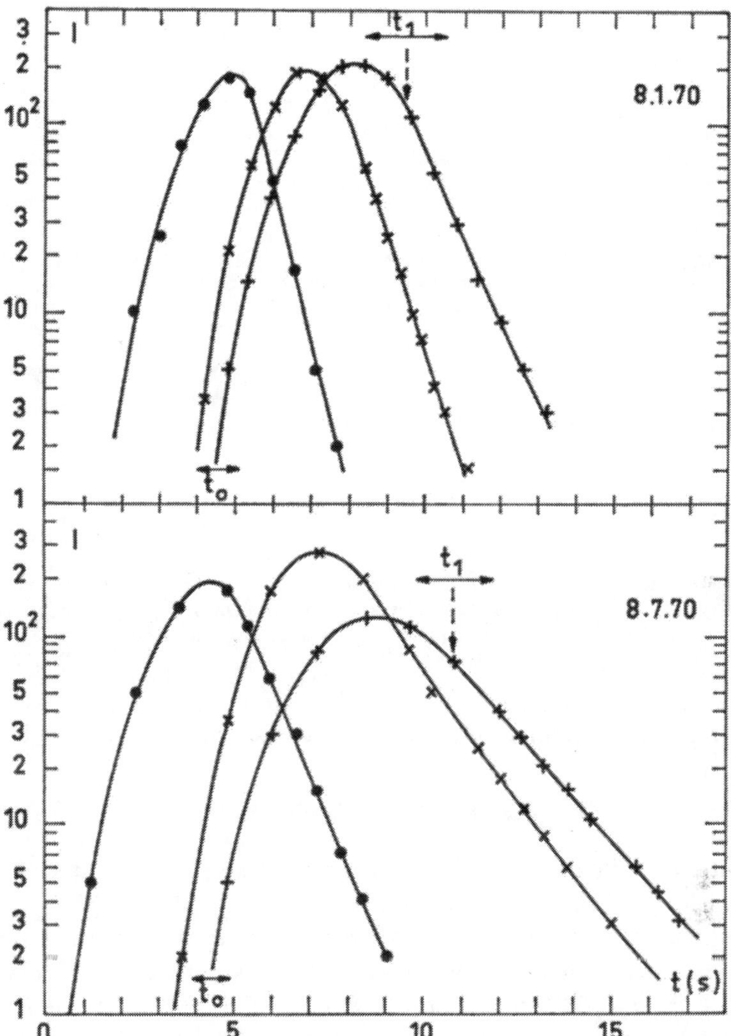

Fig. 15. Examples of type III time profiles plotted on a log I – linear t scale. The intensities I are in arbitrary units. The symbols indicate observing frequencies as follows: ● 60 MHz; × 37 MHz; + 29 MHz (Aubier and Boischot, 1972).

The work of Steinberg *et al.* (1971) and Riddle (1972a) suggests that scattering effect also can contribute to the observed decay rate. At 169 MHz Steinberg *et al.* calculated a decay rate $\leqslant 0.2$ s for a pulse of radiation emitted from a point source and scattered by small-scale irregularities. If interpreted in terms of collisional damping alone this corresponds to an electron temperature $\leqslant 2 \times 10^6$ K. Hence published estimates of electron temperature based on the measured decay rates of type III bursts, ignoring scattering, can at best be taken as upper limits. Other effects, such as stream dispersion, may contribute to the initial decay rate of the type III burst (Aubier and

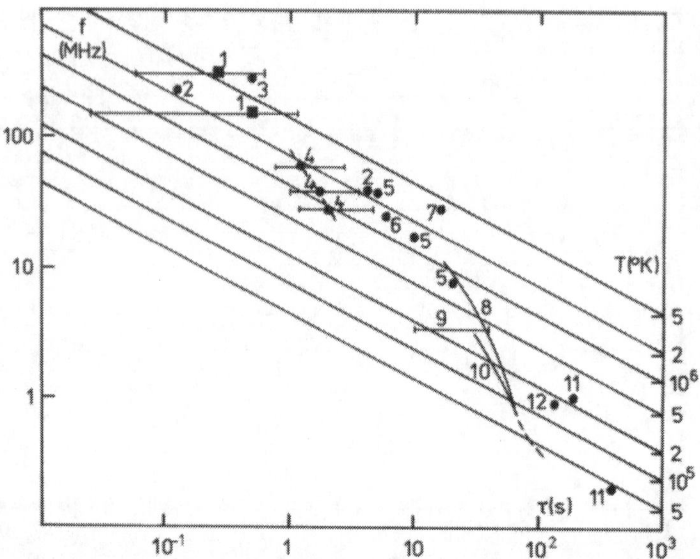

Fig. 16. Observed decay time constants, τ, vs frequency. The straight lines correspond to constant temperatures computed from the equation relating decay constant to electron temperature (see text of Section 10). For references see Figure 12 of Aubier and Boischot (1972).

Boischot, 1972). Finally, it is noted that the assumption that plasma waves are damped by electron-ion collisions and not by some other mechanism is till open to question (Smith, 1974a, b).

11. High-Resolution Spectra

High-resolution studies of type III spectra (Elgarøy, 1961; Ellis and McCulloch, 1966, 1967; Warwick and Dulk, 1969; and de la Noë and Boischot, 1972) have shown that whilst most type III bursts have a similar appearance on both high-resolution and low-resolution spectrograms, some bursts do not. These unusual bursts often consist of narrow bands ($\leqslant 100$ kHz) with low drift rates. Ellis (1969) and de la Noë and Boischot (1972) have found the bands to be nearly 100% circularly polarized. Chains of bands with a leading edge drift rate similar to normal type III bursts are called type IIIb (Figure 17). These bursts are thought to be precursors of normal type III bursts (de la Noë and Boischot, 1972) but from examples studied by the author in low-resolution spectra it appears they may be the fundamental components of harmonic type III bursts.

Other high-resolution studies (De Groot, 1970; Philip, 1969) have found an apparent association between the occurrence of a type III burst at lower frequencies and pulsating quasi-periodic structures at higher frequencies. De Groot (1970) refers to associations with type I bursts, (bright spots) and broad-band structures in (type IV) continuum bursts. Philip (1969) refers to fast-drift spike bursts which may be associated with type III bursts occurring at lower frequencies. The spike bursts have very short durations, ~ 0.1 s (at frequencies ~ 200 MHz), relatively broad bandwidths,

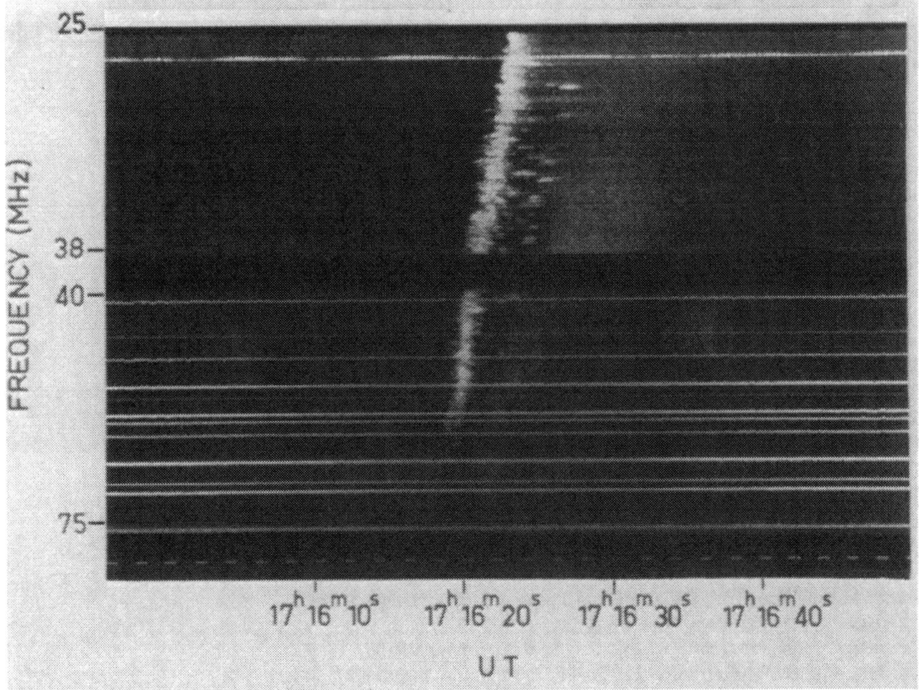

Fig. 17. Type IIIb precursor of a type III burst on 1970 August 8 observed over the 25 to 75 MHz frequency range with a high-sensitivity spectrograph (de la Noë and Boischot, 1972).

$\leqslant$30 MHz, and are almost 100% circularly polarized (Markeev and Chernov, 1971). We note that Fokker *et al.* (1974) suggest that spike bursts may be simply fine structure in a broadband continuum burst.

Another interesting relationship is that between drift pair bursts and decametre type III bursts (Roberts, 1958; de la Noë and Pederson, 1971). However, so far none of the features described above has been observed with equipment with high spatial resolution.* Until this is done interpretation based on spectral associations alone must be treated with caution.

12. Conclusion

Recent ground-based observations of the properties of type III bursts, particularly high-spatial resolution observations, are compatible with the earlier models of type III propagation along open field lines. The establishment of scattering as an important process in determining the position and angular size of the observed source has cast doubts on the validity of previous estimates of coronal densities and temperature.

* Recently Caroubalos *et al.* (1973, 1974), and Mercier and Rosenberg (1973) have combined high time resolution interferometer and spectral data in order to study source characteristics of 'paired' type III bursts and have found that the two members of the pair have sources with identical properties.

The concept of type III propagation along the axis of a coronal streamer is still very appealing but as yet has little supporting evidence.

References

Akabane, K. and Cohen, M. H.: 1961, *Astrophys. J.* **133**, 258.
Alvarez, H. and Haddock, F. T.: 1973, *Solar Phys.* **29**, 197.
Anderson, K. A. and Lin, R. P.: 1966, *Phys. Rev. Letters* **16**, 1121.
Aubier, M. and Boischot, A.: 1972, *Astron. Astrophys.* **19**, 343.
Axisa, F., Martres, M. J., Pick, M., and Soru-Escaut, I.: 1973, *Solar Phys.* **29**, 163.
Bhonsle, R. V. and McNarry, L. R.: 1964, *Astrophys. J.* **139**, 1312.
Billings, D. E.: 1966, *A Guide to the Solar Corona*, Academic Press, N.Y.
Boischot, A., de la Noë, J., du Chaffant, M., and Rosolen, C.: 1971, *Compt. Rend. Acad. Sci. (France)* **272**, 166.
Bougeret, J. L., Caroubalos, C., Mercier, C., and Pick, M.: 1970, *Astron. Astrophys.* **6**, 406.
Caroubalos, C. and Steinberg, J. L.: 1974, this volume, p. 239.
Caroubalos, C., Heyvaerts, J., Pick, M., and Trottet, G.: 1974, this volume, p. 243.
Caroubalos, C., Pick, M., Rosenberg, H., and Slottje, C.: 1973, *Solar Phys.* **30**, 473.
Chernov, G. P., Chertok, I. M., Formichev, V. V., and Markeev, A. K.: 1972, *Solar Phys.* **24**, 215.
Chin, Y. C., Lusignan, B. B., and Fung, P. C. W.: 1971, *Solar Phys.* **16**, 135.
Daene, H. and Voigt, W.: 1964, *Publ. Astrophys. Obs. Potsdam* **31**, 58.
De Groot, T.: 1970, *Solar Phys.* **14**, 176.
De la Noë, J. and Boischot, A.: 1972, *Astron. Astrophys.* **20**, 55.
De la Noë, J. and Møller Pedersen, B.: 1971, *Astron. Astrophys.* **12**, 371.
De la Noë, J., Boischot, A., and Aubier, M.: 1973, in R. Ramaty and R. G. Stone (eds.), *Proc. Symposium on High Energy Phenomena on the Sun* at Maryland, 28–30 Sept. 1972, Goddard Space Flight Center Preprint X-693-73-193, p. 602.
Dodge, J. C.: 1972, Ph. D. Thesis, University of Colorado.
Dodge, J. C.: 1973, 'Polarization Spectra of Type III Radio Bursts', paper presented at the *IAU Symposium No. 57 on Coronal Disturbances*, Surfers Paradise, Australia, 7–11 Sept. 1973.
Elgarøy, Ø.: 1961, *Astrophys. Norv.* **7**, 123.
Ellis, G. R. A.: 1969, *Australian J. Phys.* **22**, 177.
Ellis, G. R. A. and McCulloch, P. M.: 1966, *Nature* **211**, 1070.
Ellis, G. R. A. and McCulloch, P. M.: 1967, *Australian J. Phys.* **20**, 583.
Fainberg, J.: 1974, this volume, p. 183.
Fainberg, J. and Stone, R. G.: 1970, *Solar Phys.* **15**, 433.
Fainberg, J., Evans, L. G., and Stone, R. G.: 1972, *Science* **178**, 743.
Fokker, A. D.: 1965, *Bull. Astron. Inst. Neth.* **18**, 111.
Fokker, A. D.: 1971, *Solar Phys.* **19**, 472.
Gleeson, L.: 1965, Ph.D. Thesis, Part 1, Monash University.
Grognard, R. J. M. and McLean, D. J.: 1973, *Solar Phys.* **29**, 149.
Haddock, F. T. and Alvarez, H.: 1973, *Solar Phys.* **29**, 183.
Harvey, C. C. and Aubier, M. G.: 1973, *Astron. Astrophys.* **22**, 1.
Kai, K.: 1970, *Solar Phys.* **11**, 456.
Kane, S. R.: 1972, *Solar Phys.* **27**, 174.
Kane, S. R.: 1974, this volume, p. 105.
Kane, S. R., Kreplin, R. W., Martres, M.-J., Pick, M., and Soru-Escaut, I.: 1974, this volume, p. 147.
Kuiper, T. B. H. and Pasachoff, J. M.: 1973, *Solar Phys.* **28**, 187.
Kundu, M. R.: 1965, *Solar Radio Astronomy*, Interscience Publishers, N.Y.
Kundu, M. R., Erickson, W. C., Jackson, P. S., and Fainberg, J.: 1970, *Solar Phys.* **14**, 394.
Labrum, N. R. and Duncan, R. A.: 1974, this volume, p. 235.
Labrum, N. R. and Stewart, R. T.: 1970, *Proc. Astron. Soc. Australia* **1**, 316.
Leblanc, Y.: 1973, *Astrophys. Letters* **14**, 41.
Leblanc, Y., Kuiper, T. B. H., and Hansen, S.: 1974, this volume, p. 227.
Lin, R. P.: 1970, *Solar Phys.* **12**, 266.

Lin, R. P.: 1974, this volume, p. 201.
Lin, R. P., Evans, L. G., and Fainberg, J.: 1973, *Astrophys. Letters* **14**, 191.
Loughead, R. E., Roberts, J. A., and McCabe, M. K.: 1957, *Australian J. Phys.* **10**, 483.
McDonald, F. B. and Van Hollebeke, M. A.: 1973, in R. Ramaty and R. G. Stone (eds.), *Proc. Symposium on High Energy Phenomena on the Sun* at Maryland, 28–30 Sept. 1972, Goddard Space Flight Center Preprint X-693-73-193, p. **404**.
McLean, D. J.: 1970, *Proc. Astron. Soc. Australia* **1**, 315.
McLean, D. J.: 1971, *Australian J. Phys.* **24**, 201.
Malville, J. M.: 1962, *Astrophys. J.* **135**, 834.
Markeev, A. K. and Chernov, G. P.: 1971, *Soviet Astron.* **14**, 835.
Martres, M. J., Pick, M., Soru-Escaut, I., and Axisa, F.: 1972, *Nature* **236**, 25.
Melrose, D. B. and Sy, W. N.: 1972, *Australian J. Phys.* **25**, 387.
Mercier, C. and Rosenberg, H.: 1974, *Solar Phys.*, in press.
Newkirk, G.: 1961, *Astrophys. J.* **133**, 983.
Philip, K. W.: 1969, *Bull. Am. Astron. Soc.* **1**, 359.
Pneuman, G. W.: 1968, *Solar Phys.* **3**, 578.
Rao, U. V. G.: 1965, *Australian J. Phys.* **18**, 283.
Riddle, A. C.: 1972a, *Proc. Astron. Soc. Australia* **2**, 98.
Riddle, A. C.: 1972b, *Proc. Astron. Soc. Australia* **2**, 148.
Roberts, J. A.: 1958, *Australian J. Phys.* **11**, 215.
Rosenberg, H.: 1973, Ph.D. Thesis, University of Utrecht.
Slottje, C.: 1974, *Astron. Astrophys.* **32**, 107.
Slysh, V. I.: 1967, *Kosmi Issled* **5**, 897; *Cosmic Res.* **5**, 759.
Smerd, S. F., Wild, J. P., and Sheridan, K. V.: 1962, *Australian J. Phys.* **15**, 180.
Smith, D. F.: 1974a, *Space Sci. Rev.* **16**, 91.
Smith, D. F.: 1974b, this volume, p. 253.
Smith, D. F. and Pneuman, G. W.: 1972, *Solar Phys.* **25**, 461.
Solar Radio Group Utrecht: 1974, *Space Sci. Rev.* **16**, 45.
Steinberg, J. L., Aubier-Giraud, M., Leblanc, Y., and Boischot, A.: 1971, *Astron. Astrophys.* **10**, 362.
Stewart, R. T.: 1965, *Australian J. Phys.* **18**, 67.
Stewart, R. T.: 1972, *Proc. Astron. Soc. Australia* **2**, 100.
Stewart, R. T. and Labrum, N. R.: 1972, *Solar Phys.* **27**, 192.
Swarup, G., Stone, P. H., and Maxwell, A.: 1960, *Astrophys. J.* **131**, 725.
Teske, R. G., Soyumer, T., and Hudson, H. S.: 1971, *Astrophys. J.* **165**, 615.
Van Allen, J. A. and Krimigis, S. M.: 1965, *J. Geophys. Res.* **70**, 5737.
Vorphal, J. A. and Zirin, H.: 1972, *Big Bear Solar Obs. Rep.* 114.
Warwick, J. W. and Dulk, G. A.: 1969, *Astrophys. J.* **158**, L123.
Weiss, A. A.: 1963, *Australian J. Phys.* **16**, 240.
Weiss, A. A. and Stewart, R. T.: 1965, *Australian J. Phys.* **18**, 143.
Weiss, A. A. and Wild, J. P.: 1964, *Australian J. Phys.* **17**, 282.
Wild, J. P.: 1950, *Australian J. Sci. Res.* **A3**, 541.
Wild, J. P.: 1970, *Proc. Astron. Soc. Australia* **1**, 365.
Wild, J. P. and Smerd, S. F.: 1972, *Ann. Rev. Astron. Astrophys.* **10**, 159.
Wild, J. P., Sheridan, K. V., and Neylan, A. A.: 1959, *Australian J. Phys.* **12**, 369.
Wild, J. P., Smerd, S. F., and Weiss, A. A.: 1963, *Ann. Rev. Astron. Astrophys.* **1**, 291.
Zaitsev, V. V., Mityakov, N. A., and Rapoport, V. O.: 1972, *Solar Phys.* **24**, 444.
Zheleznyakov, V. V. and Zaitsev, V. V.: 1970, *Soviet Astron. AJ* **14**, 250.

SOLAR RADIO BURSTS AT LOW FREQUENCIES

J. FAINBERG

Goddard Space Flight Center, Greenbelt, Md. 20771, U.S.A

Abstract. Properties of solar radio bursts observed by spacecraft at frequencies below several MHz are reviewed. In this frequency range most of the observed bursts are type III events (associated with particles) but several cases of type II emission (associated with shocks) have been reported. The analyses which lead to emission levels of type III solar bursts out to beyond 1 AU from the Sun also indicate that the low frequency radiation is observed at the harmonic of the emission region plasma frequency. Simultaneous particle and radio measurements imply that the bursts are generated near the leading edge of impulsive streams of solar electrons with energies extending from several hundred keV to several keV. Recent experiments measuring the direction of arrival of the radio emission allow the exciter particles to be tracked along the interplanetary magnetic field from regions near the Sun out to 1 AU.

1. Introduction

Satellite measurements of traveling solar radio bursts extend the range of observations of the origin of these phenomena from 5 $R_{\odot}$ on out to past 1 AU. In these regions the coronal properties change from the rigid rotation conditions near the solar surface and take on the flow characteristics of the solar wind.

Type III solar radio bursts generated by energetic particles moving outward along open magnetic field lines provide a very important and unique source of information about physical conditions along the path of the exciter particles. In many cases these radio measurements provide an important link between observations of phenomena occurring near the solar surface with extensive *in situ* observations of solar wind, energetic particles, and magnetic fields at 1 AU.

In this paper I will review some of the general properties of type III bursts in the two-decade frequency range extending from several MHz down to 30 kHz. A principal problem in this observational range has been the determination of an average emission level scale in order to relate the physical information derived from the radio bursts to specific coronal distances. I will discuss the radio studies relating to this and will then illustrate how, with such an emission level scale and with measurements of radio arrival directions, it has been possible to track the motion of the exciter particles along the interplanetary magnetic field out to 1 AU.

2. Properties of Type III Bursts

Type III bursts generally occur in groups of overlapping bursts. At low frequencies, where the radiation lasts for some tens of minutes, single bursts are not common except during quiet times. Although isolated single bursts occur rarely, it is generally assumed that complex events can be described as combinations of individual bursts. However, it may be that in such complex events other phenomena such as type II or type IV radiation are involved.

Gordon Newkirk, Jr. (ed.), Coronal Disturbances, 183–200. *All Rights Reserved.*

Figure 1 shows a simple type III burst observed on the RAE-1 spacecraft (Weber *et al.*, 1971). The lower figure shows intensity-time profiles at 6 frequencies. The upper part shows the same burst observed by a multichannel receiver displayed in a frequency-time contour plot of 1 dB intensity intervals.

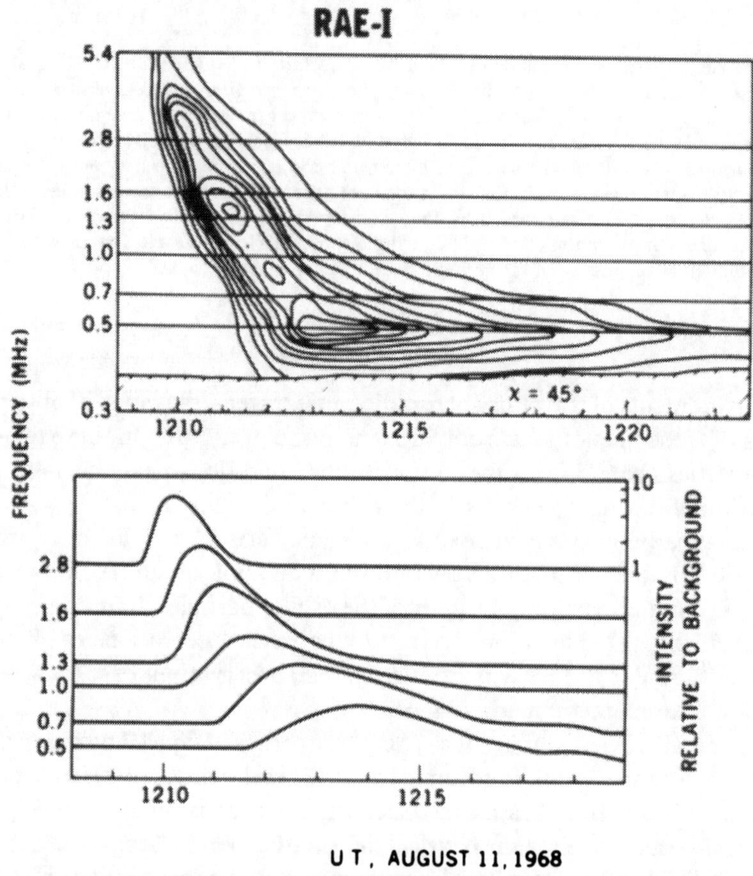

Fig. 1. Dynamic spectrum of a type III burst at hectometer wavelengths observed with the RAE-1 satellite. The upper figure shows a computer-developed dynamic spectrum from 5.4 to 0.3 MHz with 1-dB intensity contours. The lower figure shows the intensity profiles at six frequencies corresponding to the horizontal lines through the upper figure.

The properties of such bursts can be described in terms of certain measurable burst parameters:

(a) Onset time – the time at which the burst rises to a detectable amount above background. This time indicates the arrival of the fastest particles in the exciter at the appropriate coronal level.

(b) Peak time – the time of burst maximum. This time indicates the arrival of the more numerous average-energy particles in the exciter at the corresponding coronal level.

(c) Decay time – Single bursts in the low-frequency region always have an exponential decay. For large bursts, this phase may extend for 5 decades of intensity and indicates a damping process with a rate of energy loss proportional to the amount of energy remaining. The decay time is usually defined as the *e*-folding time during the exponential decay. This quantity has been interpreted as the decay of plasma waves due to electron-ion collisions. There are several problems with this. Temperatures derived from observed decay times at very low frequencies are an order of magnitude lower than expected. Alvarez and Haddock (1973) have suggested that an additional electron-ion collisional energy exchange introduced by Wolff *et al.* (1971) to explain electron-proton temperatures at 1 AU will also account for the radio decay rates. Zaitsev *et al.* (1972) have suggested that Landau damping of plasma waves by the tail of the exciter will affect the decay rate and they have calculated burst profiles from a one-dimensional analysis. Although the predicted burst shapes seem to agree well with observations of small bursts this is not the case for large bursts where the theoretical profiles have decays much more rapid than the observed exponental decays. In addition, Harvey and Aubier (1973) have recently shown that in the solar wind where the plasma waves are convected into regions of decreasing density, the waves should experience a sudden abrupt attenuation due to Landau damping in the medium, but this is also not observed. It may be that some burst plasma waves can escape this fate by propagating towards the Sun to compensate for the outward solar wind convection. An additional interesting result which may not have physical significance is that the number of oscillation cycles within the decay time remains remarkably constant over the full range of type III observations.

(d) Excitation time – the time from the burst onset to when the decay first becomes uniformly exponential. The excitation time is taken (Aubier and Boischot, 1972) to be the effective duration of the burst exciter at a particular plasma level. The increase in excitation time at lower frequencies can be interpreted in terms of the growth in exciter length due to the dispersion of velocities in the exciter.

(e) Drift rate (onset or peak) – frequency drift rate between closely-spaced frequencies. This important quantity depends on the separation between the two emission levels, the velocity of the exciter, and the path difference to the observer. Drift times have been used often to derive information about emission levels.

(f) Cutoff frequencies. The dynamic spectra of most bursts have a high- and low-frequency limit. In satellite observations there is a tendency for the upper frequencies to be cut off for western events and for the low frequencies to be cut off for eastern events.

Other burst parameters which can be determined are burst intensity, burst size and burst arrival direction.

3. Storms of Type III Bursts

Although large type III events are well associated with flares, the great proportion of observable type III's are associated with storm periods (Fainberg and Stone, 1970a) lasting for intervals of several days to over one solar rotation period. Figure 2 shows one such storm period in August 1968 about 4 days before central meridian passage (CMP) of the active region. Note the steady succession of drifting events and their drift rates. In Figure 3 the same storm is shown near CMP. Note the more rapid drift rates in this case. This is due to the fact that the low frequencies have a shorter

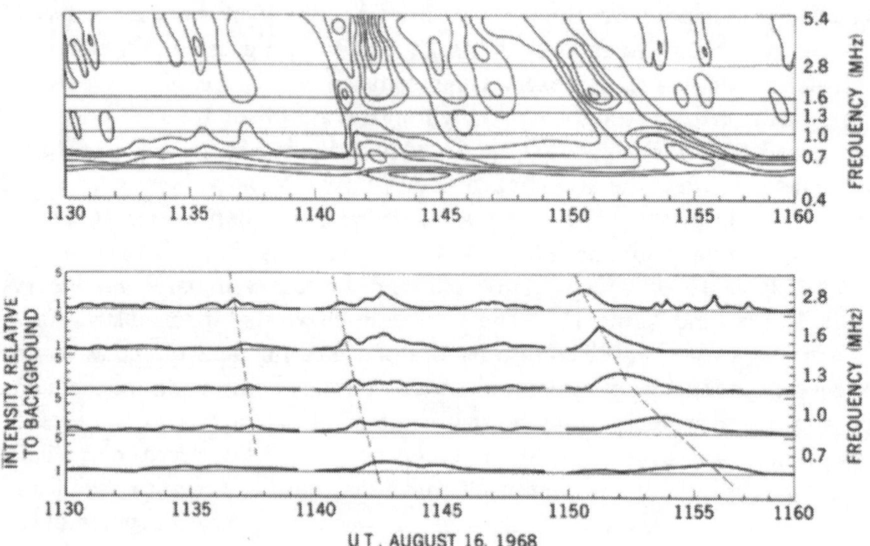

Fig. 2. A 30-min segment of data of a solar storm of type III bursts 4 days prior to CMP. The drift in frequency, limited burst bandwidth, and large numbers of small bursts can be seen.

radiation travel path than the higher frequencies – a time of flight effect discussed by Wild *et al.* (1959). The increased activity is probably due to refractive focusing of electromagnetic waves in the radial direction. The occurrence rate of one burst per 10 s suggests that several hundred thousand storm bursts may be released during one solar rotation.

In addition to bursts there is also a continuum component to the storms (Fainberg and Stone, 1971a). Figure 4 shows this for the August 1968 storm. The points plotted are the minimum background values for each 10-min interval during this period and represent the levels between bursts. These storm periods are also well associated with decametric storms and with metric type I storms (Sakurai, 1971; Stewart and Labrum, 1972).

I would like to close my discussion of storms by showing the variation of drift rates during the storm in Figure 5. Here, for each day of a 14-day period, the number distribution of observed drift times as a function of drift time is plotted. It is clear that a systematic variation of drift times occurs which depends on the orientation of the active region with respect to central meridian. By utilizing a three-parameter drift-time function depending on exciter speed, level separation, and CMP time, a least-

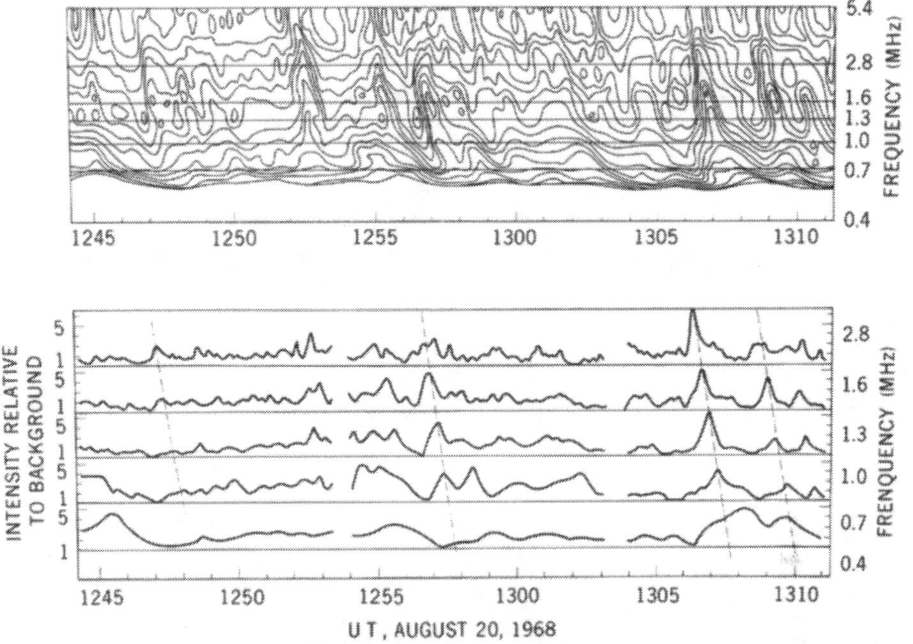

Fig. 3. A 30-min segment of data from the same storm as Figure 2, but near CMP. The higher occurrence rate and faster drift rate of the bursts are evident.

squares fitting procedure was followed to derive best fits of these quantities simultaneously from 2500 bursts. The scale so derived is presented in Figure 6. In this figure the bottom line represents the best estimate of coronal densities near solar minimum made by Newkirk (1967). The region between 10 $R_\odot$ and 215 $R_\odot$ is essentially an interpolation between values determined by light scattering and values from *in situ* measurements.

The radio data are plotted assuming that the observed radio frequency is at the plasma frequency. Reasons for believing that this assumption is questionable will be presented later. Slysh's (1967) point measured at 1 MHz results from a spin modulation plus lunar occultation measurement on several bursts observed by Luna 11 and 12. This technique determined the burst arrival direction from the spin modulation minima. The dashed line results from an extensive drift analysis of bursts observed by Hartz (1969) on the Alouette satellites. In this analysis, Hartz assumed an exciter speed of 0.35c and derived a set of level separations from the observed burst drift times. He also found his scale to be consistent with a pressure balance argument applicable to streamers using temperatures derived from burst decay rates. A similar analysis was made with other bursts observed on the ATS satellite by Alexander *et al.* (1969). The RAE points result from a drift-rate analysis of Fainberg and Stone (1970b, 1971b) in which exciter speed was also derived.

All of these analyses of type III emission levels, when converted to densities, indicate that the emission regions are enhanced in density by a factor of about 10. Close

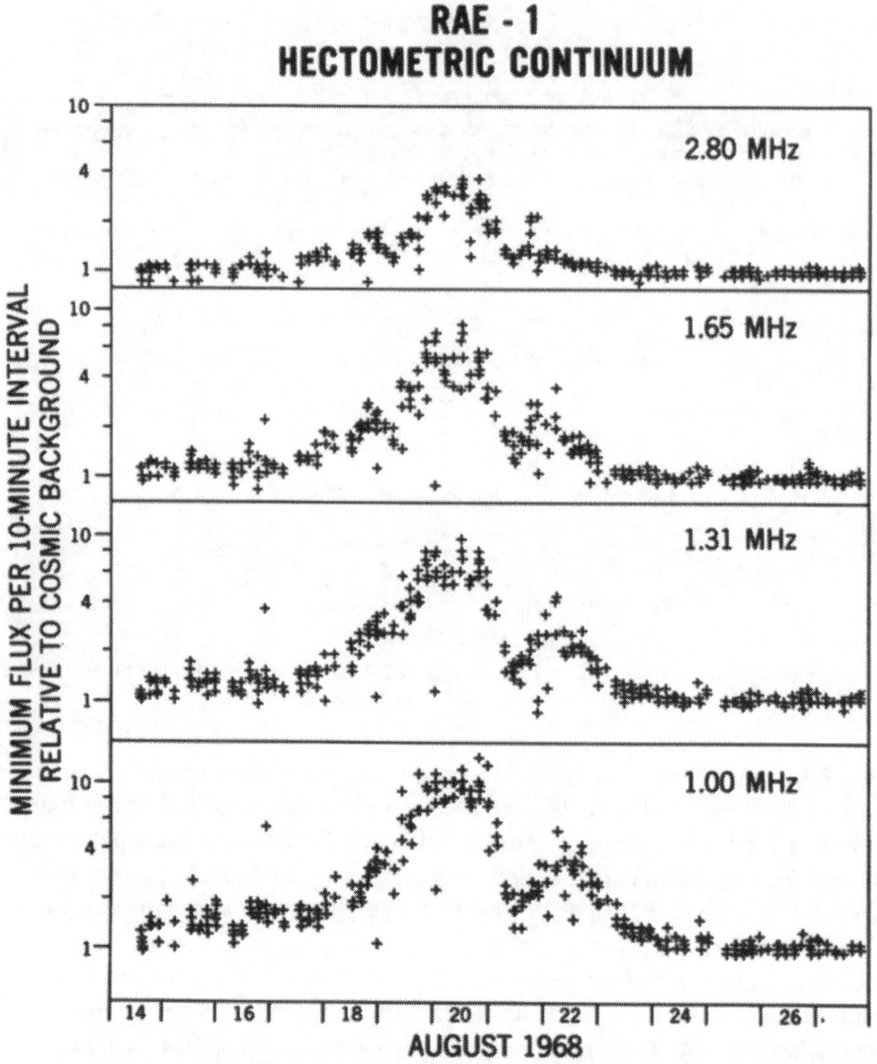

Fig. 4. The minimum flux values for 10-min periods of the August, 1968 storm as a function of heliographic longitude (day number) of the associated active region. The temporary decrease in storm activity on August 21 was also observed at metric frequencies. The minimum flux at the higher frequencies is essentially the level between bursts and represents the continuum component.

to the Sun, such enhancements are fairly common and could explain why type III emission can propagate to the Earth from limb events. However, optical measurements of density at increasing distances from the Sun do not support the assumption that enhancements occur often enough to account for the frequent occurrence of type IIIs.

With the launching of the IMP-6 satellite in 1971 the first results measured over a range of closely-spaced frequencies became available from a quiet spacecraft located

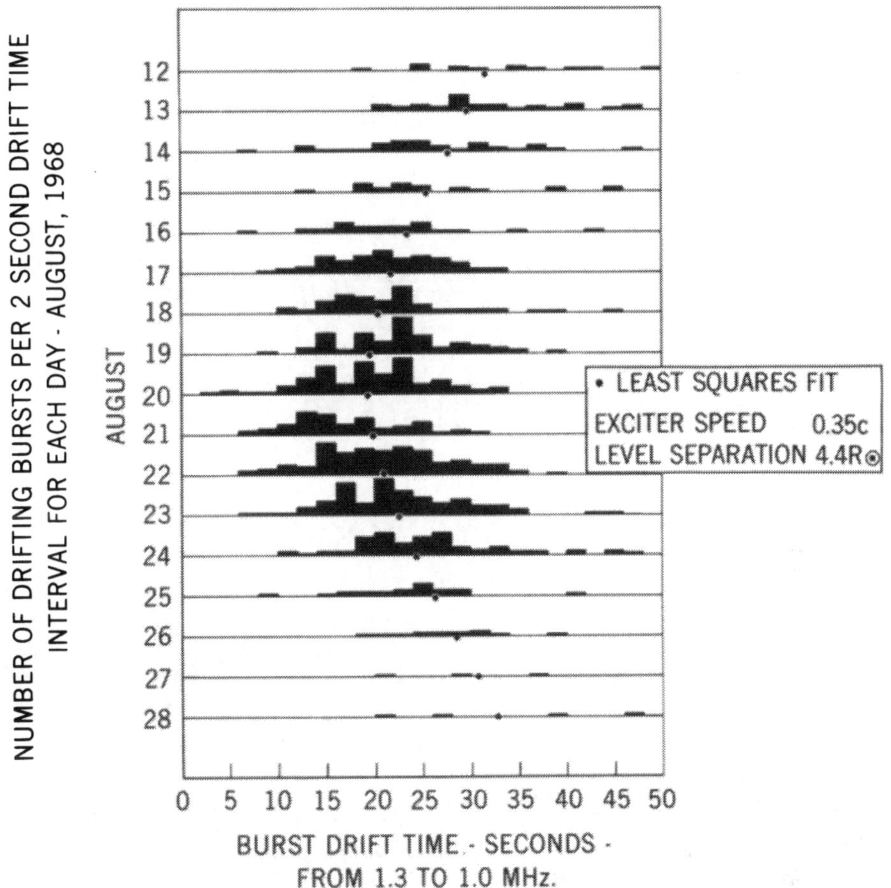

Fig. 5. The dependence of burst drift rate between 1.3 and 1.0 MHz on heliographic longitude for the August, 1968 storm. The least-squares fit, indicated by black circles in the figure, gives a level separation of 4.4 $R_{\odot}$ and an average exciter speed of 0.35 c.

well out in the solar wind. Figure 7 (Fainberg *et al.*, 1972) illustrates a typical single event going down to low frequencies. Here we have plotted the intensity-time profiles of the burst envelopes observed at a number of frequencies by the GSFC experiment on IMP-6. Note that the vertical scale is logarithmic; this burst is over 40 dB above background at mid-frequencies. This large western event had a high-frequency cutoff near 1 MHz and was not observed by ground equipment at Clark Lake Radio Observatory in the decameter range.

The insert in this figure illustrates the observed modulation pattern caused by the rotation of the dipole antenna in the spin plane (which is the same as the ecliptic plane). The minima of this pattern occur when the dipole axis passes closest to the source direction; the maxima occur 90 deg later when the antenna is broadside to the source direction. From an accurate knowledge of antenna orientation from optical sensors, we can determine the direction of arrival of the radiation with respect to the Sun–Earth

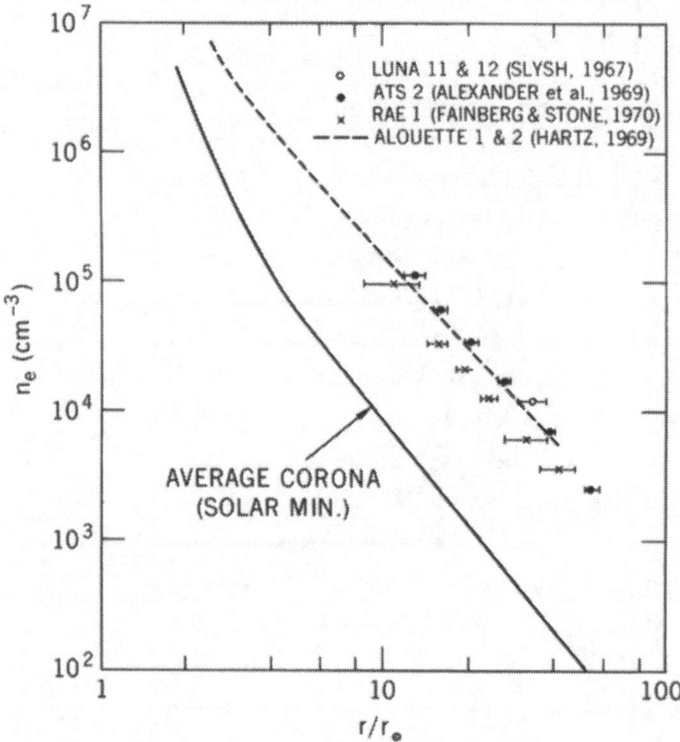

Fig. 6. Electron density vs distance from the Sun in the outer corona. The solid curve represents the values given by Newkirk (1967) as best estimates during solar minimum. The values given by the radio measurements are from single bursts and are placed assuming the type III radiation was observed at the fundamental of the local plasma frequency. They should be displaced downward a factor of 4 if the radiation was observed at twice the plasma frequency.

line. We have developed a very powerful data-processing technique which utilizes the entire waveform to derive the radio arrival direction. An error analysis indicates that with several minutes of data we can determine burst arrival directions to within 1 deg for small sources. It is clear that from a study of limb events the radio limb of the Sun can be determined at a wide range of frequencies. To interpret these results fully, it will be necessary to include effects of scattering and refraction discussed by Steinberg (1972).

In addition to burst arrival direction, we can also measure the depth of modulation accurately. This quantity, M, which we call the modulation index we define as

$$M = \frac{I_{max} - I_{min}}{I_{max} + I_{min}},$$

where I_{max} and I_{min} are the maximum and minimum intensities observed during one spin period. M has values between 0 and 1. A fully modulated burst with modulation index of 1 occurs only when a point source is located in the spin plane so that the antenna nulls pass through the source. A modulation index less than one results from a

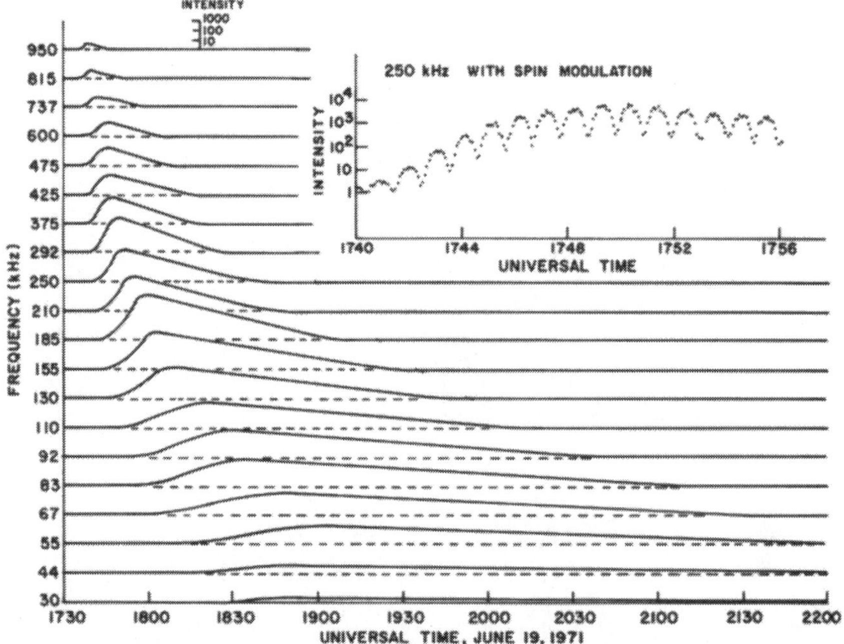

Fig. 7. A type III burst observed between 1 MHz and 30 kHz by the IMP-6 radio experiment. The insert figure illustrates the observed spin modulation at a frequency of 250 kHz, while for the main figure, only the burst envelopes are shown for clarity.

distributed source centered in the spin plane or from a smaller source located out of the plane. Within this range of ambiguity, which we have explored, we can determine some information on source size.

Figure 8 shows the results of this data-processing technique applied to the burst of Figure 7. The positions of the Earth and Sun are shown in the ecliptic plane. From spin modulation we determined the source direction to be along the lines labeled by the frequency of observation. The open circles on this figure are the distances of closest approach to the Sun which form a lower bound on emission levels for the burst. An extrapolation of the emission level scale derived in RAE storm analysis yields intersections of the emission lines with the emission surfaces represented by the solid circles. In examining many limb events, we have found that this scale is a reasonable one to use. It is understood that further refinements in such a scale will occur when the effects of scattering and refraction (Steinberg, 1972) are taken into account.

For this burst we have found that the modulation index becomes essentially zero at 55 kHz, implying that the radiation was nondirective and arrived isotropically to the antennas, so that this emission originated close to 1 AU. Note the spiral configuration of the burst trajectory resulting from the interplanetary magnetic field.

Figure 9 shows these results plotted on a density scale. The lower line is an interpolation between light-scattering results at 10 $R_\odot$ and long term *in situ* measurements at 1 AU. The upper line is the RAE scale. The open circles are the minimum emission

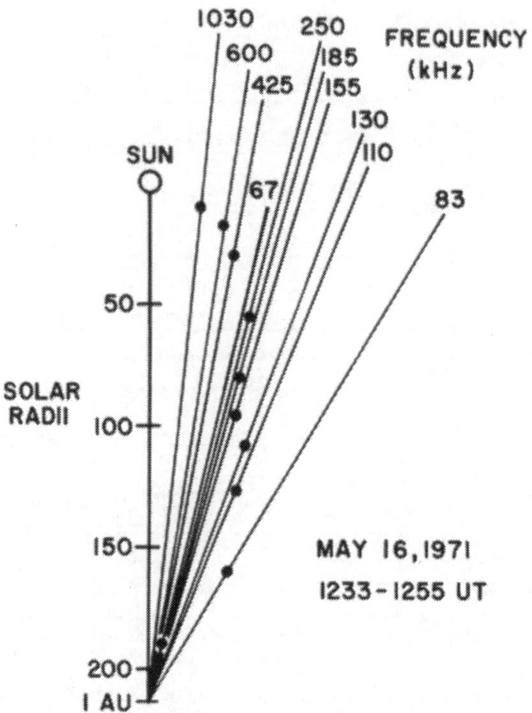

Fig. 8. Directions of arrival (in the ecliptic plane) determined from the spin modulation for the burst in Figure 7. The trajectory shown by the black dots is fixed by using the RAE-1 emission level scale (see text). The trajectory shown by the open circles is determined by the distance of closest approach permitted by the arrival direction.

levels permitted by the distance of closest approach. Again, we see a density enhancement by an order of magnitude, even out to 1 AU. Direct measurements of such enhancements are very rare and this strongly motivates one to seek other interpretations. The most obvious one is that the low-frequency bursts are observed at the harmonic of the plasma frequency, which then drops the radio density levels on this diagram by a factor of 4. In many thousands of type III bursts observed by RAE and IMP-6 with good frequency resolution from 5 MHz down, we have never seen a convincing fundamental-harmonic pair of type III bursts and this suggests that we are looking primarily at one class or the other. The densities derived from the emission levels favor the harmonic interpretation.

Recently Haddock and Alvarez (1973) have interpreted many low-frequency complex type III events observed on OGO-5 as having fundamental and harmonic components. They find there is a transition frequency below several MHz where the observed burst changes from predominantly fundamental radiation at higher frequencies to predominantly harmonic radiation at lower frequencies. However, multicomponent complex events can be deconvolved in different ways; in the absence of published profiles it is not clear how uniquely the data support this interesting conclusion.

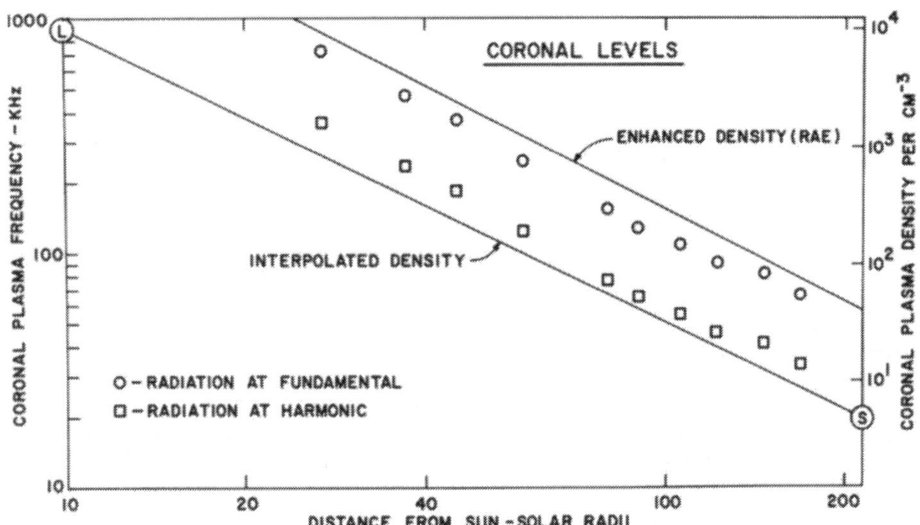

Fig. 9. Interplanetary density and plasma frequency scale as a function of distance in solar radii. The interpolated density scale is given by a line between the value determined by light-scattering observations, labeled L, and that from space probe measurements, labeled S. The circles are the density scale corresponding to the closest approach trajectory shown in Figure 8, assuming radiation occurs at the plasma frequency. These points map into the squares if radiation occurs at the second harmonic of the plasma frequency.

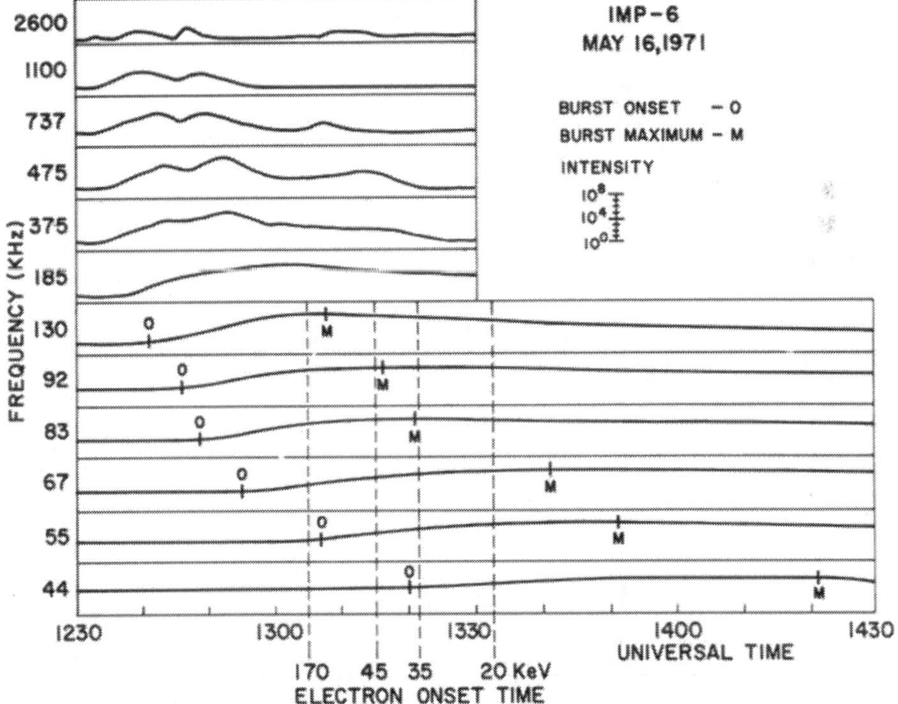

Fig. 10. Intensity-time profiles (spin modulation removed) of a type III burst observed on 16 May 1971 for frequencies from 2600 kHz to 44 kHz. The loss of spin modulation at 55 kHz places this level at 1 AU. Note the observed electron onset times.

Additional evidence for the harmonic interpretation at low frequencies comes from IMP-6 observations of type II radiation which is clearly observed at both the fundamental and harmonic frequencies (Malitson *et al.*, 1973a). An analysis of one event reported by Malitson *et al.* (1973b) strongly indicates that the position of the type II fundamental agrees with the RAE type III scale only if the III's are observed at the harmonic of the plasma frequency. In addition to these arguments there is the association of radio burst onset time with energetic electrons reported by Lin *et al.* (1973). Figure 10 shows the comparison for one event in May, 1971. For this event, the loss of modulation at 55 kHz indicates that this level was near 1 AU. The series of electron onset times measured by Lin shows that the radio burst onset agrees with the onset of 170-keV electrons. As the arrival of the more numerous slower electrons occurs, the radio burst increases in intensity with the burst maximum corresponding to the onset of approximately 10-keV electrons. These data strongly suggest that the beam energy conversion causing the radio burst takes place at the leading edge of the electron stream where velocity separation of electrons in transit causes a positive slope on the differential energy spectrum (Zaitsev *et al.*, 1972; Lin *et al.*, 1973). Figure 11 shows the trajectory of this radio burst. Again, the dots are the intersections of the measured arrival directions with emission-level surfaces determined from the RAE scale. We see that this is again a western event with the burst trajectories out-

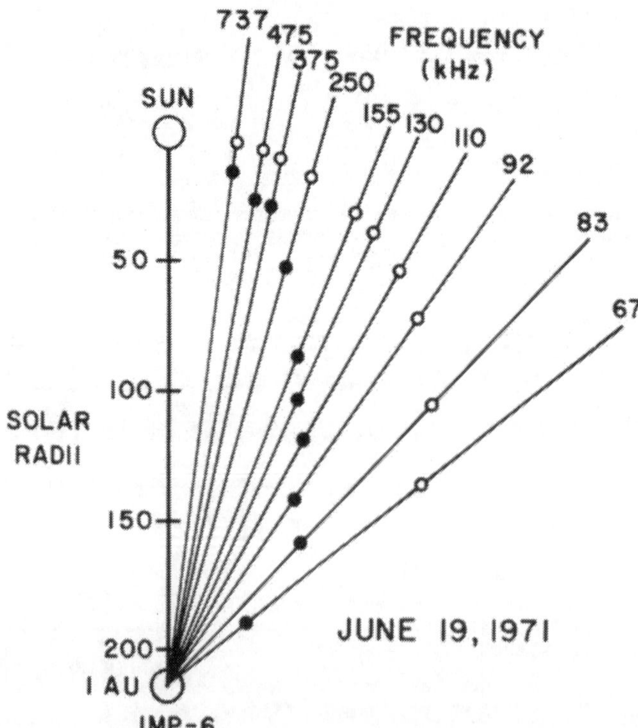

Fig. 11. Average trajectory of the type III burst for which simultaneous electron measurements were made. See Figure 10.

lining the spiral curvature of the magnetic field. This figure is a summary over the history of the burst. In order to look for dynamic changes we have developed techniques for computer processing the IMP-6 data to generate movies of solar bursts in progress.

Figure 12 shows the format of the plots. The plane of the figure is the ecliptic with

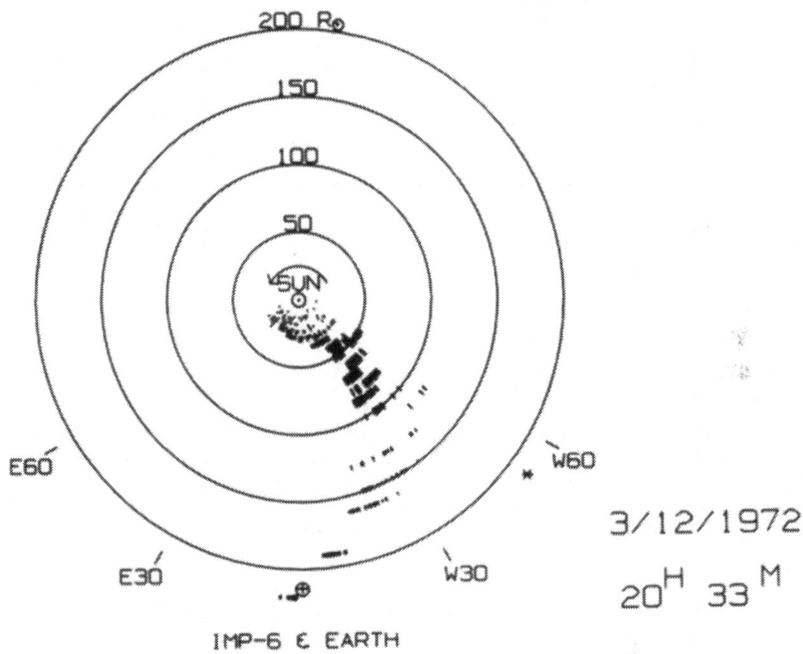

Fig. 12. Format of a computer-generated plot of data from IMP-6. The plane of the figure is the ecliptic with IMP-6 and Earth located at bottom – ⊕. For each minute of data, the intersection of the measured type III arrival directions with the spherical emission level is determined, and a radial line segment with length proportional to log intensity is plotted. This figure is a collection of data from 32 frequencies taken at 5-s intervals for a 2-min period.

west to the right and east to the left. A grid of four concentric circles separated by 50 $R_\odot$ is plotted. A fifth innermost circle is at 5 $R_\odot$, just under the arrow showing the direction of solar rotation; we note how these satellite observations complement ground-based observations by pointing out that ground-based measurements deal with phenomena occurring within the 5 $R_\odot$ circle.

For each minute of spin-modulated data we determine the arrival direction. We then calculate the intersection of this direction with the spherical emission level given by the RAE scale and plot a radial line segment whose length is proportional to the logarithm of the burst intensity above background. Each frame of the movie is a collection of these intersections for a certain period of the burst – usually for 5 min. Successive movie frames differ by 5 s so that there is a steady progression in time. In this single frame we see a solar burst in the process of moving out from the Sun.

Figure 13 shows a sequence of frames separated by 2 min taken from this movie.

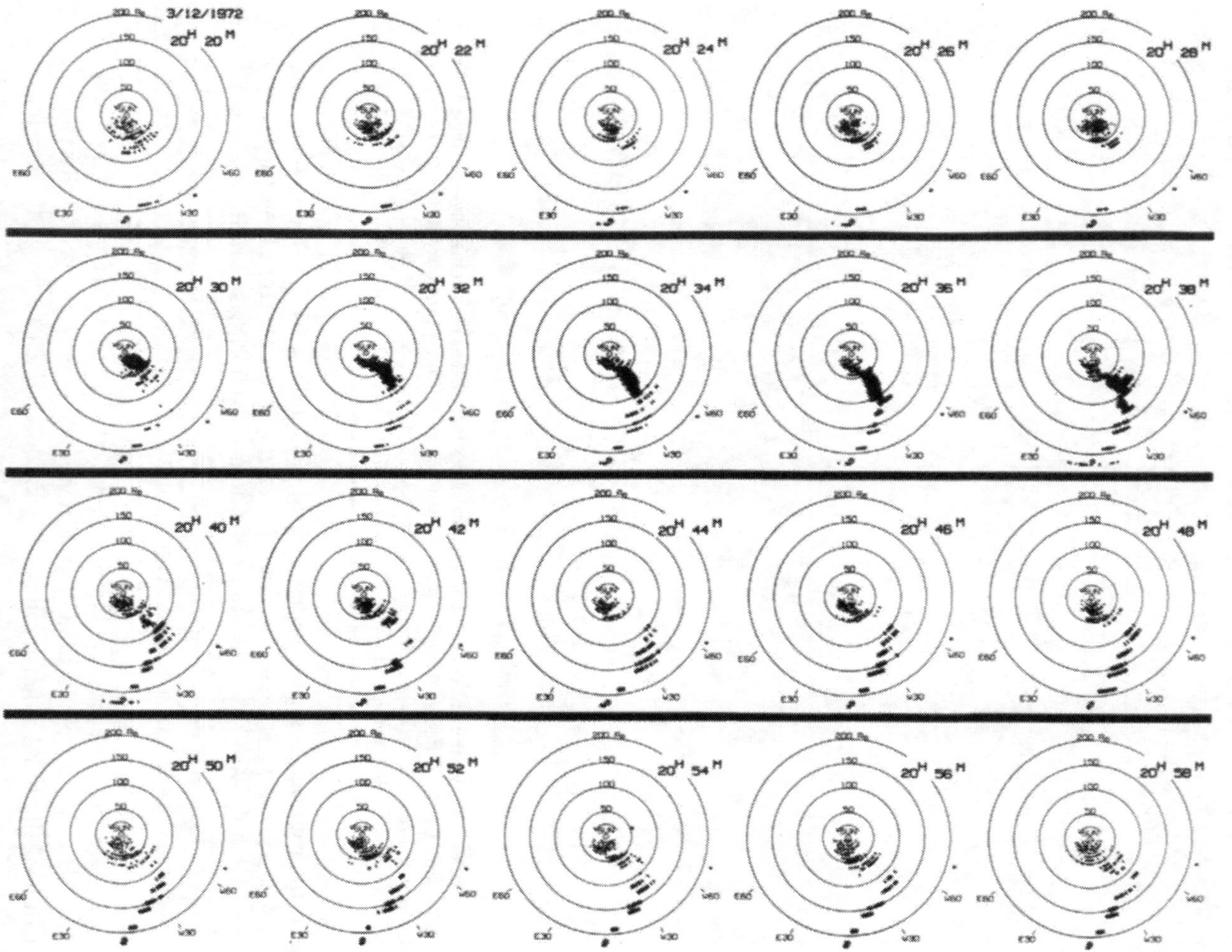

Fig. 13. Frames separated by 2-min intervals during a large type III burst. A low-level solar storm visible over the first $\frac{1}{3}$ AU is present before and after the burst.

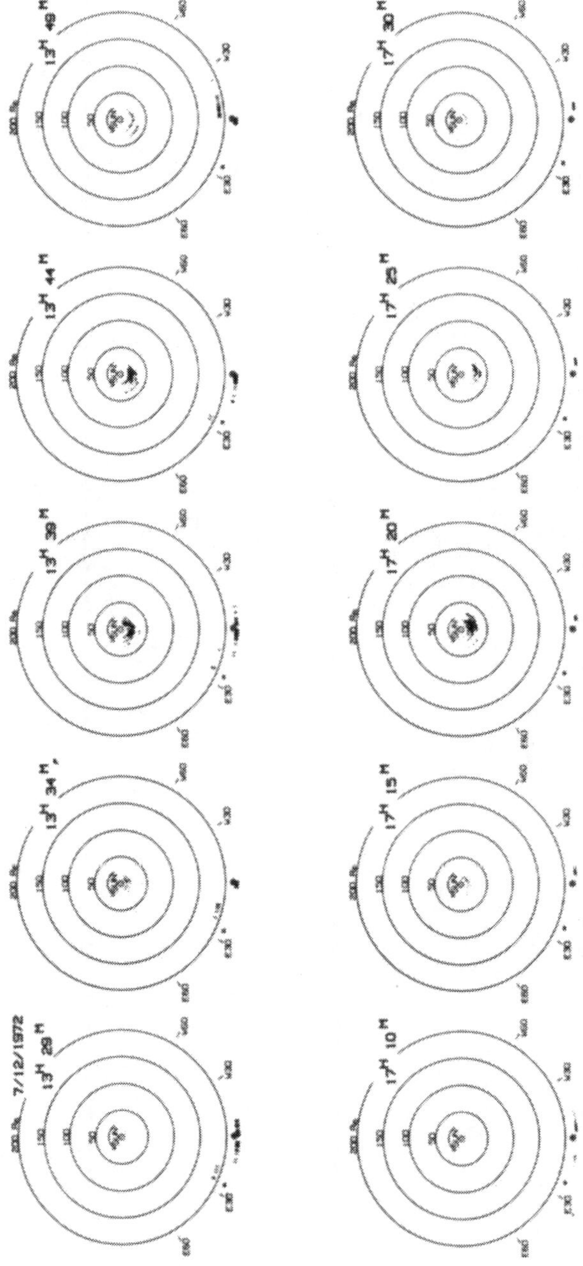

Fig. 14. Two type III bursts showing a low-frequency cutoff. The bursts are not visible beyond 40 $R_\odot$.

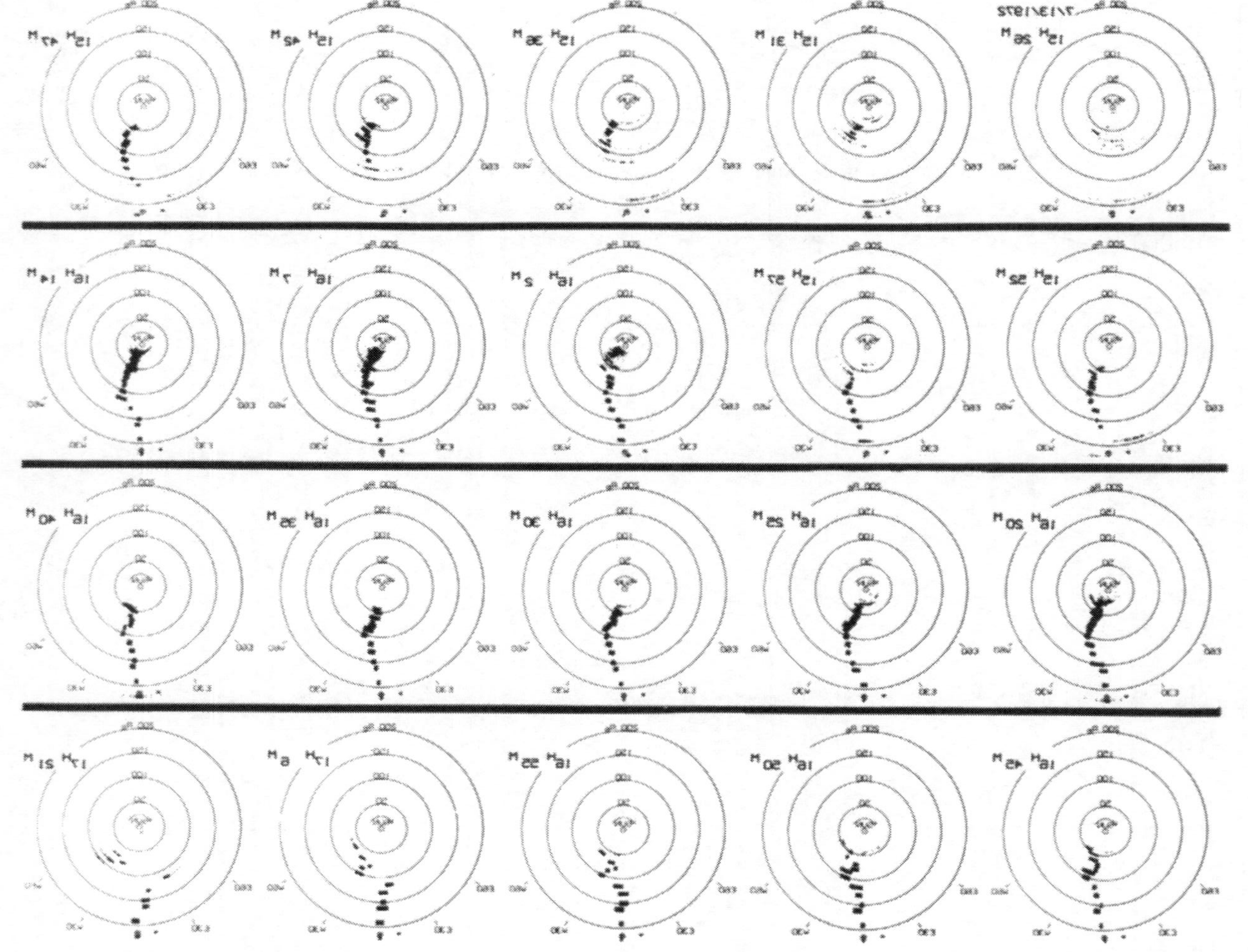

Fig. 15. Two type III bursts. The first, starting near 1531 UT, shows no high-frequency components although it is quite large. The second burst, starting near 1602 UT, does have high-frequency components.

Prior to the burst there is a long-lasting solar storm in progress, consisting of many overlapping small bursts visible out to 70 $R_{\odot}$. At 2026 UT, a type III becomes visible near the Sun and the emission region progresses outward as the exciter electrons propagate along the interplanetary magnetic field. At 2036 UT, the main part of the exciter is at 50–100 $R_{\odot}$ while the fastest particles have already reached 1 AU. The spiral structure of the interplanetary magnetic field is outlined by this burst.

Very often, type III bursts have a limited frequency range over which they are visible. Figure 14 shows a sequence of frames from 2 bursts on 12 July 1972 which are normal type III events, but without low-frequency components. Such bursts are fairly common, especially during storms. They may indicate interplanetary magnetic field conditions which cause the burst exciter packet to become very diffuse as it moves out.

Figure 15 shows an example of a type III burst without high-frequency components. This burst first becomes visible near 45 $R_{\odot}$ at 1531 UT and moves out to 1 AU. At 1602 UT, a second type III burst starts much closer to the Sun and progresses outward along a similar trajectory. It may be that the first burst is an example of electron acceleration by magnetic field reconnection at large distances from the Sun. It is clear that a wide range of interplanetary phenomena can be explored by these radio techniques.

There is very great interest in out-of-the-ecliptic configurations. The second RAE spacecraft was launched in July of this year into lunar orbit. For the first several weeks it was in a spinning mode with just the dipoles extended. We placed the axis of spin in the ecliptic plane and perpendicular to the Sun–Earth line so that we could determine arrival directions out of the ecliptic. We should get additional information from lunar occultation. The only solar burst we have examined is consistent with a trajectory out from the Sun inclined about 10° to the ecliptic. The next spinning spacecraft will be Helios A which will go in to about $\frac{1}{4}$ AU from the Sun and will be launched in 1974.

Goddard also has a joint experiment with Steinberg's group at Meudon on the IME spacecraft, set for 1978. This will involve the synthesis of a spinning tilted dipole so that the two angles defining the arrival direction can be determined; in addition, source size can be found. This will allow the tracking of solar bursts out of the ecliptic and will yield the radio determination of magnetic field configurations extending outside the ecliptic plane.

The ideal situation would be to have two spinning spacecraft simultaneously measuring the same burst from an extended baseline. This would permit triangulation to determine source location and would allow a direct determination to be made of emission levels as well as source directivity. With fewer space launches being planned, it is not clear when such an opportunity might occur.

References

Alexander, J. K., Malitson, H. H., and Stone, R. G.: 1969, *Solar Phys.* **8**, 388.

Alvarez, H. and Haddock, F. T.: 1973, in R. Ramaty and R. G. Stone (eds.), *High Energy Phenomena on the Sun*, NASA SP-342, Washington.

Aubier, M. and Boischot, A.: 1972, *Astron. Astrophys.* **19**, 343.
Fainberg, J. and Stone, R. G.: 1970a, *Solar Phys.* **15**, 222.
Fainberg, J. and Stone, R. G.: 1970b, *Solar Phys.* **15**, 433.
Fainberg, J. and Stone, R. G.: 1971a, *Astrophys. J.* **164**, L123.
Fainberg, J. and Stone, R. G.: 1971b, *Solar Phys.* **17**, 392.
Fainberg, J., Evans, L. G., and Stone, R. G.: 1972, *Science* **178**, 743.
Haddock, F. T. and Alvarez, H.: 1973, *Solar Phys.* **29**, 183.
Hartz, T. R.: 1969, *Planetary Space Sci.* **17**, 267.
Harvey, C. C. and Aubier, M. G.: 1973, *Astron. Astrophys.* **22**, 1.
Lin, R. P., Evans, L. G., and Fainberg, J.: 1973, *Astrophys. Letters* **14**, 191.
Malitson, H. H., Fainberg, J., and Stone, R. G.: 1973a, *Astrophys. Letters* **14**, 111.
Malitson, H. H., Fainberg, J., and Stone, R. G.: 1973b, *Astrophys. J.* **183**, L35.
Newkirk, G., Jr.: 1967, *Ann. Rev. Astron. Astrophys.* **5**, 213.
Sakurai, K.: 1971, *Solar Phys.* **16**, 125.
Slysh, V. I.: 1967, *Kosm. Issled.* **5**, 897; *Cosm. Res.* **5**, 759.
Steinberg, J. L.: 1972, *Astron. Astrophys.* **18**, 382.
Stewart, R. T. and Labrum, N. R.: 1972, *Solar Phys.* **27**, 192.
Weber, R. R., Alexander, J. K., and Stone, R. G.: 1971, *Radio Sci.* **6**, 1085.
Wild, J. P., Sheridan, K. V., and Neylan, A. A.: 1959, *Australian J. Phys.* **12**, 369.
Wolff, C. L., Brandt, J. C., and Southwick, R. G.: 1971, *Astrophys. J.* **165**, 18.
Zaitsev, V. V., Mityakov, N. A., and Rapoport, V. O.: 1972, *Solar Phys.* **24**, 444.

THE FLASH PHASE OF SOLAR FLARES: SATELLITE OBSERVATIONS OF ELECTRONS

R. P. LIN

Space Sciences Laboratory, University of California, Berkeley, Calif. 94720, U.S.A.

Abstract. Satellite observations of solar electrons bearing on flare particle acceleration and the generation of radio and X-ray emission are reviewed. The observations support a two stage acceleration process for electrons, one stage commonly occurring at the flare flash phase and accelerating electrons up to ~ 100 keV, and a second stage occurring only in large proton flares and accelerating electrons up to relativistic energies. The location of the acceleration region appears to be no lower than the lower corona.

The accelerated non-relativistic electrons generate type III radio burst emission as they escape from the Sun. Direct spacecraft observations of the type III emission generated near 1 AU and the energetic electrons, provide quantitative information on the characteristics of the electrons exciting type III emission, the production of plasma waves, and the conversion from plasma waves to electromagnetic radiation.

1. Introduction

In the past decade spacecraft penetrating beyond the Earth's magnetosphere have been able to directly observe the low energy particle emissions of the Sun. Low energy solar particles are very frequently emitted from the Sun, especially non-relativistic electrons (see Table I) which commonly originate in small importance 1 flares or subflares. Electrons up to $\sim 10^8$ eV in energy have been observed (Datlowe, 1971) from larger flares. Energetic electrons appear to contain the bulk of the flare energy

TABLE I

Solar particle events in an active year

Number of solar flares[a]	$\sim$16,000
Number of non-relativistic electron events[b]	$\sim$ 400
Number of energetic proton events[b]	$\sim$ 70

[a] Normalized for the whole Sun = number observed $\times$ 2.
[b] Normalized by cone of emission of $\sim 70°$ for non-relativistic electron and $\sim 100°$ for proton events.

in those flares where they are accelerated (Lin and Hudson, 1971; Syrotvatskii and Shmeleva, 1972) and they are responsible for most of the observed flare energetic X-ray and radio emission. The particle observations, combined with improved observations of solar electromagnetic radiation (much of which has also been provided by spacecraft) define a detailed physical picture of energetic electrons in small solar flares. Here we review the results obtained from direct spacecraft sampling of electrons from flares pertaining to the acceleration of electrons at the sun and to the generation of type III radio emission at 1 AU.

Gordon Newkirk, Jr. (ed.), Coronal Disturbances, 201–223.

2. Interplanetary Propagation

The electron observations at 1 AU are for the most part limited to the particle intensity above a given energy threshold. The intensity vs time profiles for flare associated events are generally consistent with an impulsive injection into the interplanetary medium. If the electrons are scattered a great deal in their propagation from the Sun to 1 AU (mean free path $= \lambda \ll 1$ AU) then the intensity-time profile will look diffusive, as in Figure 1. If, on the other hand, very little scattering occurs ($\lambda \gtrsim 1$ AU) then the

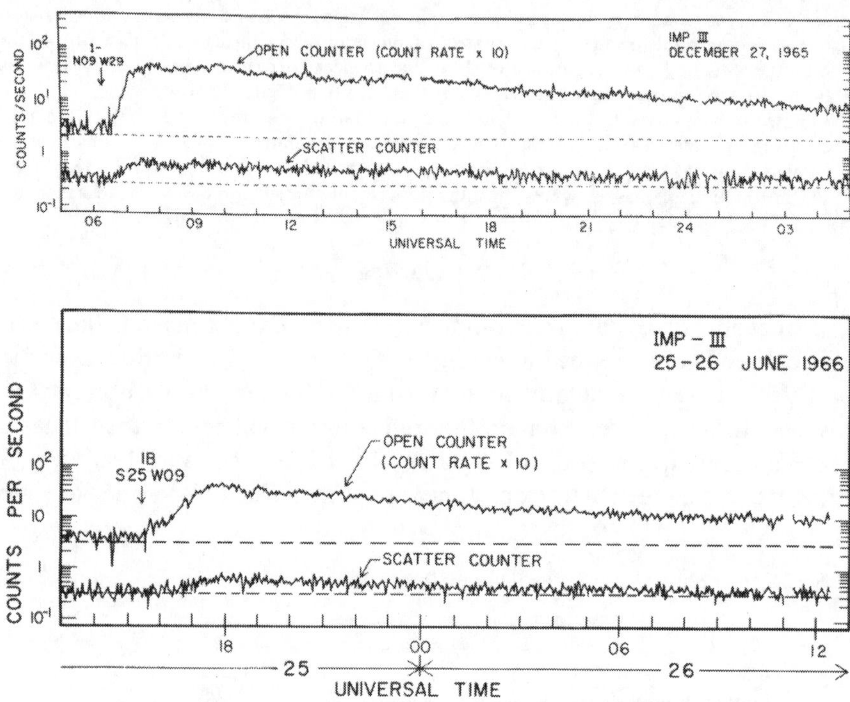

Fig. 1. Two diffusive electron events. The scatter counter is only sensitive to >45 keV electrons while the open counter counts both >40 keV electrons and >0.5 MeV protons. The 27 December 1965 event shows a rapid rise during onset which is observed for many electron events.

intensity-time profile will consist of a rapid increase and decrease as in Figure 2. Both kinds of events, diffusive and scatter-free (Lin, 1970), and events in between as well, are observed, with scatter-free events ($\lambda > 1$ AU) numbering ~20% of the total.

It might seem, then, that it would be very difficult to derive the characteristics of the injected electrons since the observations at 1 AU are greatly affected by the amount of scattering and how it varies with energy, etc. Actually, however, the *maximum flux* is remarkably insensitive to the details of the scattering (Lin, 1971), so long as there is enough scattering to give diffusive profiles. We can obtain the relationship between the maximum flux and the number of particles emitted for an event which can be

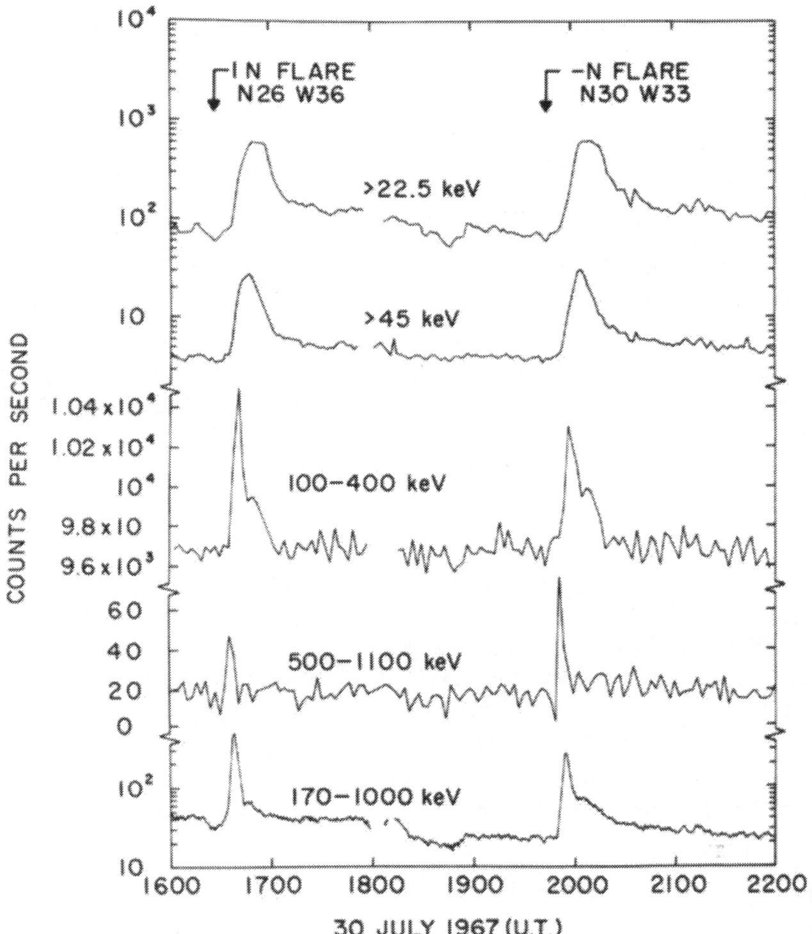

Fig. 2. Two scatter-free impulsive solar electron events from McMath plage 8905 (Wang *et al.*, 1971). The sharp initial peak in the high energy channels place an upper limit on the duration of the electron injection into the interplanetary medium of $\lesssim 3$ min. Note the long decay following the initial peak in the low energy channels.

described by a diffusion equation (Parker, 1963)

$$\frac{\partial \varrho}{\partial t} = \frac{1}{r^{\alpha}} \frac{\partial}{\partial r}\left(M r^{\alpha+\beta} \frac{\partial \varrho}{\partial r}\right), \tag{1}$$

where the diffusion coefficient $D = Mr^{\beta}$, M and β are parameters. Here ϱ is the particle density at position r, time t, of energy E and α is the parameter specifying the dimension of the space to be used. From Krimigis (1965), we obtain for an assumed isotropic flux $J = \varrho v/4\pi$,

$$J(r, t) = \frac{Nv \exp\left[-\frac{1}{Mt} \frac{r^{2-\beta}}{(2-\beta)^2}\right]}{4\pi (2-\beta)^{(2\alpha+\beta)/(2-\beta)}\, \Gamma[\alpha+1)/(2-\beta)]} \left(\frac{1}{Mt}\right)^{(\alpha+1)/(2-\beta)}, \tag{2}$$

where N = number of particles emitted per unit solid angle, and v = particle velocity. At time of maximum we obtain

$$J_{max}(r, t_{max}) = \\ = \frac{Nv \exp\left[-\frac{\alpha-1}{2-\beta}\right]}{4\pi(2-\beta)^{(2\alpha+\beta)/(2-\beta)}\,\Gamma[(\alpha+1)/(2-\beta)]}\left(\frac{(2-\beta)(\alpha+1)}{r^{2-\beta}}\right)^{(\alpha+1)/(2-\beta)}. \tag{3}$$

Note that the relationship between N and J_{max} is independent of M. Thus, as long as the diffusion coefficients for the different energy particles have the same spatial dependence (same β) regardless of the value of M, the constant of proportionality between J_{max} and Nv remains the same. In this case the shape of the flux spectrum derived from the maximum flux at each energy observed at 1 AU is exactly that of the emitted particles.

Some computations have been made of the validity of this method as λ increases (Lin *et al.*, 1973b). These indicate that up to $\lambda \lesssim 0.3$ AU this constant relationship holds.

3. Location of the Acceleration Region

The electrons escaping to the interplanetary medium will lose energy during their passage through the solar atmosphere overlying the acceleration region. A straightforward calculation, assuming rectilinear upward path through fully ionized hydrogen to 1 AU without regard to deflections or other energy loss mechanisms (such as wave particle interactions or radio emission), (Lin, 1973) shows that an initial power law spectrum at height h at the Sun

$$\frac{dn}{dE_1} = AE_1^{-\delta} \quad \text{with } A, \delta \text{ constants} \tag{4}$$

becomes a peaked spectrum

$$\frac{du}{dE_2} = \frac{AE_2}{(E_2^2 + 2k)^{(\delta+1)/2}} \tag{5}$$

with peak at

$$E_{2m} = \left(\frac{2k}{\delta}\right)^{1/2} \tag{6}$$

and $k = -2.6 \times 10^{-18} \int_h^{1\text{ AU}} n_i(x)\,dx$, where n_i is in cm^{-3}, h in cm.

In actuality the helical paths of the electrons along the field line and deflections will increase the path length traversed by the electrons in escaping to 1 AU, so that the E_{2m} given by Equation 6 will be a lower limit. A few low energy spectra of electrons observed at 1 AU have become available from recent spacecraft observations (Figure 3). All of these spectra extend smoothly in a power law to below ~6 keV, and on occasion to lower energies. Using Equation (6) we find that a peak $\lesssim 6$ keV implies that the total path length is given by

$$\int_{h}^{1\,\mathrm{AU}} n_i(x)\,\mathrm{d}x \lesssim 3.5 \times 10^{19}\ \mathrm{cm}^{-2}.$$

This corresponds to $\lesssim 60\ \mu\mathrm{g\ cm}^{-2}$ of hydrogen, equivalent to an ambient density at the acceleration region of certainly less than $\sim 10^{10}\ \mathrm{cm}^{-3}$. This density corresponds to a height of $\gtrsim 2 \times 10^4$ km above the photosphere for a 10 X Baumbach-Allen active region density model.

We wish to re-emphasize the fact that this estimater is a *lower* limit to the actual

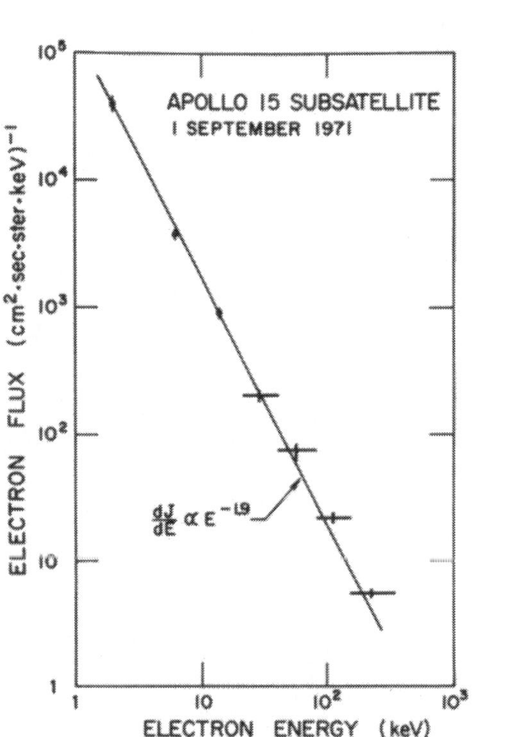

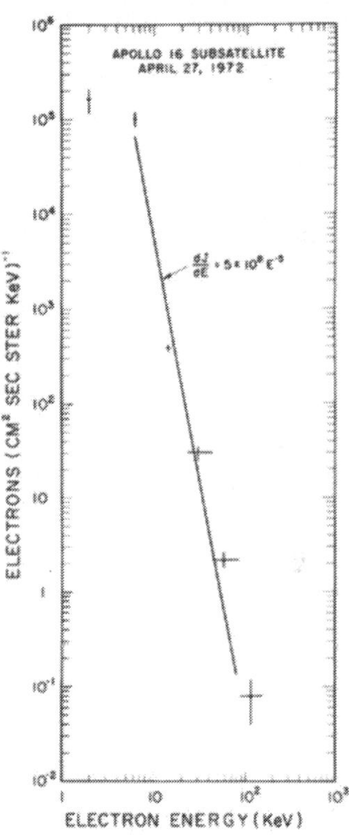

Fig. 3. Two electron energy spectra extending to low energies. The 1 September event is accompanied by energetic protons while the 27 April event (from Lin *et al.*, 1973) is not. Both spectra extend smoothly in a power law to below ~6 keV.

height of acceleration since the effects which were not taken into account would tend to increase the minimum energy of the peak. Clearly electron acceleration must have occurred in the lower corona. Although only a few events have been observed to energies below ~20 keV, in no events has a turnover been observed at higher energies. Thus electron acceleration at the flash phase appears to be a coronal phenomenon, at least for events observed to emit electrons into the interplanetary medium.

This location and ambient density is consistent with the observed starting frequencies (~200–1000 MHz) of type III bursts, and is also consistent with the occasional observation of an electron event at 1 AU without detectable X-ray emission (Kane and Lin, 1972). Presumably in those events the magnetic field structure in the vicinity of the acceleration region is such as to prevent the electrons from entering dense regions where a detectable X-ray flux would be produced.

4. Accelerated vs Escaping Electron Energy Spectra

Non-thermal X-rays give direct information on the electrons at the Sun. The X-ray spectrum can be directly related to the instantaneous X-ray producing electron spectrum (Brown, 1971; Kane and Anderson, 1970). This relationship can be written for a power law X-ray spectrum as

$$\frac{\mathrm{d}n}{dE}=3.85\times 10^{41}\,\gamma(\gamma-1)^2\,B(\gamma-\tfrac{1}{2},\tfrac{3}{2})\frac{AE^{-\gamma+1/2}}{n_iV}\ \mathrm{cm}^{-3}\,\mathrm{keV}^{-1}, \tag{7}$$

where

$$\frac{\mathrm{d}J(h\nu)}{\mathrm{d}(h\nu)}=A(h\nu)^{-\gamma}.$$

$B(x, y)$ is the beta function, n_i the ambient ion density, V the volume of the X-ray region, and A, γ constants.

The relationship of the instantaneous X-ray producing electron spectrom to the accelerated electron spectrum depends on the evolution of the electrons subsequent to acceleration. Suppose the electrons are accelerated in one region and produce the bulk of the observed X-rays in another region (these two regions may be one and the same but for the sake of generality we will allow them to be different). The evolution of the electron distribution, $N(E, t)=V(\mathrm{d}n/\mathrm{d}E)$, where $V=$ volume in the X-ray emitting region, can be described by the equation

$$\frac{\partial N(E,t)}{\partial t}=F(E,t)-\frac{N(E,t)}{\tau_e(E)}-\frac{\partial}{\partial E}\left[N(E,t)\frac{\mathrm{d}E}{\mathrm{d}t}\right], \tag{8}$$

where $F(E, t)$ is the input source of electrons $\mathrm{keV}^{-1}\,\mathrm{s}^{-1}$, $N(E, t)/\tau_e(E)$ is the number of electrons escaping the region $\mathrm{s}^{-1}\,\mathrm{keV}^{-1}$, and the third term describes energy loss processes for the electrons. Note that X-ray observations define $N(E, t)$ subject to a choice of ambient density n_i (Equation (7)). Thus this equation can be solved for $F(E, t)$, given n_i and given the form of $\tau_e(E)$ if only collisional energy losses are assumed to be important.

To a good approximation we can consider $N(E, t)$ as constant over some time interval, Δt, and zero outside that interval. This removes the time dependence of the equation. Additionally we shall consider only the power law case, $N(E, t)=BE^{-\delta}$, so that inserting for dE/dt the energy loss in ionized hydrogen (Trubnikov, 1965)

$$\frac{dE}{dt} = -4.9 \times 10^{-9}\, n_i E^{-1/2} (\text{keV s}^{-1}), \tag{9}$$

where n_i = ambient density in cm^{-3} and E is in keV, Equation (8) becomes

$$F(E) = N(E)\left[\frac{1}{\tau_e(E)} + \frac{4.9 \times 10^{-9}\, n_i(\delta + \frac{1}{2})}{E^{3/2}}\right]. \tag{10}$$

We have computed the anticipated energy dependence of $F(E, t)$ compared to $N(E, t)$ and $dJ(h\nu)/d(h\nu)$, and the energy dependence of the escaping electrons for two extremes:

(1) where the escape term is much larger than the collisional energy loss term. This situation is the *thin-target* approximation for X-ray emission.

(2) where the collisional energy loss term is much larger than the escape term. This situation is the *thick-target* approximation for X-ray emission.

We have used two obvious choices for the energy dependence of τ_e, although other forms might be appropriate. These two are: (1) τ_e = constant, and (2) $\tau_e \propto 1/E^{1/2}$, i.e., proportional to the scale size of the X-ray region divided by the particle velocity. The results are summarized in Table II.

TABLE II

Spectral dependence of electrons and X-rays

	Thick target	Thin target
Spectrum of X-rays	$\frac{dJ(h\nu)}{d(h\nu)} = A(h\nu)^{-\gamma}$	$\frac{dJ(h\nu)}{d(h\nu)} = A(h\nu)^{-\gamma}$
Spectrum of electrons in X-ray emitting region $N(E) \propto \frac{dne}{dE} \propto E^{-\delta}$	$\delta = \gamma - \frac{1}{2}$	$\delta = \gamma - \frac{1}{2}$
Spectrum of accelerated electrons $F(E) \propto E^{-\delta_a}$	$\delta_a = \gamma + 1$	$\delta_a = \gamma - \frac{1}{2}$ for τ_e = constant $\delta_a = \gamma - 1$ for $\tau_e \propto E^{-1/2}$
Spectrum of electrons escaping from the X-ray region, $S(E) \propto E^{-\delta_e}$	$\delta_e = \gamma - \frac{1}{2}$ for τ_e = constant $\delta_e = \gamma - 1$ for $\tau_e \propto E^{-1/2}$	$\delta_e = \delta_a =$ $\gamma - \frac{1}{2}$ for τ_e = constant $\gamma - 1$ for $\tau_e \propto E^{-1/2}$

The spectrum of electrons escaping from the X-ray region is not necessarily the spectrum of the electrons escaping to the interplanetary medium. The electrons need not escape to the interplanetary medium to be lost from the X-ray region; they may also escape to the low density, $n_i \lesssim 10^9\ \text{cm}^{-3}$, upper corona, where the flux of X-rays they produce will be below the threshold of current X-ray detectors. Also the acceleration region may be much higher in the solar atmosphere than the X-ray region, and the electrons observed in space may have come directly from the accelerated population (i.e., $\delta_e = \delta_a$).

Datlowe and Lin (1973) noted that it is possible to distinguish between thick and thin target cases under the assumption that the spectrum of electrons observed in the interplanetary medium is representative of the accelerated electron spectrum (i.e., $\delta_e = \delta_a$, see Table II). For a flare event where high energy resolution measurements were available for the electrons and X-rays above 20 keV (see Figure 4), the result was $\delta_a = \gamma - \frac{1}{2}$, favoring thin target. Other X-ray electron events studied where only measurements with poor energy resolution were available are also generally consistent

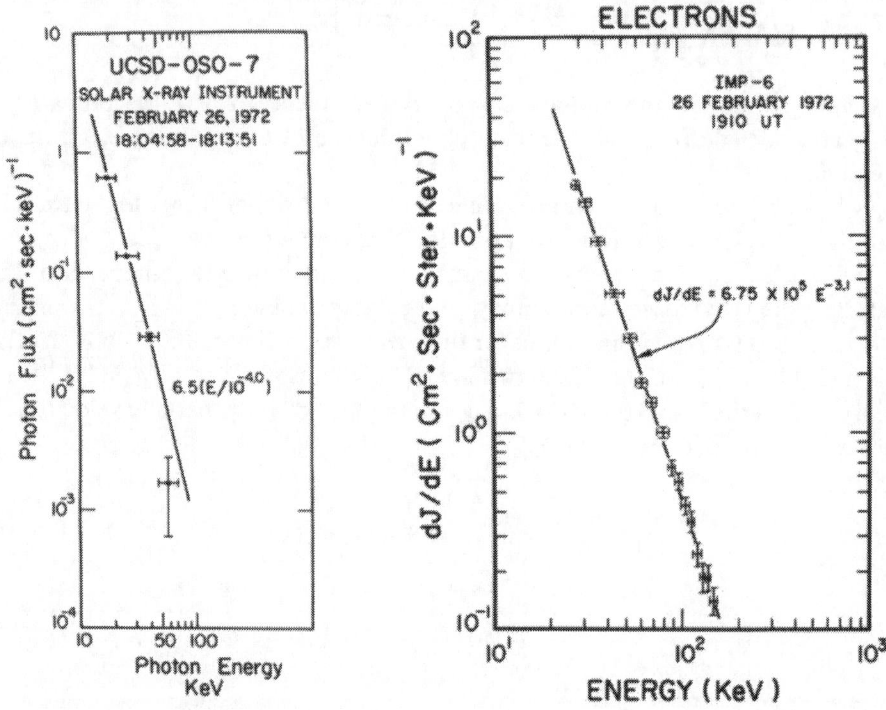

Fig. 4. The spectra of hard X-rays and electrons observed at 1 AU for the same flare event. The photons fit a power law spectrum $\mathrm{d}J(h\nu)/\mathrm{d}(h\nu) = A(h\nu)^{-\gamma}$ where $\gamma = 4.0 \pm 0.3$, while the electrons fit a spectrum $\mathrm{d}J/\mathrm{d}E = 6.75 \times 10^5\ E^{-3.1}$. Since $\mathrm{d}n/\mathrm{d}E = v\ \mathrm{d}J/\mathrm{d}E$ where v is the electron velocity, the electron fit a *density* spectrum $\mathrm{d}n/\mathrm{d}E \propto E^{-\delta}$ with $\delta = 3.6 \pm 0.1$. These two spectra are consistent with thin target emission under the assumption the escaping electrons have the same spectrum as the accelerated electrons.

(see Lin and Hudson, 1971; Kane and Lin, 1972) with a thin target model. The thin target case is also consistent with the location of the acceleration region ($n_i \lesssim 10^{10}\ \mathrm{cm}^{-3}$) derived from considerations of the low energy electron spectrum observed at 1 AU.

In favor of thick target processes we note that if non-relativistic electrons penetrate to the dense ($n_i \gtrsim 10^{12}\ \mathrm{cm}^{-3}$) regions of the chromosphere-corona boundary and below, they could produce the observed EUV and perhaps provide the energy for heating the Hα flare region through collisional loss (and possibly even heat the white light flare region) (Hudson, 1972). The close time coincidence between the hard X-ray spike and the EUV spike (Kane and Donnelly, 1971) is consistent with such an interpretation. At those densities the thick-target approximation would certainly be appropriate.

There are several possible ways of reconciling the observations in support of thick and thin target. One possibility is that the electrons injected into the interplanetary medium may have a spectrum modified from that of the electrons initially accelerated. From comparisons of the total energy in accelerated electrons derived from the X-ray observations and the total energy in escaping electrons, Lin and Hudson (1971) found that the escape efficiency is only ~0.1 to 1%. Thus the probability of escape of the electrons into the interplanetary medium may be a function of electron energy. We do not favor this possibility in view of (a) the relatively small amounts of matter which are traversed by the escaping electrons, and (b) the otherwise coincidental agreement between the X-ray and electron spectra.

A second possibility is that electrons of low energies, say below ~10 keV, are described by the thick target approximation while higher energy electrons are in an essentially thin target situation (Kane, 1973). This dichotomy could arise, for example, if the electrons are accelerated and contained by a magnetic 'bottle' in a low density, $n_i \lesssim 10^{10}$ cm^{-3}, region. Electrons only appear in high density, $n_i \gtrsim 10^{10}$ cm^{-3}, regions near the feet of the magnetic bottle if they are scattered into the loss cone. Since the amount of scattering is a strongly decreasing function of energy, essentially only the low energy electrons will be dumped into the loss cone. This interpretation is consistent with the observations which show that the correspondence between rising portion of the EUV emission and the rising portion of the non-thermal X-rays is best for the lowest energy, ~10 keV, X-rays.

A third possibility is that the acceleration of the escaping electrons is separate from the acceleration of the electron producing X-rays. However the time of injection of the electrons into the interplanetary medium, which can be obtained accurately ($\lesssim$10 min) by analyses of the velocity dispersion observed during the onset of electron events at 1 AU, is clearly between the Hα onset and maximum, i.e., at the time of the flash phase X-ray and radio flare phenomena. The duration of the injection is inferred to be $\lesssim$3 min from the duration of the highly scatter-free events. Furthermore, escaping electrons generate type III emission. Type III emission occurs exactly at the time of X-ray bursts to within seconds when both are observed from flares (Kane, 1972).

These considerations indicate that the >20 keV electron spectrum observed at 1 AU is probably an essentially undistorted sample of the electrons accelerated in the flare.

5. Two Stage Acceleration

Two types of electron spectra are observed (Lin, 1970). For events which are unaccompanied by energetic protons* and relativistic electrons, i.e., pure non-relativistic electron events, the electron spectrum can be fit to a power law with exponent from ~2 to 5, usually with a steepening to >5 at ~100–200 keV (Figure 5). Events which are accompanied by energetic protons usually have electron spectra which extend

* Above a threshold of ~0.3 (cm^2 s ster)$^{-1}$ above 10 MeV.

smoothly in a power law to relativistic energies (Figure 6). These two types of spectra, one with a 'cut-off' and one without, suggest two stages of acceleration, one a flash phase acceleration of mainly just ~5–100 keV electrons, and the second an acceleration of protons and electrons to high, even relativistic energies which occurs only in some flares. This concept is further supported by spacecraft observations which in-

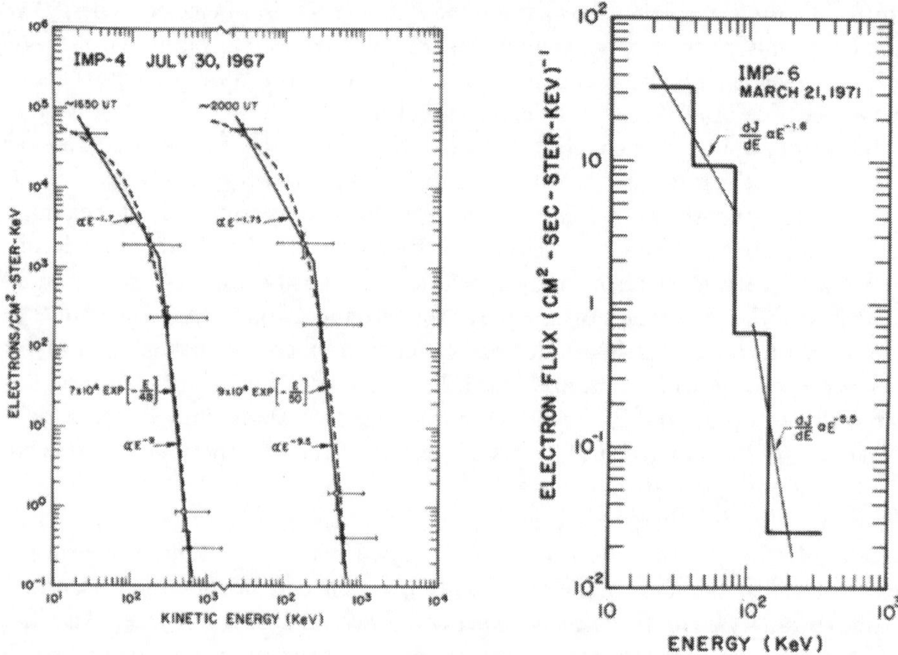

Fig. 5. The electron energy spectra of the events of Figure 2 (from Wang *et al.*, 1971) illustrating the steepening in the spectrum above ~100–200 keV typical of pure electron events. Note the wide variation in flux from the 30 July 1967 events to the 21 March 1971 event.

dicate that relativistic electrons and energetic protons are usually injected into the interplanetary medium $\gtrsim$10 min *after* the non-relativistic electrons (Figure 7) (Sullivan, 1974; Simnett, 1974; Lin and Anderson, 1967).

Two stages are sometimes observed in the hard X-ray event accompanying energetic proton flares (Figure 8) (Frost and Dennis, 1971). The X-ray energy spectrum shows a cut-off at ~100 keV for the flash phase but no energy cut-off even to the limits of their observation ($\gtrsim$250 keV) in the long second phase. That phase starts at the onset of the type II burst. The radio, X-ray, and particle observations are generally consistent with the acceleration of particles in the second state by the type II shock front in the *corona* by a stochastic Fermi-type mechanism. The X-ray event of 30 March 1969 (Figure 8) was associated with a behind-the-limb flare (~W 110°) so the origin of the X-ray burst must be in the corona. Interplanetary shocks and the Earth's bow shock both accelerate particles, electrons and nuclei, up to energies of ~10^2 keV (Fan *et al.*, 1964; Anderson, 1965; McGuire *et al.*, 1972) and ~10 MeV (Palmeira *et al.*, 1971) re-

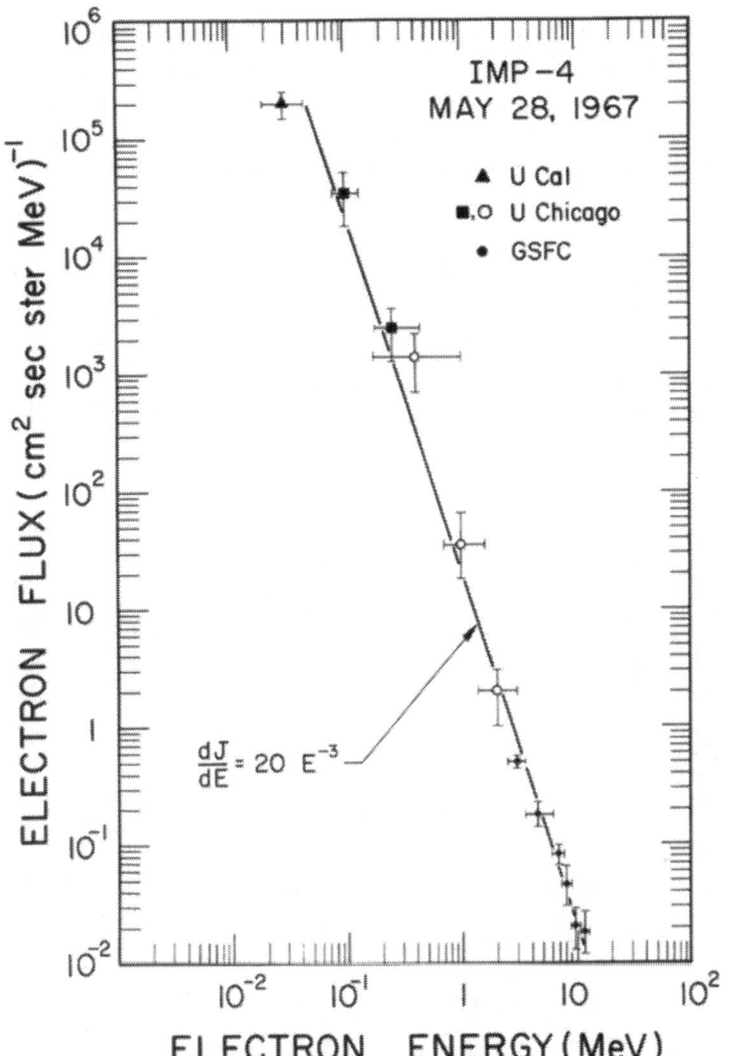

Fig. 6. The differential electron energy spectrum for the mixed electronproton event of 28 May 1967, compiled from four different detector systems aboard IMP-4. The lowest energy point is obtained from Geiger Müller detector observations (Lin, 1970). The University of Chicago points are from solid-state detector telescopes (Sullivan, 1973), and the points above ~2 MeV are from the Goddard Space Flight Center range and energy loss scintillation detectors (Simnett, 1971). The points fit to a single power over three decades in energy, even though they are from several different detectors with different view directions.

spectively. In the much higher magnetic fields and densities near the Sun it seems likely that substantially higher particle energies will be attained.

6. Generation of Type III Radio Emission

It is well established that energetic electrons generate most (if not all) of the non-thermal radio emission observed from the Sun. Detailed calculations of gyro-synchrotron and synchrotron emission of energetic electrons in solar magnetic fields give generally good agreement with observations of impulsive microwave bursts and type IV emission. However until recently the various theoretical treatment of type III emission differed as to the exciter, whether electrons or protons or waves, and numbers

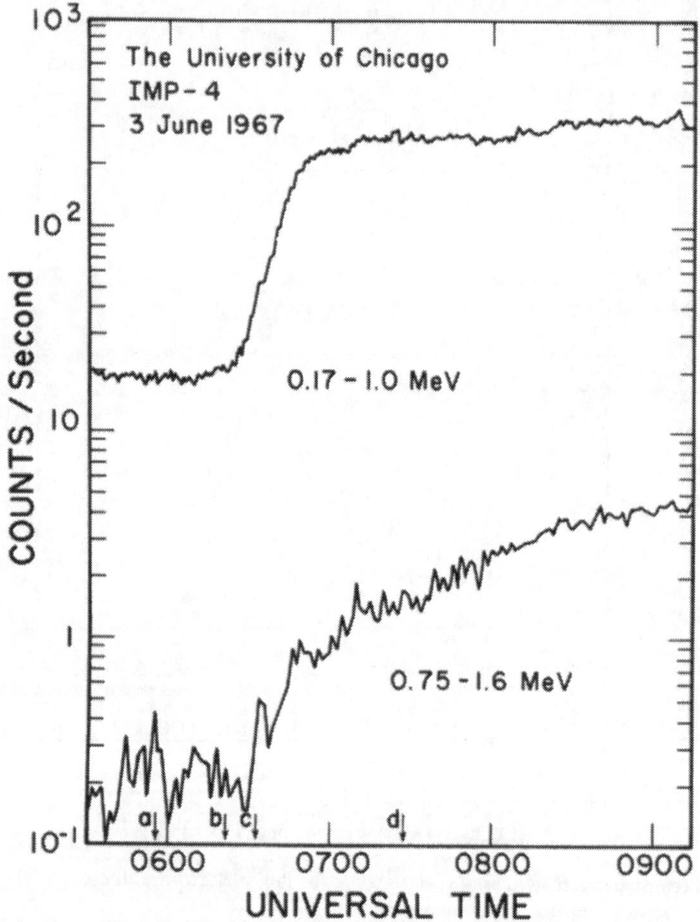

Fig. 7. The delay of the relativistic (0.75–1.6 MeV) electrons vs the non-relativistic electrons (0.17–1 MeV) is illustrated here. The onset of the soft 2–12 Å flare X-rays is denoted by (a), the 0.17–1.0 MeV electron onset by (b), the 0.75–1.6 MeV onset by (c), and the 9.6–18 MeV proton onset by (d) (taken from Sullivan, 1973).

of particles needed varied over 10–12 orders of magnitude (Evans *et al.*, 1971). Thus direct observations of the exciter, and simultaneously the type III emission generated by them at 1 AU, are of critical importance in establishing a firm theoretical base for the emission processs.

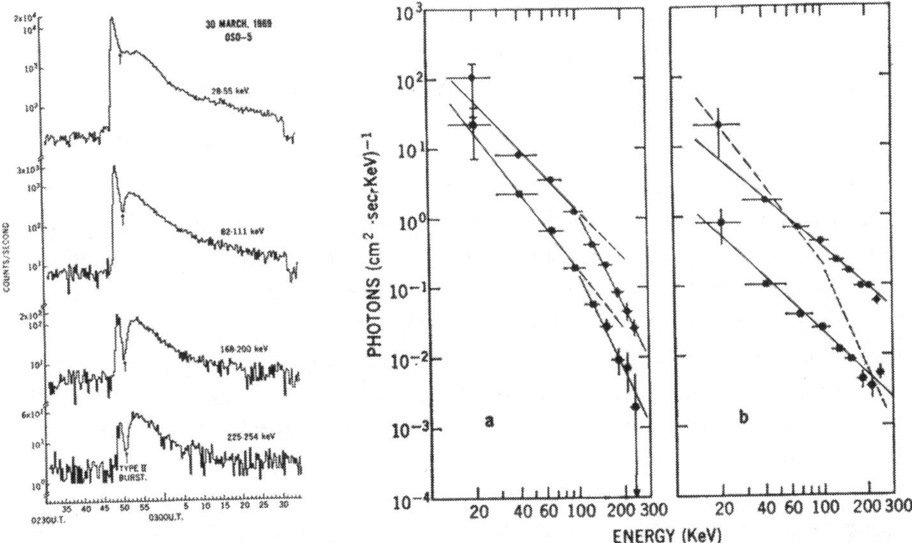

Fig. 8. The two-stage hard X-ray burst of 30 March 1969 (from Frost and Dennis, 1971), which was followed by an intense mixed electron-proton event observed at 1 AU. This X-ray event presents clear evidence for two-stage acceleration. Photon energy spectra during the initial X-ray burst (shown in the middle panel marked 'a') have a spectrum steepening above ~100 keV, fairly typical of flash phase events. The spectra during the second phase (shown in the right panel marked 'b') which starts at the time of intense type II emission (Smerd, 1970) shows a smooth and very hard spectrum to the upper limits of the X-ray detector's energy range (~300 keV).

6.1. Physical Mechanism

Type III solar radio bursts are the most common type of impulsive phenomena observed from the Sun. These bursts are characterized by a rapid frequency drift from high to low frequencies, and occasionally by the presence of two bands of emission, one at approximately twice the frequency of the other (see reviews by Wild, *et al.*, 1963; Wild and Smerd, 1972). A theoretical basis for the plasma hypothesis for type III solar radio bursts (Wild, 1950) was first introduced by Ginzburg and Zheleznyakov (1958), and although it has been developed and refined in the intervening years, the basic ideas have remained unchanged (see review by Smith, 1973). A group of fast particles injected near the Sun generate longitudinal electron plasma waves at frequencies near the local plasma frequency as they pass through the coronal plasma. These plasma waves then scatter off ion density fluctuations to produce electromagnetic radiation near the plasma frequency (fundamental), and off other plasma waves to produce emission at twice the plasma frequency (2nd harmonic). As the fast particles go upward in the corona and into the interplanetary medium the radio emission will drift from high to low frequencies. Typical drift rates for these bursts indicate velocities of ~0.3–0.5 c for the particles where c is the speed of light.

The plasma waves are generated through a coherent Cerenkov plasma process. In order to produce plasma waves more rapidly than they are damped, the velocity distribution of the fast particles must have a positive slope, that is, a peak must exist in the non-thermal particle velocity distribution. Observations of the characteristics of type III bursts and the particles that excite them will thus provide a test of basic beamplasma and mode-mode coupling theory over a wide range of plasma conditions.

6.2. THE TYPE III BURST EXCITER

Wild *et al.* (1954) were first to suggest that energetic protons might be the exciters of type III emission. More recently Smith (1970) summarized the theoretical difficulties of stabilizing a spatially unbounded and homogeneous electron stream, and noted that a proton stream can be stabilized. Smith (1970) suggested that the ~20–100 MeV protons which might produce the burst at the Sun were so few in number that after diffusion in the interplanetary medium their fluxes would be too low to observe at 1 AU. However, the observations of type III bursts generated near 1 AU imply that substantial fluxes of protons sufficient to produce emission should be observed *at 1 AU*. Such fluxes are not generally observed except in large proton events which are quite rare compared to electron events.

The theoretical difficulties for electron streams can apparently be overcome by considering a spatially bounded stream with inhomogenities in the front and back (Zaitsev *et al.*, 1972). In addition direct observations of the electron velocity distributions at 1 AU show that peaked distributions do exist, contrary to the theoretical predictions otherwise.

A very highly significant correlation, almost one to one, exists between intense kilometric wavelength type III's and >20 keV electrons observed at 1 AU from flares located in the western solar hemisphere (Alvarez *et al.*, 1972). In addition storms of weak type III bursts are observed at hectometric wavelengths which appear to be closely related to type I storms at metric wavelengths (Fainberg and Stone, 1970). These type III storms are accompanied by non-impulsive co-rotating $\gtrsim$20 keV electron fluxes observed at 1 AU.

Recently Frank and Gurnett (1972) and Lin *et al.* (1973a) have reported observations at 1 AU of energetic electrons and type III burst emission at the low frequencies characteristic of the near 1 AU plasma environment. Frank and Gurnett (Figure 9) did not observe radio emission simultaneously with the arrival of the 5–6 keV electrons, which, in their interpretation, are the exciters of the emission. Rather they assumed that the radiation is generated primarily at the fundamental, i.e., local plasma frequency, and that the lowest frequency radiation they observe, ~31 kHz, originates some distance away from 1 AU. Approximately 2600 s after the onset of the 31 kHz emission the ~6 keV electrons are observed to arrive. Although the fundamental emission generated at 1 AU (~20 kHz) is not observed, the authors note that the calculated time of onset of the fundamental emission is in agreement with the arrival of the ~6 keV electrons ($v \approx 0.15\, c$). Since the drift rates of bursts near the Sun indicate velocities of ~0.3 c, they conclude that deceleration of the electrons may be substantial.

Lin *et al.* (1973a) located the position of the type III burst emission at each fre-

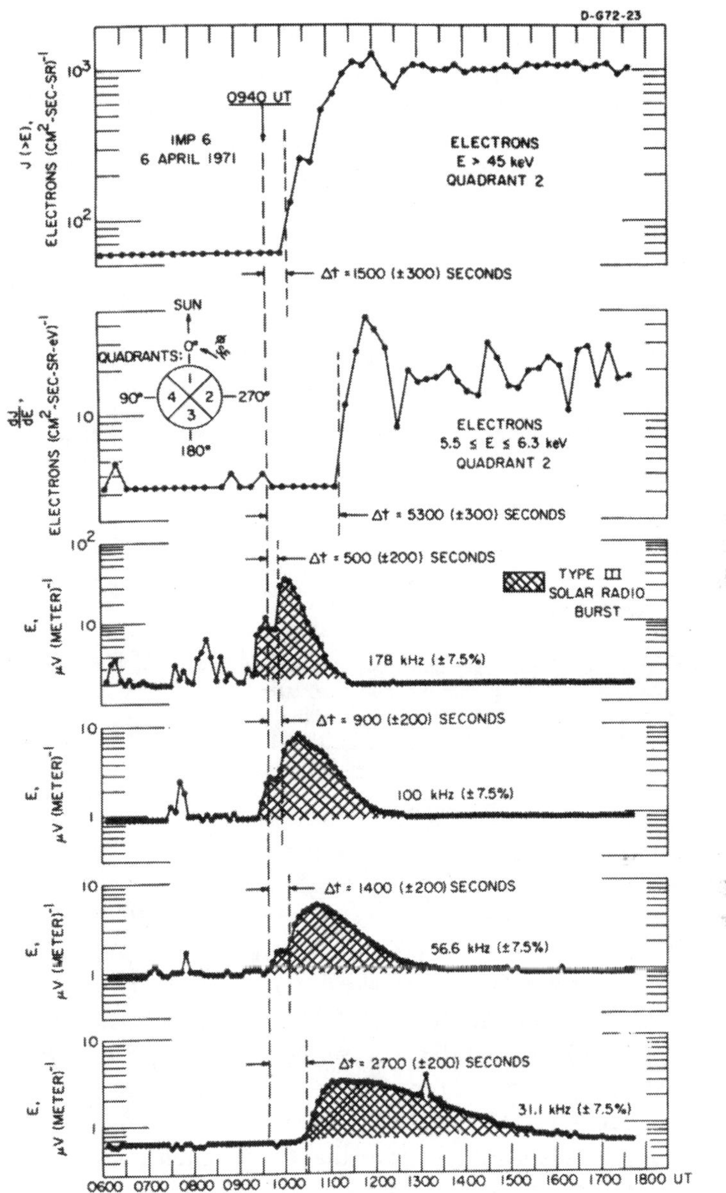

Fig. 9. Simultaneous electron and type III burst observations at 1 AU (from Frank and Gurnett, 1972). The local electron plasma frequency is ~20 kHz, but no oscillations were observed at that frequency. If the radio emission is at the fundamental of the plasma frequency then the 5.5–6.3 keV electrons would be likely candidates for the burst exciter. If, on the other hand, the radio emission is at the second harmonic of the plasma frequency then higher energy, $\gtrsim 10$ keV, electrons would coincide with the emission (see text).

quency (Figure 10) from the spin modulation of the radio signal. The emission originating at 1 AU was then compared with the particle data. Their results show (Figure 11) that the onset of the emission located at 1 AU corresponds to the arrival of electrons of ~100 keV energy. The subsequent build up of the radiation corresponds to the arrival of lower energy electrons, until maximum is reached when ~10 keV

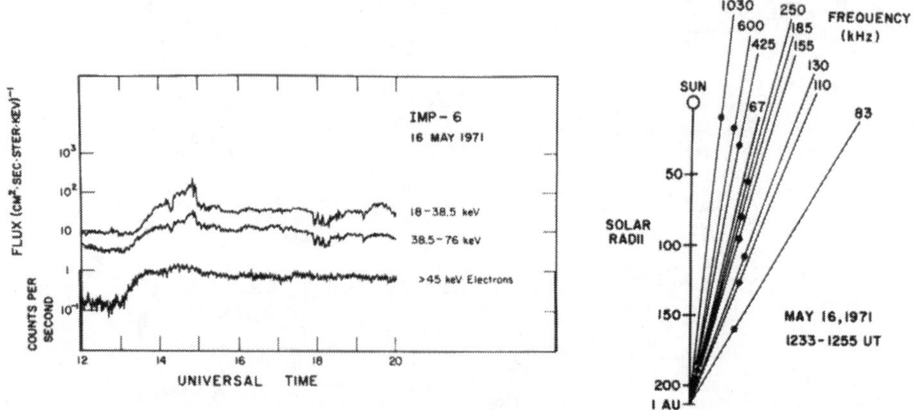

Fig. 10. (a) The May 16, 1971 event. The upper two channels are sensitive to low energy protons as well as electrons while the lower channel is sensitive only to electrons. Some upstream terrestrial protons are observed ~1430–1500 UT, a time well after the period of analysis. (b) The trajectory in the interplanetary medium of the type III burst of 16 May 1971, determined from the spin modulation of the observed radio signal at different frequencies.

electrons first arrive. By the time 6 keV electrons arrive the emission is already decaying rapidly.

Since the propagation of these electrons may differ markedly from event to event, depending on the changeable scattering characteristics of the interplanetary medium, two events were compared. One event was scatter-free, the other diffusive. The frequency drift rate of the interplanetary type III bursts was more rapid for the scatter-free event and less rapid for the diffusive event, corresponding nicely to the difference in the computed distance traveled for the first arriving electrons of 1.4 AU in the scatter-free event and 1.7 AU for the diffusive event.

The evidence from radio studies indicate that the second harmonic emission rather than fundamental is predominant for low frequency type III radiation (Fainberg *et al.*, 1972; Smith, 1972; Malitson *et al.*, 1973; Haddock and Alvarez, 1973). Thus, the frequencies of the near 1 AU radiation are ~twice the local plasma frequency. Under the assumption of second harmonic radiation the observations of Frank and Gurnett would be in close agreement with Lin *et al.* (1973a).

Even under the second harmonic hypothesis some *apparent* deceleration is observed (Fainberg *et al.*, 1972). This can be attributed to scattering of the electrons as they propagate outward in the interplanetary medium. Such scattering will lower the apparent velocity along a smooth spiral field line.

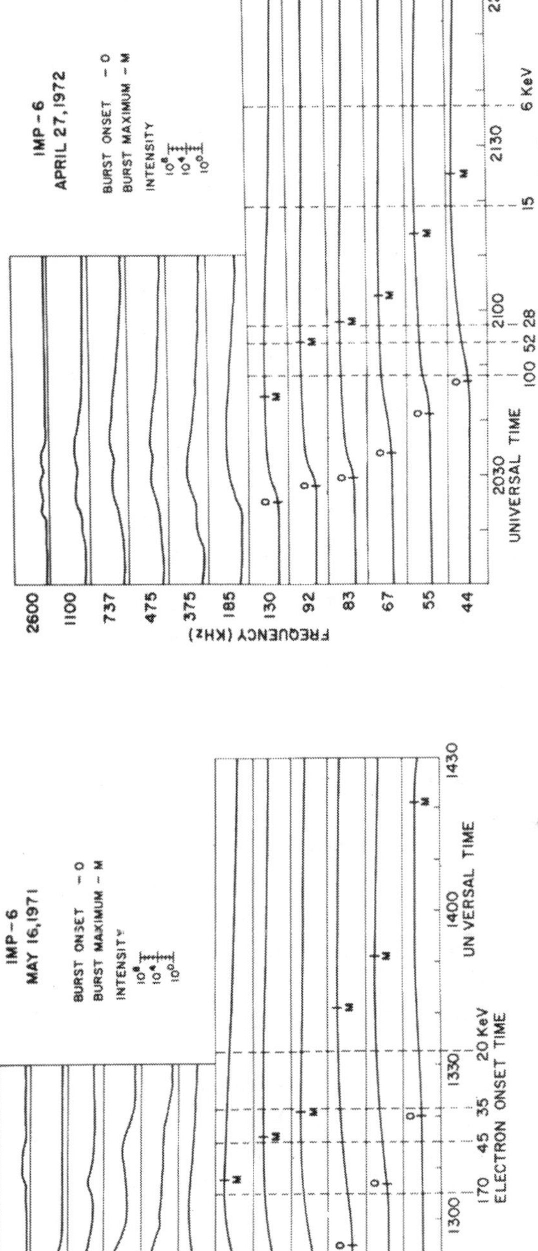

Fig. 11. (a) The type III burst of 16 May 1971 observed at frequencies from 2.6 MHz to 44 kHz. The 55 kHz emission originates closest to 1 AU (see Figure 10). (b) The type III burst of 27 April 1972. Here the 44 kHz emission is closest to 1 AU.

6.3. Comparison with Theories

Using calculated incoherent Cerenkov emission efficiency (Cohen, 1958) and a maximum estimated value for the efficiency of coupling to EM radiation, we find that the emission from incoherent Cerenkov processes calculated for the observed electron fluxes is insufficient to account for the observed emission.

The peaked distribution needed to generate coherent Cerenkov radiation is clearly present in the electrons (Figure 12). This peaked distribution arises from velocity

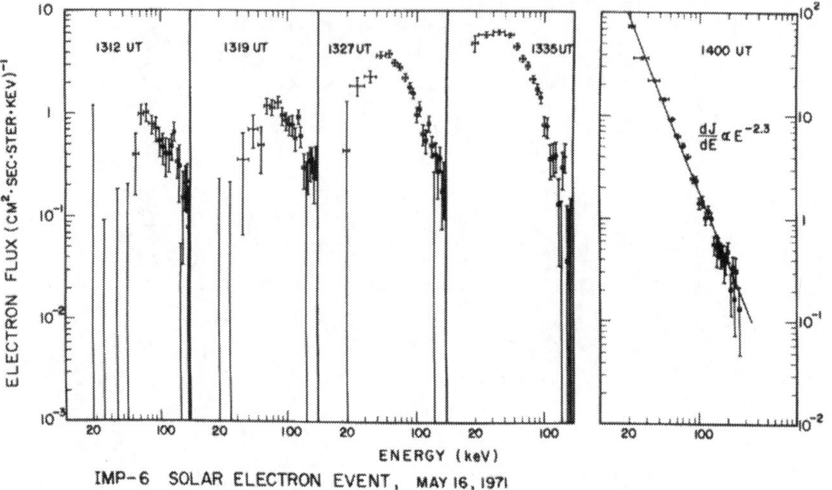

Fig. 12. The energy spectrum of the electrons in the 16 May 1971 event at different times during the onset.

dispersion rather than an initially peaked distribution of injected electrons at the Sun, since the injection spectrum at the Sun is observed to extend down to $\lesssim 5$ keV without a peak.

The radio emission is observed to increase in intensity until the peak in the electron distribution falls below ~ 15 keV. The radio burst begins a constant exponential decay, independent of the behavior of the electron flux, after the peak in the electron distribution goes below ~ 10 keV.

It should be noted that the coherent Celenkov processes, at least in these observations, are not so strong as to substantially modify the peaked distribution. That distribution evolves essentially as would be anticipated from velocity dispersion alone. The implication is that the time scale for relaxation of the beam through the wave-particle interaction is long compared to the time scale for evolution of the distribution by velocity dispersion. This is contrary to the expectation of most theories for type III emission.

One factor not considered in most theoretical treatments which may be relevant in explaining the long time scale for relaxation is the angular distribution of the electrons. The average distribution for the 16 May 1971 event during the period of generation of type III emission is shown in Figure 13. The distribution is not just outward along the field line as assumed by most theoretical models, but instead is relatively isotropic, with a maximum to minimum ratio of ~2:1. This angular distribution presumably arises from the pitch-angle scattering of the electrons by irregularities in the interplanetary magnetic field.

Table III lists some characteristics of the low frequency type III burst at 1 AU

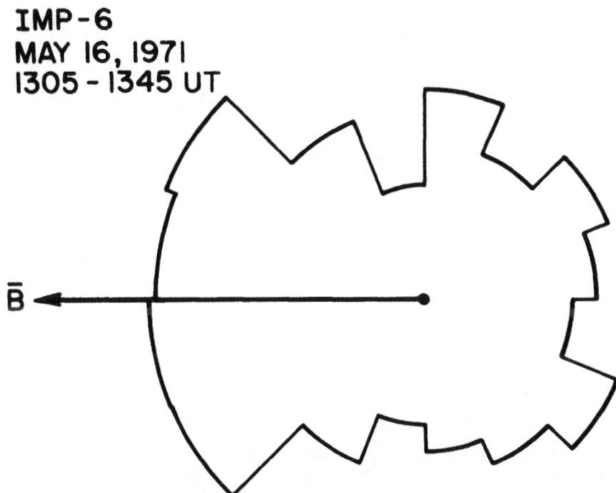

Fig. 13. The pitch angle distribution of the >45 keV electrons during the onset of the 16 May 1971 event. The approximate direction of B is indicated by the arrow.

TABLE III

Relevant parameters of type III bursts and burst exciters at 1 AU

(1) Type III burst	
(a) Typical frequencies of observation (2nd harmonic)	~40–60 kHz
(b) Intensity of emission	$\sim 3 \times 10^{-17}$ W m^{-2} Hz
(c) Cross-sectional area of source	$\sim 10^{26}$ cm^2
(d) Average emissivity of source	$\sim 6 \times 10^{-24}$ erg cm^{-3} s^{-1}
(2) Fast electron exciters	
(a) Typical energy of electrons	~10–100 keV
(b) Density of fast electron at burst maximum	$\sim 10^{-6}$ cm^{-3}
(c) Total number of fast electrons producing emission	$\sim 10^{33}$
(d) Average rate of energy loss to electromagnetic emission (per electron)	$\sim 4 \times 10^{-9}$ keV s^{-1}
(e) Average interparticle distance	$\sim 10^2$ cm
(3) Ambient plasma medium (solar wind)	
(a) Mean densities	5–10 cm^{-3}
(b) Mean temperatures – electron	$\sim 1.2 \times 10^5$ K
– proton	$\sim 7 \times 10^4$ K
(c) Debye length	$\sim 10^3$ cm

and some parameters of the emission process. It is clear from these preliminary studies that a quantitative plasma theory for type III bursts can be obtained through further observations of this kind. Such observations and theory, besides illuminating problems of plasma physics otherwise impossible to study, will also be applicable for the quantitative interpretation not only of solar flare radio phenomena, but possibly of galactic and extragalactic radio emission as well.

Acknowledgements

I wish to acknowledge discussions with Drs S. R. Kane and K. A. Anderson. This research was funded in part by NASA grant NGL-05-003-017.

References

Alvarez, H., Haddock, F., and Lin, R. P.: 1972, *Solar Phys.* **26**, 468.
Anderson, K. A.: 1965, *Proc. 9th International Conf. on Cosmic Rays*, London, p. 520.
Brown, J. C.: 1971, *Solar Phys.* **18**, 450.
Cohen, M. H.: 1958, *Phys. Rev.* **123**, 711.
Datlow, D.: 1971, *Solar Phys.* **17**, 436.
Datlowe, D. and Lin, R. P.: 1973, *Solar Phys.* **32**, 459.
Evans, L. G., Fainberg, J., and Stone, R. G.: 1971, *Solar Phys.* **21**, 1968.
Fainberg, J., Evans, L. G., and Stone, R. G.: 1972, *Science* **178**, 743.
Fainberg, J. and Stone, R. G.: 1971, *Astrophys. J.* **164**, L123.
Fan, C. Y., Gloeckler, G., and Simpson, J. A.: 1964, *Phys. Rev. Letters* **13**, 149.
Frank, L. A. and Gurnett, D. A.: 1972, *Solar Phys.* **27**, 446.
Frost, K. J. and Dennis, B. R.: 1971, *Astrophys. J.* **165**, 655.
Ginzburg, V. L. and Zheleznyakov, V. V.: 1958, *Soviet Astron.* **AJ2**, 653.
Haddock, F. T. and Alvarez, H.: 1973, *Solar Phys.* **29**, 183.
Hudson, H. S.: 1972, *Solar Phys.* **24**, 414.
Kane, S. R.: 1972, *Solar Phys.* **27**, 174.
Kane, S. R.: 1973, *Bull. Am. Astron. Soc.* **5**, 274.
Kane, S. R. and Anderson, K. A.: 1970, *Astrophys. J.* **162**, 1003.
Kane, S. R. and Donnelly, R. F.: 1971, *Astrophys. J.* **164**, 151.
Kane, S. R. and Lin, R. P.: 1972, *Solar Phys.* **23**, 457.
Krimigis, S. M.: 1965, *J. Geophys. Res.* **70**, 2943.
Lin, R. P.: 1970a, *J. Geophys. Res.* **75**, 2583.
Lin, R. P.: 1970b, *Solar Phys.* **12**, 209.
Lin, R. P.: 1971, *Conf. Papers, 12th International Conf. on Cosmic Rays*, Hobart, Tasmania, Australia, **5**, 1805.
Lin, R. P.: 1973, in R. Ramaty and R. G. Stone (eds.), *Proc. Symposium on High Energy Phenomena on the Sun*, Goddard Space Flight Center, Greenbelt, Maryland, September 1972, p. 439.
Lin, R. P. and Anderson, K. A.: 1967, *Solar Phys.* **1**, 446.
Lin, R. P. and Hudson, H. S.: 1971, *Solar Phys.* **17**, 412.
Lin, R. P., Evans, L. G., and Fainberg, J.: 1973a, *Astrophys. Letters* **14**, 191.
Lin, R. P., Wang, J. R., and Fisk, L. A.: 1974, *J. Geophys. Res.*, in preparation.
Malitson, H. H., Fainberg, J., and Stone, R. G.: 1973, *Astrophys. Letters* **14**, 111.
McGuire, R. E., Anderson, K. A., Chase, L. M., Lin, R. P., and McCoy, J. E.: 1972, *Trans. AGU* **53**, 1086.
Palmeira, R. A. R., Allum, F. R., and Rao, U. R.: 1971, *Solar Phys.* **21**, 204.
Parker, E. N.: 1963, *Interplanetary Dynamical Processes*, Interscience Publishers, Chapter 8.
Simnett, G. M.: 1971, *Solar Phys.* **20**, 448.
Simnett, G. M.: 1974, *Space Sci. Rev.* **16**, 257.
Smerd, S. F.: 1970, *Proc. Astron. Soc. Australia* **1**, 305.
Smith, D. F.: 1970, *Solar Phys.* **15**, 202.

Smith, D. F.: 1972, in R. Ramaty and R. G. Stone (eds.), *Proc. Symposium on High Energy Phenomena on the Sun*, Goddard Space Flight Center, Greenbelt, Maryland, September, 1972, p. 558.
Sullivan, J. D.: 1974, *J. Geophys. Res.*, to be published.
Syrovatskii, S. I. and Shmeleva, O. P.: 1972, *Soviet Astron. AJ* **16**, 273.
Trubnikov, B. A.: 1965, *Rev. Plasma Phys.* **1**, 105.
Wang, J. R., Fisk, L. A., and Lin, R. P.: 1971, *Conf. Papers, 12th International Conference on Cosmic Rays*, Hobart, Tasmania, Australia, **2**, 438.
Wild, J. P.: 1950, *Australian J. Sci. Res.* **A2**, 541.
Wild, J. P., Murray, J. D., and Rowe, W. C.: 1954, *Australian J. Phys.* **7**, 439.
Wild, J. P., Smerd, S. F., and Weiss, A. A.: 1963, *Ann. Rev. Astron. Astrophys.* **1**, 291.
Wild, J. P. and Smerd, S. F.: 1972, *Ann. Rev. Astron. Astrophys.* **10**, 159.
Zaitsev, V. V., Mityakov, N. A., and Rapoport, V. O.: 1972, *Solar Phys.* **24**, 444.

DISCUSSION

(After Three Review Papers by Stewart, Fainberg, and Lin)

Smith: What is your opinion about the constancy of the exciter speed? We heard Dr Stewart this morning saying it is constant; you wrote that there is a deceleration.

Fainberg: In the range of altitudes 50 $R_\odot$ to the Earth's orbit, there is a deceleration, but it might not be a real variation of velocity but the result of changes in the pitch angle distribution. It is also possible that the conversion efficiency to EM waves changes with electron energy as the exciters move through different coronal regions.

Smith (to Fainberg): What do you think of the result of Haddock and Alvarez where they see the transition from the dominance of fundamental to second harmonic radiation?

Fainberg: We have specialized in simple events. The others have so many components that it is very difficult to keep track of those which go to low frequencies or not. It is always very difficult to trace individual features. Haddock and Alvarez flew their equipment on a very noisy spacecraft and had to study only strong bursts which always go in groups. Their dynamic range was limited and their events often reached saturation. I cannot have an opinion without seeing the data.

Smith (to Lin): The reason the electron stream does not relax quickly is that its density is very low relative to the background plasma so that the relaxation time is very long. The almost isotropic distribution of electrons may explain the decay of the burst below 10 keV since *any* isotropic velocity distribution is stable relative to the production of plasma waves.

Dryer (to Fainberg): The IMP 6 results strongly suggest that the type III electrons may be used to map the interplanetary magnetic field. Is it possible to use this technique to map discontinuities such as shocks (emitted earlier)?

Fainberg: Yes, as long as type III's are visible. We can see that all type III's are not coming along nice spirals. But to study these disturbances systematically we would need a continuous type III output.

Sturrock: Is it true that type III's have not been correlated with the second stage of acceleration? If so, what are we to infer?

Smith: The electrons accelerated in the second phase are accelerated to relativistic energies which means that they are much more effectively stabilized by non-linear processes than lower energy electrons.

Wild: In reply to Dr Sturrock's question as to why type III's occur during the first phase of acceleration and not the second, the answer may be that, in fact, they *are* observed in both phases. It is possible that second phase acceleration is caused by the interaction of the type II shock wave with magnetic flux tubes. In fact, one quite often sees type III's or type III-like features (herringbone effect) emanating from type II bursts.

Pneuman (to Fainberg): The type II observations you showed in your movie seemed to show emission only in front of the disturbance propagated outward. Now if the type II burst is produced by a hydromagnetic shock moving out from the flare site, this shock wave would be expected to move out more or less in all directions. Then one should expect to see a sort of spherical geometry. Do you have any explanation as to why this does not seem to be the case?

Fainberg: The reason is that in the movie we show only the centroid of the emitting region. You cannot infer the size of this region from these pictures.

Zirin (to Lin): You said you do not see any Coulomb scattering effects but only 10^{-3} of the particles arrive near the Earth. Where do the other electrons go?

Lin: Presumably most of the energetic electrons are trapped high in the corona at ambient densities too low to produce observable levels of hard X-ray emission and only $\sim 1\%$ of the electrons escape to be observed at 1 AU.

Rosenberg (to Stewart): We have been told that only harmonics are observed at low frequencies. What is the ratio of the number of fundamental to harmonic bursts in ground-based observations?

Stewart: I cannot quote numbers. It is very difficult to see pairs.

Rosenberg (to Fainberg): You showed a slide with the rise and decay times of low frequency type III showing a constant ratio. The same is true also for higher frequencies and the curve might be extended to ~ 200 MHz as noted by De Groot many years ago.

Schmidt (to Fainberg): Have you tried to correlate the shape of the spiral pattern followed by type III's in the range of altitudes 0 to 50 $R_\odot$ with the velocity of solar wind?

Fainberg: We looked for that, but we do not have enough resolution to see the spiral pattern in the range 0–50 $R_\odot$; but the technique is in principle capable of doing that.

Erickson (to Fainberg): You stated that the motion picture and slides gave only the position of the centroid of type III emission. Could you give an estimate of the actual size of the emitting region at, say, 0.5 AU?

Fainberg: About 20 to 30° as seen from the Sun.

Wild (to Sturrock): One of the important points raised this morning is the doubt cast on the more or less accepted doctrine that type III electrons travel along neutral sheets – doubt cast by the paper of Smith and Pneuman – which says that this is impossible because of the existence of transverse fields. Perhaps Dr Sturrock would care to give a second opinion on this conclusion.

Sturrock: It is certainly a problem to reconcile these conflicting views. However, I think that reconnection of field lines could produce type III electrons most easily if current sheets are very thin. If this is so, the filling out of these current sheets would allow free outflow of electrons and this could resolve the paradox.

Rosenberg: We have redone the same calculations as Pneuman and Smith. We find a much smaller transverse magnetic field than they do.

Smith: We have also redone these computations and found the same result as we published. We said that electrons cannot travel *on* the axis of a streamer for very long but they can travel there for $\frac{1}{2}R_\odot$ – then go out and remain close to the axis.

Pneuman (to Rosenberg): Any computation of a model neutral sheet relies, of course, upon the value of the electrical conductivity in the corona about which very little is known. Other considerations lead us to believe that the conductivity is considerably lower than one would normally estimate. If this is the case, then the transverse field would be even larger than either you or I would estimate. My second point is that we must clearly differentiate distinct parts of a streamer configuration – that associated with the density enhancement and the very much smaller neutral sheet in the center of the streamer. Thus, the fact that electrons cannot travel in the sheet does not exclude the possibility that streamers could be the location of type III emission, i.e. the electrons could travel outside the sheet but still in the streamer.

Stewart (to Kane): Can Dr Kane say at what altitude electrons can be accelerated by shock waves?

Kane: The observations I presented are not related to the acceleration of electrons in shock waves. The observations do indicate that during the flash phase the acceleration of electrons occurs at an altitude where the ion density is about 10^9 cm^{-3}.

HIGH RESOLUTION STUDIES OF TYPE III SOLAR RADIO BURSTS

C. CHIUDERI and R. GIACHETTI
Osservatorio Astrofisico di Arcetri, Firenze, Italy

C. MERCIER
Observatoire de Paris, Meudon, France

and

H. ROSENBERG* and C. SLOTTJE
Sterrewacht 'Sonnenborgh', Utrecht, The Netherlands

Abstract (*Solar Phys.*). High spectral, temporal and spatial resolution observations were obtained with the 60-channel Utrecht solar radio spectrograph (160–320 MHz) and the 169 MHz Nançay solar radioheliograph. From a large number of type III bursts the average height was found to be 0.37 solar radius above the photosphere, corresponding to approximately the Newkirk streamer density, if the bursts are emitted at the harmonic of the local plasma frequency. No center-to-limb variation, nor east–west asymmetry was observed. All double bursts, double humped bursts, precursor-type III had exactly the same position and general shape for both members of the pair. From this it was concluded that fundamental-harmonic pairs are very rare at frequencies above 160 MHz (Mercier and Rosenberg, 1974).

An analysis of the fine structure in decametric type IIIb–type III pairs indicates that the type IIIb burst is related to the fundamental of the plasma frequency, while the type III burst is emitted at the harmonic of the local plasma frequency. The type IIIb fine structure is explained by emission at the same frequency of the plasma frequency and the lowest harmonics of the electron gyro-frequency, through the coupling of plasma waves and low harmonics electron Bernstein waves. The derived densities and magnetic field strengths satisfy lateral pressure equilibrium across the neutral sheet in a coronal streamer. This is regarded as circumstantial evidence for the occurrence of type III bursts inside neutral sheets in coronal streamers (Rosenberg, 1973).

It is shown that a 100 keV electron beam can generate Bernstein waves under coronal conditions. Only the lowest two or three harmonics are rendered unstable, as is observed in the fine structures of type IIIb bursts (Chiuderi *et al.*, 1974).

Recent spectral observations of type III bursts between 160 and 320 MHz show circular polarization up to 100% during the rise of the type III burst. The degree of polarization falls off exponentially with a time constant of approximately 0^{s}.1 (Slottje, 1974).

* Presently at Harvard College Observatory, Cambridge, Mass., U.S.A.

Gordon Newkirk, Jr. (ed.), Coronal Disturbances, 225–226. All Rights Reserved.

References

Chiuderi, C., Giachetti, R., and Rosenberg, H.: 1974, to be published.
Mercier, C. and Rosenberg, H.: 1974, to be published.
Rosenberg, H.: 1973, Thesis, Utrecht.
Slottje, C.: 1974, *Astron. Astrophys.* **32**, 107.

COMMENTS

Leblanc: You are referring to electron density models of a streamer given by Newkirk in 1958 (2 times the quiet corona). I wish to state that there is a large scatter between quiet corona models (Leblanc, Leroy, and Lecautet: 1973, in press, *Solar Phys.*). These models refer to the period 1970–1972, and we show that they differ from Newkirk's model of the last maximum. Moreover, one streamer model differs from the other. You find that the pairs of type III's are in the same positions, and you conclude that the hypothesis of pairs of fundamental and harmonic components is excluded. I feel that your arguments are not strong enough because when taking scattering into account (Riddle and Leblanc), we could find fundamental and harmonic at the same position. Even the size criterion is not sufficient due to uncertainties in the measurements.

Rosenberg: Our model is Newkirk's 1952 streamer model and not a quiet corona model. We find pairs all over the disk. If there were scattering effects there should be a center-limb effect.

Newkirk: To your knowledge, is there any evidence for curvature in the trajectories of type III bursts as might be expected if they propagate up the legs of streamers on their way up to the nearly radial current sheet?

Rosenberg: As yet, no. Only two-dimensional, multi-frequency observations can show this. Perhaps, with the three-frequency Culgoora array we might see this.

Stewart: We have never seen any trace of highly polarized type III's at Culgoora.

Rosenberg: To see them you need a high time resolution.

CORONAL DENSITY STRUCTURES IN REGIONS OF TYPE III ACTIVITY

YOLANDE LEBLANC
Observatoire de Meudon, France

THOMAS B. H. KUIPER*
University of Maryland, U.S.A.

and

SHIRLEY HANSEN
High Altitude Observatory, U.S.A.

Abstract (*Solar Phys.*). We identify from daily *K*-corona observations the coronal regions located above type III burst sources. The radio observations were made with the east–west log-periodic array at Clark Lake Observatory. East–west positions of 260 isolated type III bursts during March and April 1971 have been measured between 20 and 65 MHz (disregarding storms of type III bursts). The *K*-corona observations are made with the HAO *K*-coronameter at Mauna Loa, Hawaii. During 1971, the heights of observation were 3.6′, 5′, 9′ and 13′ above the limb.

Kuiper (1973) analyzed the radio observations and showed that type III bursts occur in discrete regions of the corona corotating with the Sun, below which Hα activity was simultaneously observed.

We have identified these regions with *K*-corona features at limb passage. It appears that nearly all type III burst regions are not located on the axis but rather at the edge of dense structures or in low density regions (Figures 1 and 2). By comparing the observed positions of the bursts at two frequencies, and assuming a radial propagation of type III exciters, we derive the electron density and the vertical gradient at 1.5 R_0. These values, corrected for scattering effects, are compared with electron density models (Figure 3) obtained from optical observations, and though the precision is poor, there is good quantitative agreement.

It is suggested that the propagation of the type III exciters does not occur along the neutral sheet of streamers which corresponds to the densest part of coronal structures but along field lines which delineate these structures and/or along radial field lines in regions of lower density.

* Present address: Jet Propulsion Laboratory, 183 B-365, California Institute of Technology, Pasadena, Calif. 91103, U.S.A.

Gordon Newkirk, Jr. (ed.), Coronal Disturbances, 227–230.

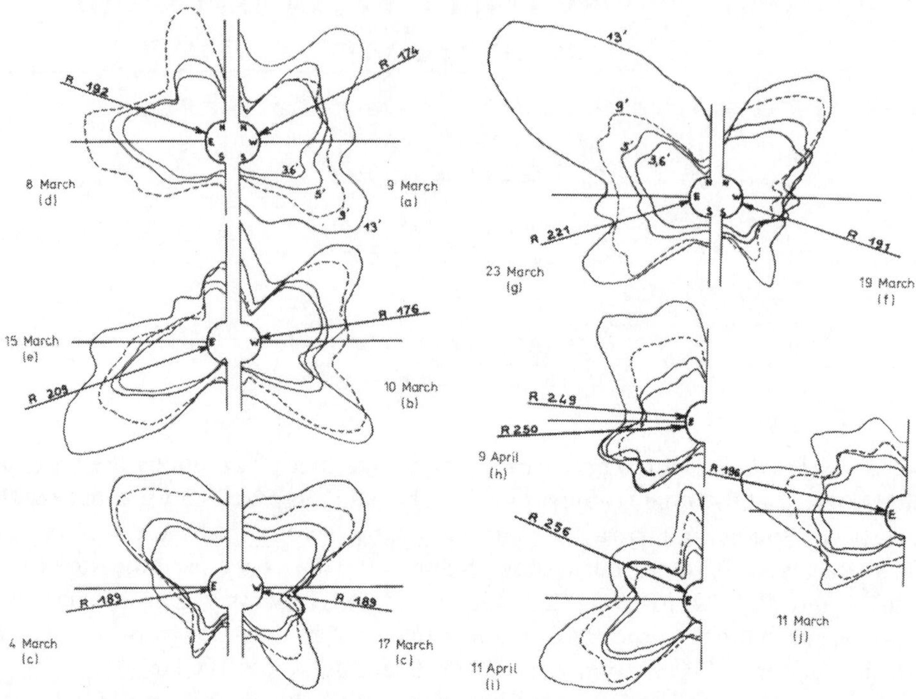

Fig. 1. *K*-corona observations: Intensity profiles obtained at four heights. The dashed line indicates the intensity at 9′ above the limb. The position (latitude) of the McMath region associated with a solar burst region is shown by an arrow. Region 196 did not show any type III activity.

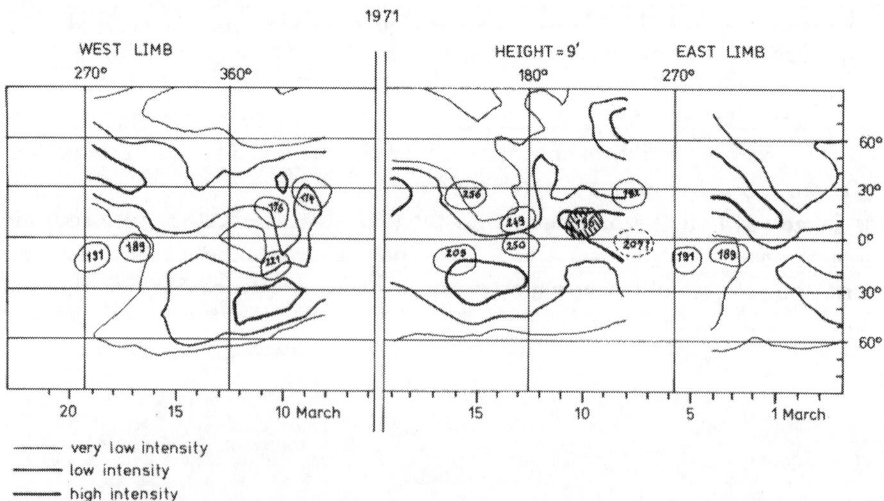

Fig. 2. *K*-corona observations: Synoptic maps, indicating positions of McMath plages (identified by last 3 digits) associated with type III regions.

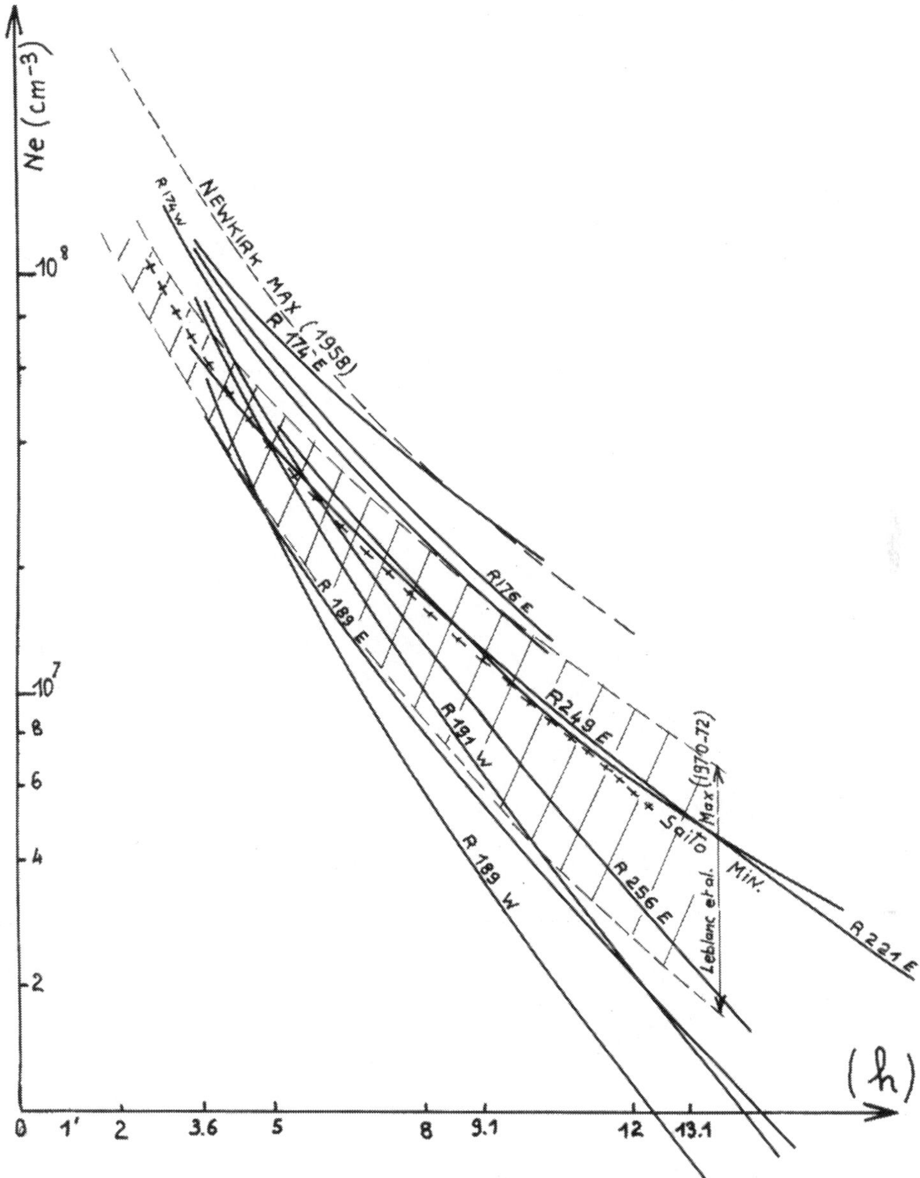

Fig. 3. Calculated electron density models. For comparison are also given: Newkirk's model (1956–58 maximum), Saito's model (average model at minimum), the Leblanc *et al.* (1973) models (1970–1972 maximum-hatched area).

COMMENTS

Kundu: This morning we heard from Ron Stewart that type III electrons travel through dense coronal regions. On the other hand, you imply that the electrons travel through low density regions. I am wondering whether you are not reaching important conclusions on the basis of insufficient data.

Leblanc: I agree that a more systematic study should be done, however for all the set of data considered here, the type III burst regions avoid the centres of large dense structures in the corona, and it is of interest to consider MacMath region 196 located in a dense structure: while it showed considerable Hα activity no type III bursts were associated with this region.

METER AND DECAMETER WAVELENGTH POSITIONS OF SOLAR RADIO BURSTS OF JULY 31–AUGUST 7, 1972

M. R. KUNDU and W. C. ERICKSON

Astronomy Program, University of Maryland, College Park, Md., U.S.A.

Abstract (*Solar Phys.*). The positional analysis of solar bursts at meter and decameter wavelengths during the period July 31–August 7, 1972 is presented. The observations were taken with two arrays – a log periodic array of 16 elements situated on an E–W base line of 3.3 km and portions of the new Clark Lake array in the form of a Tee (an E–W arm of 32 log spiral antennas and a N–S arm of 16 similar antennas). The new array operates over the frequency range 10–120 MHz and has angular resolutions of approximately 3′.5 at 100 MHz and 8′.5 at 40 MHz in the E–W direction.

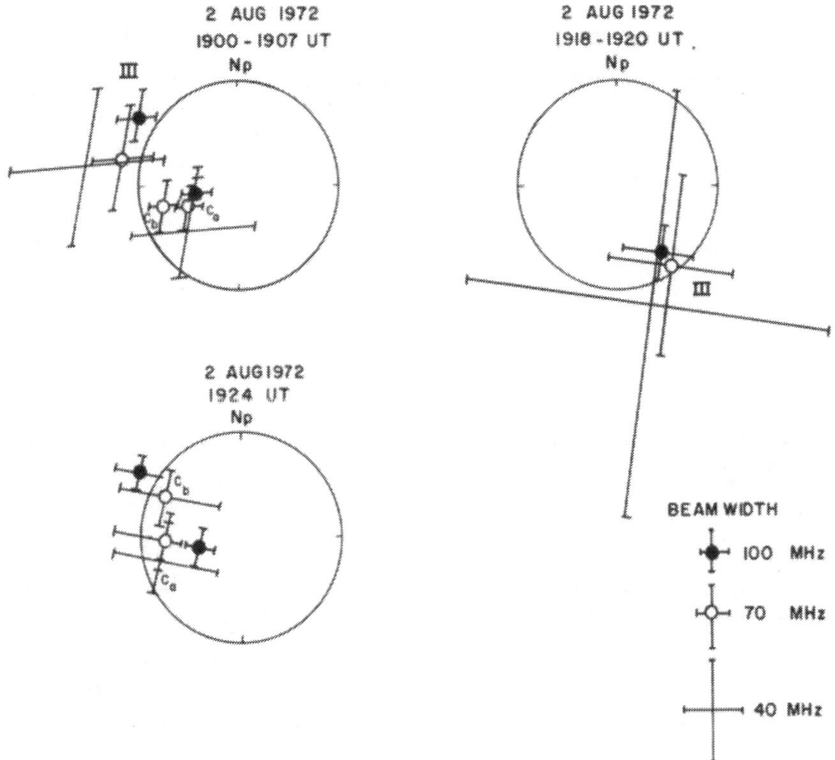

Fig. 1. Positions of type III and continuum bursts for August 2, 1972. The big circle represents the optical disk. The radio positions were obtained from a combination of both E–W and N–S data. These positions are given by the radius vector from the Sun's center and a position angle measured to the east from the north pole of the Sun. The E–W and N–S sizes of the burst sources are indicated by the lengths of the lines aligned with the appropriate angle.

Gordon Newkirk, Jr. (ed.), Coronal Disturbances, 231–233. *All Rights Reserved.*

In the N–S direction the instrument has a resolution of 6′ at 100 MHz and 15′ at 40 MHz. During this period the activity was often so complex that we represent it only by 'snapshots' taken at representative times at three representative frequency ranges 90–110 MHz, 60–80 MHz and 30–50 MHz. Most of the activity during this period was associated with the active regions McMath 11976 and 11970. The radio-

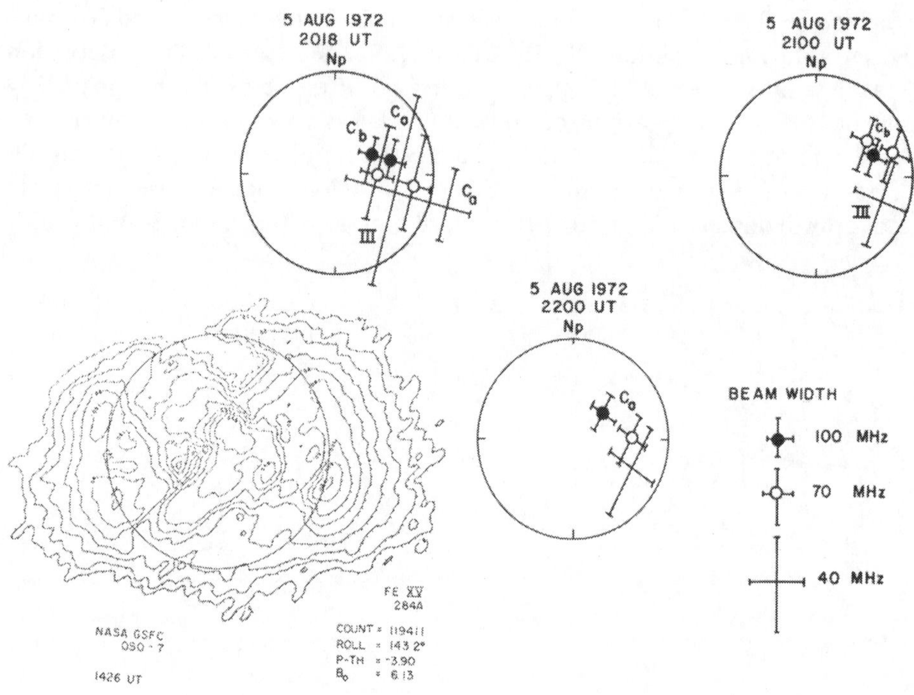

Fig. 2. Positions of bursts for August 5, together with the Fe xv line (284 Å) map obtained by NASA-GSFC OSO-7. The radio positions are indicated in the same way as for August 2 (Figure 1).

emissive regions were also closely associated with the Fe xv (284 Å) coronal maps produced by NASA-GSFC from OSO-7 observations (Figures 1, 2 and 3). Except near the CMP of the region 11976, two regions of continuum emission were often observed – one a relatively smooth continuum and the other a continuum superimposed with many type III's and other fine structure. It seems possible to interpret these continua in terms of plasma waves originating from two sources located at different heights or with different electron density gradients. The angular size of type III sources seems to increase with decreasing frequency. This implies that the open field lines along which the type III electrons travel have larger angular extent at greater height.

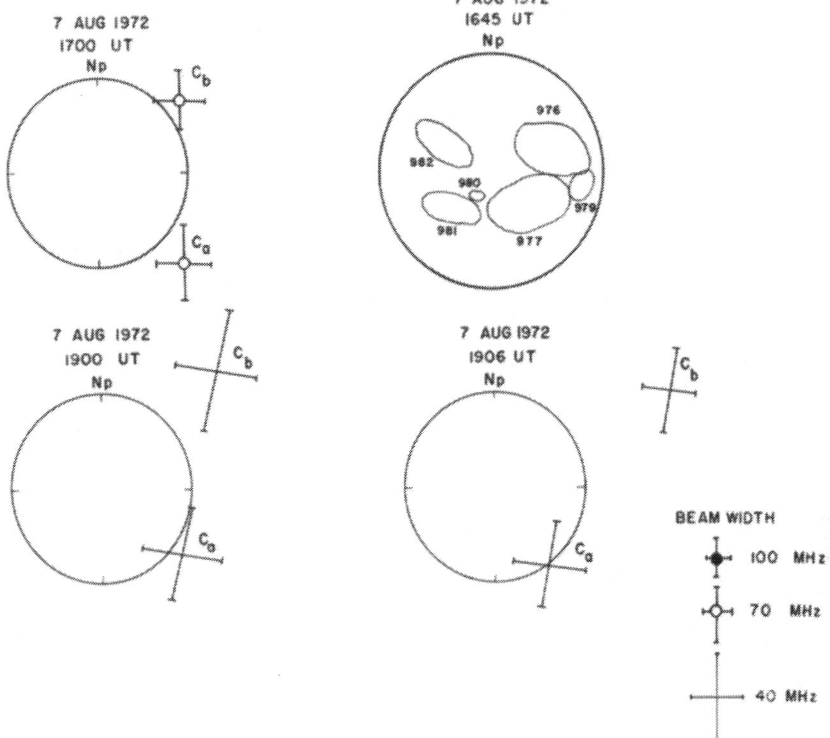

Fig. 3. Positions of bursts for August 7, together with the McMath-Hulbert Calcium report. The positions are indicated in the same was as for August 2 (Figure 1). Note the displaced position of the continuum source C_b at 1906 UT relative to its position at 1900 UT, which indicates an ejection with a velocity of about 1400 km s^{-1} from the parent continuum source.

COMMENTS

Pick: Did you observe any sudden variation of the diameter of type III's with decreasing frequency?

Erickson: The size generally increases with wavelength but that is not always so.

Zirin: Have you seen something connected with the huge brightening of the flare on that day (August 2, 1972)?

Erickson: The radiation became darker and darker.

SOURCE STRUCTURE IN METRE-WAVE TYPE V SOLAR BURSTS

N. R. LABRUM
Division of Radiophysics, CSIRO, Sydney, Australia

and

R. A. DUNCAN
Division of Mathematical Statistics, CSIRO, Sydney, Australia

Abstract (*Astrophys. Letters*). The type V burst has been defined as a wideband continuum which sometimes appears for a minute or so following a type III burst (Wild *et al.*, 1959b). It is now generally accepted that type III bursts arise from plasma waves set up by electrons escaping with velocity $\sim c/3$ along open magnetic field lines (Wild *et al.*, 1959a; Stewart, 1965); the most widely accepted explanation of type V continua is that they arise from plasma waves set up by electrons of similar velocity which have become trapped in a coronal magnetic loop (Weiss and Stewart, 1975). On this hypothesis the plasma waves are set up by two opposing electron streams in the trapping region, and from this consideration Zheleznyakov and Zaitsev (1968) have concluded that type V emission should be predominantly at the second harmonic of the local plasma frequency. In this paper we describe and discuss some two-dimensional observations of source positions of type III–V events which were obtained at 80 MHz on the Culgoora radioheliograph.

The 14 type III–V events summarized in Table I, associated with 14 different solar active regions, were observed at 80 MHz with the radioheliograph during the period 1968 May to 1969 December. All 14 events were listed as type III–V in the Radiophysics spectrum catalogue, but only 3 had classic III–V spectra resembling that in Figure 1(a). The remainder graded down to bursts only a little more diffuse than a normal type III or *U* burst. However, the sensitivity of the heliograph is greater than that of the spectrograph, and in some of these less definite events a second heliograph source, which we believe to be type V, was recorded. The less definite events are listed separately in the second line of Table I.

TABLE I

Displacement of type V from type III source position – based on 14 events observed during the period 1968 May to 1969 December

Spectrograph appearance	Number of events			
	Outward	Tangential	None	Inward
Classic III, V	3	0	0	0
Diffuse III or *U*	4	2	5	0

Gordon Newkirk, Jr. (ed.), Coronal Disturbances, 235–238.

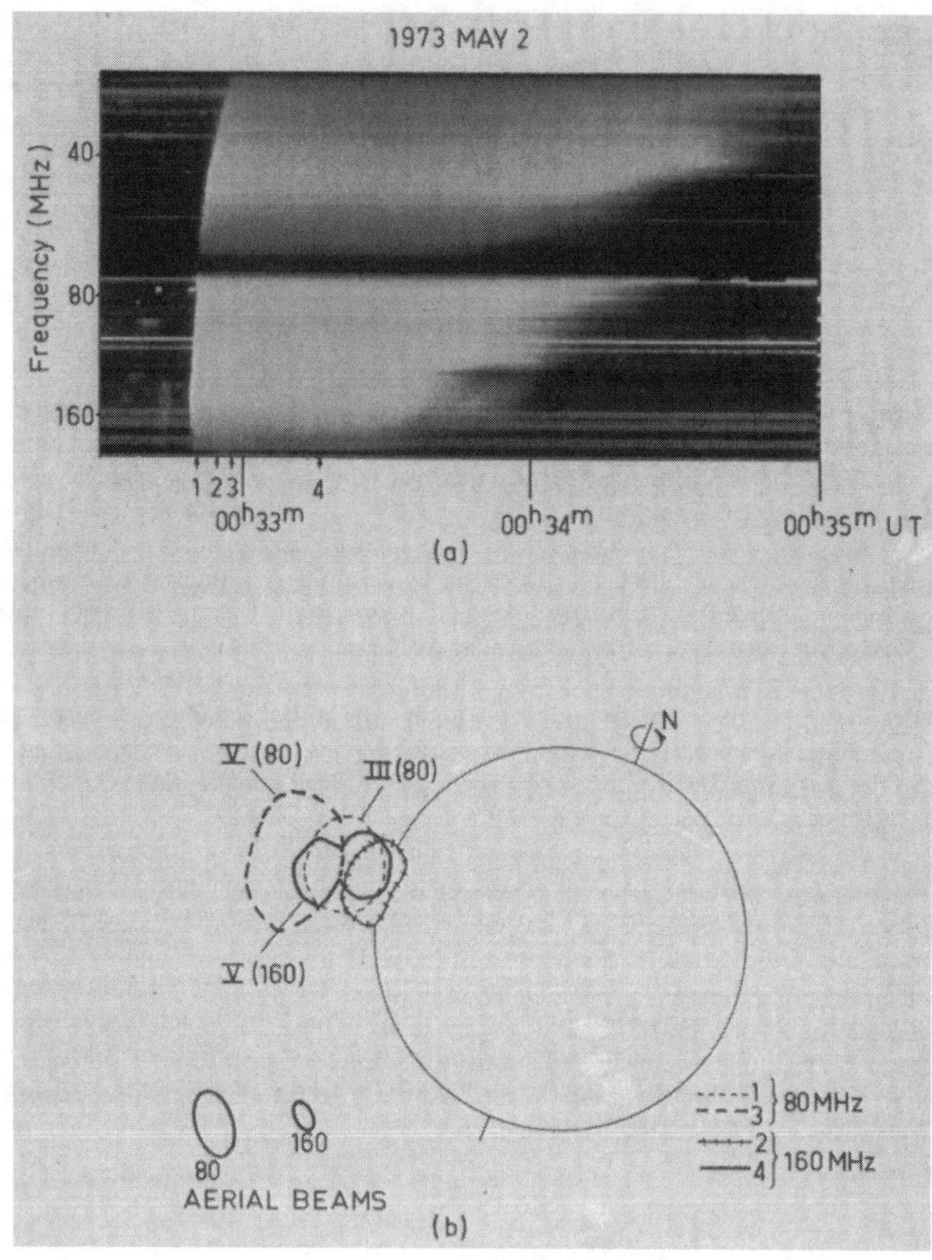

Fig. 1. Type III–V burst, 1973 May 2.
(a) Spectrum. (b) Source positions at 80 and 160 MHz at times indicated.

The angular size of the type V source was usually similar to that of the type III – about 6′ to half maximum intensity at 80 MHz. With the exception of a single event, the degree of circular polarization of both sources was low – less than 10%.

The most interesting characteristic of the radioheliograms is the frequent displacement of the type V from the type III source position. This displacement was always either tangential or radially outward; in no case was the type V source closer to the disk centre than the type III source (Table I). Figure 1 is a more recent example of a type III–V event, with radioheliograms at both 80 and 160 MHz. At 80 MHz both the type III and the type V source components were recorded, and the outward displacement of the type V from the type III source position is clearly seen. The dispersion of type V height with frequency is also evident (Figure 1b).

The complete absence of inward displacements seems to rule out the possibility of explaining the observed outward displacements (Table I) in terms of tangential displacements and accidents of perspective. It therefore seems necessary to explain them in terms of emission and propagation.

At first sight the outward displacements might seem explicable in terms of type III emission at the fundamental plasma frequency, and Zheleznyakov and Zaitsev's (1968) suggestion of type V emission at the second harmonic of the plasma frequency. This simple explanation, however, conflicts with other evidence. Stewart (1972) has observed the fundamental and second harmonic of individual type III bursts almost to coincide in position, and Steinberg *et al.* (1971) and Riddle (1972) have explained this effect in terms of coronal refraction and scattering. These results make the question of emission at the fundamental or second harmonic irrelevant to a discussion of apparent source heights. Thus, if we retain the belief that type V emission arises through a plasma wave mechanism, if we accept Stewart's (1972) observations and their explanation, and if we suppose that the propagation characteristics of type V emission resemble those of type III emission, it becomes necessary to postulate that in the type V source region the coronal plasma density is enhanced.

Acknowledgements

We are indebted to Dr S. F. Smerd and Dr J. A. Roberts for suggestions on the presentation of this paper.

References

Riddle, A. C.: 1972, *Proc. Astron. Soc. Australia* **2**, 98.

Steinberg, J. L., Aubier-Giraud, M., Leblanc, Y., and Boischot, A.: 1971, *Astron. Astrophys.* **10**, 362.

Stewart, R. T.: 1965, *Australian J. Phys.* **18**, 67.

Stewart, R. T.: 1972, *Proc. Astron. Soc. Australia* **2**, 100.

Weiss, A. A. and Stewart, R. T.: 1965, *Australian J. Phys.* **18**, 143.

Wild, J. P., Sheridan, K. V., and Neylan, A. A.: 1959a, *Australian J. Phys.* **12**, 369.

Wild, J. P., Sheridan, K. V., and Trent, G. H.: 1959b, *Paris Symp. Radio Astron.*, Stanford Univ. Press, p. 176.

Zheleznyakov, V. V. and Zaitsev, V. V.: 1968, *Astron. Zh.* **45**, 19; (English translation) *Soviet Astron. AJ* **12**, 14 (1969).

COMMENTS

Smith: In the case assumed by Zheleznyakov and Zaitsev of two counter streaming electron streams, plasma waves will be generated very effectively. In this case, the fundamental will be produced more effectively than the harmonic because it is produced by induced scattering which is effectively a much more non-linear process than for the second harmonic. This would explain the outward displacement of the type V burst when combined with results of scattering calculations.

Sturrock: If only 10 out of 14 cases show that type V is a greater height than the type III source, this seems hardly significant statistically.

Duncan: We have now twice as many cases.

Sturrock: It seems that the relative heights, if real, could be understood if the coronal density is higher in the closed field regions than in the open field regions.

Uchida: It seems that the trailing edge of your type V is quite irregular. Is it real?

Labrum: Yes.

Melrose: In connection with the suggestion that the type V sources might appear higher than second harmonic type III's because of preferential backward emission of the latter, I would like to point out that the theories predicting backward emission are outdated. Smerd, Wild, and Sheridan (1962, *Australian J. Phys.* **12**, 369) assumed coalescence of plasma waves generated by the stream with thermal plasma waves, but this leads only to thermal emission. Zheleznyakov and Zaitsev (1970, *Soviet Astron. AJ* **16**, 67) assumed that scattered plasma waves coalesce with the initial plasma waves, but they ignored the decrease in wave number which accompanies the scattering of plasma waves. Coalescence of plasma waves with larger and smaller wave numbers leads to preferential emission in the direction of the larger wave number. Which is the forward direction?

Rosenberg: How can you reconcile both type *U* and type V bursts as being due to particles trapped in a closed magnetic field bottle? I would then expect a *U* followed by a V or something like that.

Newkirk: The rarity of type *U* bursts and the fact that type III and V's are often associated may be due to the configuration of the ambient magnetic field. We have calculated the fates of the energetic particles emitted above some 46 major flares. About 30% escape the Sun directly, 70% go into mirroring orbits (type V?), and only 0.2% impact the chromosphere at the conjugate point. These latter particles might be identified with *U* bursts.

DIRECT MEASUREMENTS OF THE DIRECTIVITY OF TYPE I AND TYPE III RADIATION AT 169 MHz

C. CAROUBALOS and J. L. STEINBERG
Observatoire de Meudon, France

Abstract. A 169 MHz radiometer fed from a Sun-pointed 8 dB antenna was flown on the Soviet probe Mars-3 in 1971–1972. This experiment called Stereo-1 is part of a cooperation program between the U.S.S.R. and France.

Analysis of the data shows that the spacecraft systems and the experiment performed satisfactorily throughout the mission and that absolute calibrations on the Earth and on board the spacecraft are reliable. The stereo angle θ between the observing directions varied from 0 to 76°; 185 h of observations have been logged including several hundreds of type III's and about 10000 type I bursts which have not yet been analyzed in detail.

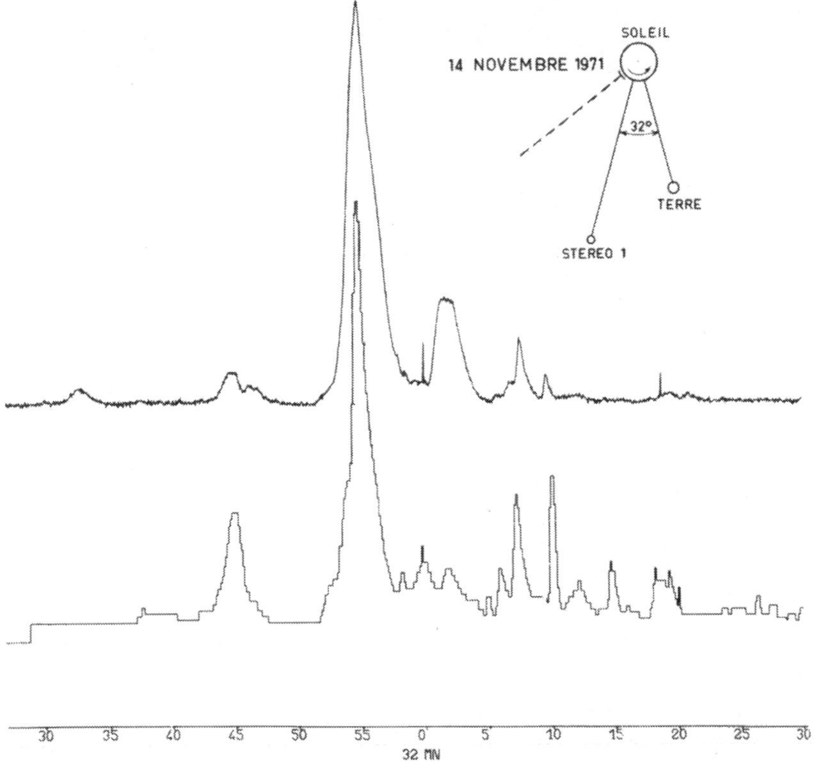

Fig. 1. A group of 169 MHz type III bursts recorded on the ground (top) and on board the Mars-3 space probe. Stereo angle was 33°. The dotted line gives the radial direction through the active region where those events took place.

Gordon Newkirk, Jr. (ed.), Coronal Disturbances, 239–242.

The first results of a partial analysis of type I and type III bursts are:

Type III

– The time profile of isolated type III's at 169 MHz is independent of the observing direction; for instance, there are only negligible propagation effects on the decay time of these events.

– This property has been used to separate individual events when they overlap; it has thus been possible to identify fundamental and harmonic pairs among the November 14, 1971 events which took place at about 70° E longitude as seen from the Earth. 80% of these events are harmonics without detectable fundamentals.

– The directivity ratio R is defined as the ratio of the event intensity in the probe direction to its intensity in the Earth direction. R is constant over individual events but can change from that event to the next. The distribution of R's over a daily observing period (shorter than 1 h) can change from day to day between two extremes:

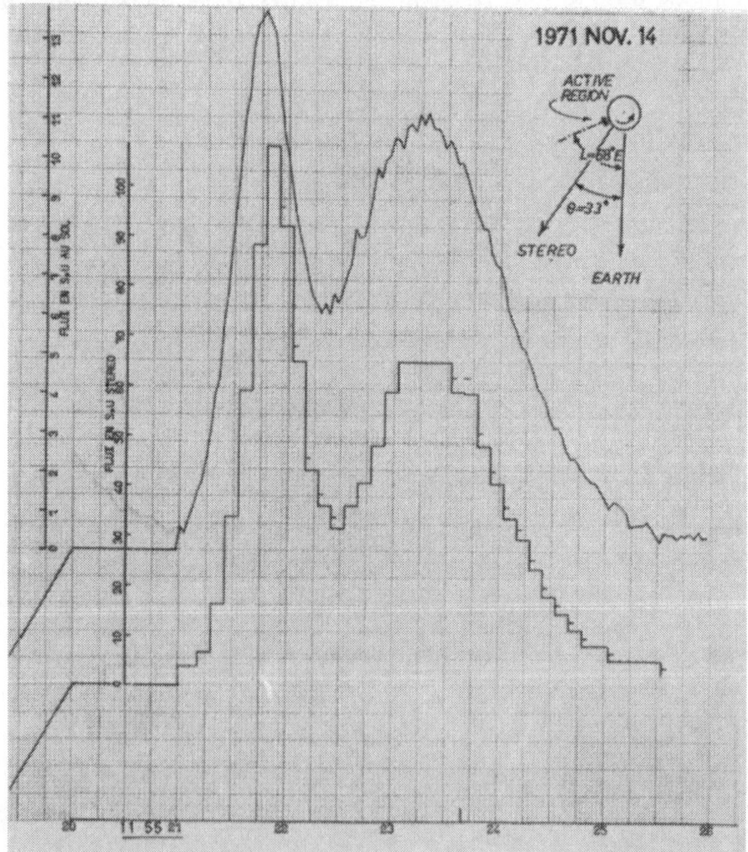

Fig. 2. A typical fundamental-harmonic pair from the 6 pairs observed on November 14, 1971. The fundamental component is always found more directive than the harmonic.

– The range of R's can be narrow around some stable mean value which depends upon the geometrical configuration (source position, lines of sight); in these cases, the source is seen close to the Sun's center from both observing stations.

– The R values are widely scattered around a mean value which can be time dependent; in most of these cases, the source is close to the limb. On November 14, 1971, fundamental components are more directive than their associated harmonic components ($\theta = 33°$) and R ranges from 1 to 50.

– The observed directivity is always higher than predicted from spherically symmetric coronal models where the scattering power is large enough to account for a noticeable part of the type III sizes and for the relative positions of the fundamental and harmonic components. For sources near the center, the difference between the computed and observed R's can be made small only if the scattering power is small and the harmonic source directive; moderate angular shifts of the computed radiation pattern are generally required. For limb sources large deviations from spherical symmetry have to be built into the models.

– It has been shown that streamer models in an otherwise spherical corona cannot fit the data better than spherical models.

Type I

Type I radiation is strongly beamed in space; the correlation between Mars-3 and

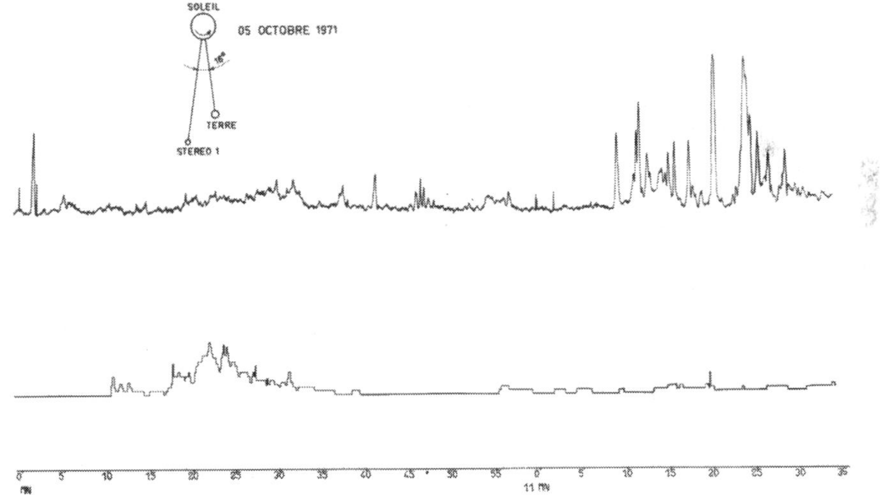

Fig. 3. Groups of 169 MHz type I bursts recorded on the ground (top) and on board the Mars-3 space probe. Decorrelation is evident with a stereo-angle of only 16°.

Earth records drops to zero for a stereo angle of $\sim 20°$. However in some cases a good correlation is observed for larger angles (i.e. 27° or even 35°).

The present preliminary report is based on an extensive analysis of only part of the available data.

Notes added in proof. On Figure 1 the Stereo angle should read 33°.

The material summarized above is now published in full in:

Caroubalos, C. and Steinberg, J. L.: 1974, *Astron. Astrophys.* **32**, 245.

Caroubalos, C., Poquerusse, M., and Steinberg, J. L.: 1974, *Astron. Astrophys.* **32**, 255.

PAIRED TYPE III BURSTS

C. CAROUBALOS, J. HEYVAERTS, M. PICK, and G. TROTTET
Meudon Observatory, France

Abstract. Radioheliographic observations of type III bursts occurring by pairs in a repetitive fashion were made with the high time resolution E–W Nançay Radio heliograph (169 MHz). The data were available in digitized form with a rate of 25 E–W profiles per second. The study is based on two days of observations: for the first one (January 14, 1971) the radio source was at 16′ W and for the second one (June 29, 1971) at only 4′ W.

As seen in Figure 1, the time separations between the two components of the pairs are found to be maintained for the pairs of a given group. These time separations are found to be respectively 0.95 ± 0.15 s and 0.3 ± 0.05 s for the first and second days of observation whereas the corresponding time spacings between pairs are 2.4 ± 0.4 and 1 ± 0.25 s.

The two components of a paired burst do not systematically differ in any space-time aspect: the durations, profiles, diameters and positions are found to be the same within the limits of the accuracy of our measurements. No specific law for the variation of the flux maxima is found, – the second component of the pair can be either weaker or stronger than the first one.

Different possible interpretations are reviewed: the hypothesis that the emission of both the fundamental and second harmonics is the cause of the pair cannot be accepted because of the similarity of the space-time properties of the source in both

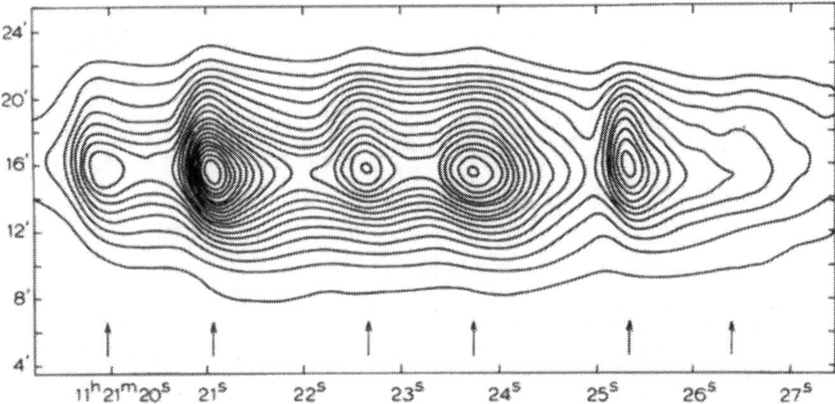

Fig. 1. January 14, 1971. E–W intensity distribution of the source of type III burst vs time observed at 169 MHz. The horizontal axis is universal time: the vertical axis indicates minutes of arc west of the central meridian. Arrows indicate maxima of elements of pairs.

Gordon Newkirk, Jr. (ed.), Coronal Disturbances, 243–244. All Rights Reserved.

components of the pair seen at the same frequency. As no consistent relationship exists between the flux of the two components, it is also hard to accept an echo effect. Furthermore, a numerical simulation of such an effect for the second harmonic performed by the Monte-Carlo method proves that it is impossible to obtain the observed paired configuration even if only a slightly scattering corona is considered.

A possible explanation is suggested here: when a high current is being established in the lower atmosphere of the Sun, an oscillating regime is possibly set up, the period of which would be defined by the electric parameters of the circuit. The current flows in fine structures in which it would produce a pinching if the plasma is dense enough. The pinched plasma column will oscillate with a frequency determined by its own characteristics, perhaps, producing a beam of fast electrons at each maximum contraction. Calculations of the oscillations of such a plasma column for a given external sinusoidal current suggest that for reasonable conditions, especially certain periods of the external current oscillations, compressions of the column are obtained at intervals of several seconds. These may split into a pair more closely spaced at a typical time interval of about a second. Further development is necessary to test the validity of such an explanation.

Reference

Caroubalos, C., Heyvaerts, J., Pick, M., and Trottet, G.: 1974, *Solar Phys.* **37**, 205.

COMMENTS

Smith: What accelerating mechanism have you in mind? You have only discussed the gross dynamics of the pinch. In fact, in all pinch experiments in the laboratory in which acceleration occurs, there are opposing fields and reconnection.

Pick: A pinched plasma column – the oscillations of which could produce beams of accelerated electrons at every maximum constriction. We did not try to present a detailed theory of the particle acceleration in the present study.

Stewart: Should not the decay rate hide everything? Maybe these events are not type III's.

Pick: At decimeter and meter wavelength, type III bursts are often very short. No confusion can exist between type III and pulsating structures. The latter are polarized and their source is not at the same position as the type III source.

Mullaly: Micro-wave bursts of duration smaller than 1 s appear to repeat themselves with a 10 s period. Maybe this could explain your meter wave observations?

Pick: Such periodicities of some seconds have effectively been observed in the microwave range. The time delay between two elements of a pair is in our case equal or less than 1 s.

Cole: I have observed similar structures in type III bursts using the electro-optic spectrograph at Culgoora.

UNUSUAL ABSORPTION OF A SOLAR TYPE II BURST BY 'SHADOW' TYPE III BURSTS

K. KAI*

Division of Radiophysics, CSIRO, Sydney, Australia

Abstract (*Proc. Astron. Soc. Australia*). The Hα-flare on 1968 August 23^{d}23^{h}45^m was followed by a short-lived type II burst. The radio spectrograph record obtained at Culgoora is shown in Figure 1. The type II burst was of split-band structure with unusual spectral features in each component of the split band. The most remarkable one is an absorption feature between 23^{h}52^{m}02^s and 23^{h}52^{m}10^s: suppression of the bright type II emission along an 'inverted *U*' (between 120 and 85 MHz).

The present event was recorded also with the 80 MHz radioheliograph at Culgoora.

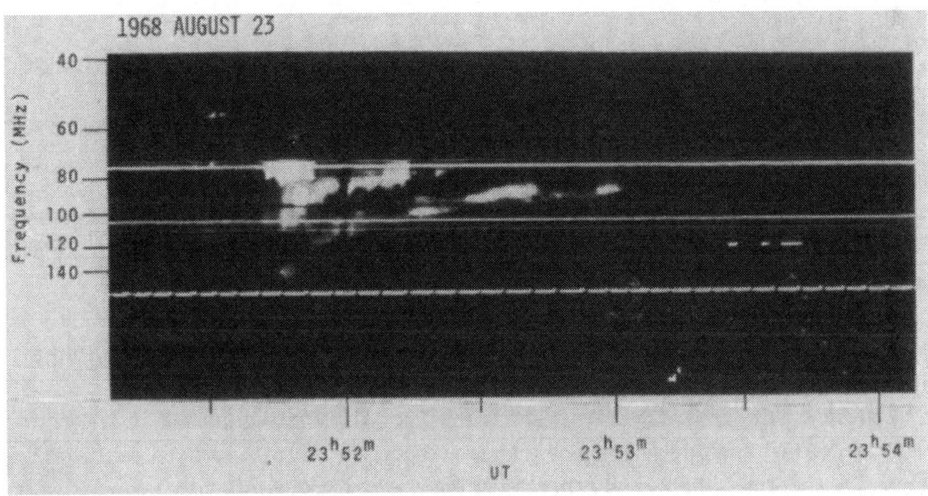

Fig. 1. Radio spectrograph record of the split-band type II burst on 1968 August 23. Note the unusual absorption features in the spectrum of the type II burst (particularly around 23^{h}52^m); which appear as 'shadow' bursts superimposed on the type II emissions.

The lower-frequency band of the type II burst crossed 80 MHz between 23^{h}51^{m}50^s and 23^{h}52^{m}20^s, the upper-frequency band later, at about 23^{h}52^{m}46^s. The observed size (to half the peak brightness) is $\sim 12' \times 7'$ for each split-band source. The separation between the upper and lower band sources is $\sim 4'$. The most dramatic absorption at 80 MHz occurred between 23^{h}52^{m}02^s and 23^{h}52^{m}04^s. The lower band of the type II emission was cut along a sharp dark lane which drifted from high to low frequencies with a drift rate (~ -10 MHz s^{-1}) similar to that of a typical type III

* On leave from Tokyo Astronomical Observatory, University of Tokyo.

Gordon Newkirk, Jr. (ed.), Coronal Disturbances, 245–248. All Rights Reserved.

burst. It is not clear, however, whether the dark lane is a shadow burst of type III or of the ascending branch of an inverted-U burst. Contour plots of the 80 MHz type II source are shown in Figure 2 for two successive seconds, just before and during the absorption. The observed maximum brightness temperature was $\sim 10^9$ K before the absorption; it then dropped by a factor of 5 to 1.8×10^8 K. It should be noted that

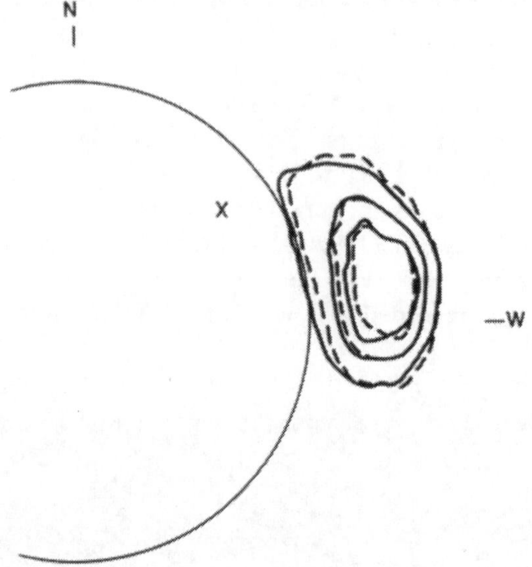

Fig. 2. Two contour plots of the 80 MHz type II burst taken successively, before and during the absorption (full line for $23^h52^m01^s$ and dashed line for $23^h52^m02^s$). The contours represent $1/\sqrt{2}$, 1/2 and 1/4 the maximum brightness.

this marked depression in brightness occurred during one second and without significant change of the source structure.

The observation of shadow inverted-U and type III bursts against the background type II emission strongly suggests that the absorption is caused by a beam of fast electrons ($v \sim c/3$) which would otherwise give rise to 'emission' of the same spectral type. Then, irrespective of the detailed mechanism, we have the following requirements for absorption. (1) The absorbing region must be on the observer's side of the type II emitting regions and of sufficient angular extent to cover the latter (both for the upper-frequency and lower-frequency band emissions, whose observed sources are separated by 4′). (2) The region must act as an absorber only when the fast electrons pass through the region. At such times its optical depth must increase to values of the order of 1. (3) The brightness temperature of the U and III bursts which are generated by the fast electrons in the absorbing region must be at least one order of magnitude lower than that of the type II burst.

A possible geometrical relation between the source regions of the type II and of the shadow bursts is schematically illustrated in Figure 3. The sources of the shadow bursts are in front of the type II sources along the line of sight: for the U-bursts the form is a large-scale magnetic loop. The heavy absorption of the type II emissions depends critically on the above, or other similarly stringent, relative positions and sizes of the emitting and absorbing regions. However, we have to recall that such events have rarely been observed.

We can offer no detailed model which would explain the absorption phenomenon.

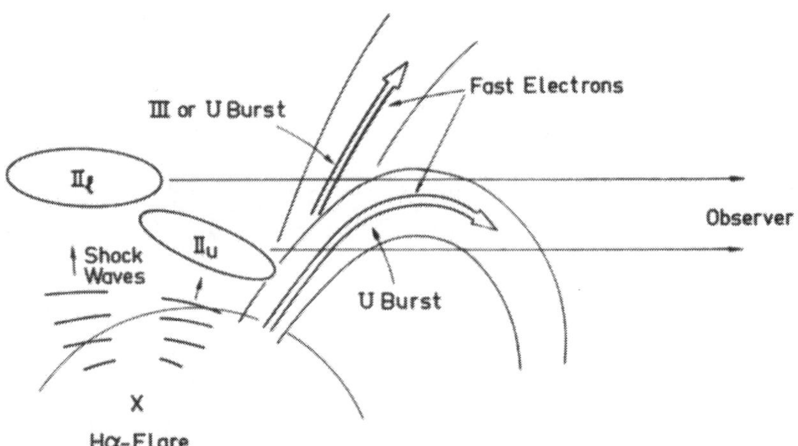

Fig. 3. A possible geometrical relation between the source regions of the type II and the shadow bursts. Here II_u and II_l indicate respectively the regions where the upper-frequency and lower-frequency bands of the type II burst are emitted.

We only suggest that the absorption or scattering takes place resonantly on plasma waves which are excited by a beam of fast electrons close to the local plasma frequency and possibly its second harmonic. In the absence of the beam, fluctuations of the background electrons scatter incident transverse waves only by negligible amounts. Once strong longitudinal waves are set up by the electron beam, they could scatter the incident waves much more efficiently.

COMMENTS

Uchida: Was there any indication or even a trace of brightening at the second harmonic on your dynamic spectrum at the moment of absorption by shadow type III probably through the interaction of plasma waves with frequency ω in the type III region caused by the particle stream and radio wave of frequency ω from the type II source.

Kai: There is no evidence of second harmonic emission. As computed by D. Melrose their intensity would be too small to be detected.

Rosenberg: How are you sure that you are dealing with a true absorption? I can show slides of the Utrecht spectrograph records where there is an awful lot of structure including 'black' type III's which could be taken as absorbing features.

Kai: I am afraid you could not see clearly the absorption along an 'inverted *U*' shape on my slide. If you looked at a better print of the spectrum, you would be convinced that we are dealing with a genuine absorbing feature.

Melrose: The theories of mine to which Dr Kai referred are as follows: The first theory is based on the assumption that the absorption is due to the inverse of the familiar plasma emission processes. Absorption at both the fundamental and the second harmonic are possible, but the latter occurs only under unacceptably extreme conditions. Absorption at the fundamental requires that the two sources overlap, i.e., fill the same volume, in order to explain the observed bandwidth. The other theory is based on the assumption that the type III stream generates ion sound waves. This theory can explain all the details of Kai's observations, but the assumption that ion sound waves are so generated is not considered likely from the viewpoint of existing theories of the propagation of electron streams.

HIGH RESOLUTION OBSERVATIONS OF GENERALIZED FAST DRIFT BURSTS

ØYSTEIN ELGARØY and PER H. ROSENKILDE
Institute of Theoretical Astrophysics, Oslo University, Norway

Abstract (*Solar Phys.*). During 1969 and 1970 groups of generalized fast drift bursts were observed on 21 days with a high resolution radio spectrograph at Oslo Solar Observatory. Totally 48 groups were detected in the frequency band 310–340 MHz. In the great majority of the cases the groups were accompanied by metre wave type III bursts at lower frequencies.

In Figure 1 the distribution of durations of single bursts as measured at a chosen frequency (318 MHz) is shown. The average burst duration amounted to 0.26 s in a sample comprising 834 bursts. For normal type III bursts observed in the same frequency band the average duration is 1 s (Elgarøy and Lyngstad, 1972).

Using data on the location of emissive regions at 169 MHz and at 408 MHz and

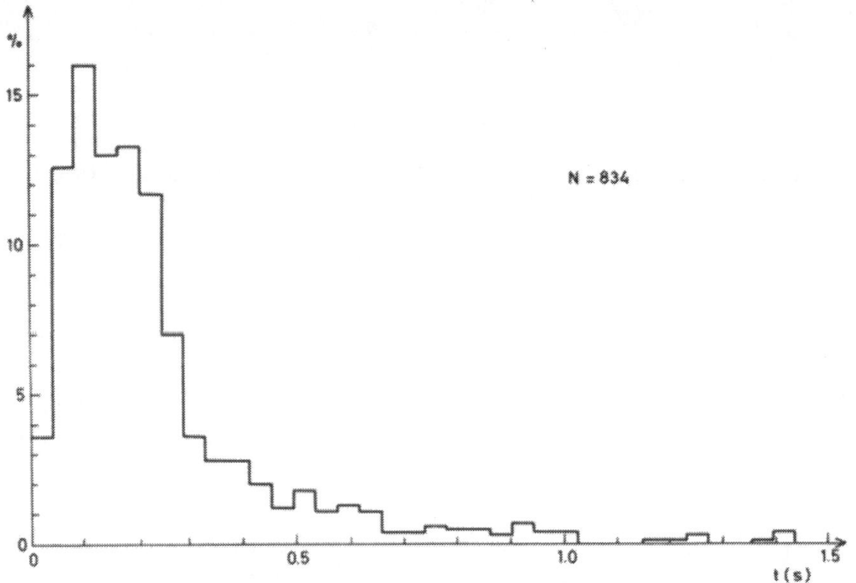

Fig. 1. Distribution of burst durations.

also data on the position of type III groups at 169 MHz from Nançay, the position of the groups of generalized fast drift bursts on the solar disk has been estimated. Plotting the average burst duration against the distance of the source from the central meridian, it is found from Figure 2 that the duration increases with increasing distance from the centre. This may indicate the presence of scattering effects.

The frequency drift velocity was very high, and positive as well as negative drifts

Gordon Newkirk, Jr. (ed.), Coronal Disturbances, 249–251.

occurred. Frequently the starting time of a burst seemed to be the same at all frequencies. The distribution of measured values of inverse frequency drift velocities for 402 bursts is shown in Figure 3. In 50% of the cases the bursts had a negative drift, 17% of the bursts showed a positive drift, and for the last 33% of the bursts the fre-

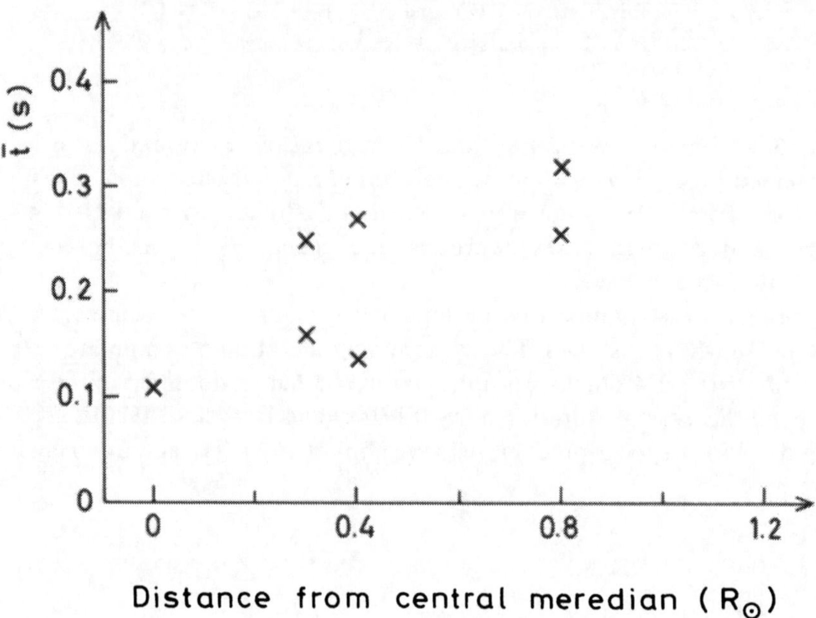

Fig. 2. Average burst duration at different distances from the central meredian.

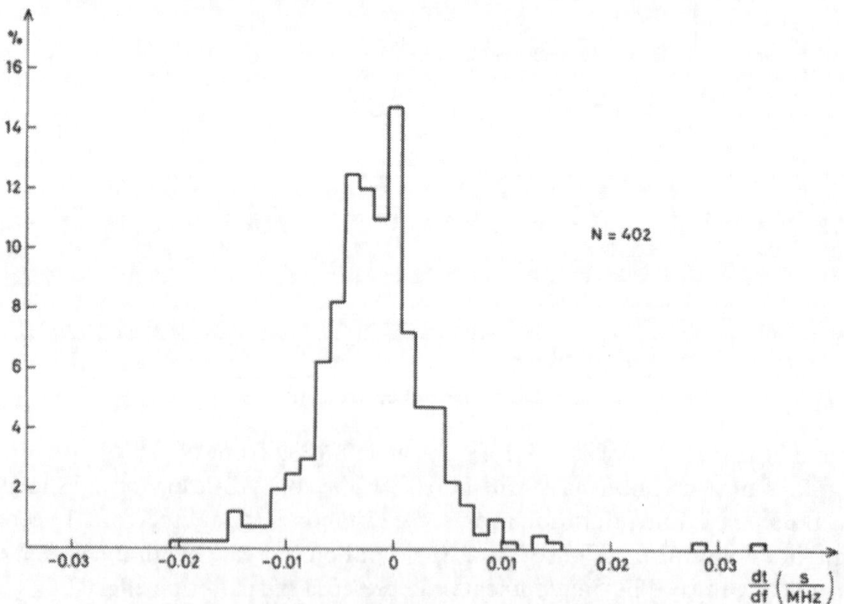

Fig. 3. Distribution of inverse frequency drift velocities ($\mathrm{d}t/\mathrm{d}f$).

quency drift was immeasurable ($|dt/df| < 0.002$ s MHz^{-1}). The distribution in Figure 3 is very different from the one found for metre wave type III bursts in the 300 MHz range (Elgarøy and Lyngstad, 1972).

About 50% of the groups of generalized fast drift bursts occurred when Explorer 37 observed X-ray bursts from the Sun.

Generalized fast drift bursts strongly resemble some other types of short lasting bursts observed with the 310–340 MHz radio spectrograph.

The fast drift bursts described here may be generated by plasma waves excited by particle streams directed towards the solar surface. The stream motion and differential group delay are then counteracting sources of frequency drift. When the first source dominates, positive frequency drift is observed whereas negative drift results when the situation is reversed. Zero frequency drift occurs when the two effects balance. The whole matter is a question of fractions of a second.

Acknowledgements

The authors are indebted to the solar group at Meudon for positional data on type III bursts at 169 MHz.

This research has been sponsored in part by the Air Force Cambridge Research Laboratories, United States Air Force under Grant AFOSR 72-2294.

References

Elgarøy, Ø. and Lyngstad, E.: 1972, *Astron. Astrophys.* **16**, 1.

MECHANISMS FOR FLASH PHASE PHENOMENA IN SOLAR FLARES

DEAN F. SMITH

High Altitude Observatory, National Center for Atmospheric Research, Boulder, Colo., U.S.A.*

Abstract. Mechanisms for explaining the various forms of particles and radiation observed during the flash phase of solar flares are reviewed under the working hypothesis that the flash phase is the time in which electrons and to a lesser degree protons are accelerated in less than one second. A succession of such accelerations is allowed to explain longer lasting or quasi-periodic phenomena. Mechanisms capable of such acceleration are reviewed and it is concluded that first-order Fermi acceleration in a reconnecting current sheet is the most likely basic process. Such acceleration, however, gives rise to a rather narrow distribution of particle velocities along a given field line which is unstable to the production of electron plasma and ion-acoustic waves. This plasma turbulence can heat the plasma to produce soft X-rays and filter the initially narrow velocity distribution to produce a power law energy distribution. Electrons travelling inward from the acceleration region produce hard X-rays by bremsstrahlung and microwave bursts by gyro-synchrotron emission. Whereas the interpretation of X-ray spectra is relatively straightforward, the interpretation of microwave spectra is difficult because the source at low frequencies can be made optically thick by several different mechanisms.

Electrons travelling further inward presumably thermalize and produce impulsive EUV and Hα emission. The theory for these emissions, although amenable to present techniques in radiative transfer, has not been worked out. Electrons travelling outward give rise to type III radio bursts by excitation of electron plasma waves and the electrons observed at the Earth. Study of the interaction of a stream of electrons with the ambient plasma shows that the electron spectra observed at the Earth do not necessarily reflect their spectrum at the acceleration region since they interact via plasma waves as well as through Coulomb collisions. The mechanisms for the conversion of plasma waves into radiation and the propagation of the radiation from its source to the observer are reviewed.

1. Introduction

We begin by defining the flash phase of a solar flare as that phase in which rapid acceleration of electrons and to a lesser extent protons and heavier nuclei occurs along with a rapid heating of part of the flare plasma. This heating may be due to collisional losses of the accelerated particles, to some part of the acceleration process or to a combination of these two processes. By rapid acceleration of electrons we have in mind a time scale of less than one second. Although it cannot be said that definitive theories exist for the conversion of low energy (1–500 keV) electron energy into all its other manifest forms, it can be said that for all impulsive phenomena which have been examined in detail, the low energy electrons do have sufficient energy. Thus a definition of the flash phase which has rapid acceleration as its basic ingredient should serve as a good working hypothesis on which to try to build a comprehensive theory. While it is clear that any theory for the flash phase must be consistent with the preflare buildup and subsequent flare phases (e.g. act as a trigger for subsequent phases), we limit ourselves to the flash phase as an entity in itself in keeping with the observations which have been presented.

* The National Center for Atmospheric Research is sponsored by the National Science Foundation.

Gordon Newkirk, Jr. (ed.), Coronal Disturbances, 253–282.

The basic problem of acceleration of electrons and other particles on the very short (< 1 s) time scales required for flash phase phenomena has received little attention in recent years. It has been shown (Smith and Priest, 1972) that the evacuation mechanisms of Alfvén and Carlqvist (1967), Carlqvist (1969), and Syrovatsky (1969, 1972) suffer from basic inconsistencies. These mechanisms were appealing for their simplicity. Takakura (1971) has shown how an electric field can be derived self-consistently by the rotation of a spherical cloud in a magnetic field – a cosmic dynamo. To generate significant electric fields by this mechanism quite large rotational velocities are required. Takakura would overcome this difficulty by placing two such dynamos in magnetic fields of opposite polarity as shown in Figure 1, i.e. in a current sheet, and build up a large current system which would have a large potential drop across it *ala* Alfvén and Carlqvist. This stored energy could be released by allowing the electric field to become large enough to excite electron plasma waves which would reduce the conductivity by several orders of magnitude in the corona. However, Takakura (1971) takes 1 MeV as the maximum energy that a particle can attain which is the potential drop across his current system. This is incorrect in a field of plasma waves since the net acceleration is determined by the microscopic encounters of the particles with the waves rather than the macroscopic potential drop. While there are many attractive features in Takakura's model the configuration of Figure 1 seems too exotic to be a commonly occurring phenomenon on the Sun.

On the other hand, one of the basic ingredients of Figure 1, namely the current sheet, should be a commonly occurring phenomenon and it is well known that large electric fields can be developed when the opposing magnetic fields are reconnecting (Sweet, 1958; Petschek, 1964; Sturrock, 1968). The basic problem here is that, as pointed out by Parker (1973), no one has been able to find a constraint from the reconnection region itself which would determine the maximum rate of reconnection.

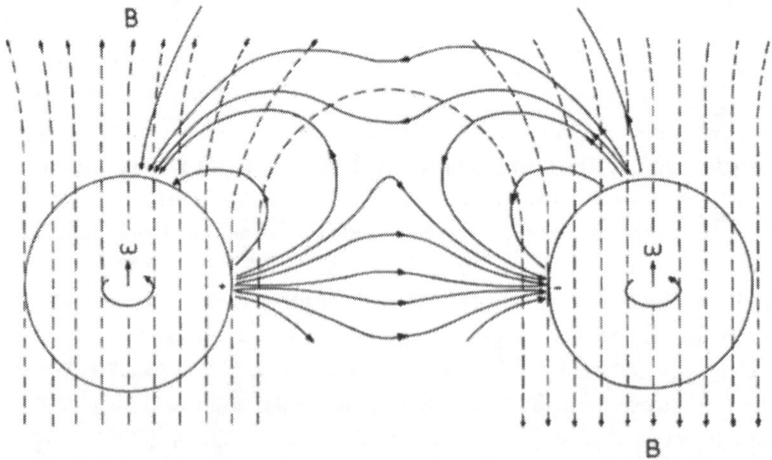

Fig. 1. Schematic picture showing lines of electric force between two spherical clouds rotating in the same direction in magnetic fields of opposite polarities (dashed lines) in Takakura's model.

Thus Yeh and Axford (1970) have surmised that the rate of reconnection is determined by boundary conditions. It should also be noted that the ideal configurations studied by Petschek (1964), Sonnerup (1970), Yeh and Axford (1970), and Coppi and Friedland (1971) are unlikely to occur in nature since in general there will be a component of magnetic field perpendicular to these two dimensional configurations, as shown in Figure 2, i.e. field lines in general will reconnect at an angle other than 180°. It can be shown by arguments similar to those used by Yeh and Axford (1970) and Parker (1973) for two dimensional configurations that the decrease in the rate of reconnection in a steady state due to this additional component of the field is also undetermined by anything in the reconnection region itself (Cowley, private communication). In the face of such indeterminacy, about all we can do is to look at indicative computer results such as a nonlinear analysis of the tearing mode instability (van Hoven and Cross, 1973), laboratory experiments (Bratenahl and Yeates, 1970) and solid evidence of reconnection in the vicinity of the Earth (Aubry *et al.*, 1970) to affirm our faith that it occurs on the Sun. Having taken that step it is natural to assert that the flash phase is that interval of time in which extremely fast reconnection has begun, but not gone so far as to preclude very coherent phenomena such as rapid acceleration of particles taking place.

Given that the reconnection geometry of Figure 2 is the simplest configuration in which acceleration can take place, one can then study the fate of individual particles in this geometry which has been started by Speiser (1965), Friedman (1969), Cowley

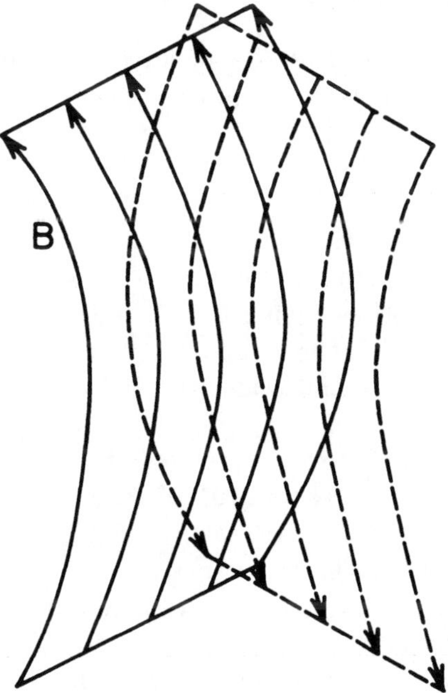

Fig. 2. Schematic drawing of how field lines with significant antiparallel components are likely to occur in nature with field components in every direction.

(1972), and Eastwood (1972, 1974). We shall review and extend these results in Section 2 and see to what extent they fit the observations.

With nonthermal particles, mostly electrons, the question of how these particles interact to give the myriad forms of radiation observed arises. If we place the height of the primary acceleration region at about 10000 km, then electrons moving down field lines will give rise to microwave bursts by gyro-synchrotron emission and hard X-rays by bremsstrahlung (Holt and Ramaty, 1969; Takakura, 1972), and impulsive EUV by free-bound, bound-bound and bremsstrahlung emission (Kane and Donnelly, 1971) and Hα kernels (Zirin and Tanaka, 1973) by bound-bound emission from excited partially ionized plasma. The theories for these emissions are reviewed in Sections 3 and 4. The electrons travelling up field lines give rise to type III radio bursts by excitation of plasma waves and the electron events observed near the Earth (Smith, 1974a). The theory for type III bursts is reviewed in Section 5. There is no theory for the propagation of low energy electrons in the interplanetary medium, but some facets of this problem are included in Section 5.

Finally there is the heating of the flare plasma at the flash phase which causes the rise of soft X-ray emission that continues into the flare proper (Kahler and Kreplin, 1970; Thomas and Teske, 1971). In the concluding section we check to see how this emission fits in to the theory reviewed and summarize what remains to be done to improve our understanding of the flash phase.

2. Acceleration of Particles in Current Sheets

We concentrate on electrons and recall the requirements provided by the observations. The acceleration mechanism must provide electrons in less than 1 s with a power law distribution.

$$\frac{dn(E_e)}{dE_e} \propto E_e^{-\delta} \tag{2.1}$$

with spectral exponent δ in the range 2–5 (Kane, 1973), where $n(E_e)$ is the number of electrons per cm^3 with energies in the range E_e to $E_e + dE_e$. The spectrum often shows a break at about 100 keV and is harder above 100 keV, especially for large flares (Kane and Anderson, 1970; Frost and Dennis, 1971). For example $\delta = 2.3$ may characterize the spectrum to 100 keV and $\delta = 4.6$ may characterize the spectrum beyond 100 keV. The total number of electrons accelerated above 22 keV is about 10^{36} (Lin and Hudson, 1971) which represents about 10% of the total flare energy (Kane, 1973). In other words the acceleration mechanism must be highly efficient. This requirement does not necessarily rule out stochastic processes, but it does imply that the only wave modes which can be involved are electron plasma waves or radiation since they are the only modes which can exist in the corona in a sufficiently lossless regime. Since the simplest manner of exciting these oscillations is by a stream of suprathermal electrons, the efficiency and time requirements imply that at least the initial stream of electrons must be accelerated directly. In other words electron plasma waves or radiation can at most act as a filter for an already directly accelerated electron stream. Consideration of collision losses (Syrovatsky and Shmeleva, 1972; Cheng, 1972; Biswas and Radhakrishnan, 1973) imply that acceleration will be most effective in the corona which may, however, extend to quite low heights (e.g., 2000 km) in some parts of an active region (Athay, private communication).

We also recall some basic requirements for particle acceleration provided by theory. From the force equation

$$\dot{\mathbf{v}} = \frac{e}{m_e}\left(\mathbf{E} + \frac{1}{c}\mathbf{v}\times\mathbf{B}\right), \tag{2.2}$$

where $\mathbf{v}$ is the particle velocity, $\mathbf{E}$ is the electric field and $\mathbf{B}$ is the magnetic field, acceleration requires either an electric field or a timedependent magnetic field. Electric space charge fields due to a charge imbalance via Poisson's equation

$$\nabla\cdot\mathbf{E} = 4\pi e(n_e - n_i), \tag{2.3}$$

where n_e and n_i the electron and ion densities, are eliminated by plasma processes on a time scale ω_{pe}^{-1}, where $\omega_{pe} = (4\pi n_e{}^2/m_e)^{1/2}$ is the plasma frequency. For a density of 3×10^3 cm^{-3}, the time scale is 4×10^{-10} s which effectively eliminates this possibility in the corona. The only other possibility is an electric field arising from a magnetic field whose lines of force are moving in the system in which acceleration takes place. This induced electric field $-(1/c)\,\mathbf{V}\times\mathbf{B}$, where $\mathbf{V}$ is the velocity of field lines in the frame in which acceleration is taking place, is perpendicular to $\mathbf{B}$ and so leads to an $\mathbf{E}\times\mathbf{B}$ drift across the field lines at a rate

$$\mathbf{v}_E = c\frac{\mathbf{E}\times\mathbf{B}}{B^2} = +\mathbf{V}. \tag{2.4}$$

However, once a particle has left the region where the induced electric field exists, which it can do after making half a gyration in the field $\mathbf{B}$, it will have acquired the velocity of the moving magnetic field line as well since the $\mathbf{E}\times\mathbf{B}$ drift is across this field line so that the net velocity gain is $2\mathbf{V}$. In other words, as pointed out by Fermi (1954), the particle effectively makes an elastic collision with the field line. It can be shown (Hayakawa *et al.*, 1964) that all types of magnetic acceleration reduce either to the Fermi mechanism or betatron acceleration (Swann, 1933). The advantage of the Fermi mechanism for our purposes is that it increases the particle velocity along field lines whereas betratron acceleration increases the velocity perpendicular to the field lines. When the velocity of a particle systematically increases, the acceleration process is referred to as a first order process.

We examine particle trajectories in reconnecting current sheets to see to what extent these requirements are satisfied. The simplest two-dimensional configuration devoid of complications such as slow shocks (Sonnerup, 1970) is shown in Figure 3. Field lines start to move in towards $y=0$ from both sides, for reasons which we shall not treat in view of the problems enumerated in the introduction, with a velocity far from the sheet v_f where the z-component of the magnetic field is B_f. As the field lines approach the neutral point they bend in towards the neutral point, are reconnected and move out in both directions along z. As shown in Figure 3, the motion of field lines gives rise to an electric field

$$E_x = \frac{v_f B_f}{c}. \tag{2.5}$$

A particle which is injected so that it is travelling exactly along this axis will be accelerated indefinitely by this field as shown by Speiser (1965). It sees an infinite

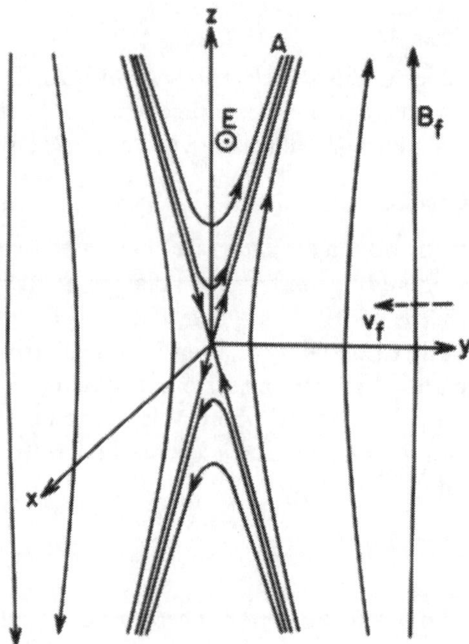

Fig. 3. Qualitative picture of the reconnection process showing the coordinate system employed. Magnetic field lines of strength B_f move toward the neutral point with velocity v_f which gives rise to an electric field E_x.

succession of first order Fermi accelerations and its final energy is limited only by the size of the system. However, the number of particles which satisfy the very stringent condition of being injected right along the neutral line is extremely small. The vast majority of particles entering the current sheet end up where there is a finite transverse field B_y. Thus we consider the fate of these particles in detail.

As first shown by Speiser (1965), one of the simplest ways to treat this problem is to make a transformation to a system moving with a field line since in this frame the electric field vanishes. We consider the field line labelled A in Figure 3. In the Sun's frame of reference, the incoming velocity of an electron is given by

$$\mathbf{v}_0 = \mathbf{v}_{\parallel 0} \pm \mathbf{v}_{c0} + \mathbf{v}_E, \tag{2.6}$$

where $\mathbf{v}_{\parallel 0}$ is the velocity parallel to $\mathbf{B}$ at time $t=0$, $\mathbf{v}_{c0}$ is the initial perpendicular velocity and $\mathbf{v}_E$ is the drift velocity given by Equation (2.4) with **B** the field strength of field line A and **E** given by Equation (2.5). Transforming to the rest frame of the field line (primed frame) which is moving with speed E_x/B_y in the z-direction, Equation (2.6) becomes

$$\mathbf{v}'_0 = \mathbf{v}'_{\parallel 0} + \mathbf{v}'_{c0}, \tag{2.7}$$

where the transformation velocity has been absorbed into $\mathbf{v}'_{\parallel 0}$. If we neglect the thermal velocity v_e of the electron because we are interested in transformation velocities somewhat higher than v_e, then the electron may simply be assumed to be streaming into the reversal region with a velocity $\mathbf{v}'_0 = \mathbf{v}'_{\parallel 0}$, where $\mathbf{v}'_{\parallel 0} \cdot \hat{z} = v'_{z0} = E_x/B_y$ and $\mathbf{v}'_{\parallel 0} \cdot \hat{y} = v'_{y0} = = E_x/B_z$. Thus the initial electron velocity is specified by E_x, B_y and B_z. Particle motion in this frame, which has been studied both analytically and numerically by Speiser (1965) and Eastwood (1972), consists simply of gyrations about the B_y and B_z fields. A particle oscillates between the reversing B_z field and is gradually turned so that it is moving in the positive z-direction by the B_y field which must be much weaker for v'_{z0} to be large. What is important for our purpose is that after half a gyration in the B_y field, a particle is ejected almost along the field line with a velocity $2v'_{z0}$ in the Sun's frame of reference. In other words, aside from a slightly more complicated motion due to the non-adiabatic nature of a current sheet, a particle is accelerated just as in a first-order Fermi process and because $B_y/B_z \ll 1$, almost along B_z. The energy gained by a particle is directly proportional to its mass as is the time for acceleration

$$\tau_A = \frac{\pi mc}{eB_y}. \tag{2.8}$$

The transverse field B_y is a very important parameter in this process since together with E_x it determines the energy gain

$$\Delta E = 2mv_{z0}^2 = \frac{2mv_f^2 B_f^2}{B_y^2}, \tag{2.9}$$

where we have used Equation (2.5). Some typical values for ΔE for electrons are given in Table I. Friedman (1969) showed that by allowing B_y to vary as $z^3/|z|$, a power law distribution of particle energies would be obtained with the index δ depending on $B_{z0} = B_f$, the plasma density n_e, the Mach number v_f/v_A where v_A is the Alfvén velocity in B_f and the spatial variations of the magnetic field. However, Friedman failed to explain how the different energy particles going along different field lines will get mixed. In fact, as pointed out by Eastwood (1972, 1974), the distribution of particle velocities along a given field line at one time is quite narrow and going in the opposite direction to the incoming particles as shown in Figure 4. This type of distribution is unstable to excitation of electron plasma waves if the distance between peaks is greater than v_e which can easily be satisfied according to Table I and ion-acoustic waves by the two-stream instability (Stringer, 1964). The growth rate for electron plasma waves in this case is about 0.4 ω_{pe} so that 20 e-folding steps take 1.8×10^{-8} s in a plasma of density 3×10^9 cm^{-3}. Thus, if the reconnection region has any reason-

TABLE I
Energy gains of electrons in a reconnecting current sheet

v_f (cm s^{-1})	B_f (G)	B_y (G)	E (keV)
		0.5	1.2
10^6	500	2.0	0.1
		5.0	0.01
		0.5	11.5
3.1×10^6	500	2.0	0.7
		5.0	0.1
		0.5	115
10^7	500	2.0	7.2
		5.0	1.2

able finite thickness like 0.1 km (Sturrock, 1968), then this instability is unavoidable in the reconnection region itself. Friedman (1969) considered such plasma turbulence to be highly likely, but he considered the turbulence to be isotropic. This led him to conclude that electrons could not be efficiently accelerated because they would be scattered and only ions with a quite high injection velocity would be unaffected, i.e. the plasma turbulence would act as a selection mechanism to allow a few ions to reach very high energies. If this indeed were the case, this mechanism would clearly be unsuitable for our purpose.

However, strongly excited electron plasma waves and ion-acoustic waves in a magnetic field which is sufficiently strong so that $\omega_{He} \approx \omega_{pe}$, where $\omega_{He} = eB/m_e c$ is the electron Larmor frequency, are much more nearly one-dimensionally distributed along the magnetic field. The reason for this behavior for electron plasma waves can be seen from the dispersion relation (Kaplan and Tsytovich, 1973)

$$\omega_p(\mathbf{k}) = (\omega_{pe}^2 + \omega_{He}^2 \sin^2 \theta + 3v_e^2 k^2)^{1/2}, \tag{2.10}$$

where θ is the angle between the direction of the magnetic field and the wave vector **k**. It is a characteristic of nonlinear interactions which must occur when the energy density in plasma waves W_p becomes sufficiently high that the frequency of the plasma

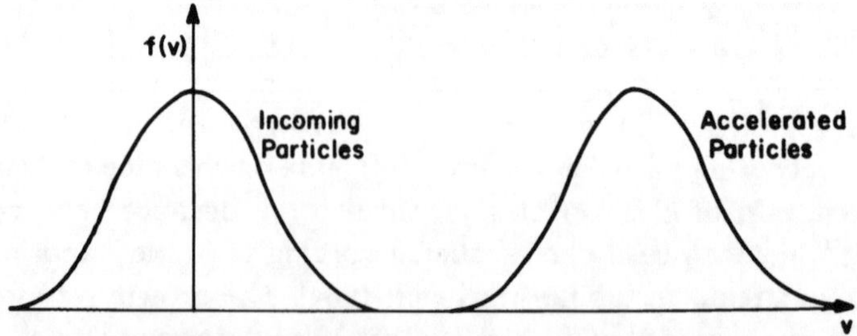

Fig. 4. The distribution of particle velocities along a given field line at one time which consists of the incoming and accelerated particles with their thermal spreads.

wave decreases somewhat in the process. In contrast to the case in the absence of a magnetic field where this decrease can only occur by a decrease in the wave number k, it can be seen from Equation (2.10) that in a magnetic field the decrease can also occur by a decrease in the angle θ. As a result plasma waves become distributed almost completely along the magnetic field. Ion-acoustic waves in a magnetic field $\omega_{He} \approx \omega_{pe}$ tend to be unaffected by the magnetic field, but for high levels of the energy density in these waves, W_s, become subject to resonance broadening (Tsytovich, 1971). This has the effect of keeping the waves close to the direction in which they were excited which is primarily along the magnetic field since the growth rate maximizes here and is consistent with experiments where the angular distribution of waves has been measured directly (Paul *et al.*, 1972).

As a consequence of this almost one dimensional plasma turbulence electrons will be preferentially accelerated along the field lines. We have in effect overcome the problem of electric space-charge fields by providing a fluctuating field. The nonlinear evolution of a spectrum of electron plasma waves in a field of ion-acoustic waves has not been worked out in detail, but the processes which will be important can be enumerated. The most important process is the induced decay of a plasma wave p into another plasma wave p' and an ion-acoustic wave s (Tsytovich, 1970)

$$p \rightarrow p' + s. \tag{2.11}$$

The wave p' has a smaller wave number k' than p since some momentum goes into the ion-acoustic wave, and thus a larger phase velocity $v'_{\rm ph} = \omega_{pe}/k'$. As a result of many decay processes of type (2.11) coupled with the fact that when the p and s waves have approximately equal wave numbers the process

$$p + s \rightarrow p' \tag{2.12}$$

which increases $v'_{\rm ph}$ will complete with process (2.11), a quasi-stationary spectrum of plasma turbulence will be set up (Pikel'ner and Tsytovich, 1968). This spectrum may have the form shown in Figure 5 with $k_0 < \omega_p/c$ so that the only part of the spectrum for which $v_{\rm ph} < c$ which can interact with particles is a power law of the form

$$W_p(k) \propto k^{-\nu}. \tag{2.13}$$

The type of particle spectrum resulting from such a distribution of waves can be studied by means of the Fokker-Planck equation (Tsytovich, 1970)

$$\frac{\mathrm{d}f_\varepsilon}{\mathrm{d}t} = D\frac{\partial^2 f_\varepsilon}{\partial \varepsilon^2} - 2D\frac{\partial}{\partial \varepsilon}\left(\frac{f_\varepsilon}{\varepsilon}\right), \tag{2.14}$$

where f_ε is the distribution of particle energies ε and D is the diffusion coefficient given by (Tsytovich, 1970)

$$D = \frac{2\pi^2 e^2 \omega_{pe}^2}{v^3} \int_{\omega_{pe}/v}^{\infty} \frac{W_p(k)}{k^3}\,\mathrm{d}k. \tag{2.15}$$

The first term on the right hand side of Equation (2.14) describes random acceleration while the second term defines systematic acceleration. For the power law spectrum of waves (2.13),

$$D \propto \varepsilon^{2(\nu-1)/2} \tag{2.16}$$

(Pikel'ner and Tsytovich, 1968) and the resulting differential electron energy spectrum

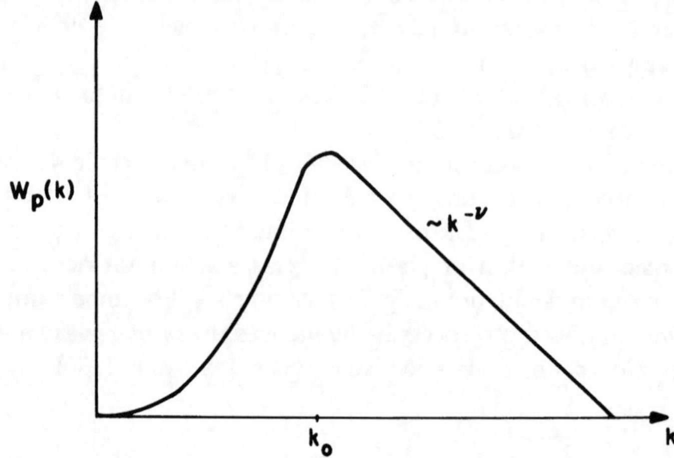

Fig. 5. A schematic of the plasma wave distribution along a field line resulting from strong nonlinear interaction of plasma waves. The wave number k_0 characterizes the beginning of the power law spectrum of waves.

has the form of Equation (2.1) with $\delta = 2$–5 for $\nu = 5$–11. Thus, although it has not been shown in detail how plasma wave spectra of the form of (2.13) with $\nu = 5$–11 are formed, we have shown that such spectra will lead to the observed electron spectrum. Moreover, since δ is determined by microscopic processes such as (2.11) rather than several macroscopic parameters as in the case of Friedman's analysis, the range of δ should be much smaller since microscopic processes tend to lead to well defined spectra (Tsytovich and Chikhachev, 1970). One of the most important effects of the 'plasma wave filter' will be to produce 100 keV electrons along field lines which originally had only 12 keV electrons.

In summary we have found that the initial direct acceleration of particles in current sheets leads to the preferential acceleration of electrons because, although ions obtain m_i/m_e times more energy, it takes them m_i/m_e more time to obtain this energy (Equation (2.8)). The accelerated electrons and ions flowing back into the incoming plasma are unstable which leads to production of electron plasma waves and ion-acoustic waves which are almost aligned along the field. Nonlinear processes lead to a spectrum of plasma waves which leads to a power law distribution of electrons. Since there are many more electrons which can resonate with the plasma waves, electrons again will be preferentially accelerated in contrast to the conclusion reached by Friedman (1969). Since the growth time for plasma waves and nonlinear processes is at most a few micro-seconds, it is readily possible to satisfy the acceleration time requirement. There is no way without a detailed analysis to determine the efficiency of such an acceleration process. However, the initial direct acceleration in the current sheet converts magnetic energy into particle kinetic energy at close to unit efficiency. The electron plasma waves are rapidly transferred to high phase velocities and thus subject only to collision damping. The ion-acoustic waves, on the other hand, are subject to very strong damping until the electron temperature T_e becomes much higher than the ion temperature T_i by damping of both wave modes. Thus an efficiency of 10% would certainly seem to be within the capabilities of this mechanism, but it remains to be shown. The observation of a small number of 50 MeV protons at the time of a type III burst (MacDonald and van Hollebeke, 1973) is consistent with this mechanism since some of the protons directly accelerated by the current sheet will be further accelerated by the plasma waves.

The increase in the electron temperature which occurs on a time scale which is long compared to the acceleration time will eventually raise the threshold for instability for plasma waves and increase the relative importance of the last term in Equation (2.10). In other words the plasma waves which are excited will become more three-dimensional and the situation envisaged by Friedman (1969) will take place where the electrons which are directly accelerated are scattered by the plasma and ion-acoustic waves and effective acceleration ceases except for a very select group of ions. Thus, while it cannot be said that we have a quantitative theory for efficient electron acceleration at the present time, the basic components of such a theory seem to be well in hand. The triggering mechanism for this process is the triggering mechanism for magnetic field reconnection and remains an open question until the conditions determining the rate of reconnection become clear. The addition of a third component of the field to Figure 3 as in Figure 2 will not significantly alter the results presented as long as neutralization of accelerated particles can occur sufficiently quickly to rapidly balance any charge imbalance. We shall defer a discussion of the height of the acceleration region to Section 5.

3. Impulsive Microwave and Hard X-Ray Emission

The nonthermal electrons travelling down field lines in a region close to or contiguous with the acceleration region give rise to microwave and hard X-ray bursts. Because each feature in the microwave intensity profile is followed by the X-ray intensity profile, most theories have attempted to explain both emissions from sources which are in total coincident (Holt and Ramaty, 1969; Takakura, 1972). Both of the cited analyses assumed that X-rays were produced by a thin-target bremsstrahlung process. The thin vs thick-target controversy is now the principle problem in interpreting X-ray spectra. The rapid rise time (~ 1 s), power law spectrum and tracking of impulsive microwave bursts which must be produced by nonthermal electrons (Ramaty and Petrosian, 1972) argue very strongly for the nonthermal nature of hard ($\gtrsim 20$ keV) X-rays. In the thick-target model electrons lose all of their energy through collisions in the X-ray emitting region so that the spectrum is a result of the equilibrium between the injection of newly accelerated electrons and the loss of electrons through collisions. In the thin-target model, electrons escape from the X-ray region in a time short compared to their collisional lifetime, and the spectrum of injected and X-ray emitting electrons is the same since collisions are negligible.

The main proponents of a the thick-target model are Brown (1972) and Hudson (1972). While their arguments are valid for large flares, they may meet with serious difficulty for the majority of flares which are small because:

(1) There are no major differences in disk flare spectra and one behind-the-limb flare spectrum which was produced high enough so that it was probably due to thin-target emission (Kane, private communication) as would be expected for thick-target emission from disk flares since γ in the relation

$$\frac{\mathrm{d}J_{h\nu}}{\mathrm{d}(h\nu)} = A(h\nu)^{-\gamma}\ \mathrm{cm}^{-2}\ \mathrm{s}^{-1}\ \mathrm{keV}^{-1} \tag{3.1}$$

decreases by a factor of about 2 for thick target emission.

(2) The electron spectra observed at the earth are consistent with thin-target emission for several measured X-ray events on a statistical basis (Kane, 1973) and for one small flare where the spectra were measured concurrently (Datlowe and Lin, 1973).

Thus we can say that the possibility of primarily thick-target emission in most flares is not favored by simple interpretations of data and concentrate on thin-target emission. Even within this realm there is the question of whether the injection of electrons into the X-ray source is impulsive or continuous. In the impulsive model (Takakura and Kai, 1966) electron injection begins at the onset of the X-ray burst and stops at the time of X-ray maximum. In the continuous injection model (Kane and Anderson, 1970; Syrovatskii and Shmeleva, 1972; Brown, 1972) the electron acceleration process determines primarily the time intensity profile of the entire X-ray burst, the time constants for trapping and/or collision loss being assumed short in comparison to the characteristic time for acceleration. At photon energies greater than about 30 keV, the continuous injection model is favored because the decay of the burst often has a profile much like the rise of the burst (Kane and Anderson, 1970), but at lower energies the spectra may be contaminated by a thermal component and it is difficult to provide evidence to support either hypothesis. Because the continuous injection model is favored at higher energies, we shall assume that it is applicable to the whole nonthermal component of interest here.

In this range X-ray production is simply due to nonthermal bremsstrahlung and leads to a spectrum of X-rays at the Earth of (Kane and Anderson, 1970)

$$\frac{\mathrm{d}J(E)}{\mathrm{d}E} \approx 5.6\times10^{-52}\frac{n_i V}{E}\int_E^{100}\frac{1}{E_e}\frac{\mathrm{d}J_e(E_e)}{\mathrm{d}E_e}\times$$

$$\times \ln\left[\left(\frac{E_e}{E}\right)^{1/2} + \left(\frac{E_e}{E} - 1\right)^{1/2}\right] \mathrm{d}E_e \text{ photons cm}^{-2}\,\text{s}^{-1}\,\text{keV}^{-1}, \qquad (3.2)$$

where $E = h\nu$ is the photon energy, n_i is the ion density and $\mathrm{d}J_e(E_e)/\mathrm{d}E_e$ is the electron spectrum which is of the form of (2.1), but with exponent $\delta' = \delta - \frac{1}{2}$. In particular, for this form of $\mathrm{d}J_e(E_e)/\mathrm{d}E_e$, the X-ray spectrum (3.2) will have the form of a power law. Although this type of analysis is a gross simplification because the anisotropy in pitch angles, inhomogeneous density structure of the source, etc. are neglected, we are confronted with the following situation. All more sophisticated analyses such as those of Brown (1972) which take into account the anisotropy and change of pitch angles in the source must do so by introducing additional parameters. These parameters taken singly inevitably have the undesirable effect of requiring a change in the electron spectral index δ'. On the other hand the possible virture of Equation (3.2) is that it predicts spectra consistent with the range of electron spectra observed at earth on a statistical basis (Kane, 1973) and on a one-one basis for one event (Datlowe and Lin, 1973). Thus, while Equation (3.2) ignores a considerable amount of basic physics, any more sophisticated approach must introduce at least two additional parameters which, at least in small flares, have opposite effects on the required electron spectrum, an approach which hardly seems better with present observations unless it can be shown that use of Equation (3.2) leads to an incorrect estimate of the number of electrons required. This may well be the case in large flares as the analysis of Brown (1972) would indicate. In the case that thick target processes are dominant, an even simpler equation than (3.2) results (Brown (1972)). The only case of difficulty is the intermediate target case where both thin and thick target processes play a role. Since thick target emission is several hundred times more efficient than thin target emission, it is likely to be dominant when present and the intermediate target case may not be a problem.

In contrast to the simplified approach useful for most X-rays, a quite involved approach is required to make the microwave and X-ray sources coincident in total (Holt and Ramaty, 1969; Takakura, 1972) and to explain the rather large variability of microwave spectra (Ramaty and Petrosian, 1972). The source of this difficulty lies largely in the fact that the microwave source is optically thick in some region whereas the X-ray source can always be considered optically thin. Since there are several possible absorption mechanisms one is forced to consider all the basic physics including the inhomogeneity of the magnetic field. Thus we limit our review to a qualitative discussion of which effects seem to be important. The emission mechanism is gyro-synchrotron emission (Ramaty, 1969). The most important absorption mechanism operative at low frequencies (1–20 GHz) in most sources is self-absorption of the emitted radiation by the emitting electrons (Holt and Ramaty, 1969; Takakura, 1972). The frequency of maximum emission ν_{sa} resulting from self-absorption of a relativistic synchrotron source was given by Slish (1963)

$$\nu_{sa} \approx 10^{12.5} (S_m/\Omega)^{2/5}\, B_{\perp}^{1/5} \qquad (3.3)$$

and was shown to agree with numerical results (Ramaty, 1969) to within 10% over the range of interest for most microwave bursts (Ramaty and Petrosian, 1972). Here S_m/Ω is the maximum brightness and $B_\perp$ is the magnetic field component perpendicular to the line of sight in the emitting region. Both Holt and Ramaty (1969) and Takakura (1972) have shown that Razin-Tsytovich suppression is negligible for densities less than or of order 10^{10} cm^{-3}. Another absorption mechanism which is important for the relatively rare microwave bursts with flat spectra is free-free absorption (Zirin *et al.*, 1971) which leads to

$$\nu_{ff} = 0.39 (EM)^{1/2}\, T_e^{-3/4}\, R^{-1}\, \Omega^{-1/2} \tag{3.4}$$

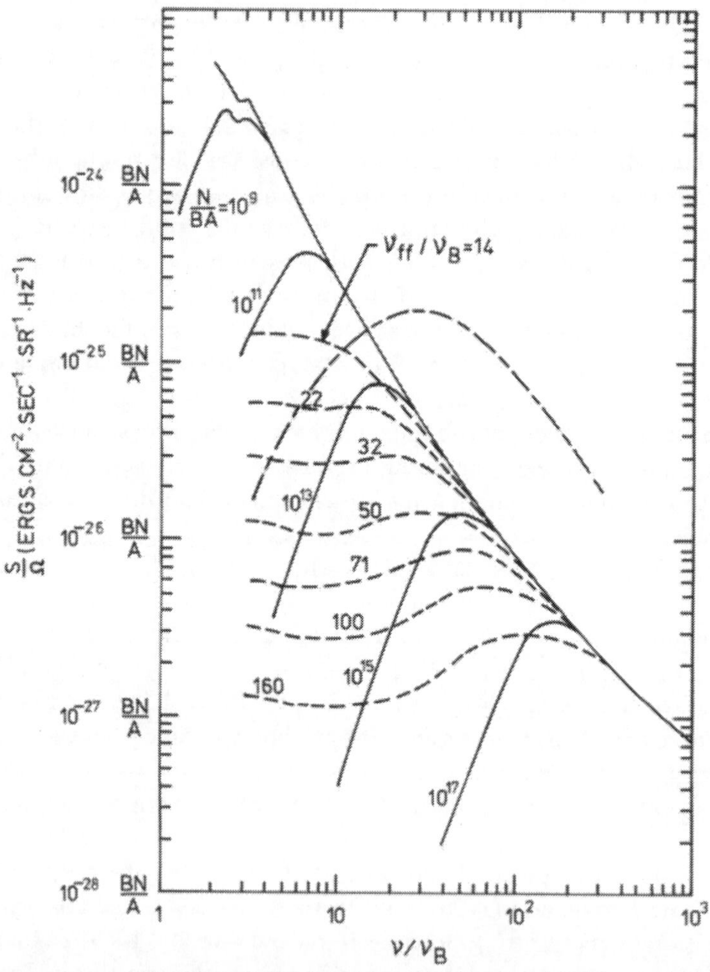

Fig. 6. Gyrosynchrotron spectra from a nonthermal electron distribution. Ramaty's and Ramaty and Petrosian's results for a source with a uniform magnetic field: *Heavy solid line*, gyrosynchrotron emissivity; *light solid line*, self-absorbed brightness; *dashed line*, free-free absorbed brightness. Takakura's result for self-absorbed brightness in a non-uniform magnetic field, *long-short dashed line*, showing the broadening effect of nonuniformity.

for the frequency corresponding to unit optical depth (Ramaty and Petrosian, 1972), where $EM = n_e^2 V$ is the thermal emission measure, R is the distance to the source of volume V and Ω is the solid angle which it subtends. Another possibility is absorption below the plasma frequency (Zirin *et al.*, 1971).

The effect of a non-uniform magnetic field, in particular a dipole field, together with self-absorption was studied by Takakura (1972). His results for a dipole of strength 2500 G are shown in Figure 6 for a height at which $B = 60$ G together with those of Ramaty and Petrosian for free-free and self-absorption in a uniform field. We see that the effect of a nonuniform magnetic field is to broaden the spectrum and make it look more like many observed spectra. Each part of the spectrum comes from a limited part of the source which decreases the apparent problem of too many electrons for the observed fluxes found by earlier workers (Peterson and Winkler, 1959; Holt and Cline, 1968) for coincident X-ray and microwave sources. Anisotropy of electrons with most of the energy along the field also acts in this direction as does a high energy cutoff (Holt and Ramaty, 1969). In short whereas X-ray production seems comparatively insensitive to source parameters in the thin-target regime, microwave production is very sensitive to these parameters and there are several possibilities for decreasing the microwave production from a given electron population. This in turn has the undesirable effect of making interpretations of microwave spectra extremely model dependent so that until other parameters in active regions become better known, they offer little possibility of increasing our knowledge of these regions. In particular, selection of one cutoff mechanism over another with qualitative arguments (Zirin *et al.*, 1971) seems dubious.

From the analysis of impulsive hard X-ray and microwave bursts (Holt and Ramaty, 1969; Takakura, 1972), we can place the following limitations on their source. The electron density $n_e \lesssim 10^{10}$ cm^{-3} and the magnetic field $B \lesssim 350$ G. The total number of electrons with energies greater than 100 keV, $N(>100 \text{ keV}) \lesssim 10^{37}$ and these electrons have a power-law spectrum with exponent $\delta' = 3-5$. The total amount of energy involved is up to 1.6×10^{30} erg, but is more typically a few times 10^{28} erg in interpretations of a small solar flare (e.g., Kane, 1973).

4. Impulsive EUV and Hα Emission

As some of the nonthermal electrons descend into the chromosphere, they excite EUV radiation (Kane and Donnelly, 1971; Donnelly *et al.*, 1973) and Hα (Zirin *et al.*, 1971; Vorpahl, 1973). Since the highest energy electrons which are subject to least scattering will descend the deepest (Brown, 1972; Hudson, 1972), we are confronted with the fact that these electrons must also produce X-rays by thick-target bremsstrahlung processes. From the results of Section 3 it can be argued that the percentage of flares in which thick-target bremsstrahlung is the dominant source of X-rays is probably small so that only a small fraction of the accelerated electrons penetrate deep into the chromosphere in most flares. The electrons which do manage to make their way are rapidly thermalized in the chromosphere due to Coulomb and ionizing collisions (Brown, 1973) which results in a much larger number of quasithermal electrons with higher effective temperatures than normal. These electrons in turn excite atoms as in a thermal plasma to produce EUV and Hα emission.

The only attempt at any quantitative calculation of the impulsive EUV or Hα emission has been made by Kane and Donnelly (1971) for the continuum part of the spectrum produced by recombination assuming the line contribution is small. Since short of 512 Å there is always an enhancement of chromospheric line emission during the flash phase (Hall and Hinteregger, 1969) whereas there is no firm evidence of any continuum enhancement (Neupert, private communication), complete neglect of the line component is unjustified (see also Kane, 1974). Thus we do not reproduce Kane and Donnelly's computation. Rather, following the approach of Brown (1973), we indicate what needs to be done. Brown derived the temperature and density profiles for a model atmosphere in quasi-equilibrium with a power law spectrum of electrons injected vertically downward treating the dominant Balmer continuum correctly in contrast to Hudson (1972) who neglected it by use of a one-level atom. However, he still treated the radiative losses incorrectly which requires use of a three-level atom (Canfield, private communication). Since a part of the chromosphere is 'boiled off' in this approach in which constant pressure is assumed, a hydrodynamic problem must be solved, but the radiative transfer can be treated as a steady-state problem.

In the case of interest for most impulsive phenomena in flares, on the other hand, the time scale for gas motion t_D which is given roughly by the time for a sound wave to traverse one density scale height H is larger than the time scale for electron injection t_B. For example, for $T=10^4$ K and $H=750$ km, $t_D \approx 50$ s whereas most impulsive injection occurs on a time scale less than t_D. Thus the assumption of constant density would be appropriate leading to simpler hydrodynamics, but the radiative transfer may have to be done as a time-dependent problem since t_B is less then the recombination or inonization time scale when $t_B \lesssim 1$ s. In other words the problem which needs to be solved for most flash phase EUV and Hα emission is the response of a static chromosphere to a sudden injection of energy by fast electrons balanced by radiative cooling.

While this has not been done, it has been shown by Kane and Donnelly (1971) that only 10^{28} erg are required for impulsive EUV emission in the 10–1030 Å range in a small flare and only 4.5×10^{25} erg is required for one Hα kernel (Vorpahl, 1973). Thus we may say that the dumping of only a small fraction of the accelerated electrons into a thick target in the low chromosphere is consistent with small flare energy requirements, but until the problem outlined above is solved, such checks are at best indicative.

5. Type III Radio Bursts and Interplanetary Electrons

A small fraction of the electrons accelerated which manage to move on open field lines give rise to type III bursts and interplanetary electrons. Although the energy involved may be small, the potential information content is large because these bursts together with their exciting electrons serve as electron probes of the corona from about 100000 km above the photosphere to the orbit of the Earth and an unknown distance beyond. The spectrum of interplanetary electrons as a function of time provides important clues on the acceleration of electrons during the flash phase.

The theory for type III bursts can be divided into three main parts: (1) The interaction of a spatially and temporally inhomogeneous electron distribution with a plasma and the resultant distribution of plasma waves. (2) The conversion of plasma waves into radiation. (3) The propagation of radiation from its source to the observer. For conciseness we refer to these as the plasma wave source, the radiation source and propagation. The theory has recently been reviewed in detail (Smith, 1974a) and a more complete account can be found there.

5.1. Plasma wave source

In the recent past this part of the problem has been the subject of great controversy largely because of incorrect or incomplete applications of theory. The observations of Stewart (1965) and Fainberg and Stone (1970) seemed to imply that the velocity of the exciting electrons was constant for several tens of solar radii. Application of the theory of quasilinear relaxation of an electron beam which is spatially homogeneous (Vedenov *et al.*, 1961; Drummond and Pines, 1962) showed that the beam should lose a significant fraction of its energy in at most 100000 km (Sturrock, 1964; Kaplan and Tsytovich, 1968; Melrose, 1970b; Smith, 1970a). In quasilinear relaxation the beam excites plasma waves with phase velocities slightly lower than the beam which in turn accelerate electrons with slightly lower phase velocities, etc. until a quasi-plateau is formed from the beam as shown in Figure 7. On the basis of this apparently large energy loss, various mechanisms were proposed to suppress this relaxation; i.e. to stabilize the beam, and the search is still in progress (Papadopolous *et al.*, 1974). The results of all attempts including that of Papadopolous *et al.* is that some rather special

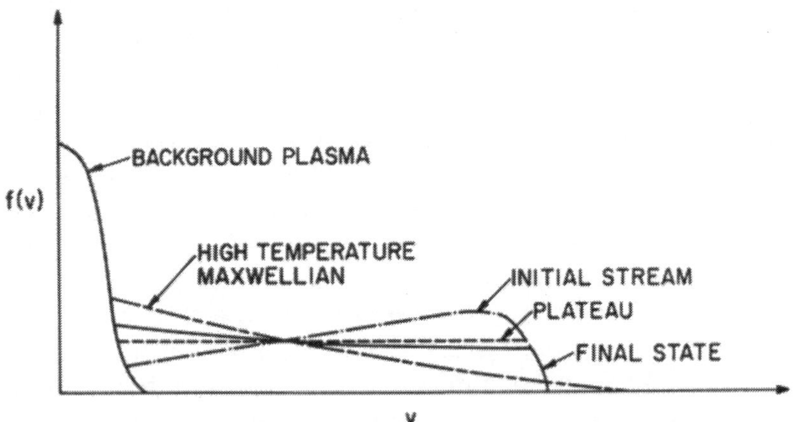

Fig. 7. Forms of the electron distribution function at various times in quasilinear relaxation. In a homogeneous plasma an initial stream (·–·–·) relaxes to a plateau (–––) and finally to a high temperature Maxwellian (-----) because of spontaneous emission processes. In the inhomogeneous relaxation characteristic of type III bursts there is insufficient time to reach a high temperature Maxwellian and the final state (———) is in between it and a plateau, closer to a plateau.

conditions have to be satisfied to stabilize an electron stream. For example, the nonlinear mechanism used by Papadopolous *et al.* is the oscillating two-stream instability which depends quite sensitively on the effective width of the stream since it only works well when the phases of all the waves are the same. Since we know from the previous sections that the stream must start out as a power law, it is likely to have a broad distribution at all times and hence fixed phase effects such as suggested by Papadopolous are unlikely to be important in the main part of the burst. Both Zheleznyakov and Zaitsev (1970a) and Smith and Fung (1971) have shown that the strongest nonlinear effect for a broad stream distribution, nonlinear Landau damping, is unable to stabilize an electron stream. Thus the stream must relax and the fact that the stream is both temporally and spatially inhomogeneous must prevent the rapid loss of energy indicated for a homogeneous stream.

Zheleznyakov and Zaitsev (1970a) were the first to consider this situation. They used a one-dimensional formalism and took collisional damping as well as Landau damping of the stream-plasma system in the direction of the stream into account. However, they integrated over the plasma wave spectrum in order to obtain a solution which is not allowable to correctly consider deceleration of the stream (Smith, 1973) or for the purpose of computing radiation near the fundamental of the plasma frequency (Smith, 1970a, 1974a). For these purposes the detailed shape of the plasma wave spectrum is important. In the low frequency regime (3 MHz and lower) Zaitsev *et al.* (1972) have constructed a one-dimensional similarity solution which only takes into account Landau damping of the stream in the direction of the stream. This solution also required integration over the plasma wave spectrum as well as complete neglect of the characteristics of the background plasma. This solution is completely lossless so that all of the energy given to the plasma waves by the stream is reabsorbed by the stream and the stream travels with constant velocity. This solution can be criticized on two major counts. (1) Although neglect of Landau damping by the background plasma may be allowable above 3 MHz because collisional damping is stronger, below 3 MHz Landau damping by the background plasma exceeds collisional damping (Harvey and Aubier, 1973) and must be included. Unfortunately a two-dimensional formalism for the plasma waves is required for this purpose (Smith, 1974a). (2) By integrating over the plasma wave spectrum the basic nature of quasi-linear relaxation which is to reduce the average velocity of the electrons involved has been neglected. As noted above, the manner in which the quasi-plateau in Figure 7 is reached is that fast electrons of velocity v_s excite plasma waves whose phase velocities v_{ph} are less than v_s. These plasma waves then accelerate some electrons to velocity $v_{\mathrm{ph}} < v_s$ which relax and produce plasma waves with phase velocities $v'_{\mathrm{ph}} < v_{\mathrm{ph}}$ and so on (Tsytovich, 1970). The resultant distribution of plasma waves (see, e.g. Nunez and Rand, 1969) thus has an average phase velocity v_{ph} considerably less than the initial average stream velocity v_s. Since the primary interaction of these waves is with electrons of the same phase velocity via the Cherenkov condition

$$\omega_p = \mathbf{k} \cdot \mathbf{v}, \tag{5.1}$$

where ω_p is the frequency of the plasma wave with wave vector **k** and **v** is the electron velocity, these waves cannot on the average give energy back to electrons with velocity v_s. In other words even in inhomogeneous quasilinear relaxation, the average velocity of the electrons must decrease. Since we now know that both the velocity of the leading edge and maximum of type III bursts do decrease (Fainberg *et al.*, 1972), this facet of quasilinear relaxation is consistent with the observations.

Because an investigation of the complete inhomogeneous quasilinear equations is likely to take some time, simple solutions in the spirit of Zaitsev *et al.* (1972), but which may provide a closer approximation to the real situation are desirable. For this reason Smith (1973) constructed a solution which overestimates the real losses suffered by a stream and which may be termed the 'completely lossy' solution. In this case which is an opposite limiting case to that of Zaitsev *et al.*, each time the stream forms a hump, the hump is cutoff and the valley filled in numerically as shown in Figure 8. It is assumed that the difference in energies of the initial and final distribu-

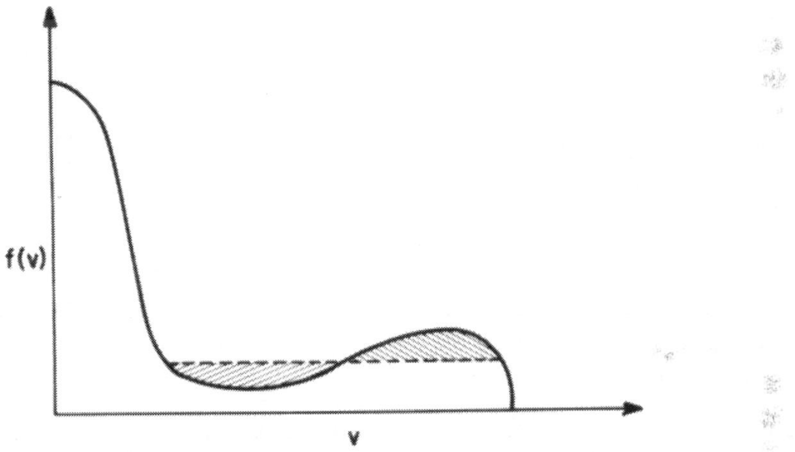

Fig. 8. Procedure of 'cutting off the hump and filling in the valley'. The original distribution (———) is converted into the final distribution (---) by cutting off the peak in the distribution (\\\) and filling in the valley (///) in such a way that the area cut off equals the area filled in.

tions is released into plasma waves and none of the energy is returned to the stream. For an initial power law distribution of electrons with $\delta' = 2.3$ which is the value for the simultaneous electron-type III event analyzed by Lin *et al.* (1973), the deceleration of the leading edge of the burst from 10–20 $R_\odot$ can be compared with the average deceleration determined experimentally (Fainberg *et al.*, 1972) as shown in Figure 9. For comparison, the result of Zaitsev *et al.* (1972) would be a horizontal line in this figure. Because the scattering properties of the interplanetary medium for low energy electrons are unknown and some of the apparent deceleration could be due to scattering in pitch angle, whether the results of Smith (1973) satisfy the observations more closely than those of Zaitsev *et al.* (1972) is an open question. The main result to be drawn is that depletion of very fast electrons occurs quite rapidly and hence a more complete analysis taking into account losses more realistically as well as following the two-dimensional axially symmetric plasma wave distribution in detail is required. The stream was not followed beyond 20 $R_{\odot}$ for this reason. The interested reader may find the equations to be solved in Smith (1974a).

Beyond this basic stream-plasma interaction, many questions remain to be answer-

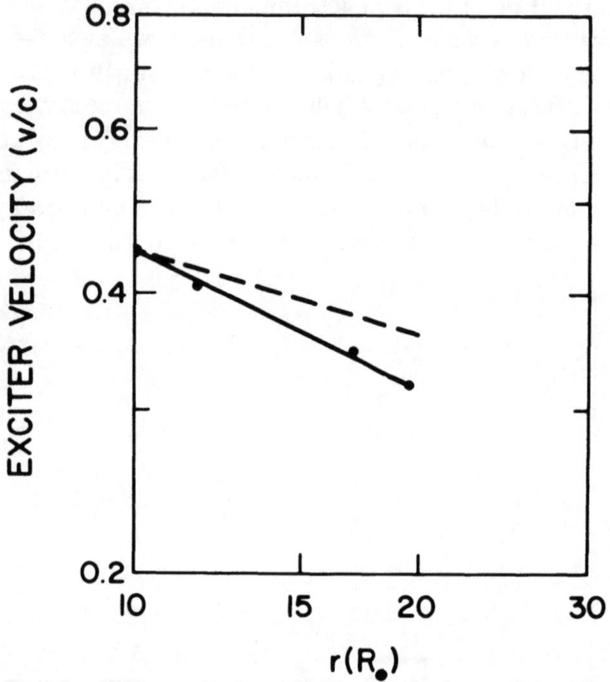

Fig. 9. Deceleration of the leading edge of a type III burst between 10 and 20 $R_{\odot}$ according to Smith's model (———) and according to the observations for a power law velocity distribution with index 4.6 and an initial stream density of 6.3×10^5 cm^{-3}, at a few thousand kilometers above the photosphere.

ed. What are the limitations of the above approach and when do nonlinear effects play a role? Although it takes about 100000 km for an initial power law to become unstable due to fast electrons overtaking slower ones (Smith, 1973), the starting frequencies of type III bursts are often much lower than the corresponding plasma frequency at this height (Stewart, 1974). A possible explanation for this is that quasi-linear relaxation may be suppressed by ion-acoustic waves or other inhomogeneities in the corona (Sturrock, 1964; Smith, 1970a, b). The occasional intermittency of type III bursts (Elgarøy, 1961; Ellis, 1969) argues that this must be possible. An alternate explanation for low starting frequencies is that electrons are accelerated high in the corona.

Another question is how does the stream become charge and current neutralized? This question was considered by Smith (1972a) for the case of a narrow velocity spread of the stream. While this is not allowable for the peak of the burst, it may be applicable to the leading edge of the burst. In the frame of the plasma, when the stream enters the plasma, the plasma reacts by accelerating electrons in the opposite direction to the stream to a velocity

$$v_a = \frac{-2n_s v_s}{n_e}, \tag{5.2}$$

where n_s is the density of the stream. As a result a current now flows in the direction opposite to the stream because the plasma has overreacted to the presence of the stream. The plasma then reacts to the presence of this current in the same manner as to the initial stream and accelerates electrons in the direction of the stream so that the initial stream is present which it reacts to again and so forth. These current oscillations which are accompanied by charge oscillations form a large amplitude plasma wave at the head of the stream which is damped by collisions. This large amplitude plasma wave may be related to the possible precursor phenomena observed by de la Noë and Boishot (1972) and others.

Finally there is the question of the macroscopic interaction of the electron stream with the inhomogeneous magnetic field and density structure of the corona. There is good evidence that electrons travel along coronal streamers in the metric band (Stewart, 1974), but no such evidence in the hectometer and kilometer band. Smith and Pneuman (1972) studied the motion of particles in a streamer model of Pneuman (1972) which included finite conductivity and added the curvature of the streamer due to solar rotation. In this model there is a small field across the streamer. They found that electrons could not travel more than about 1 $R_\odot$ in the current sheet along the streamer axis and have the properties required for type III bursts because of the finite transverse field. Thus the electrons causing type III bursts, which are most likely accelerated in current sheets as in Section 2, must drift out of these sheets due to the curvature of the streamer. Some aspects of the fine structure of type III bursts may also be due to the large scale inhomogeneous structure of the corona.

5.2. Radiation source

The basic processes for radiation near the fundamental and second harmonic of the plasma frequency are by now well known. Near the fundamental the process is the scattering of a plasma wave p on the polarization cloud of an ion i (electron density fluctuation) to produce a transverse electromagnetic wave t

$$p + i \rightarrow t + i', \tag{5.3}$$

with the polarization cloud going to a new state i' due to the absorption of some energy and momentum. Process (5.3) can be either spontaneous or induced and the absorption coefficient can be negative, i.e. the radiation can be amplified. It was shown by Smith (1970a) that the fundamental must be amplified to obtain observed ratios of fundamental/second harmonic power. Thus the fundamental is produced by a process which is effectively much more nonlinear than for the second harmonic. Zheleznyakov and Zaitsev (1970b) concluded that the fundamental would not be amplified using an inapplicable formula from Tsytovich (1970). The radiation which results from the amplification process has been examined in a detailed model of the source region at the 80 MHz level based on approximate plasma wave spectra (Smith, 1974b).

Near the second harmonic the radiation process is the combination of two excited plasma waves p and p' to produce a wave $t(2\omega_p)$

$$p+p'\rightarrow t(2\omega_p). \tag{5.4}$$

In contrast to process (5.3), the absorption coefficient for process (5.4) can only be positive. By considering this absorption, Melrose (1970a) and Smith and Sturrock (1971) showed that both of the combining plasma waves must be excited, i.e. have high effective temperatures, in order to obtain nonthermal radiation. Energy and momentum conservation require that the two plasma waves meet almost 'head-on'. Since the plasma waves produced by quasilinear relaxation are directed primarily in the direction of the stream, some secondary plasma waves must be produced moving in the opposite direction by scattering on the polarization clouds of ions

$$p+i\rightarrow p'+i'. \tag{5.5}$$

Again this process can be either spontaneous or induced. Generally the energy density in plasma waves is not sufficiently high for the induced process to play an important role. The emission and absorption coefficients for this case were calculated by Melrose (1970a) and Smith (1972b). The emission pattern for $v_s \lesssim 0.5\ c$ is a quadrupole.

At high frequencies ($\sim$100 MHz) the simultaneous observation of fundamental and second harmonic emission can provide important information about the source as shown by Smith (1970a). From the second harmonic power the energy density in plasma waves W_p required for a given source area A can be determined. By tracing rays in model sources of varying areas A with corresponding W_p and taking into account amplification of the fundamental, the area A and energy density W_p can be determined uniquely for a given source geometry. The results of such an analysis for a strong source at the 80 MHz level with $P(\omega_p)/P(2\omega_p)\approx 10$, where P is source emissivity, are that the source is about 14000 km or 0.3 min in diameter, that $W_p \approx 3\times 10^{-7}$ erg cm^{-3} and that all of the important rays come out the sides of the model cylinder travelling outward within an angular range of 26° (Smith, 1974b). However, scattering of the radiation in the source was neglected in deriving these results.

At low frequencies ($\sim$300 KHz) the energy density in plasma waves becomes so small that effective amplification of the fundamental is no longer possible and emission near the second harmonic becomes dominant. In this case there is no way to determine the source parameters from the radiation itself.

Melrose and Sy (1972) have recently considered the radiation mechanisms (5.3–4) in the presence of a magnetic field. For the case of a weak magnetic field, i.e. $\omega_{He}^2 \ll \omega_{pe}^2$, the only effect of the presence of the field is to give a slight or complete circular polarization to both fundamental and harmonic radiations. These results coupled with the observed degree of circular polarization could be used to set an upper limit to the magnetic field in the source if there were no mode coupling outside the source. However, the analysis of Dodge (1972) indicates that there may be significant mode coupling (see below).

5.3. Propagation

The radiation leaving the source is scattered, refracted, polarized and absorbed on its way to the observer. All of the recent work on this part of the problem has neglected the polarization of radiation and concentrated on the effect of scattering by random density inhomogeneities in a refracting absorbing medium. The method of treating this problem is to trace rays and to let them suffer random changes in direction after travelling a step size in a refracting absorbing plasma (Steinberg *et al.*, 1971; Riddle, 1972a, b). Since the results of this procedure have been summarized by Stewart (1974), and Smith (1974a) we only comment on some of Stewart's possible interpretations. Riddle (1974) has considered the case of anisotropic inhomogeneities in which the scale length $h=a\varrho$ in directions normal to the radius and $h=ab\varrho$ in the radial direction. With $b=5$, Riddle found no significant differences for a point source compared with the isotropic inhomogeneity case. Riddle (1974) also showed that with scattering in a streamer taken into account the differences between isotropic emission, preferential emission backwards (Zheleznyakov and Zaitsev, 1970b) and preferential emission sideways (Smerd *et al.*, 1962) for the second harmonic would not be detectable observationally for a point source. Thus there is no reason to consider the possibility of preferential backward emission without considering a finite size source. It is quite possible that the intrinsic source size increases during a burst due to a larger number of field lines being 'lit-up' by high-energy electrons. It is also possible that the effective fundamental source is smaller than the second harmonic source because amplification by plasma waves is only effective in the center of the source where W_p is sufficiently high.

Consideration of finite size sources, time varying sources and different size fundamental and second harmonic sources appear as logical next steps for this part of the problem.

The observations of Dodge (1972) and Grognard and McLean (1973) on the polarization of type III bursts imply that the polarizing region must be a few solar radii above the radiation source at metric and decametric frequencies. This result is also in accord with that of Riddle (1974) who showed that the linear polarization observed is unlikely to represent conditions at the source because of the varied paths and path lengths by which radiation reaches the observer. The only known mechanism to convert the circularly polarized emission expected from the radiation source into primarily polarized radiation is mode coupling within regions of quasi-transverse magnetic field (Cohen, 1960; Zheleznyakov and Zlotnik, 1964). Cohen's mechanism for linear polarization requires that one of the incident magnetoionic modes be stronger than the other and the degree of linear polarization depends directly on the ratio of intensities. Thus strong linear or highly elliptical polarization would require that the quasi-transverse region be illuminated almost completely by *o*-mode radiation which is possible under the condition (Melrose and Sy, 1972)

$$v_{ph} > \max\{v_e(3\omega_{pe}/\omega_{He})^{1/2},\ v_i(2\omega_{pe}/\omega_{He})\}, \tag{5.6}$$

where v_i is the ion thermal velocity.

5.4. Electron propagation in the interplanetary medium

Besides gradient drifts, etc. in large scale field inhomogeneities, electrons will also be scattered by magnetic field fluctuations. Unfortunately, analytic approximations are inapplicable in the energy range 10–100 keV so that no adequate theory for such propagation exists as for higher energy electrons (Jokopii, 1971). It appears that the best approach to the problem is a combination of numerical modelling and empirical results. Once the complete quasilinear equations (Smith, 1974a) which neglect pitch angle scattering are solved numerically, the amount of deceleration due to quasi-linear relaxation will be known. Any additional apparent deceleration could then be attributed to scattering of electrons in pitch angle by interplanetary magnetic field fluctuations.

Kane and Lin (1972) and Lin (1973) have concluded that the electron acceleration region must have a density $n_e \lesssim 2\times 10^9$ cm^{-3} by noting that power law spectra observed at the earth have no turnover at low energies before ~6 keV. This deduction is based on the premise that the only interaction that electrons encounter in escaping from the Sun in a socalled 'scatter-free' event is the Coulomb interaction. However, we have seen in Section 2 and again in subsection 5.1 that other quasilinear and non-linear interactions involving plasma waves are possible and that these can reconstitute a power law spectrum out of a non-power law spectrum. Moreover, for the non-linear interactions of Section 2 very little radiation would be expected by the mechanisms of subsection 5.2 because the plasma waves are rapidly transformed to the region of k-space where the radiation produced has very low group velocities and so is absorbed before it has a chance to leave the source (Kaplan and Tsytovich, 1973). For the case where nonlinear interactions are not sufficiently strong it is possible that some of the decimetric emission (Kundu, 1965) is associated with such plasma turbulence. Thus there is no compelling reason for accepting the argument of Kane and Lin (1972) and Lin (1973), and aside from placing the electron acceleration region in the low corona where the magnetic field is still quite high for reasonable efficiency, its height is an open question. The numbers used in the first two sections of this review of $h \approx 10000$ km above the photosphere with $n_e \approx 3\times 10^9$ cm^{-3} and $B \approx 500$ G are simply guesses which satisfy the rather loose requirements stated above.

6. Discussion

We have reviewed our current knowledge of the mechanisms responsible for the non-thermal and impulsive quasi-thermal phenomena occurring during the flash phase of a flare. The relation of these mechanisms to the main thermal phenomenon going on during the flash phase, namely the rise of the soft X-ray burst is of interest. It is often observed that the rate of rise of the soft X-ray flux maximizes at the time of the most impulsive hard X-ray or microwave spike and that the flux continues to rise into the flare proper after the hard X-ray spike is over, but at a slower rate (Neupert, 1968). Thus the acceleration mechanism for electrons should be characterized by maximum heating being coincident with maximum acceleration. This is a natural consequence of the sequence of mechanisms proposed for acceleration in Section 2. Namely, the amount of energy going into heat in the reconnection process will be maximum when reconnection and hence acceleration is most efficient. The production of electron plasma and ion-acoustic waves by the accelerated particles will be most efficient when the acceleration is most efficient. The ion-acoustic waves in turn are heavily damped on a time scale $\tau_i \approx (m_i/4\pi n_i e^2)^{1/2}$ until T_e becomes much larger than T_i and the energy of the waves is rapidly converted into electron thermal energy. As noted in Section 2, the increase of electron temperature will tend to decrease the efficiency of acceleration by electron plasma waves. Since it will also make scattering through large angles more efficient, we know from the results of Friedman (1969) that the increase in electron temperature will make the direct first-order Fermi acceleration ineffective for all but a very select group of fast ions. In other words the process of rapid acceleration in a reconnection geometry is self-quenching by nature and this is consistent with the observations. However, both before and after efficient acceleration the current sheet is capable of acceleration of particles to energies slightly above their thermal values by slow reconnection. This energy is rapidly converted into heat since the collision mean free path of these slightly nonthermal particles is small, especially in a turbulent plasma. This fact allows an explanation for the fact that highly nonthermal phenomena such as type III bursts appear to rise out of a plasma of increasing temperature as indicated by soft X-rays (Teske *et al.*, 1971) and that the soft X-ray flux continues to rise after the impulsive phenomena are over (Neupert, 1968).

If we now ask what needs to be done to improve the theories of flash phase mechanisms, it is not hard to find several good problems. The parts of the acceleration mechanism outlined in Section 2 need to be worked out in detail. To start with, what determines the rate of reconnection? Can the acceleration of particles act like a siphon to suck field lines and more particles into the acceleration region or is everything determined by conditions far from the reconnection region? What is the spatial distribution of plasma turbulence as a function of time and can it provide power law spectra with the indices observed? What are the fates of particles brought in on incoming field lines? How does the temporal development of plasma turbulence affect the efficiency of the first-order Fermi process?

In the hard X-ray interpretations are we being too naive in favoring dismissing the possibility of a significant thick-target contribution in most flares? Better observations of the location of the hard X-ray source will help to answer this question. In the interpretation of impulsive microwave bursts can we find some relatively simple means of determining which absorption mechanism is dominant in a given situation? The impulsive EUV and Hα emission during the flash phase is perhaps the problem most ripe for solution. A large amount of data has recently become available (Kane, 1974), but there is no consistent model to which it can be fit. The construction of such a model will inevitably involve a detailed calculation of the radiative response of the chromosphere in hydrogen to a flux of impulsive nonthermal electrons as outlined in Section 4. The problem of impulsive EUV lines and continuum of atoms other than hydrogen can then be attacked since the hydrogen lines and continuum determine the basic state of the chromosphere.

While the type III burst problem has probably seen more progress than other problems because of its relative simplicity, much remains to be done. The most pressing problem is the numerical solution of realistic quasilinear equations which are inhomogeneous in time and space. With this problem solved and the resultant realistic plasma wave spectra, the radiation source can be defined much better. With a better radiation source, the problems of scattering and polarization of radiation can be treated more realistically, etc. With a complete numerical solution of the quasilinear equations, the amount of apparent deceleration due to pitchangle scattering and thus the diffusion coefficients for low-energy electrons in the interplanetary medium can be determined. These can then be used to treat the stream-relaxation more realistically and thus refined by successive approximations. With this information in hand, we could define the required electron source much more precisely.

For conciseness and to avoid too much speculation this review has followed a rather narrow line. Certainly many other phenomena are most likely initiated at the flash phase such as the white light and CN flare, the formation of type II burst sources and interplanetary shock waves, the acceleration of particles to relativistic energies and possibly the formation of type IV burst sources. There are also many known interrelations between various phenomena such as a relation between Hα absorption features and type III bursts (Axisa *et al.*, 1973; Kuiper and Pasachoff, 1973). However, only rudimentary qualitative ideas exist for many of these phenomena and the rest are more properly treated along with the main phase of flares.

Acknowledgements

The author would like to express his gratitude for the stimulating discussions he has had with a number of co-workers and especially to Drs R. Canfield, S. Kane, W. Neupert and K. Tanaka.

References

Alfvén, H. and Carlqvist, P.: 1967, *Solar Phys.* **1**, 220.
Aubry, M. P., Russell, C. T., and Kivelson, M. G.: 1970, *J. Geophys. Res.* **75**, 7018.
Axisa, F., Martres, M. J., Peck, M., and Soru-Escaut: 1973, *Solar Phys.* **29**, 163.
Biswas, S. and Radhakrishnan, B.: 1973, *Solar Phys.* **28**, 211.
Bratenahl, A. and Yeates, C. M.: 1970, *Phys. Fluids* **13**, 2696.
Brown, J. C.: 1972, *Solar Phys.* **26**, 441.
Brown, J. C.: 1973, *Solar Phys.* **31**, 143.
Carlqvist, P.: 1969, *Solar Phys.* **7**, 377.
Cheng, C.-C.: 1972, *Solar Phys.* **22**, 178.

Cohen, M. H.: 1960, *Astrophys. J.* **131**, 664.
Coppi, B. and Friedland, A.: 1971, *Astrophys. J.* **169**, 379.
Cowley, S. W. H.: 1972, *Cosmic Electrodyn.* **11**, 112.
Datlowe, D. W. and Lin, R. P.: 1973, *Solar Phys.* **32**, 459.
de la Noë, J. and Boishot, A.: 1972, *Astron. Astrophys.* **20**, 55.
Dodge, J. C.: 1972, Ph.D. Thesis, Univ. of Colorado (unpublished).
Donnelly, R. F., Wood, A. T., and Noyes, R. W.: 1973, *Solar Phys.* **29**, 107.
Drummond, W. E. and Pines, D.: 1962, *Nuclear Fusion Suppl.*, Part 3, p. 1023.
Eastwood, J. W.: 1972, *Planetary Space Sci.* **20**, 1555.
Eastwood, J. W.: 1974, *Planetary Space Sci.* (in press).
Elgarøy, Ø.: 1961, *Astrophys. Norv.* **7**, 123.
Ellis, G. R. A.: 1969, *Australian J. Phys.* **22**, 177.
Fainberg, J. and Stone, R. G.: 1970, *Solar Phys.* **15**, 443.
Fainberg, J., Evans, L. G., and Stone, R. G.: 1972, *Science* **178**, 743.
Fermi, E.: 1954, *Astrophys. J.* **119**, 1.
Friedman, M.: 1969, *Phys. Rev.* **182**, 1408.
Frost, K. J. and Dennis, B. R.: 1971, *Astrophys. J.* **165**, 655.
Grognard, R. J. M. and McLean, D. J.: 1973, *Solar Phys.* **29**, 149.
Hall, L. A. and Hinteregger, H. E.: 1969, in C. de Jager and Z. Švestka (eds.), 'Solar Flares and Space Research', *COSPAR Symp.*, p. 81.
Harvey, C. C. and Aubier, M. G.: 1973, *Astron. Astrophys.* **22**, 1.
Hayakawa, S., Nishimura, J., Obayashi, H., and Sato, H.: 1964, *Prog. Theor. Phys. Suppl.* No. 30, 86.
Holt, S. S. and Cline, T. L.: 1968, *Astrophys. J.* **154**, 1027.
Holt, S. S. and Ramaty, R.: 1969, *Solar Phys.* **8**, 119.
Hudson, H. S.: 1972, *Solar Phys.* **24**, 414.
Jokipii, J. R.: 1971, *Rev. Geophys. Space Sci.* **9**, 27.
Kahler, S. W. and Kreplin, R. W.: 1971, *Astrophys. J.* **168**, 531.
Kane, S. R.: 1973, in R. Ramaty and R. G. Stone (eds.), *High Energy Phenomena on the Sun*, NASA SP-342, p. 55.
Kane, S. R.: 1974, this volume, p. .
Kane, S. R. and Anderson, K. A.: 1970, *Astrophys. J.* **162**, 1003.
Kane, S. R. and Donnelly, R. F.: 1971, *Astrophys. J.* **164**, 151.
Kane, S. R. and Lin, R. P.: 1972, *Solar Phys.* **23**, 457.
Kaplan, S. A. and Tsytovich, V. N.: 1968, *Soviet Astron. AJ* **11**, 956.
Kaplan, S. A. and Tsytovich, V. N.: 1973, *Plasma Astrophysics*, Pergamon, London.
Kuiper, T. B. H. and Pasachoff, J. M.: 1973, *Solar Phys.* **28**, 187.
Kundu, M. R.: 1965, *Solar Radio Astronomy*, Wiley Interscience, New York.
Lin, R. P.: 1973, in R. Ramaty and R. G. Stone (eds.), *High Energy Phenomena on the Sun*, NASA SP-342, p. 439.
Lin, R. P. and Hudson, H. S.: 1971, *Solar Phys.* **17**, 412.
Lin, R. P., Evans, L. G., and Fainberg, J.: 1973, *Astrophys. Letters* **14**, 191.
MacDonald, F. B. and van Hollebeke, M. A.: 1973, in R. Ramaty and R. G. Stone (eds.), *High Energy Phenomena on the Sun*, NASA SP-342, p. 404.
Melrose, D. B.: 1970a, *Australian J. Phys.* **23**, 871.
Melrose, D. B.: 1970b, *Australian J. Phys.* **23**, 885.
Melrose, D. B. and Sy, W. N.: 1972, *Australian J. Phys.* **25**, 387.
Neupert, W. N.: 1968, *Astrophys. J.* **153**, L59.
Nunez, P. and Rand, S.: 1969, *Phys. Fluids* **12**, 1666.
Papadopolous, K., Goldstein, M. L., and Smith, R. A.: 1974, *Astrophys. J.* **190**, 175.
Parker, E. N.: 1973, *Astrophys. J.* **180**, 247.
Paul, J. W. M., Daughney, C. C., Holmes, L. S., Rumsby, P. T., Craig, A. D., Murray, E. L., Summers, D. D. R., and Beaulieu, J.: 1972, *Plasma Physics and Controlled Nuclear Fusion Research*, IAEA, Vienna, Vol. 3, p. 251.
Peterson, L. E. and Winkler, J. R.: 1959, *J. Geophys. Res.* **64**, 697.
Petschek, H. E.: 1964, in W. N. Hess (ed.), *The Physics of Solar Flares*, NASA SP-50, p. 425.
Pikel'ner, S. B. and Tsytovich, V. N.: 1969, *Soviet Phys. JETP* **28**, 507.
Pneuman, G. W.: 1972, *Solar Phys.* **23**, 223.

Ramaty, R.: 1969, *Astrophys. J.* **158**, 753.
Ramaty, R. and Petrosian, V.: 1972, *Astrophys. J.* **178**, 241.
Riddle, A. C.: 1972a, *Proc. Astron. Soc. Australia* **2**, 98.
Riddle, A. C.: 1972b, *Proc. Astron. Soc. Australia* **2**, 148.
Riddle, A. C.: 1974, *Solar Phys.* **35**, 153.
Slish, V. I.: 1963, *Nature* **199**, 682.
Smerd, S. F., Wild, J. P., and Sheridan, K. V.: 1962, *Australian J. Phys.* **15**, 180.
Smith, D. F.: 1970a, *Adv. Astron. Astrophys.* **7**, 147.
Smith, D. F.: 1970b, *Solar Phys.* **15**, 202.
Smith, D. F.: 1972a, *Solar Phys.* **23**, 191.
Smith, D. F.: 1972b, *Astrophys. J.* **174**, 121.
Smith, D. F.: 1973, *Solar Phys.* **33**, 213.
Smith, D. F.: 1974a, *Space Sci. Rev.* **16**, 91.
Smith, D. F.: 1974b, *Solar Phys.* **34**, 393.
Smith, D. F. and Fung, P. C. W.: 1971, *J. Plasma Phys.* **5**, 1.
Smith, D. F. and Sturrock, P. A.: 1971, *Astrophys. Space Sci.* **12**, 411.
Smith, D. F. and Pneuman, G. W.: 1972, *Solar Phys.* **25**, 461.
Smith, D. F. and Priest, E. R.: 1972, *Astrophys. J.* **176**, 487.
Sonnerup, B. U. O.: 1970, *J. Plasma Phys.* **4**, 161.
Speiser, T.: 1965, *J. Geophys. Res.* **70**, 4219.
Steinberg, J. L., Aubier-Giraud, M., Leblanc, Y., and Boischot, A.: 1971, *Astron. Astrophys.* **10**, 362.
Stewart, R. T.: 1965, *Australian J. Phys.* **18**, 67.
Stewart, R. T.: 1974, this volume, p. 161.
Stringer, T. E.: 1964, *Plasma Phys.* **6**, 267.
Sturrock, P. A.: 1964, in W. N. Hess (ed.), *The Physics of Solar Flares*, NASA SP-50, p. 357.
Sturrock, P. A.: 1968, in K. O. Kiepenheuer (ed.), 'Structure and Development of Solar Active Regions', *IAU Symp.* **35**, 471.
Swann, W. F. G.: 1933, *Phys. Rev.* **43**, 188.
Sweet, P. A.: 1958, in B. Lehnert (ed.), 'Electromagnetic Phenomena in Cosmical Physics', *IAU Symp.* **6**, 123.
Syrovatsky, S. I.: 1969, in C. de Jager and Z. Švestka (eds.), 'Solar Flares and Space Research', *COSPAR Symp.*, p. 346.
Syrovatsky, S. I.: 1972, in R. Dyer (ed.), *Solar Terrestrial Physics*, Part I, Reidel, Dordrecht, p. 119.
Syrovatsky, S. I. and Shmeleva, O. P.: 1972, *Soviet Astron. AJ* **16**, 273.
Takakura, T.: 1971, *Solar Phys.* **19**, 186.
Takakura, T.: 1972, *Solar Phys.* **26**, 151.
Takakura, T. and Kai, K.: 1966, *Publ. Astron. Soc. Japan* **18**, 57.
Teske, R. G., Soyumer, T., and Hudson, H. S.: 1971, *Astrophys. J.* **165**, 615.
Thomas, R. J. and Teske, R. G.: 1971, *Solar Phys.* **16**, 431.
Tsytovich, V. N.: 1970, *Nonlinear Processes in a Plasma*, Plenum Press, New York.
Tsytovich, V. N.: 1971, *Plasma Phys.* **13**, 741.
Tsytovich, V. N. and Chikhachev, A. S.: 1970, *Soviet Astron. AJ* **14**, 385.
van Hoven, G. and Cross, M. A.: 1973, *Phys. Rev.* **A7**, 1347.
Vedenov, A. A., Velikhov, E. P., and Sagdeev, R. Z.: 1961, *Nucl. Fusion* **1**, 82.
Vorpahl, J.: 1973, *Solar Phys.* **26**, 397.
Yeh, T. and Axford, W. I.: 1970, *J. Plasma Phys.* **4**, 207.
Zaitsev, V. V., Mityakov, N. A., and Rapoport, V. O.: 1972, *Solar Phys.* **24**, 444.
Zheleznyakov, V. V. and Zlotnik, E. Ya.: 1964, *Soviet Astron. AJ* **7**, 485.
Zheleznyakov, V. V. and Zaitsev, V. V.: 1970a, *Soviet Astron. AJ* **14**, 47.
Zheleznyakov, V. V. and Zaitsev, V. V.: 1970b, *Soviet Astron. AJ* **14**, 250.
Zirin, H., Pruss, G., and Vorpahl, J.: 1971, *Solar Phys.* **19**, 463.
Zirin, H. and Tanaka, K.: 1973, *Solar Phys.* **32**, 173.

DISCUSSION

Wild: Can you account for the quasi-periodicity of groups of type III bursts?

Smith: No, I can't in this simple model. However, one can imagine a sheet with some extent and a hydromagnetic wave moving down the sheet causing reconnection to be enhanced as it passes, which might explain the quasi-periodicity.

Mayfield: Can you give an estimate of the lifetime of these current sheets?

Smith: No, because I haven't studied their formation and we don't know what determines the rate of reconnection which will lead to their decay.

Lin: You used 500 G in your estimates. This seems like a high value for a height of 10000 km.

Smith: As I noted the acceleration is proportional to B^2. I used 500 G because if we have a dipole 10000 km long of strength 500 G, then it will still be 500 G at a height of 10000 km.

Dryer: Could a jump in the magnetic field (as in a fast MHD shock) produce accelerated electrons which could produce type III bursts?

Smith: No, the field should be antiparallel as in a separate body of plasma with its own field interacting with the ambient field.

Dryer: As in a plasma piston, perhaps?

Smith: Yes.

Rosenberg: Kuperus and I looked into neutral sheet configurations. Taking a flux tube and twisting its end points, it becomes kink unstable, but stabilizes nonlinearly with a kink. Going on, you end up with a narrow, very long spiralled neutral band (giving the possibility for fast reconnection). The other advantage is the great amount of magnetic energy present with relatively small field strengths.

Smith: I have treated the simplest configuration here. One of the problems with this configuration is that you need a large area of sheet to release a significant amount of energy. Any way you can twist up a sheet into a three-dimensional configuration is likely to be better than the configuration in my talk.

Brown: I have two comments: (1) I agree with your criticisms of the radiative loss model in my 1973 paper but as regards the question of constant pressure or density, I considered the former because it is appropriate in large events, e.g. Cline *et al.* (1967) which *e*-folded in 100 s as against a 50 s dynamical time. When you modify my analysis to constant density, the higher density greatly enhances the radiative loss (even neglecting Hα) and it becomes hard to get the flare hot enough. (2) It is important to point out that models of flare heating by electrons, widely mentioned here today, strongly depend on how low the steep electron spectra extend in energy since all their energy resides at the low end. This limit is not merely instrumentally hard to determine, but is mixed with the thermal X-ray contribution which must therefore be considered carefully, as I hope to do in Part III.

Smith: I agree with you completely on your second comment. Regarding your first comment, it still seems to me that whether constant density or constant pressure is appropriate depends on how much energy is delivered how quickly. It is quite possible that in small flares we are seeing essentially direct conversion of electron energy into EUV radiation with no significant heating of the chromosphere.

Kane: In Takakura's model the energy is transported from the acceleration region to the EUV and optical source through heat conduction. What time constant do you expect for such an energy transport?

Smith: I believe the fine structure in impulsive EUV and hard X-ray data rule out any heat conduction model because the time constant would be too long.

McLean: With regard to your explanation of linear polarization in type III bursts, I should like to ask if you do not find this inconsistent with the observation of circular polarization in type I emission?

Smith: No, because the two types of bursts are thought to be produced in different regions of the corona. Melrose might wish to comment.

McLean: The problem is made more difficult by the fact that these bursts are seen over a wide range of longitude and so we should certainly expect to see both types of burst through the same parts of the corona, yet for one type of burst circular polarization is converted into linear and not for the other.

Melrose: It is thought that the difference between type I and type III is associated with the much stronger field strengths for the former. If one wishes to explain linear polarization in type III, but none in type I, by mode coupling at a *QT* region, this region must be low in the corona because otherwise *both* type I and type III would be linearly polarized. However with a *QT* region low in the corona the Faraday rotation between the *QT* region and Earth should wash out the linear polarization. Linear polarization by this mechanism seems unacceptable.

Stewart: How do you explain the high level in the corona at which type III's start in type I–type III storms?

Smith: There are two possibilities: (1) The electrons are generated low, but because the corona is turbulent to a certain level, generation of plasma waves is supressed. (2) Electrons are accelerated in current sheets high in the corona.

Krall: There is a recent result from Papadopoulos *et al.* which provides an alternate mechanism from either Zaitsev's or Tsytovich's for stabilizing an electron beam. This mechanism is based on the fact that electron plasma waves in the presence of a beam can decay unstably into waves which are not resonant with the beam. This prevents the formation of the quasilinear plateau (Zaitsev's idea) as well as giving an ion-wave coupling stronger than that calculated in Tsytovich's model.

Smith: I am familiar with the work of Papadopoulos *et al.* on stabilizing a narrow stream by the oscillating two-stream instability. However, this is a fixed-phase effect which wouldn't work for a wide stream as expected from a stream which starts out from a power law. I think using the simplest approach is best unless it can be shown to lead to inconsistent results, i.e. quasilinear relaxation.

A THEORY OF TYPE III SOLAR RADIO BURSTS

V. V. ZAITSEV, N. A. MITYAKOV, and V. O. RAPOPORT
Radiophysical Research Institute, Gorkii, U.S.S.R.

(Presented by D. B. Melrose)

Abstract. This paper is an extension of an earlier paper by Zaitsev *et al.* (1972).

In the earlier paper a theory with the following properties was constructed for the propagation of type III streams:

(i) The injection of electrons accelerated in the flare region is 'explosive', i.e. of negligible duration.

(ii) Plasma waves are generated at the front of the stream where quasi-linear relaxation occurs.

(iii) The plasma waves are either (a) damped due to collisions, or (b) reabsorbed by the following electrons in the stream.

(iv) The stream is maintained by the faster electrons (i.e. those which gain energy as a result of the quasi-linear relaxation) escaping from the front of the stream.

In the present paper assumption (i) is relaxed to allow either *explosive* or *continuous* injection of electrons. The results include the following:

(1) At high frequencies ($\gtrsim$20 to 30 MHz) collisions dominate in the damping (assumption (iiia)) while at low frequencies ($\lesssim$5 MHz) collisions are negligible (assumption (iiib)).

(2) The model with explosive injection and collisional damping implies an unacceptably rapid deceleration of the stream.

(3) The model with continuous injection and collisional damping involves an increase in the energy density in plasma waves (and so of the observed radiation) followed by a decay as t^{-6}.

(4) The appropriate model at low frequencies involves continuous injection from the region (e.g. at 5 to 10 $R_{\odot}$) which separates the two limiting cases in (1).

(5) This model involves a smooth increase in the energy density in plasma waves followed by a decay as t^{-3}.

The theoretical time profiles for low frequency bursts (at the second harmonic) are compared with those observed by Fainberg and Stone (1970), and reasonable agreement is found.

In addition, type III bursts with quasi-oscillatory decay are attributed to stabilized (in the sense of Zaitsev *et al.*, 1972) proton streams with the oscillation associated with the combined effects of collisional damping and induced scattering of the plasma waves.

Gordon Newkirk, Jr. (ed.), Coronal Disturbances, 283.

A RELATIONSHIP BETWEEN THE BRIGHTNESS TEMPERATURES FOR TYPE III BURSTS

D. B. MELROSE*

Division of Radiophysics, CSIRO, Sydney, Australia

Abstract (*Solar Phys.*). The widely accepted emission mechanisms for type III bursts involve at least two stages. The first stage is the generation of Langmuir waves by the inferred stream of electrons. Emission at the fundamental frequency arises when these waves are scattered by thermal ions. Emission at the second harmonic arises when two Langmuir waves coalesce; however, the coalescence is possible only after an intermediate stage in which the distribution of Langmuir waves evolves towards isotropy due to scattering by thermal ions.

Quantitative theories for these emission processes are unsatisfactory because uncertainties in both stages compound each other. However, the predicted brightness temperatures T_1 and T_2, at the fundamental and the second harmonic respectively, at a given plasma frequency f_p depends only on the spectrum of Langmuir waves and on coronal parameters. After making a plausible choice of the form of the spectrum of Langmuir waves, it is possible to eliminate the parameter describing their level of excitation (e.g. their energy density) to find a relation between T_1, T_2 and f_p. Comparison of this predicted relation with the observed quantities should avoid uncertainties associated with the theory of the generation of the Langmuir waves by the stream.

Theory predicts

$$T_1 = (T_1)_0 (e^{\tau} - 1), \qquad T_2 = (T_2)_0 \tau^2, \tag{1}$$

with

$$(T_1)_0 \simeq 10^9 \text{ K}, \qquad (T_2)_0 \simeq \frac{10^{21} L_2}{f_p^2 L_1^2} \text{ K}, \tag{2}$$

and

$$\tau \simeq 3 \times 10^{-7}\, W^l L_1 / f_p, \tag{3}$$

where f_p is in megahertz, W^l is the energy density in Langmuir waves in ergs per cubic centrimetre, and L_1 and L_2 are the path lengths in centimetres over which generation of the fundamental and second harmonic respectively occur. In the above quantitative estimates the temperature of the corona has been taken as 10^6 K, the velocity of the stream v_s and the velocity spread Δv_s have been taken as $v_s = c/3$ and $\Delta v_s = v_s/3$ and the phase velocities of the Langmuir waves have been assumed to be in this same range. In the estimate of T_2 the Langmuir waves have been assumed isotropic; the

* On leave from Dept. of Theoretical Physics, Faculty of Science, The Australian National University, Canberra.

Gordon Newkirk, Jr. (ed.), Coronal Disturbances, 285–287.

characteristic time-scale on which they become isotropic is

$$t \sim (3\times10^{11}\,W^l/f_p)^{-1}\ \text{s}, \tag{4}$$

where the units are as above. With $W^l \gtrsim 3\times10^{-10}$ for $T_1 > 10^9$ K ($\tau \geqslant 1$) at $f_p = 100$ MHz this gives $t \lesssim 1$ s, i.e. less than the duration of a type III burst (at fixed frequency at 100 MHz).

The characteristic lengths can be estimated from the bandwidth of the emission mechanisms and the characteristic distance $L_N = f_p|\text{grad}\, f_p|^{-1}$ ($\sim 3\times10^{10}$ cm at $f_p = 80$ MHz) over which the plasma frequency changes. The bandwidth over which emission occurs is very narrow, e.g. $\Delta f/f \sim 10^{-3}$, at both the fundamental and the second harmonic. Because of this reasonable estimates would be $L_1 \sim L_2 \sim 3\times10^7$ to 10^8 cm.

Eliminating τ between the Equations (1) leads to a relationship between T_1 and T_2 for fixed f_p. This relationship is plotted in Figure 1 for $L_1 = L_2 = 10^8$ cm and $f_p =$

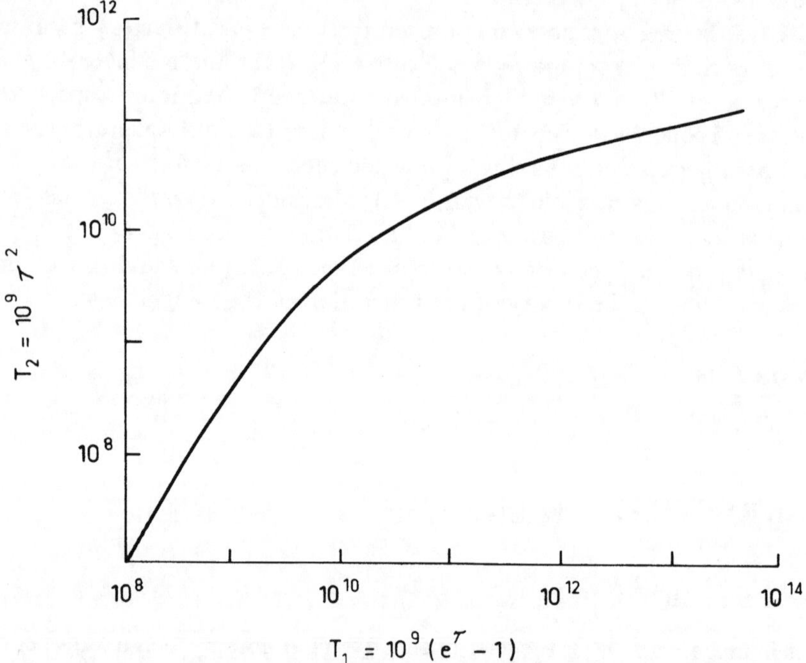

Fig. 1. The brightness temperature at the second harmonic (T_2) is plotted as a function of the brightness temperature at the fundamental (T_1) for $f_p = 100$ MHz and $L_1 = L_2 + 10^8$ cm.

$= 100$ MHz. It should be emphasized that the relevant comparison is at fixed plasma frequency and not, for example, of the fundamental and second harmonic at a single frequency.

The relationship predicted here is the simplest one compatible with theory. It should be anticipated that the relation between the observed brightness temperatures will be found to deviate from the predicted relation. Various effects which have been neglected would favour specific kinds of deviation. Thus the qualitative form of the difference between the predicted and any observed relation should provide further insight into the detailed properties of the emission process.

DISCUSSION

Leblanc: If one wants to compare the brightness temperature of the fundamental and the harmonic emission, one must take into account scattering effects, since the fundamental suffers much more absorption than the harmonic.

Melrose: I agree. The effect you mention, namely collisional damping, favours the second harmonic. There are other effects, listed in my paper, some of which would favour the fundamental. The theory I have presented is the 'zeroth' order one. To improve on it we need to know whether it is consistent with observation and, if it is not, in which direction theory and observations differ.

Rosenberg: Wouldn't you expect that, if induced scattering plays an important role, the burst would be highly polarized?

Melrose: I would not say that I expect it or not, but it is possible. Dr Sy's paper later in the session is concerned with one mechanism whereby induced scattering leads to highly polarized radiation.

THE THIRD HARMONIC IN SOLAR RADIO BURSTS

V. V. ZHELEZNYAKOV and E. Ya. ZLOTNIK
Radiophysical Research Institute, Gorkii, U.S.S.R.

(Presented by D. B. Melrose)

Abstract. Generation of third harmonic plasma emission by coalescence of second harmonic radiation with a plasma wave is more favourable than by direct coalescence of three plasma waves. The predicted ratio of the intensities at the third (III) and second (II) harmonics is $I_\omega^{\mathrm{III}}/I_\omega^{\mathrm{II}} \sim \omega_L W_l L/Nmc^3$, where ω_L is the plasma frequency, W_l is the energy density in plasma waves, L the linear dimensions of the source and N the ambient electron number density. The intensity at the second harmonic is

$$I_\omega^{\mathrm{II}} \simeq e^2 W_l^2 L_N / m^2 c^2 \omega_L^2,$$

where $L_N(\sim 10^{10}$ cm at $\omega_L \sim 2\pi \times 10^8\ \mathrm{s}^{-1})$ is the typical dimension of coronal inhomogeneities.

It is suggested that the third harmonic in a *U*-burst observed by Haddock and Takakura and in type V emission reported by Benz (1973) were due to this process. For the latter event the theory predicts $I^{\mathrm{III}}/I^{\mathrm{II}} \sim 3 \times 10^{-2}$ compared with the observed ratio of 10^{-1}.

Gordon Newkirk, Jr. (ed.), Coronal Disturbances, 289–290.

DISCUSSION

(After Two Abstracts of Zaitsev et al. and Zheleznyakov and Zlotnik)

Rosenberg: (1) Continuous injection seems difficult to reconcile with X-ray data. Is that true? (2) The fall-off with E^{-6} or E^{-3} seems to contradict the observed exponential decay. (3) Has anybody observed third harmonics? In my opinion the two cases mentioned should be used with care.

Melrose: As I understand it, continuous injection would give type III burst the same duration at all frequencies up to the 30 MHz the authors mention. The power law decay is considered satisfactory by the authors.

Smith: There is nothing difficult about a short continuous injection of a few tenths of a second. Indeed, physically the injection time must be finite. Zaitsev *et al.* appear to obtain agreement with the data on decay because they considered a limited frequency range and drew sufficiently large dots around the data points.

Takakura: Zaitsev *et al.* have neglected the induced scattering of plasma waves in the resonance region into the non-resonance region. However, if we adopt their solution and take the observed flux of solar electrons at the Earth, the time scale for induced scattering is less than the growth time of the plasma waves in the resonance region everywhere between the Sun and the Earth. Therefore their result is inconsistent. I have modified their result taking into account induced scattering, but there is no time to go into the details.

Smith: A problem with the induced scattering time of Tsytovich is that it needs to be multiplied by a factor of about 600 for a narrow spectrum of waves according to the numerical analysis which I carried out with Fung and by an even larger factor for a broad spectrum of waves. With this correction, induced scattering would not lead to a significant change in the angular distribution of waves for a broad spectrum of waves.

Melrose: The fact that fundamental radiation must be amplified to obtain observed brightness temperatures requires that induced scattering of plasma waves also be important since the process is the same for both waves.

Wild: We have looked for a third harmonic for many years. It often looks possible, but there is no convincing case.

Kane: There is evidence for continuous acceleration or injection of electrons in the observed hard X-ray emission and its time variation. This is independent of the model assumed, i.e. a thin or thick target model for the X-ray source.

FINE STRUCTURE IN TYPE IV SOLAR RADIO BURSTS

C. CAROUBALOS and M. PICK
Observatoire de Paris, Meudon, France

C. CHIUDERI and R. GIACHETTI
Osservatorio Astrofisico di Arcetri, Firenze, Italy

and

H. ROSENBERG* and C. SLOTTJE
Sterrewacht 'Sonnenborgh', Utrecht, The Netherlands

Abstract (*Solar Phys.*). The fine structure in solar type IV radio bursts was studied using the 169 MHz Nançay radioheliograph and the 60 channel radiospectrograph at Utrecht (160–320 MHz). The observed fine structure includes pulsating structure, zebra patterns (parallel drifting bands) and intermediate drift bursts. All are considered as modulation of high frequency radiation by low frequency oscillations or as the result of up conversion of low frequency oscillations to higher frequencies (Rosenberg, 1973).

Definite proof is presented that the pulsating structure is produced by a source which is imbedded in the continuum source and which is considerably smaller than the continuum source (Caroubalos *et al.*, 1973).

The zebra patterns can be understood as due to the coupling of electron Bernstein waves and upper hybrid waves. These waves become unstable if a hot electron cloud is trapped inside a magnetic flux tube filled with cooler coronal plasma (Chiuderi *et al.*, 1974). A source size of 1000 km and an energy density in the plasma waves and Bernstein waves equal to 10^{-3} of the thermal coronal energy density is compatible with the observations. It is concluded that stationary type IV radiation is not synchrotron radiation but magnetoplasma radiation strongly dependent upon the magnetic field (Chiuderi *et al.*, 1973).

References

Caroubalos, C., Pick, M., Rosenberg, H., and Slottje, C.: 1973, *Solar Phys.* **30**, 473.
Chiuderi, C., Giachetti, R., and Rosenberg, H.: 1973, *Solar Phys.* **33**, 225.
Chiuderi, C., Giachetti, R., and Rosenberg, H.: 1974, to be published.
Rosenberg, H.: 1973, Thesis, Utrecht.

DISCUSSION

Wild: Perhaps an alternative and simpler explanation of the Zebra pattern on dynamic spectra would be to postulate a propagation effect in which the radiation reaches the observer along two well-defined paths, causing an interference pattern. A similar bi-prism phenomenon is observed on the dynamic spectrum of radio-star scintillations, though perhaps in this case the 'bi-prism' would be in the solar atmosphere. One would only need a path-length difference of, e.g., 10 m to explain a frequency separation between adjacent stripes of 30 MHz.

* Presently at Harvard College Observatory, Cambridge, Mass.

Sy: (1) If an instability is involved, you would expect different intensities in the harmonics. (2) You assume the Bernstein waves to be isotropic which is doubtful.

Rosenberg: (1) Intensities are observed to be comparable in the laboratory. Harmonics are thought to feed each other nonlinearly. (2) We used an isotropic distribution for simplicity, but only a small portion of it is important for coalescence.

A THEORY OF TYPE I SOLAR RADIO BURSTS

W. N.-C. SY
Division of Radiophysics, CSIRO, Sydney, Australia

Abstract (*Proc. Astron. Soc. Australia*). A theory is developed to account for the observed properties of type I storm bursts in terms of plasma radiation – that is, electromagnetic radiation at the electron plasma frequency resulting from the non-linear scattering of electron plasma waves on plasma ions. Now the average brightness temperature of a type I source is greater than 10^9 K, or even higher if, because of coronal scattering, the apparent source size is larger than the true source size. For brightness temperatures as high as 10^9 K the non-linear scattering must be of the induced kind in which electromagnetic radiation below the frequency of the electron

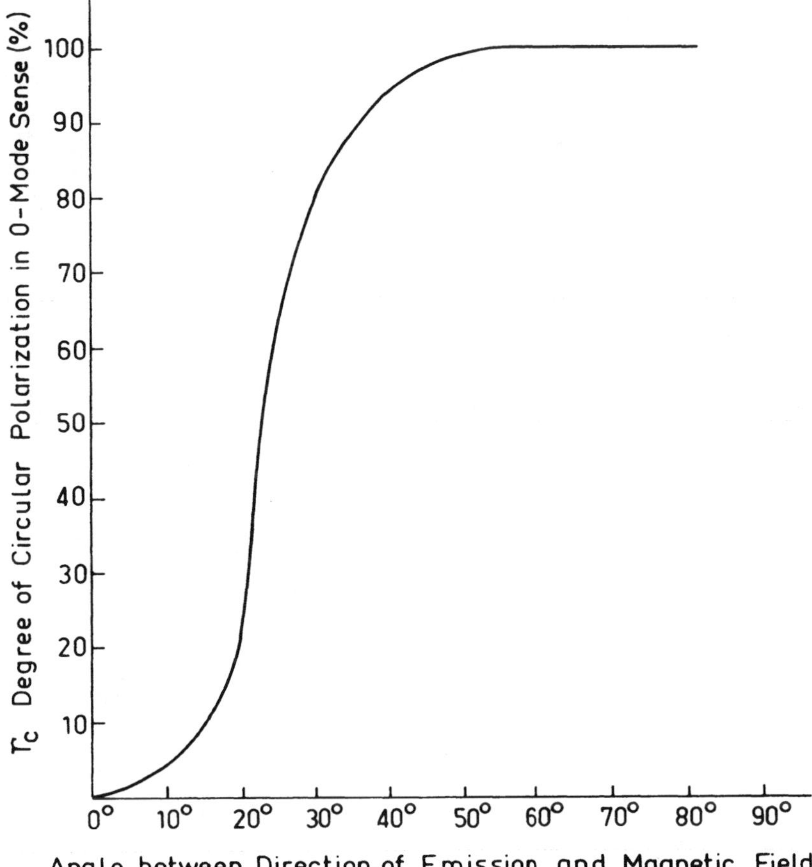

Fig. 1. The degree of circular polarization of amplified plasma radiation as a function of the angle between the directions of emission and of the magnetic field calculated for a simple model of a type I burst source.

Gordon Newkirk, Jr. (ed.), Coronal Disturbances, 293–294. All Rights Reserved.

plasma waves is amplified. For such radiation to be strongly circularly polarized in the *o*-mode, as observed in type I bursts, requires that the amplification be more effective in the *o*-mode than in the *x*-mode (Figure 1). This is found to be so for plasma waves excited by electrons travelling parallel to the magnetic field. The electric field of the plasma waves is then also parallel to the magnetic field. The non-linear scattering is more efficient for that magnetoionic mode which has the greater component of electric field in the same direction. This mode is the *o*-mode.

DISCUSSION

Melrose: Your idea is interesting. How large does the optical depth have to be for the effect of the preferential amplification of the ordinary mode to be important?

Sy: The significant amplification of *o*-mode over *x*-mode depends on the angle of emission of the radiation with respect to the background magnetic field. At an emission angle of 45° or greater the optical depth needs to be 4 or greater to give close to 100% polarization.

SYNCHROTRON RADIATION IN DIRECTIONS CLOSE TO MAGNETIC-FIELD LINES

K. C. WESTFOLD

Monash University, Australia

Abstract (*Astrophys. Space Sci.*). It is characteristic of the radiation from a particle of mass m bearing a charge e moving with ultrarelativistic velocity $\boldsymbol{\beta} c$ in a magnetic field of induction $\mathbf{B}_0$ that the bulk of the emission is confined to a small cone of directions within $O(\xi)$ of the direction of motion $\boldsymbol{\tau}$ which makes a constant angle α with $\mathbf{B}_0$, and that for any direction $\mathbf{n}$ at an angle θ with $\mathbf{B}_0$ such emission is observed as harmonics of the fundamental frequency $f_{B_0}\xi/a$, where f_{B_0} is the nonrelativistic cyclotron frequency $eB_0/2\pi m$, $\xi=\sqrt{(1-\beta^2)}\ll 1$, and

$$a=1-\beta\cos\alpha\cos\theta.$$

It follows that the radiation from a distribution of particles in general appears as a series of lines, broadened with respect to the dependence of the distribution on both pitch angle and energy $E=mc^2/\xi$.

In directions away from the field lines, $\sin\theta=O(1)$, $a\simeq\sin^2\theta$, and the bulk of the emission is in the high-order harmonics $n=O(\xi^{-3})$. Although the broadening with respect to pitch angle is slight, the observed radiation is quasicontinuous.

In directions close to the field lines, for which $\sin\theta=O(\xi)$, $a\simeq\frac{1}{2}(\xi^2+\sin^2\alpha+\sin^2\theta)=$ $=O(\xi^2)$, so that pitch-angle broadening is significant and there is a significant contribution to the radiation from low-order harmonics. Specifically, if $N(1/\xi)\,\mathrm{d}(1/\xi)$ is the number density of particles in the energy range $(E, E+\mathrm{d}E)$ and $2\pi\phi(\alpha)\sin\alpha\,\mathrm{d}\alpha$ is the fraction of these in the pitch-angle range $(\alpha, \alpha+\mathrm{d}\alpha)$, the emissivity tensor corresponding to the nth harmonic is, to a first approximation, distributed with respect to the frequency f according to the formulae

$$\boldsymbol{\eta}_n(\mathbf{n})=\int_0^{nf_{B_0}/\sin\theta}\boldsymbol{\eta}_{nf}(\mathbf{n})\,\mathrm{d}f,$$

$$\boldsymbol{\eta}_{nf}(\mathbf{n})=\frac{2\pi n^2y^3}{f_{B_0}}\int_{ny-\sqrt{(n^2y^2-\sin^2\theta)}}^{ny+\sqrt{(n^2y^2-\sin^2\theta)}} N\left(\frac{1}{\xi}\right)\phi(\alpha)\langle\mathbf{P}_n(\mathbf{n})\rangle\,\mathrm{d}\xi,$$

where $y=f_{B_0}/f$ and is of magnitude $O(\xi/n)$, and, in terms of base vectors $\mathbf{i}_1$ along the projection of $\mathbf{B}_0$ on the plane transverse to $\mathbf{n}=\mathbf{i}_3$ and $\mathbf{i}_2=\mathbf{i}_3\times\mathbf{i}_1$, the emission polarization tensor

$$\langle\mathbf{P}_n(\mathbf{n})\rangle=\frac{\mu e^2cf_{B_0}^2}{2n^2y^4\xi^2}\left[\left(\frac{ny\xi-\sin^2\theta}{\sin\theta}\right)^2 J_n^2\left(\frac{\sin\alpha\,\sin\theta}{y\xi}\right)\mathbf{i}_1\mathbf{i}_1\mp\right.$$

Gordon Newkirk, Jr. (ed.), Coronal Disturbances, 295–297.

$$\mp i \frac{ny\xi - \sin^2\theta}{\sin\theta} \sin\alpha\, J_n\left(\frac{\sin\alpha \sin\theta}{y\xi}\right) J_n'\left(\frac{\sin\alpha \sin\theta}{y\xi}\right)(\mathbf{i}_1\mathbf{i}_2 - \mathbf{i}_2\mathbf{i}_1) +$$
$$+ \sin^2\alpha\, J_n'^2\left(\frac{\sin\alpha \sin\theta}{y\xi}\right)\mathbf{i}_2\mathbf{i}_2\Bigg],$$

according as $\cos\theta \gtrless O$. The pitch angle α is given in terms of ξ and y by the relation

$$\sin\alpha = \sqrt{(2ny\xi - \xi^2 - \sin^2\theta)}.$$

Since the argument of the Bessel functions is $n \times O(1)$, it is not possible to make a further simplification without making further assumptions. For emission in directions close to the field lines it is appropriate to take $\sin\theta \ll \xi$ and to expand the formulae in powers of $(\sin\theta)/\xi$. Then, for $n=1$, the magnitude of the expression in square brackets is $O(\xi^2)$, and for each subsequent harmonic it is reduced by the factor

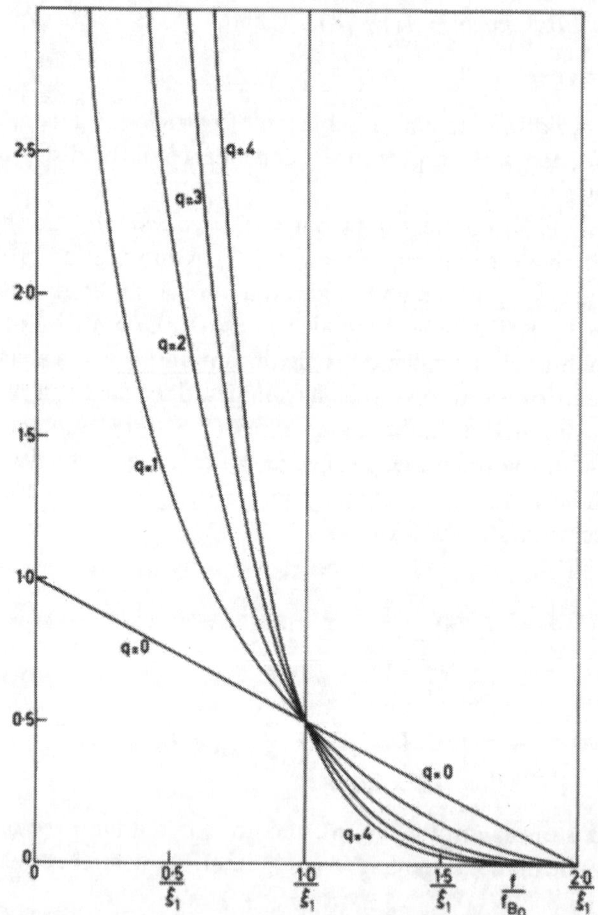

Fig. 1. The emissivity parameter $\eta_f/(\pi \mathcal{N} \mu e^2 c f_{B_0} a_q \xi_1^{q+1})$ for monoenergetic particles with pitch-angle distribution $\phi(\alpha) = a_q \sin^q\alpha$.

$O((\sin^2\theta)/\xi^2)$. The approximations have been carried through for the first two harmonics, giving formulae for the total emissivity tensor

$$\boldsymbol{\eta}_f(\mathbf{n})=\sum_n \boldsymbol{\eta}_{nf}(\mathbf{n})$$

of consistent accuracy and showing that, contrary to what has previously been reported, the degree of circular polarization is $1+O((\sin^4\theta)/\xi^4)$, RH for $\cos\theta>O$ and LH for $\cos\theta<O$.

In particular, for a monoenergetic distribution for which $N(1/\xi)=\mathcal{N}\delta(1/\xi-1/\xi_1)$, we find as a first approximation to the emissivity

$$\eta_f(\mathbf{n})=\pi\mathcal{N}\mu e^2 cf_{B_0}\,\phi(\alpha_1)\,\xi_1\left(1-\frac{\xi_1 f}{2f_{B_0}}\right),$$

where the pitch angle α_1 corresponding to the frequency f is given by

$$\frac{\sin\alpha_1}{\xi_1}=\left(\frac{2f_{B_0}}{\xi_1 f}-1\right)^{1/2}.$$

A particular case of the dependence of emissivity on pitch-angle distribution is exhibited in Figure 1 where we have taken

$$\phi(\alpha)=a_q\sin^q\alpha,$$

a distribution which varies from isotropic with respect to velocity when $q=1$ to a concentration of flat helical trajectories about the value $\alpha=\frac{1}{2}\pi$ when q is increased, resulting in reduced emission close to $\theta=0$ and π. The portion of the curves for which $O\leqslant f/f_{B_0}<1/s\xi_1$ correspond to $(\sin\alpha_1)/\xi_1>(2s-1)^{1/2}$, and should be ignored when s is so large that the requirement that $(\sin\alpha_1)/\xi_1$ should be $O(1)$ is breached. They should all fall sharply to zero at $f/f_{B_0}=\frac{1}{2}\xi_1$.

PART III

SHOCK WAVES AND PLASMA EJECTION

SHOCK WAVES AND THE EJECTION OF MATTER FROM THE SUN: RADIO EVIDENCE

D. J. McLEAN
Division of Radiophysics, CSIRO, Sydney, Australia

Abstract. The passage of shock waves and ejected matter through the solar corona can produce type II and type IV radio bursts. This paper reviews the observations of these types of bursts and their interpretation, with particular emphasis on recent work.

1. Introduction

The first radio evidence of disturbances travelling out through the solar corona came from the observation that some metre-wave bursts were progressively more delayed at lower observing frequencies. With the introduction of the first radio spectrograph by Wild and his colleagues these bursts were recognized as a distinctive type of solar radio burst which became known as a type II burst. The interpretation of this type of burst, since confirmed in a variety of ways, is that a disturbance travelling out through the corona excites the local plasma at each height to radiate at a frequency close to f_p, the local plasma frequency, and also at twice that frequency. Since the electron density law for the corona is known, at least approximately, the variation of frequency with time can be interpreted as an increase of height with time; velocities ranging from about 200 km s^{-1} up to 2000 km s^{-1} or more are deduced. Since these velocities are appreciably greater than the sound velocity in the corona (~ 150 km s^{-1}) and greater than or of the order of the probable value of the Alfvén velocity in the corona, it is generally accepted that the disturbance responsible for a type II burst is a collisionless magnetohydrodynamic shock wave.

Important characteristics of the spectra of these bursts are: (a) the relatively narrow bandwidth of drifting ridges of emission, sometimes as little as a few megahertz, although sometimes very much broader; (b) the frequent occurrence of fundamental and second harmonic components (with almost identical features); and (c) the splitting of both the fundamental and harmonic components into two bands ~ 10 MHz apart for the fundamental, twice as much for the harmonic. For an example see Figure 1. Band-splitting is observed in many type II bursts, but not all.

The most direct confirmation of the plasma hypothesis comes from position measurements with a swept-frequency interferometer by Wild *et al.* (1959) and Weiss (1963); for bursts near the limb of the Sun they found that low frequencies apparently originated higher in the corona than high frequencies. At any one frequency the source did not move. Another piece of confirmatory evidence has come from simultaneous optical and radio observations. A number of cases have been reported of eruptive prominences seen rising above the solar limb at the same time and with about the

Gordon Newkirk, Jr. (ed.), Coronal Disturbances, 301–321.

same speed as type II bursts observed in the radio spectrum (Giovanelli and Roberts, 1959; Warwick, 1965; McCabe and Fisher, 1970).

At about the time of the first spectral results (Wild and McCready, 1950; Wild, 1950), Payne-Scott and Little (1952), observing at a single frequency with a two-element interferometer, detected sources moving outward through the corona at velocities of 500 to 3000 km s^{-1}. Some years later, using observations made with the Nançay interferometer at 169 MHz, Boischot (1958) studied similar sources moving at several hundred kilometres per second out to great distances from the Sun. He named these bursts type IV and described them as a broadband emission varying smoothly in time, typically preceded by a flare and a type II burst. Since then it has been recognized that a solar flare event can include a number of broadband continuum phenomena. Because these have all been termed type IV bursts we must use the name 'moving type IV burst' for the radio bursts of relevance to this paper.

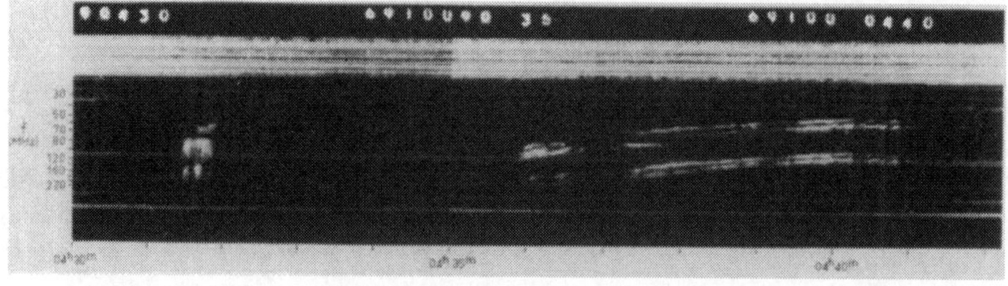

Fig. 1. Examples of a dynamic radio spectrum of a type II burst showing harmonic structure and split bands. At any point on the spectrum the whiteness of the record indicates the intensity of radiation received at that time frequency. Signals at constant frequency are terrestrial interference. (From Smerd *et al.*, 1973.)

Studies of the Earth's magnetic field often reveal periods of perturbation. The sudden commencement of many of these 'geomagnetic storms' is attributed to a disturbance from the Sun impinging on the Earth's magnetosphere. In view of their movement out from the Sun it was natural to investigate type II and type IV bursts and the flares which produce them as a probable cause of geomagnetic storms. Indeed a strong correlation was found. However, I leave further discussion of this subject and of more direct observations of shock waves in interplanetary space for Dr Hundhausen.

Films of the Sun's disk photographed in the wings of Hα occasionally reveal curved fronts, spreading out across the chromosphere from a flare at about 1000 km s^{-1}. This type of event, sometimes known as a Moreton wave, has been interpreted as the skirt of a coronal shock wave made visible by the perturbations it produces in the chromospheric material it passes over (Uchida, 1968; Meyer, 1968). In a number of cases the flares which produced Moreton waves also produced type II bursts (Wild, 1969a, b; Uchida *et al.*, 1973; Uchida, 1973; Smith and Harvey, 1971), and it has been suggested that these two phenomena may be different aspects of the same shock front. I expect that Dr Bruzek will include Moreton waves in his discussion.

2. Type II Bursts

In recent years the Culgoora radioheliograph has literally added a new dimension to observational solar radio astronomy, and most of the new data on type II bursts has come from this instrument. However, before we examine these data in detail, it is as well to recognize the major difficulties encountered in the interpretation of any observations of the positions of type II bursts.

Consider the well-known formula

$$\mu^2 = 1 - f_p^2/f^2,$$

which determines the refractive index μ of an electromagnetic wave of frequency f in a plasma where the plasma frequency is f_p and the magnetic field is so weak that its effects can be ignored (a good approximation for the corona). Clearly near the source of a type II burst the fundamental component ($f \simeq f_p$) will be severely refracted, and even the effect on the second harmonic will not be negligible ($f \simeq 2f_p$). In fact rays traced in simple smooth coronal models show that from the Earth we should not be able to observe fundamental components, except at the centre of the disk, and that near the limb even the second harmonic source should be invisible. Yet limb flares sometimes produce type II bursts for which both fundamental and second-harmonic components are recorded. It has long been assumed (Roberts, 1959) that small-scale inhomogeneities of the corona scatter the radiation above a source into directions along which it can reach the Earth.

The difficulty then is to relate the apparent position and size of a source to the true position and size. The simplest assumption, first introduced by Shain and Higgins (1959) and widely adopted since, is that the net effects of scattering and subsequent refraction tend to cancel, leaving the apparent position close to the true position.

Unfortunately we have little information about the amplitude and scale of the density inhomogeneities which determine the scattering properties of the solar corona. Observations of the angular sizes and scintillation of radio sources occulted by the corona have been used to investigate scattering further from the Sun in the region of the solar wind. By extrapolating these results it is possible for us to make a guess at the scattering properties lower in the corona. Numerical work based on this estimate has been published recently which provides useful statistical data on the probable effects of scattering (Fokker, 1965; Fokker and Rutten, 1967; Steinberg *et al.*, 1971; Aubier *et al.*, 1971; Steinberg, 1972; Caroubalos *et al.*, 1972; Riddle, 1972a, b, 1974; Leblanc, 1973). So far most of these calculations have been carried out for a spherical corona, although Riddle (1972b, 1974) has considered a source on the axis of a coronal streamer. The results of all this work are to suggest that the simple assumption introduced by Shain and Higgins (1959) is misleading. In fact, the apparent position of a fundamental source near the limb, radiating at 80 MHz will lie outside the true position by $\gtrsim 2'$, whereas the second harmonic source will appear inside its true position by a similar amount. Although these calculations are only based on guesses about the coronal inhomogeneities, their predictions are consistent (assuming a point emitter) with observations of the size of type III bursts, and the observed relative positions of fundamental and second-harmonic sources.

A number of earlier deductions were based on the assumption that the apparent position exactly equals the true position: these results should be re-evaluated. They include the high-density electron density law deduced from burst position observations (see e.g. Wild *et al.*, 1963) and the 'backward' emission hypothesis proposed by Smerd *et al.* (1962). Furthermore, in view of the remaining uncertainties, it is hazardous to make any interpretation of the data if that interpretation could be upset by shifting the source position by about 2′.

With these uncertainties in mind we shall now examine some of the results which have emerged from studies of the data from the Culgoora radioheliograph, and consider what clues these results give about the nature of the type II shock front. Because special circumstances have given extra significance to a few particular observations, these will receive special attention.

2.1. Extent

It has been suggested (Wild, 1969a, b; Uchida *et al.*, 1973) that when type II bursts and Moreton waves are observed simultaneously they are different manifestations of a single shock wave. If this is the case we should expect the type II source to subtend a very large angle at the flare. Alternatively, since at one frequency we see only the region of intersection of the moving shock front with one plasma level in the corona, we might expect to see a source which moved across the Sun as the outward motion of the shock shifted the point of intersection across the Sun.

Type II sources observed at 80 MHz are often very extensive, $\frac{1}{2} R_{\odot}$ or more is typical (Wild, 1970a). Since this is much greater than the spread due to scattering, estimated either from the theoretical work quoted above or from observations of the apparent sizes of other types of burst, there is no reason to doubt that these large sizes are real. On occasions too, progressive shifts of a type II burst position have been observed and interpreted as due to shock intersecting the 80 MHz plasma level at an angle (Dulk, 1970a; Riddle, 1970a; Smerd, 1970). The total shift, 4′ or more, is probably sufficient to eliminate the suspicion that it may be due to complicated propagation effects.

2.2. Variability

On occasions the brightness distribution of type II sources has been observed to vary very rapidly, with a time scale of about 1 s. So far these fluctuations have not been identified with any spectral features, although Wild (1969a) has suggested that these variations might be related to herringbone structure in type II bursts, which also has a very short time scale (Roberts, 1959). Since herringbone structure may be the key which relates the theories of type II and type III emission (Smith, 1971), further study of this matter is desirable.

2.3. Split bands and bright lanes

Type II burst spectra often show fine structure consisting of bright peaks ~10 MHz apart which follow the general drift of the burst. Often two very similar components are observed (e.g. Figure 1) and these are termed 'split bands'. On other occasions the brightness and drift-rate variations of the peaks do not appear correlated and there may be more than two of them; these we term 'bright lanes'. An example is shown in Figure 2. Split bands have received much more attention in the literature than lanes.

There are currently two possible theories of split bands (the reasons for discarding

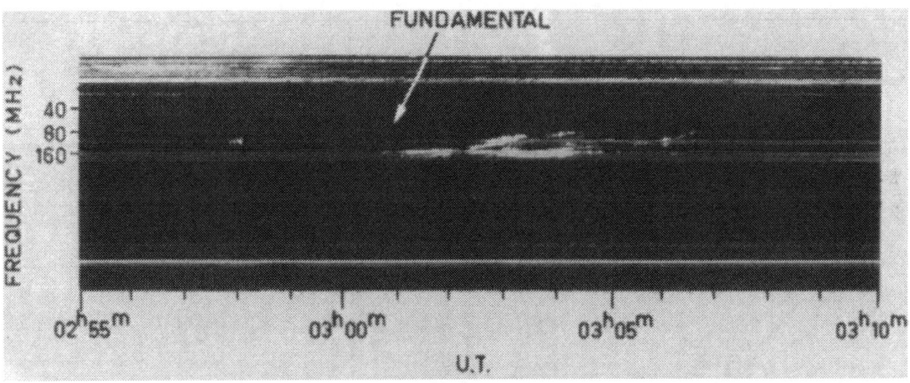

Fig. 2. Spectrum of the type II burst of 1969 October 3 showing multiple bright lanes in the harmonic component. (From Dulk, 1971.)

the earlier 'magnetic' and 'Doppler' theories are reviewed by Wild and Smerd (1972)). One theory due to McLean (1967) proposes in essence that parts of the shock front which are parallel to the surfaces of equal electron density will emit intensely at a single frequency whereas the emission from other parts of the shock front will be spread thinly across a range of frequencies. In a simple streamer structure (Newkirk, 1961) there are likely to be two different parts of the shock front which emit intensely at two slightly different frequencies, the amount of splitting being determined by the geometry of the corona.

Smerd *et al.* (1973) proposes rather that the two bands of a split-band burst correspond to emission from in front of and behind the shock front. Since the density behind a shock is greater than in front the upper-frequency band comes from behind the lower-frequency band from in front. In this theory the amount of splitting is a measure of the Mach number of the shock; Smerd *et al.* find values in the range $M_A \simeq 1.2$ to 1.7.

Observations of slightly different positions of the two components of split band bursts have been interpreted as evidence in favour of McLean's (1967) theory (Dulk, 1971; Labrum, 1969; Wild and Smerd, 1972). However Smerd *et al.* (1973) point out that we should also expect a shift if both bands were emitted from a single source;

the source will shift appreciably between the times of observation of the two bands at a given frequency, and also for the fundamental component the propagation effects will be different for the two frequency bands.

Unfortunately the number of clear cases of split bands observable at a single frequency (80 MHz) is very small. It is to be hoped that future observations at 43, 80 and 160 MHz with the modified Culgoora radioheliograph may help to choose between these theories.

Figure 3 (from Dulk, 1971) shows the 80 MHz positions of the multiple bright-lane

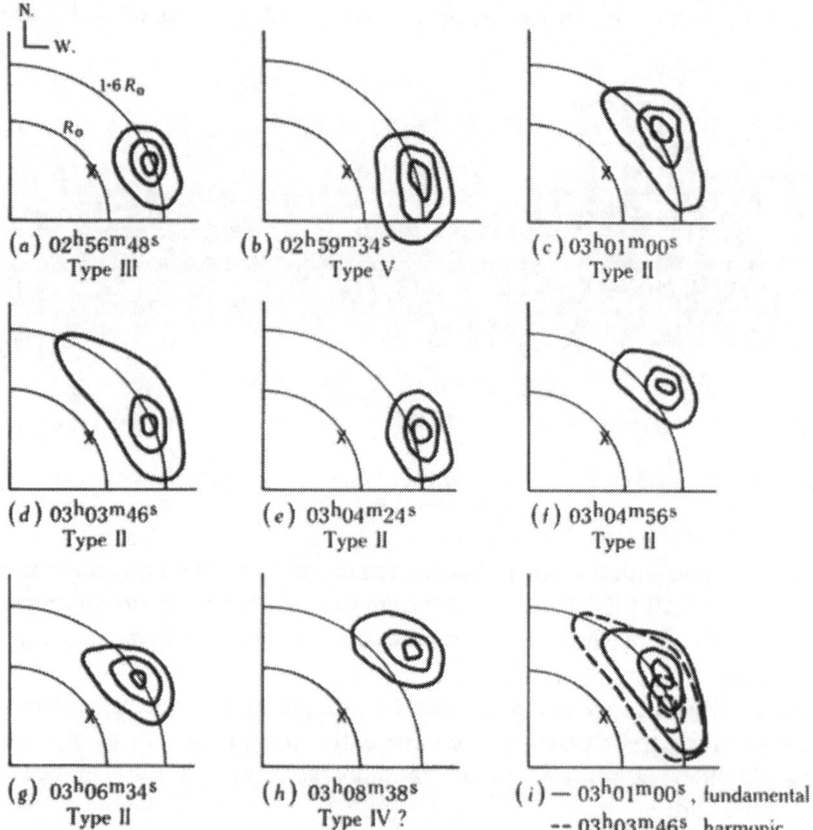

Fig. 3. Culgoora radioheliograph 80 MHz contours for a number of times during the type II burst of Figure 2. The inner and outer circular quadrants have radii 1 $R_\odot$ and 1.6 $R_\odot$ respectively. (From Dulk, 1971.)

burst illustrated by Figure 2. In these and other cases (Riddle, 1970a; Kai, 1969a) the large shifts between the positions of the different bright lines leave little doubt that different parts of the shock front cause the different lanes. We shall return to the problem of patchy shock fronts shortly.

Weiss (1963) noted that position observations of many type II bursts reveal the existence of multiple shocks, propagating in different directions. In some cases it is possible that these multiple shocks are in fact parts of one shock front propagating with different speeds in different directions (Kai, 1969a). It seems quite plausible that other events, such as the very complex event described by Riddle and Seridan (1971), might also fit this interpretation. As multiple bright lanes are a very common feature of type II bursts this interpretation would imply that radiation from a number of different parts of a shock front is typical of this phenomenon.

2.4. Reflection and refraction or guiding

We know the coronal magnetic field and density distributions to be highly inhomogeneous, and it is only to be expected that these inhomogeneities affect the propagation of type-II-producing shock fronts. Furthermore theoreticians do not agree on the nature of these shocks, or, in particular, on the direction relative to the magnetic field in which they propagate (see Dulk *et al.* (1971a) for details).

Two events have contributed to our understanding of this matter. In one case Kai (1969a) drew attention to the fact that the position of components of a type II burst indicated that the shock front had been stopped or reflected by the strong field 'behind' the flare; apparently the type II shock front could not propagate across the region of strong field.

The other case, 1969 March 30, described by Smerd (1970), has been discussed at great length in the literature. The event is believed to have originated in an active region about 20° behind the limb. The flare was also responsible for an important solar proton event. The type II bursts produced in this event were preceded by a burst which outlined half the solar limb (Figure 4): three type II bursts occurred in widely separated locations. (We note the possibility of an extended, patchy shock front consistent with the discussion in the preceding sub-sections.) However, the most significant deduction from this observation is that the shock front must have followed a curved path from the flare to at least two of the sources. Smerd (1970) concludes that the shock was either guided along the coronal magnetic field or refracted by suitable structure in the corona. Furthermore, Dulk *et al.* (1971a) have shown that the shock path was along the coronal magnetic field, whatever the cause of the curved path. The curved path and the propagation along magnetic field lines both seem to be of considerable significance for the theoretical description of type II shock waves.

Dulk *et al.* (1971a) examined the data from a number of other bursts for information about the direction of motion relative to the magnetic field but were unable to draw any firm conclusions. However, they did give other arguments which suggest that type II shocks may travel approximately along the magnetic field: often the position of a type II burst coincides with the positions of associated type III bursts, and the latter are certainly guided along the magnetic field. In addition, type II bursts which extend to very low frequencies (less than 20 MHz, say) must be moving approximately radially outward in a region where the magnetic field is also approximately radial, combed out by the solar wind.

When we recall that the best-developed type II theories assume shocks travelling perpendicular to the magnetic field, the importance of these conclusions is evident.

2.5. Interpretation

Many of the observations discussed in the previous subsections reveal a patchy structure in type II sources. Why do these bursts brighten in some places and miss other regions between the bright parts? It may be because:

(a) refraction of shock waves by magnetic and density structures of the corona

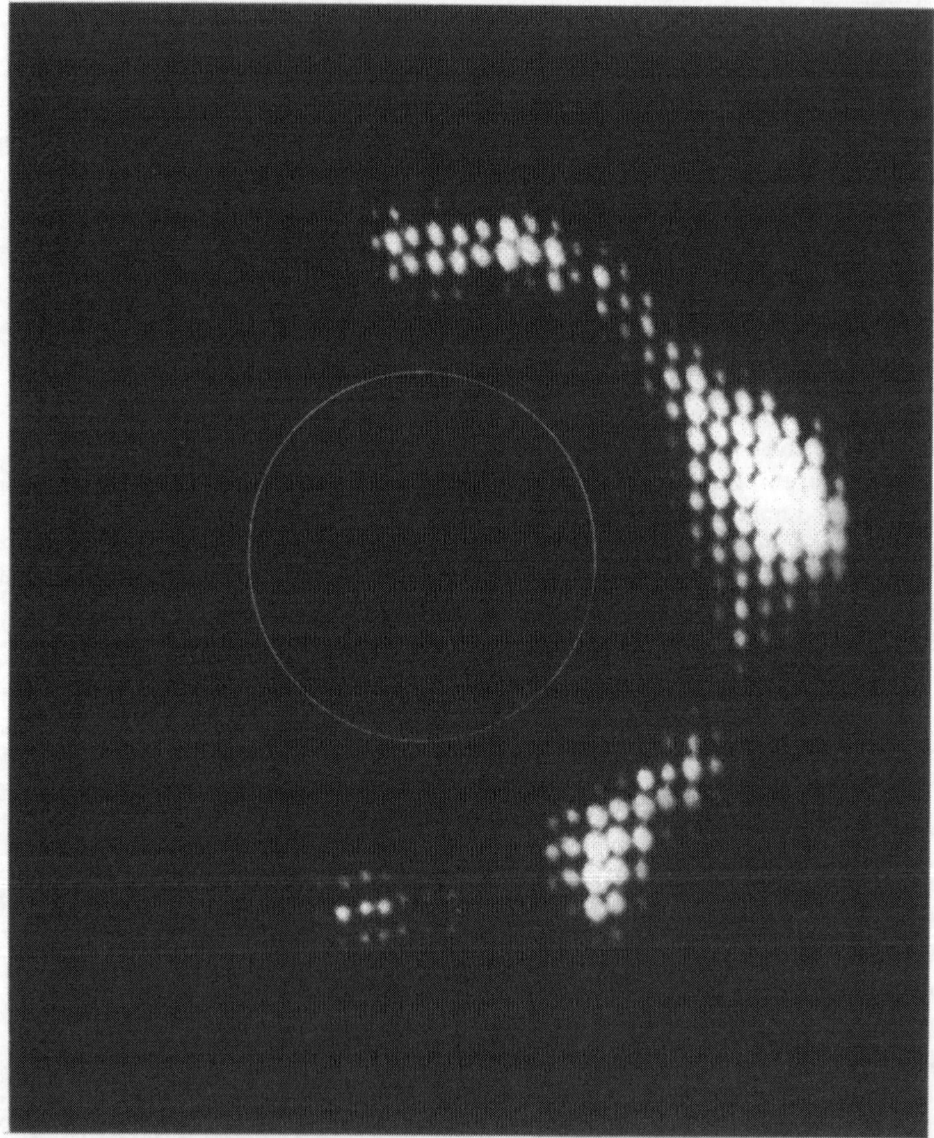

Fig. 4. A radioheligram from the event of 1969 March 30 caused by a flare behind the limb. Later in this event type II bursts appeared in three different places. (From Smerd, 1970.)

channels the shock energy into regions of low Alfvén velocity (e.g. Uchida, 1968, 1970, 1973; Uchida *et al.*, 1973; Smerd, 1970; Kai, 1969a);

(b) the shock is strongest where the Alfvén velocity is lowest and therefore presumably radiates the strongest emission from there (e.g. Dulk and Smerd, 1971);

(c) regions where the shock front is normal to the density gradient radiate intensively at one frequency, whereas the radiation from other regions is spread more thinly across the spectrum (McLean, 1967);

(d) deviations of the corona from spherical symmetry modify the propagation of the electromagnetic radiation to contribute to the uneven appearance of type II sources. However, the fundamental radiation would be affected much more than the second harmonic (e.g. Riddle, 1970a).

These four possible effects are not mutually exclusive, in fact, it seems quite probable that each plays a role. When we succeed in evaluating the relative importance of these different effects, observations of the structure of type II bursts should add to our knowledge of the underlying coronal structure.

Discussing (b), Wild and Smerd (1972) not that regions of maximum density and minimum magnetic field exist along the essentially radial core of coronal streamers; they cite this in favour of type II shocks travelling essentially parallel to the magnetic field.

The approach to (a) being developed by Uchida shows great promise of explaining Culgoora radioheliograph observations (Uchida, 1973). In this approach the shock is treated like a small amplitude magnetohydrodynamic fast-mode wave, propagating at all angles to the magnetic field.

3. Moving Type IV Bursts

A moving type IV can really only be distinguished from the various other phenomena known as 'type IV' with positional observations. Even so there is likely to be doubt about events seen near the centre of the disk.

Fortunately the outward motion of these sources carries them clear of the region of low refractive index. Consequently observations of the position, size and brightness distribution of these bursts are more readily interpreted than for type II bursts. For 80 MHz observations of 12 moving sources, Smerd and Dulk (1971) found that the average distance to which a burst could be followed was about 4 $R_\odot$, where $\mu \gtrsim 0.98$.

The first result of consequence to come from two-dimensional observations with the Culgoora radioheliograph is that the term 'moving type IV' appears to cover a variety of different phenomena – all different manifestations of explosive energy release in flares. The following four types have been recognized.

3.1. Magnetic Arch

Wild (1969c) described the prototype magnetic arch. Essentially it consisted of three sources: an unpolarized source (with polarized edges) marking the apex of the arch and two other sources, strongly circularly polarized in opposite senses, which apparently marked the intersections of the arch with the 80 MHz plasma level. These sources moved apart in a manner consistent with an arch rising in the corona while its feet spread out.

Of the 25 moving type IV sources so far observed at Culgoora and reported in the literature, at most six cases might fit this model, although none of the other cases has the compelling simplicity of the prototype.

3.2. Advancing front

The only burst definitely of this type so far observed is the burst of 1968 October 23/24 reported by Kai (1970). However, Kai (1973, private communication) finds that two or three other bursts have features in common with the prototype. Figure 5 illustrates

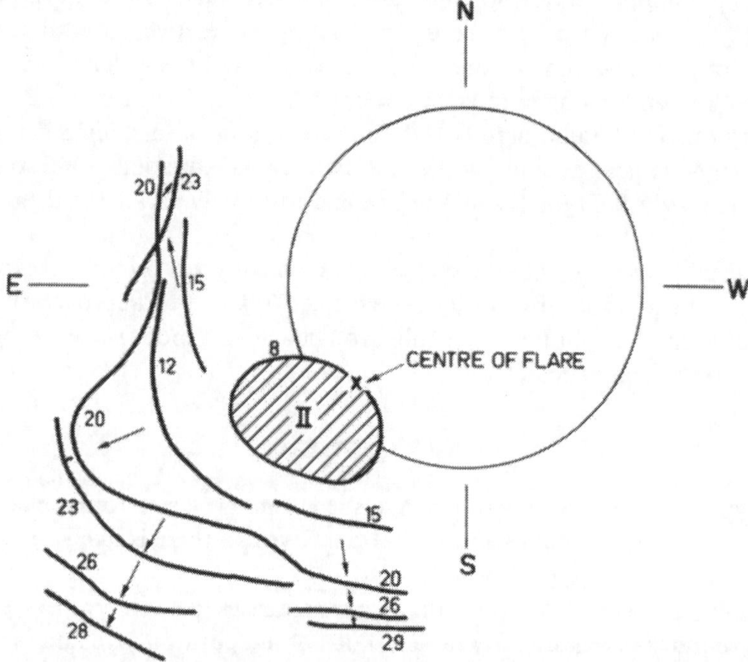

Fig. 5. Summary of radioheliograph observations of the advancing front type IV burst observed on 1968 October 23/24. The shaded area indicates the region of the two type II bursts. The thick lines indicate the ridge line of the extended type IV source for successive times, which are indicated in minutes after the onset of the flare (23^h51^m). (From Kai, 1970.)

the geometry of this burst. Two type II bursts occurred above a flare a few degrees from the limb. After the second of these, an extended arc-shaped source appeared above the limb and expanded outward, in a continuation of the general outward motion implied by the type II burst. The speeds of the second type II burst, and of the arch were both about 1400 km s^{-1} (Figure 6). The longest-lived section of the arch reached a maximum distance of 2.9 $R_\odot$ from the centre of the Sun. Throughout this event the source was only weakly polarized (<20% LH) – in contrast to the magnetic arch.

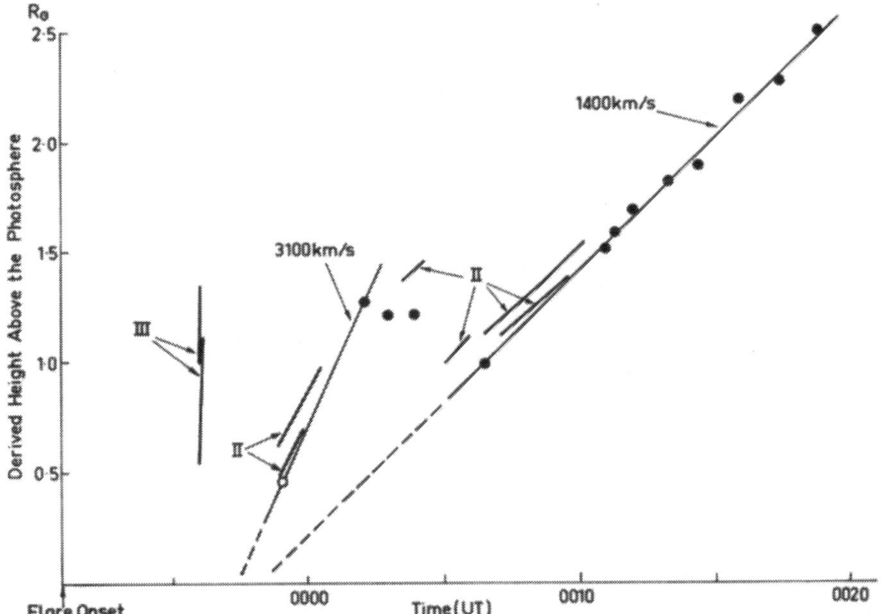

Fig. 6. For the event of 1968 October 23/24 (see Figure 5), comparison of the heights of the type II bursts (derived from the spectrum) and the height of the type IV burst from the heliograph observations. (From Kai, 1970.)

The large extent perpendicular to the direction of motion, the continuity with the type II burst and the lack of polarization appear to be the distinctive features of this burst.

3.3. Jet

Another type of burst, of which only one clear example has yet been observed, is the jet described by Riddle and Sheridan (1971). Following a flare and two complicated type II bursts, which appeared at a succession of different positions (Figure 7) the type IV source appeared as a row of bright patches forming a jet. Most of the individual brightenings moved out along the jet, as shown by Figure 8. On a number of occasions the components of the jet brightened in rapid succession, each one doubling its brightness for a second or two. This whole sequence lasted about 7 s. These brightenings have been interpreted as due to bursts of electrons travelling with speeds ~0.5 c along the jet and radiating brightly from each source. This interpretation resembles the type IV model of Warwick (1968).

Both the unusual source geometry and the sequential brightenings suggests that this jet is a distinct type and not just a variant of the most common type, 'the isolated source', which will be described in the following paragraph.

3.4. Isolated source

This term, introduced by Smerd and Dulk (1971), also covers many cases in which

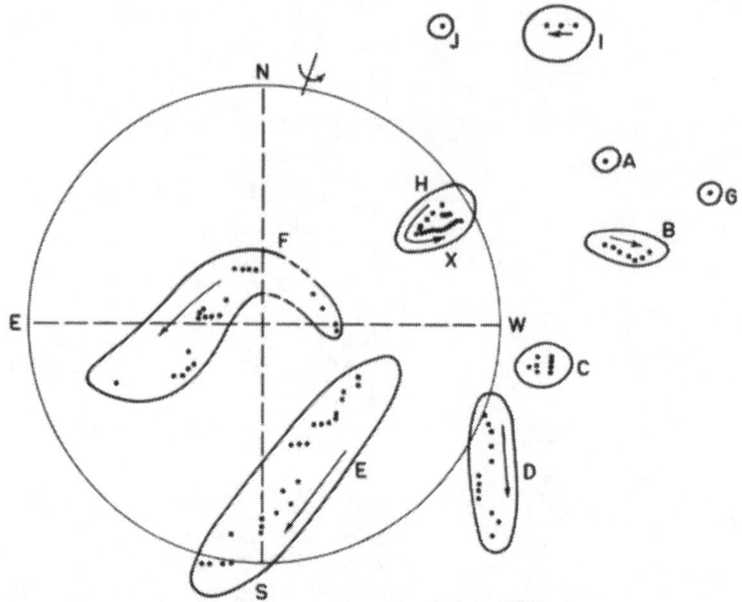

Fig. 7. Centroid positions of the many sources observed during the early phase of the event of 1971 January 24/25. Most of the positions refer to type II bursts. The letters indicate approximately the sequence in which the sources appeared and the arrows indicate the direction of apparent movements. (From Riddle and Sheridan, 1971.)

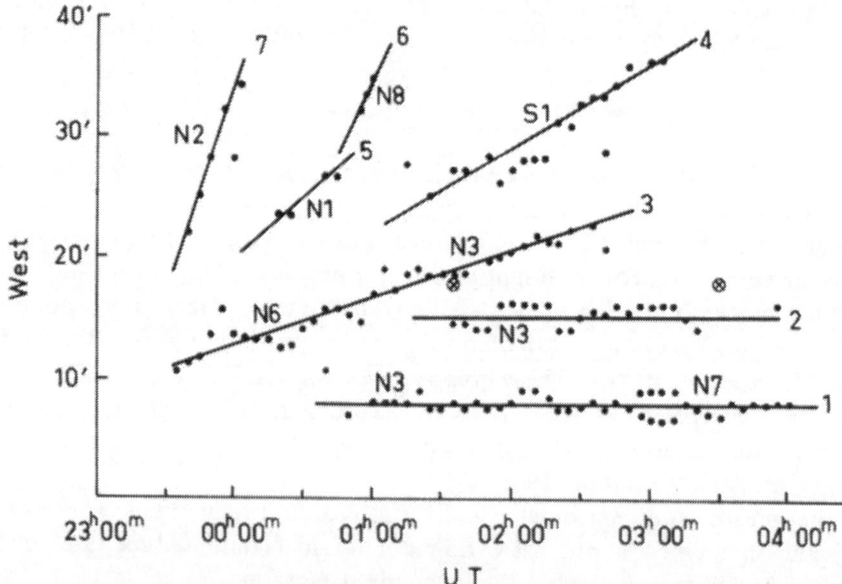

Fig. 8. Position west of the solar centre as a function of time for the sources (numbered 1 to 7) of the jet observed 1971 January 24/25 immediately after the type II burst of Figure 7. The north-south coordinate in minutes of arc is indicated by symbols such as N3. Two type III bursts are indicated by a circled cross. (From Riddle and Sheridan, 1971.)

the source is observed to break up into a number of sources moving with similar velocities. The majority of moving type IV bursts appear to fit this classification; we shall look briefly at two examples.

Sheridan (1970) reported observations of a type IV burst out to a record distance of more than 6 $R_{\odot}$ (exceeding the 5 $R_{\odot}$ reported by Riddle (1970b) for 'Westward Ho'). The event started with a flare seen on the disk, and an ejected prominence seen beyond the limb in Hα. Shortly afterwards a stationary unpolarized source (flare continuum) appeared above the position of the prominence. After this had faded another, polarized, source appeared in about the same position and moved out at a steady speed of about 290 km s^{-1}. As it moved the degree of polarization increased from 30% RH to about 90% RH. Late in the event the source split into four, all RH polarized and all moving slowly apart. By the time the furthest source reached 6 $R_{\odot}$ they were all fading rapidly and could not be followed any further.

For a second example I have chosen the burst of 1970 April 19, described by Dulk and Altschuler (1971). From Figure 9 we can see the general sequence of events. The

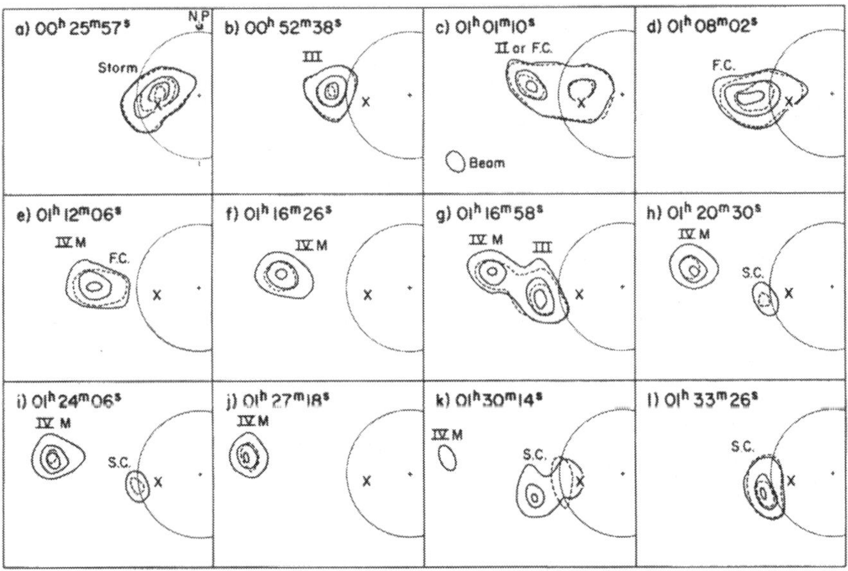

Fig. 9. Contours of radio brightness from the Culgoora radioheliograms for the event of 1970 April 29. The flare position is shown by a cross. Dashed and full lines refer to RH and LH circular polarization respectively, both normalized to the brightest point in the LH image. The symbols II, III, IVM, FC and SC denote the types of bursts: type II, type III, moving type IV, flare continuum and storm continuum respectively. (From Dulk and Altschuler, 1971.)

flare occurred at 56° E, 10° S starting at 00^{h}47^m UT. A group of type III bursts, a possible type II burst and a flare continuum occurred above the flare. The flare continuum remained approximately stationary for a few minutes, then started to move out at an average projected velocity of 880 km s^{-1}. A storm continuum source (sometimes called stationary type IV) appeared later above the flare site. Figure 10 shows the path of the moving type IV source and also the slight motion of the storm continuum source. Figure 11 shows the variation with time of the flux density and the degree of circular polarization of the main sources. These curves are characteristic of this type of burst, particularly the rapid fading of the type IV burst (in this case three decades in 12 min) and the progressive increase of the degree of polarization up to a maximum of about 80%. The rapid fading limits our ability to follow these sources to great distances from the Sun, but generally there is no indication of slowing down

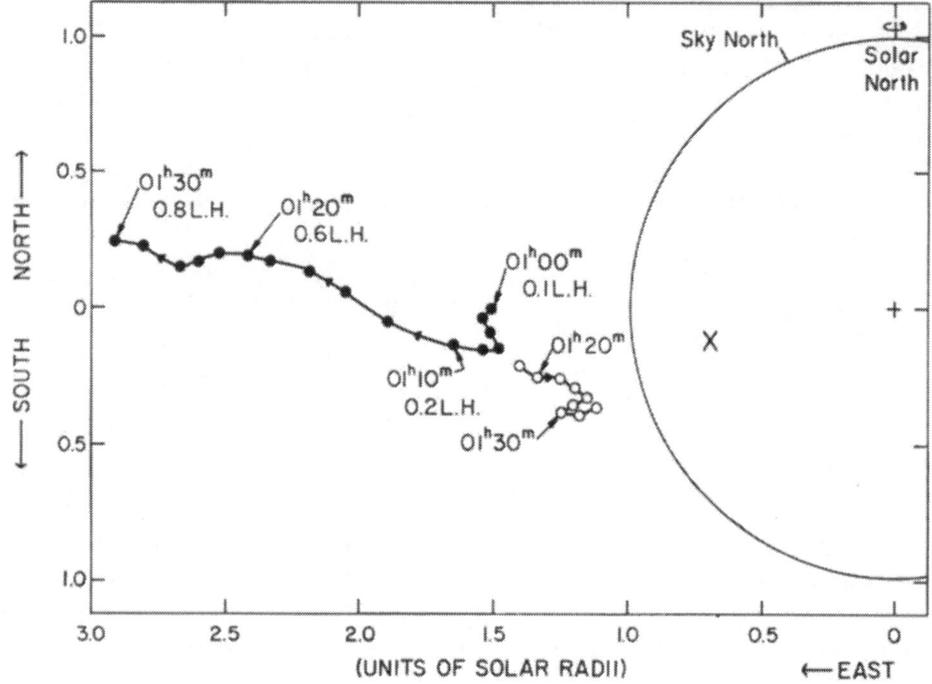

Fig. 10. The path of the centroid of the moving type IV bursts (filled circles) and of the continuum (open circles) for the event of Figure 9. The flare position is marked by a cross. (From Dulk and Altschuler, 1971.)

of the sources as they fade. The appearance of the storm continuum source late in the event is also typical, although the short life of this particular continuum source is atypical – several hours' duration is more normal.

One point of conflict with Boischot's (1958) original observations should be noted. He described sources as moving out and then falling back to a position close to the Sun. From the observations just described and a number like them it seems possible that Boischot really observed a moving source which kept moving out but faded rapidly, and a stationary source which appeared at about the same time as the moving source faded. With one position every 4 min in one dimension only it might be difficult to distinguish between these two possible interpretations.

Although the essentially linear motion of the two examples I have chosen is

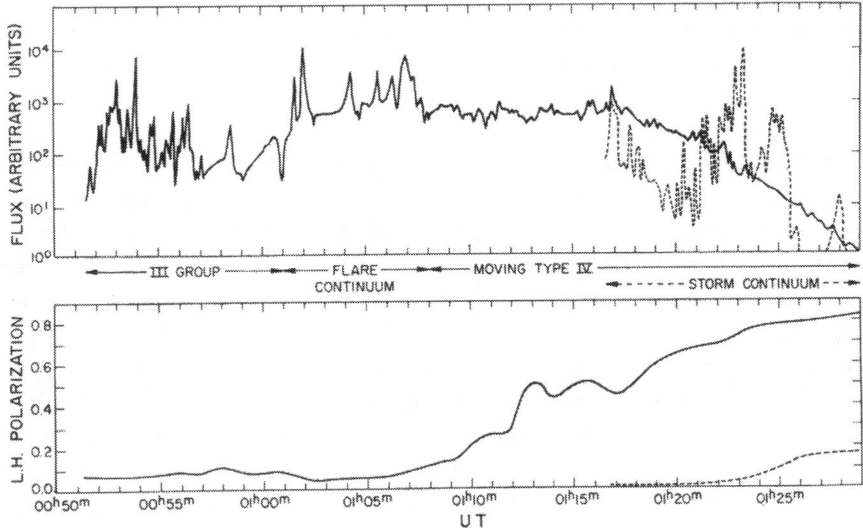

Fig. 11. Flux density and polarization of the sources of Figures 9 and 10, recorded at 80 MHz with the Culgoora radioheliograph. The dashed curves refer to the storm continuum. The other burst types are as indicated. (From Dulk and Altschuler, 1971.)

typical of most 'isolated-source' bursts there have been cases where the trajectories showed significant curvature (e.g. Smerd, 1971) or even where the movement was largely around the limb rather than radially outward (Dulk *et al.*, 1971b).

3.4.1. *The French Burst*

Although four categories seem adequate for the observations made at 80 MHz with the Culgoora instrument, there is one observation, made at Nançay at both 169 and 408 MHz, which appears to merit separate consideration. The special features of this burst (Boischot and Clavelier, 1968) are that it started suddenly and simultaneously at both observing frequencies at a great height above the photosphere (1 $R_\odot$). Both frequencies were emitted from the same source, which moved to a height of 2 $R_\odot$ above the photosphere. The spectrum of the burst had very sharp cut-offs on both the high-and low-frequency edges.

The interpretation of these features given by Boischot and Daigne (1968) and Lacombe and Mangeney (1969) is as follows. The common position for the two frequencies is consistent with the normal synchrotron interpretation, although the sharp low frequency cut-off is taken to indicate that the plasma inside the source had suppressed the low frequencies (Razin-Tsytovich effect). The sudden start at a great height is interpreted as indicating that the electrons were accelerated in situ at this height. Lacombe and Mangeney (1969) developed a shock wave model to explain this. In their model the shock wave was capable of accelerating electrons to MeV energies for only a critical range of Mach numbers, which only occurred between the heights of 1 and 2 $R_\odot$. However, Smith (1971) has criticized this model. He finds that general heating of the electron gas will occur, rather than acceleration of a select few electrons. Mangeney (1974) has since developed another theory for accelerating relativistic electrons, this time in a parallel shock, which avoids this criticism.

3.4.2. *Interpretation*

Smerd and Dulk (1971) found that it is not always possible to decide the classification of a particular event with certainty. Nonetheless the four classes just described appear to be adequate to cover all the bursts so far observed at Culgoora. (I have added the jet to the three classes considered by Smerd and Dulk; the classification of the French event as an advancing front relies on rather uncertain theoretical considerations, as discussed above.)

Since these very different phenomena all represent some sort of disturbance moving out through the corona it should be instructive to examine the similarities and differences. Since all have sources which move out to a considerable distance from the Sun – well above the 80 MHz plasma level – it has generally been accepted since Boischot and Denisse (1957) that the only possible emission mechanism is synchrotron emission from energetic electrons spiralling in a magnetic field. However, as first pointed out by Kai (1969b), the very strong circular polarization observed in the late stages of many isolated sources is a difficulty with this interpretation.

A relativistic electron, velocity βc, radiates most strongly in about the γ^3-th harmonic of its frequency of gyration, f_g/γ, where the Lorentz factor γ is $(1-\beta^2)^{-1/2}$ and f_g is the gyro frequency for non-relativistic electrons (f_g(MHz)$=2.8B$ (gauss)). But radiation in high harmonics is generally linearly polarized whereas low harmonic radiation appears strongly circularly polarized if viewed approximately along the magnetic field. Hence for strong polarization we must assume subrelativistic electrons, i.e. electrons with $\gamma \lesssim 2$. But this leads to a major difficulty; these electrons will not radiate much at frequencies above $4f_g$, and so to explain observation at 80 MHz we must suppose that $f_g \gtrsim 20$ MHz or $B \gtrsim 6$ G inside the source. Particularly at the extreme distance of 6 $R_\odot$ for the record holder (Sheridan, 1970), this field is well above all the best estimates of coronal magnetic field strength (e.g. Newkirk, 1971).

Hence there seems little choice. Until someone can propose a reasonable alternative to synchrotron emission of the radiation it seems that the moving sources must carry their own strong magnetic field with them. Of course, there is no suggestion that this structure should be stable. On the contrary, a number of authors have studied the evolution of expanding sources (e.g. Dulk, 1970b, 1973; Schmahl, 1972) and I believe Dr Dulk will be presenting a recent model of this type which successfully explains many of the observed characteristics, particularly the rapid fading and the simultaneous rise in polarization.

However, a second problem still remains to be faced: for the very high degrees of circular polarization recorded the source must be viewed along the magnetic field to within say 45°. This is a statistically unlikely orientation which should only occur in about 30% of cases, yet polarization degrees higher than 70% are reported for well over half the isolated sources so far recorded. This orientation is even more unlikely if we consider that the synchrotron emissivity of a source containing an isotropic distribution of electrons is peaked in the plane normal to the magnetic field. An anisotropic electron distribution would shift the emission peak away from the direction where the polarization is least, but it is believed that such distributions are unstable to whistler emission. Even stronger magnetic fields (and less energetic electrons) could be invoked, but the field strengths are already embarrassingly high.

Of course the problem of explaining the frequent observation of very high degrees of circular polarization would probably apply equally to any other emission mechanism which might be proposed as an alternative to synchrotron emission. Taken in isolation it seems to require a peak of emission along the direction of the magnetic field.

In the case of the magnetic arch, it seems obvious to follow Wild (1969c) and assume a different emission mechanism for the polarized feet of the arch, which appear near the plasma level and the unpolarized apex.

Similarly the shock wave character of Kai's (1970) expanding arch seems obvious both from its large extent at right angles to the direction of motion (Figure 5) and its continuity with the type II burst. Nonetheless the great height reached by the longest-lived part of the arch seems to rule out any mechanism other than synchrotron emission.

There is no mystery about the end of type IV bursts. As mentioned already, their rapid fading makes them impossiblė to follow further, although there is generally no indication of deceleration as they fade. The current idea is that the fading is a consequence of the expansion of the source. However, the breaking up into a number of separate sources of the same (e.g. Sheridan, 1970; Smerd, 1971) or opposite (e.g. Riddle, 1970b) senses of polarization is an interesting complication which has not yet received much attention.

3.4.1. *A Link With Type I Storms*

I should like now to present briefly a recent observation made at Culgoora with the radioheliograph observing at both 80 and 160 MHz. This event, described by McLean (1973) has a number of interesting features which can be summarized as follows (see Figure 12).

(a) The event was initiated by a prominence which was observed to rise above the limb.

(b) The first phase of the radio event was a type I storm observed with the Culgoora spectrograph. As the prominence rose higher in the corona the type I storm spread to lower frequencies. The 80 and 160 MHz emission during the type I storm came from different heights above the limb, presumably determined by the plasma levels.

(c) The storm moved some distance around the limb, then faded. As it faded a faint moving type IV source appeared, with both frequencies coming from near the last observed position of the 80 MHz storm.

(d) The moving type IV source moved out along a radial path. The sources at the two frequencies did not coincide exactly.

I have brought this observation in at this stage to show that type I emission may

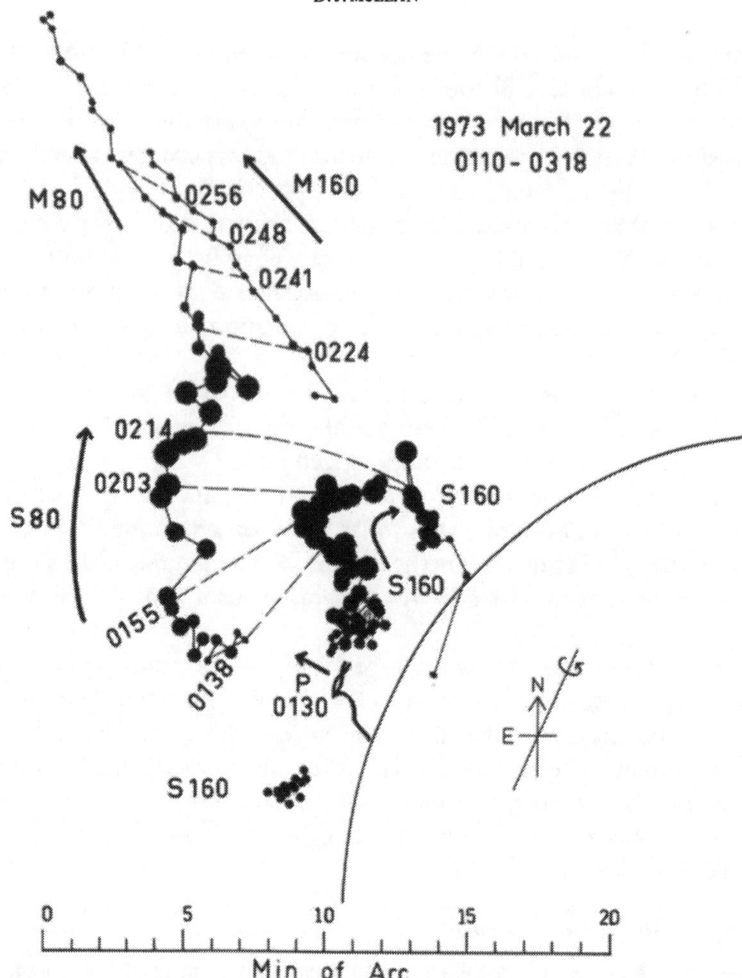

Fig. 12. Paths of the centroids of sources observed during the event of 1973 March 22. Each dot represents the position of the centroid of a radio source; the area of the dot is proportional to the flux density. S80 and S160 are the 80 and 160 MHz components of the first, storm part of the event, and M80 and M160 the components of the moving source. The appearance of the ascending prominence at 01^h30^m UT has also been shown (labelled P). Some of the almost simultaneous observations at the two frequencies have been joined by broken lines and the time of these observations are indicated. Successive positions at each frequency have been joined by full lines, and the direction of motion indicated by heavy arrows. To avoid confusion only one record in 56 has been used for this figure (i.e. the dots are at intervals of 56 s). (From McLean, 1973.)

also be associated with the ejection of matter from the Sun. We have had other, less direct evidence of this for some time. Le Squeren (1963) found that noise storms started or increased in activity about 1 h after a flare. So called stationary type IV bursts are probably just examples of this result. In addition, Wild and Zirin (1956) observed another probable type I event associated with a rising prominence. However, this aspect of the observations does not appear to have been included in any theoretical work.

4. Conclusion

Radio observations of the Sun reveal two main phenomena indicative of ejection from solar flares: type II and moving type IV bursts. Although the broad outlines of the interpretation of these phenomena are well established we have much to learn before we can really claim to understand them.

References

Aubier, M., Leblanc, Y., and Boischot, A.: 1971, *Astron. Astrophys.* **12**, 435.
Boischot, A.: 1958, *Ann. Astrophys.* **21**, 273.
Boischot, A. and Daigne, G.: 1968, *Ann. Astrophys.* **31**, 531.
Boischot, A. and Denisse, J.-F.: 1967, *Compt. Rend. Acad. Sci.* **245**, 2194.
Boischot, A. and Clavelier, B.: 1968, *Ann. Astrophys.* **31**, 445.
Caroubalos, C., Aubier, M., Leblanc, Y., and Steinberg, J. L.: 1972, *Astron. Astrophys.* **16**, 374.
Dulk, G. A.: 1970a, *Proc. Astron. Soc. Australia* **1**, 308.
Dulk, G. A.: 1970b, *Proc. Astron. Soc. Australia* **1**, 372.
Dulk, G. A.: 1971, *Australian J. Phys.* **24**, 177.
Dulk, G. A.: 1973, *Solar Phys.* **32**, 491.
Dulk, G. A. and Altschuler, M. D.: 1971, *Solar Phys.* **20**, 438.
Dulk, G. A. and Smerd, S. F.: 1971, *Australian J. Phys.* **24**, 185.
Dulk, G. A., Altschuler, M. D., and Smerd, S. F.: 1971a, *Astrophys. Letters* **8**, 235.
Dulk, G. A., Stewart, R. T., Black, H. C., and Johns, I. A.: 1971b, *Australian J. Phys.* **24**, 239.
Fokker, A. D.: 1965, *Bull. Astron. Inst. Neth.* **18**, 111.
Fokker, A. D. and Rutten, R. J.: 1967, *Bull. Astron. Inst. Neth.* **19**, 254.
Giovanelli, R. G. and Roberts, J. A.: 1959, *Australian J. Phys.* **11**, 353.
Kai, K.: 1969a, *Solar Phys.* **10**, 460.
Kai, K.: 1969b, *Proc. Astron. Soc. Australia* **1**, 189.
Kai, K.: 1970, *Solar Phys.* **11**, 310.
Labrum, N. R.: 1969, *Proc. Astron. Soc. Australia* **1**, 191.
Lacombe, C. and Mangeney, A.: 1969, *Astron. Astrophys.* **1**, 325.
Leblanc, Y.: 1973, *Astrophys. Letters* **14**, 41.
McCabe, M. K. and Fisher, R. R.: 1970, *Solar Phys.* **14**, 212.
McLean, D. J.: 1967, *Proc. Astron. Soc. Australia* **1**, 47.
McLean, D. J.: 1973, *Astron. Soc. Australia* **2**, 222.
Mangeney, A.: 1974, this volume, p. 387.
Meyer, F.: 1968, in K. O. Kiepenheuer (ed.), 'Structure and Development of Solar Active Regions', *IAU Symp.* **35**, 485.
Newkirk, G. Jr.: 1971, in C. Macris (ed.), *Physics of the Solar Corona*, D. Reidel Publ. Co., Dordrecht, Holland, p. 66.
Newkirk, G., Jr.: 1961, *Astrophys. J.* **133**, 183.
Payne-Scott, R. and Little, A. G.: 1952, *Australian J. Sci. Res.* **A5**, 32.
Riddle, A. C.: 1970a, *Proc. Astron. Soc. Australia* **1**, 310.
Riddle, A. C.: 1970b, *Solar Phys.* **13**, 448.
Riddle, A. C.: 1972a, *Proc. Astron. Soc. Australia* **2**, 98.
Riddle, A. C.: 1972b, *Proc. Astron. Soc. Australia* **2**, 148.
Riddle, A. C.: 1974, *Solar Phys.* **35**, 153.
Riddle, A. C. and Sheridan, K. V.: 1971, *Proc. Astron. Soc. Australia* **2**, 62.
Roberts, J. A.: 1959, *Australian J. Phys.* **12**, 327.
Schmahl, E. J.: 1972, *Proc. Astron. Soc. Australia* **2**, 95.
Shain, C. A. and Higgins, C. S.: 1959, *Australian J. Phys.* **12**, 357.
Sheridan, K. V.: 1970, *Proc. Astron. Soc. Australia* **1**, 376.
Smerd, S. F.: 1970, *Proc. Astron. Soc. Australia* **1**, 305.
Smerd, S. F.: 1971, *Australian J. Phys.* **24**, 229.

Smerd, S. F., and Dulk, G. A.: 1971, in R. Howard (ed.), 'Solar Magnetic Fields', *IAU Symp.* **43**, 616.
Smerd, S. F., Sheridan, K. V., and Stewart, R. T.: 1974, this volume, p. 389.
Smerd, S. F., Wild, J. P., and Sheridan, K. V.: 1962, *Australian J. Phys.* **15**, 180.
Smith, D. F.: 1971, *Astrophys. J.* **170**, 559.
Smith, S. F. and Harvey, K. L.: 1971, in C. Macris (ed.), *Physics of the Solar Corona*, D. Reidel Publ. Co., Dordrecht, Holland, p. 156.
Steinberg, J.-L.: 1972, *Astron. Astrophys.* **18**, 382.
Steinberg, J.-L., Aubier-Giraud, M., Leblanc, Y., and Boischot, A.: 1971, *Astron. Astrophys.* **10**, 362.
Le Squeren, A.-M.: 1963, *Ann. Astrophys.* **26**, 97.
Uchida, Y.: 1968, *Solar Phys.* **4**, 30.
Uchida, Y.: 1970, *Publ. Astron. Soc. Japan* **22**, 341.
Uchida, Y.: 1973, in R. Ramaty and R. G. Stone (eds.), *Proc. Symposium on High Energy Phenomena on the Sun*, Sept. 1972, NASA-GSFC, Greenbelt, Mass., p. 577, preprint.
Uchida, Y., Altschuler, M. D., and Newkirk, G., Jr.: 1973, *Solar Phys.* **28**, 495.
Warwick, J. W.: 1965, in J. Aarons (ed.), *Solar System Radio Astronomy*, Plenum Press, N.Y., p. 131.
Warwick, J. W.: 1968, *Solar Phys.* **5**, 111.
Weiss, A. A.: 1963, *Australian J. Phys.* **16**, 240.
Wild, J. P.: 1950, *Australian J. Sci. Res.* **A3**, 399.
Wild, J. P.: 1969a, in D. A. Tidman and D. C. Wentzel (eds.), *Plasma Instabilities in Astrophysics*, Gordon and Breach, N.Y., pp. 119–138.
Wild, J. P.: 1969b, *Proc. Astron. Soc. Australia* **1**, 181.
Wild, J. P.: 1969c, *Solar Phys.* **9**, 260.
Wild, J. P.: 1970a, *Proc. Astron. Soc. Australia* **1**, 365.
Wild, J. P. and McCready, L. L.: 1950, *Australian J. Sci. Res.* **A3**, 387.
Wild, J. P. and Smerd, S. F.: 1972, *Ann. Rev. Astron. Astrophys.* **10**, 159.
Wild, J. P. and Zirin, H.: 1956, *Australian J. Phys.* **9**, 315.
Wild, J. P., Sheridan, K. V., and Trent, G. H.: 1959, in R. N. Bracewell (ed.), *IAU-URSI Symposium on Radio Astronomy, Paris 1958*, Stamford Univ. Press, p. 176.
Wild, J. P., Smerd, S. F., and Weiss, A. A.: 1963, *Ann. Rev. Astron. Astrophys.* **1**, 291.

DISCUSSION

Sturrock: A simple interpretation of the blast-wave theory of type II bursts leads one to expect radiation over a wide range of frequency, namely the range of plasma frequency corresponding to the wide range of height covered at any one instant by the blast wave. This is very different from the typically narrow frequency band of a type II burst.

McLean: Emission occurs preferentially where the shock front is parallel to the electron density contours. Also, as Uchida has shown, the emission is not really a spherical wave but occurs predominantly in regions where Alfvén velocity is low.

Smith: Also, on this point, it is not inconsistent that we start out with a blast wave because we only get type II emission when special conditions are satisfied, such as the Mach number being high enough.

On another matter, an alternative explanation for 'winking' phenomena in type II bursts is that special conditions are required for emission such as the angle between the plane of the shock and the magnetic field lying in a narrow range. Is there any observational evidence to favor the hypothesis that 'winking' is related to 'herring-bone' over the above explanation?

McLean: As far as I know, no.

Kai: I would like to comment on some points in the shock wave – moving IV relation. I have reached the following conclusion from a statistical study of about 30 IV(M) which were observed at Culgoora. There are at least two kinds of IV(M): high velocity ($\gtrsim 1000$ km s^{-1}) ones, which are unpolarized and rarely observed and have close association with shock waves, and the low-velocity (~ 400 km s^{-1}) type IV's, which are strongly polarized and have no direct association with shock waves. The majority of IV(M) fall into this second class. I suggest, therefore, (and this is supported by other observational evidence) that most IV(M) are caused by low-velocity material which possibly moves along open magnetic field lines above a closed helmet structure of magnetic fields. Secondly, IV(M) are not as energetic events as we previously thought. Even minor sub-flares can often produce IV(M) bursts, IV(M) are hardly associated with the proton events measured in interplanetary space.

Maxwell: Can you clarify something about the March 30, 1969 burst? The original emission was over 180°. Was the burst due to type IIIs or, if not, what was its origin?

Smerd: The early wide spread of the 80 MHz burst sources in the 1969 March 30 event can be explained if the radiating electrons reached the source regions along similarly widespread magnetic-field lines from an explosive center behind the west limb.

Maxwell: Was this burst unusual?

Smerd: Dulk *et al.* (1971a) showed that the large-scale coronal field (as computed from the observed photospheric field) was remarkably constant over three solar rotations – including that of March 1969 – in spite of a number of major flare events during that period. The large-scale field was characterized by an unusually widespread network from the explosive center to 'all points on the Sun'.

Uchida: I understand that the harmonic-band copies most of the structures in the fundamental. Is the herringbone structure in the fundamental also copied in the harmonic band? I think this point is pretty important in understanding the production mechanism of harmonics, whatever it may be.

Smerd: Yes.

McLean: The major features certainly occur together in both fundamental and harmonic; fine details are probably repeated too.

OPTICAL EVIDENCE FOR PLASMA EJECTIONS AND WAVES IN THE SOLAR CORONA

A. BRUZEK

Fraunhofer Institut, Freiburg, i.Br., F.R.G.

Abstract. Plasma ejections and waves in the solar corona are almost exclusively flare associated phenomena. Ejections of relatively cool and dense plasma are frequently observed in Hα whereas observations in coronal light (visible, EUV- and X-radiation) are still rather scarce. Occurrence of coronal waves is so far best known from their effects on the Hα chromosphere and, of course, from the production of radio bursts. Only in relatively few cases have observations been made in coronal lines and in coronal continuum by ground based as well as by satellite borne equipment. We may expect, however, that the white light coronagraph and the X-ray telescopes on board of the Skylab will detect quite a number of events in front of the solar disk and high in the solar corona and will considerably increase and improve our imperfect knowledge and understanding of coronal ejections and waves as it is presented in this review.

1. Plasma Ejections

Three large classes of ejections may be distinguished according to their origin and their appearance in Hα: (1.1) surges, (1.2) sprays, (1.3) eruptive prominences.

1.1. Surges

The simplest type of ejection occurring in the solar corona is the classical surge which grows out of the chromosphere with velocities 100–200 km s^{-1} in the form of a straight or slightly curvilinear spike or streamer remaining connected with the chromosphere and attaining a length of 10^5 km or more. In general, surges shrink again along the path of formation and disappear after a lifetime of 10–30 min; the material either fades or it returns into the chromosphere along the trajectory of ascent. However, in some cases material was observed moving along a large arch, ascending in one branch of the arch and eventually descending along the other one (Bruzek, 1969, Figure 4). On the solar disk surges appear in absorption, in their initial phase sometimes in emission; they originate close to spots and pores. Sometimes several surges form a group resembling a veil or a fan.

The trajectory of the moving plasma and its collimation into a small solid angle indicate that surges are confined by a more or less radial magnetic field. Roy (1973b) computed the configuration of the coronal magnetic field at the position of surges in the current free approximation from measured longitudinal photospheric fields and indeed found good agreement with the observed directions of surges. Magnetic fields in surges have been measured by Harvey (1969) and field strengths of the order 50 G decreasing with height in the surge and with its length have been found.

A number of authors (Giovanelli and McCabe, 1958; Gopasyuk *et al.*, 1963; Bruzek, 1969) observing in the center of Hα reported that more than 90% of the surges emanate from brightenings in Hα which are classified as flares, the majority of them however only as subflares, even in cases of very large surges. These surge-associated

Gordon Newkirk, Jr. (ed.), Coronal Disturbances, 323–332.

flares appear as bright points or blobs, sometimes rather *U*- or ring-shaped with a typical diameter of 25000 km. The flare first expands somewhat in height until at maximum intensity the surge rapidly grows out; the flare fades simultaneously and has virtually disappeared at the time of maximum development of the surge. At the limb, the surge assumes often flare brightness in its initial stage of development and shows the characteristic flare spectrum rather than a prominence spectrum. The brightness in the surge decreases with height and time as if the flare plasma relaxes after it is shot out from the chromosphere. This development indicates that it is the flare which provides the material for the surge, or rather that the flare is transformed into the surge. Apparently, the flare at the base of the surge and the surge are two phases of the 'flare-surge event'. That the growth of the surge is true mass motion and not due to a propagating condensation of coronal material is inferred from the Doppler effects observed on the solar disk.

It is remarkable that, in general, no brightening occurs at the base of the surge during the inflow of material into the chromosphere which takes place at speeds up to 150 km s^{-1}. This observation indicates that the infall of material does not necessarily produce a flare as Hyder (1967) suggested in his impact hypothesis.

It must be pointed out here that the majority of ejections associated with *larger* flares (importance 1 to 4) are not classical surges although many observers call them surges. They belong rather to classes (2) or (3) or to some intermediate type.

In a recent investigation Roy (1973a) found almost all surges of his sample associated with bombs and only a few of them with flares or subflares. This disagreement with the previous authors is apparently due to the fact that the majority of Roy's surges were very small (155 out of his 182 were <35000 km, i.e. importance S) whereas, for instance, the sample which I had studied (Bruzek, 1969) included only surges >40000 km (i.e. importance $\geqslant 1$). That suggests that, in general, very small surges are associated with bombs, large ones, however, preferably with (sub-)flares. One should keep in mind, however, that the majority of bombs is associated with rather small, surgelike absorption features which may be of different nature than 'proper' surges. Nevertheless Roy (1973b) found the same intensity-time relation between surge development and bomb development as it was described above for flares and associated surges, and he interpreted it also in terms of growing of the surge out of the bomb! This indicates that the same ejection mechanism is effective for bomb associated and flare associated surges.

One more characteristics of surges has to be mentioned which is most conspicuous with the smaller surges starting at penumbral borders; that is the tendency to recur many times (at a rate ≈ 1 h^{-1}) at the same position (Bruzek, 1969). A very spectacular sequence of countless surge repetitions over a period of >24 h has been described by Fortini and Torelli (1968). Also large surges may appear several times at the same place and in the same shape.

Studying the dynamics of his largest surges Roy (1973b, c) found that the rising motion of the surge material, as measured at its top, is accelerated as far as 10000–50000 km above the surface where a velocity of 100–175 km s^{-1} may be attained after

5–10 min. The following desceleration of the ascending material is larger than that produced by gravitation alone, and the downward acceleration of the returning material is less than gravitational acceleration. These observations imply that (1) an accelerating force is still acting on the surge after its material has left the chromosphere; the surge is not a pure ejection with its material moving freely in the corona; (2) the motion of the material is opposed by a kind of friction.

Various mechanisms have been proposed for the initial acceleration of the surge (Livshits and Pickel'ner, 1964; Uchida, 1969; Pickel'ner, 1969; Altschuler *et al.*, 1968; Lilliquist *et al.*, 1971) which cannot be reviewed here. There is observational evidence that the required energy may be provided by the underlying magnetic field: Rust (1968) and Roy (1973a, b) have observed that the magnetic flux in satellite magnetic polarities in surge active regions decreased at a rate of $1-3\times 15$ Mx s^{-1}. Roy found for a series of well observed bombsurge events that the energy possibly provided by the changing magnetic flux as well as the thermal energy in the associated bombs and the kinetic energy of the surges were of the same order of magnitude, 5×10^{27} erg. The corresponding value for large flare-surges would be about 10^{30} erg!

The braking force observed during ascent and descent of the surge material is proportional to its velocity (Roy, 1973b, 1974). It can be interpreted as the Lorentz force acting if the material does not move exactly along the magnetic lines of force so that a small field component is perpendicular to the mass motion. A field component $B_{\perp}\approx 0.01$ G would already account for the observed effect. That means that a slight deviation from the force-free configuration, a slight twisting of the field otherwise unnoticeable would exist. It may be noted that in some large surges a helical flow has been observed.

1.2. Sprays

Another type of ejecta, the sprays, are much more vigorous than surges. They emanate also from flares, often from large flares with an explosive phase ('explosive' in this context means a rapid brightening and expansion of the flare). At the solar limb, sprays are preceded by a flare which rapidly forms a bright hill or even a large flare-bright prominence which suddenly disrupts and expels the spray material at high velocities. The trajectories of the moving plasma are spread over a large volume and not confined to a narrow cone as in surges. The spray plasma often appears in flare-bright clumps as if parts of the flare plasma were torn off. This may indicate that the flare plasma was highly turbulent before its ejection. This development suggests, as was pointed out by Smith (1968) and Gold (1968) that the flare plasma originally constrained by a closed magnetic field stretches the lines of force as its internal kinetic energy density increases until eventually the field bursts open and the plasma escapes. Nevertheless, McCabe (1971) was able to show for a typical spray event that the associated flare occurred clearly at the base of an existing open field structure and that the spray material moved outwards along the field lines. The ejected plasma attains its maximum velocity of up to 1000 km s^{-1} within a few minutes. After the initial very high acceleration (a few km s^{-2}) the ejected material moves slightly decelerated by gravitation; only a small fraction is seen to return to the solar surface, the main part of the material fades or escapes from the Sun due to its high velocity.

Observations of sprays on the solar disk are scarce and their identification is difficult. This is not a surprise if we consider the high velocities of the material and the narrow passband of the filters commonly used in the Hα patrol observations. Even for strongly inclined ejections the radial velocity component may amount to several hundred km s^{-1} corresponding to a Doppler shift 5 Å or more. So, probably the majority of sprays occurring in front of the disk simply escape observation through filters centered on Hα or shifted a few Å off the line center. A line shifting as far as about −10 Å off Hα would be required for an effective spray patrol.

The available disk observations in the center of Hα – which necessarily are inadequate – show sprays, or what is believed to correspond to sprays, as rapidly expanding material which starts bright and changes to dark, extending over a fairly wide arc. The material does not show always the clumpiness of the limb spray, it rather looks like a bright or dark semi-transparent cloud or veil. These expanding features may be confused with flare wave disturbances which have the same velocity of propagation (see below).

On the other hand, it is difficult in many cases (limb as well as disk events) to decide if we are observing a spray or a surge because there are ejections which are either intermediate in velocity and structure between classical sprays and classical surges or do not fit at all into the simple surge-spray classification scheme.

Besides the flare sprays which consist of ejected flare plasma there is another type which contains material from erupting prominences (see next section).

1.3. Eruptive prominences

Quite another class of plasma ejections in the solar corona are the eruptive prominences. These are a flare associated phenomenon, at least in the more vigorous types, but the moving material is originally prominence plasma.

The eruption of a filament or prominence consists in a concurrent growth and rise which takes place at an increasing rate until the prominence bursts open and/or disappears high in the corona. The observed maximum velocities may exceed the escape velocity. A part of the erupted material may, however, return to the solar surface. We may distinguish three different types according to the particular circumstances of occurrence, i.e. to the relationship to flares and to the temporal development. The three types are: (a) During the flash and maximum phase of a nearby flare an active region filament, even a very small one, may grow and rise rapidly and move away from the flare as if blown away – and sometimes also disrupted – by the flare; in some cases the plasma is finally driven away as a spray. The whole phenomenon lasts about 15 min. It seems likely that it is produced by a flare-produced shock or by a kind of 'flare wind' i.e. by a stream of particles accelerated by the flare process. (b) An active region filament may start to ascend and expand slowly but also at an accelerated rate several tens of minutes prior to a flare. It finally attains a velocity of several hundred km s^{-1} and disappears high in the corona just at the onset or during the flash of a flare which forms two bright ribbons on both sides of the disappeared filament. The duration of this eruption process is up to one hour. This is the classical DB-flare event. The eruption (the 'Disparition Brusque') is a preflare phenomenon probably induced by an instability of the magnetic field preceding flare occurrence. (c) Quiescent prominences outside active regions sometimes ascend and finally disappear in the course of several hours. The rising motion starts with a few km s^{-1} but may attain also high velocities in the final phase. In many cases the eruption is followed by flarelike brightenings (Bruzek, 1957; Hyder, 1967) or even major flares (Dodson and Hedeman, 1970). There are indications that these eruptions in some cases are initiated by a disturbance emanating from a distant active region either from newly emerging magnetic flux or from a flare. In the majority of cases, however, there is no clear evidence for an external cause. It has to be assumed that the eruption is due to an instability of the prominence-supporting magnetic field.

1.4. Observations in coronal light

In the preceding sections the ejections of dense and cool plasma visible in Hα have been considered. In a few instances ejections have been observed also in higher temperature lines: Neupert mentioned the observation of a surge on the solar limb in He II λ304 on OSO-7; Kirshner and Noyes (1971) reported on the observation of an ejection seen in the C III line λ977 with the OSO-6 spectroheliograph. The amount of the C III emission was found consistent with the assumption that the C III ions occupy sheets with thickness 100 km surrounding the cooler Hα emitting threads. This transition sheet is believed to be formed by heat conduction from the hot corona.

Observations of fast changes ('transients') in visible coronal lines are difficult and rather scanty. An extensive observing material in the green coronal line λ5303 has been collected at the Sacramento Peak Observatory, New Mexico in the years 1957 through 1972. In recent studies of this unique material Dunn (1971) and DeMastus *et al.* (1973) detected 30 fast transient events, 18 out of them were classified as ejections. They are described as expulsions of material from the lower corona or the chromosphere which appear as ascending clouds either large and diffuse or small and concentrated; several were definitely surgelike (Kleczek and Hansen, 1962). Almost all observed coronal ejections were found to be associated with Hα surges or eruptive prominences.

A spectacular ejection of coronal plasma clouds visible in white light has been observed by OSO-7 (Brueckner, these proceedings).

2. Coronal Waves

We have to expect that explosive flares and fast flare ejecta – which have supersonic or superalfvenic velocities – produce shocklike wave disturbances in the solar corona. Well known are the type II radio bursts which are evidence for the propagation of a flare-produced shock wave through the higher corona. In recent years evidence for waves has been found also in optical observations of the corona in monochromatic and in white light. Moreover, flare-related chromospheric disturbances have been observed in Hα which provide indirect evidence for the propagation of a wave or shock in the corona. These are: (a) the oscillation of a distant filament (winking filament); (b) a bright or dark chromospheric disturbance rapidly moving away from the flare (flare wave). The majority of these Hα phenomena have been observed at the Lockheed Solar Observatory and a comprehensive study has been presented recently by Smith and Harvey (1971). The Hα disturbances will be discussed first.

2.1. Evidence from Hα observations

The flare wave, in many cases, becomes visible only outside the associated active region propagating into a limited cone of about 90° with velocities 400–1000 km s^{-1} to distances of up to 500000 km. If the motion is extrapolated back into the flare region it is found that the disturbance starts during the explosive phase of the flare. In well developed events, such as the classical cases of 20 September 1963 and 28 August 1966 the disturbances are led by a curved dark or bright front depending on the wavelength of observation. If observed at Hα−0.5 Å a narrow bright front is followed by a broader, more diffuse, dark feature; at Hα+0.5 Å bright and dark is exchanged. In the line center of Hα usually a faint bright front is seen. This brightening and darkening is interpreted as due to a Doppler shift produced by a propagating down-up motion of the dark chromospheric elements; the velocity amplitude is about 30 km s^{-1}. This motion may be a depression produced by a passing wave front followed by a relaxation. Actually, the motion need not to be strictly vertical, it may have a horizontal component and rather be a tilting motion of the chromospheric elements (spicules, mottles). If the flare wave reaches a filament a similar effect occurs: the filament first becomes redshifted (and disappears in the line center) that is depressed, then blueshifted, again redshifted and so on for 2–5 times. That means, the filament is excited to a damped oscillatory motion by the passing flare wave. Periods of oscillation in the range 6–40 min were found (Dodson and Hedeman, 1964; Smith and Ramsey, 1966). Again, the motion may rather be an increase and decrease of inclination as suggested by Kleczek and Kuperus (1969). Limb observations of oscillating prominences would clarify the exact type of motion, but are not available so far.

In a number of cases such filament oscillations were observed following a flare without the arrival of a visible flare wave. If we assume that these oscillations are also excited by a disturbance emanating from the explosive flare we find the same velocity of propagation and we may conclude that the same type of disturbance is acting even if it has no observable effects on the chromosphere.

Smith and Harvey (1971) concur with the above interpretation of the Hα flare disturbance as a depression and relaxation of chromospheric elements (Dodson and Hedeman, 1968) for only about 50% of the events included in their study. They contend that nine cases – according to their different characteristics – should be interpreted as material ejected into the corona rather than as a Doppler shift of chromospheric elements. Following the description given by the authors one is led to the conclusion that these ejections probably are sprays. It is interesting to note that in all but one of the above nine cases filament oscillations were observed inferring a flare disturbance travelling with the typical wave velocity (500–750 km s^{-1}). Thus, although no typical chromospheric wave was observed, the filament was reached by a flare disturbance.

These observations show once more the difficulties to recognize the true nature of the chromospheric disturbances just from observations in a narrow wavelength range around Hα.

Theoretical interpretations of the Hα flare wave in terms of a flare produced MHD coronal wave have been given by Anderson (1966), Meyer (1968), and Uchida (1968). Uchida suggested that the flare emits a MHD fast-mode wavefront which expands into the solar corona. The observed travelling chromospheric disturbance is excited at the intersection of this wave front with the chromosphere which sweeps along the chromosphere ('sweeping skirt' model). Recently, Uchida *et al.* (1973) elaborated this basic hypothesis and applied it to selected observed events taking into account the actual physical conditions prevailing above the disturbance. They computed the propagation of a fast-mode MHD wavefront through the solar corona using the distribution of the coronal magnetic fields as computed from measured photospheric fields and the coronal electron densities as derived from *K*-coronameter data. They found that the progression of the fast wave mode and its intersection with the chromosphere agreed remarkably well with the observed positions of the front of the Hα disturbance. Unfortunately, the two cases for which detailed results are published belong to the nine events which Smith and Harvey considered as ejections and not as a wave disturbance. It is therefore at least doubtful whether the results are conclusive in these two particular cases.

2.2. Coronal observations

Shocks produced in the corona by flares and ejected plasma should, of course, become evident in the corona itself by some characteristic effects. The best known coronal phenomenon produced by a shock wave are the type II radiobursts. Sprays, on the other hand, are frequently associated with type II bursts (McCabe, 1971; Stewart *et al.*, 1974; Riddle *et al.*, 1974), i.e. they are actually associated with a shock. There is also strong evidence that the coronal wave inferred from the chromospheric flare disturbance just discussed is identical with the shock producing the type II burst (Wild, 1969). The statistical correlation of ejections and chromospheric waves with type II bursts, however, is rather poor. Only 30% of spray flares (Smith, 1968) and 18 out of 50 Hα flare disturbances (Smith and Harvey, 1971) were found associated with a type II burst. A possible explanation for this poor correlation will be given below.

As for optical coronal observations, Dunn (1971) found about 20 fast events in the Sacramento Peak green line coronal film which may either interpreted as the visible propagation of a shock into the corona or as the effect of a travelling shock on coronal structures. Almost all of them were associated with Hα flares or ejections. Unfortunately, so far no coronal observation is available at the time and above the position of an observed chromospheric flare wave, that means, we have not yet a direct optical identification of the coronal wave which produces the Hα chromospheric disturbance.

Different types of green line fast events have been observed. The first class includes fast expanding arches or shells emanating from a flare (see Orrall and Smith, 1961) and accelerated expanding arches which explode at flare onset quite analogue to the behaviour of preflare erupting prominences (Bruzek and DeMastus, 1970). The second class includes the features classified by Dunn (1971) as Realignments and Disruptions of existing coronal structurs. The classical case of a disruption is Evans' whip (Evans, 1957; Dunn, 1971). A somewhat similar case occurred in the 22 February 1967 event (Bruzek and DeMastus, 1970). Realignments are described as large arches seen to snap into new positions almost instantaneously which is suggested to be due to a shock.

Only about 50% of the events classified as coronal expansions were associated with type II bursts although the expansions may be considered as shocks which are expected to propagate further into the corona. In this context an observation of DeMastus *et al.* (1973) may be important. They found that some of the moving coronal features are trapped that is, their motion was more or less suddenly stopped in lower levels of the corona – probably by the magnetic field – so that they could not reach the level where the type II shock appears. This may also be the explanation for the poor statistical correlation of fast Hα events (ejections and waves) with type II bursts: shocks produced by or associated with Hα events may not reach the type II level in all cases.

Also in the white light (electron) corona different types of fast changes have been detected in *K*-coronameter observations (Hansen *et al.*, 1973). They were found associated with fast Hα or 5303 Å events; in particular, Hα sprays were associated with white light expanding arches and an erupting prominence was associated with a white light disruption.

References

Altschuler, M. D., Lilliquist, C. G., and Nakagawa, Y.: 1968, *Solar Phys.* **5**, 366.

Anderson, G. E.: 1966, Ph.D. Thesis, Univ. Colorado, Boulder.

Bruzek, A.: 1957, *Z. Astrophys.* **42**, 76.

Bruzek, A.: 1968, in K. O. Kiepenheuer (ed.), 'Structure and Development of Solar Active Regions', *IAU Symp.* **35**, 126.

Bruzek, A.: 1969, in C. de Jager and Z. Švestka (eds.), 'Solar Flares and Space Research', *COSPAR Symp.*, 61.

Bruzek, A. and DeMastus, H. L.: 1970, *Solar Phys.* **12**, 447.

DeMastus, H. L., Wagner, W. J., and Robinson, R. D.: 1973, in A. J. Hundhausen (ed.), *Proc. Conf. on Flare Produced Shock Waves*, Boulder 1974, p. 17.

Dodson, H. W. and Hedeman, R. E.: 1964, in W. Hess (ed.), *AAS-NASA Symposium on Physics of Solar Flares*, NASA SP-50, p. 15.

Dodson, H. W. and Hedeman, R. E.: 1968, *Solar Phys.* **4**, 229.

Dodson, H. W. and Hedeman, R. E.: 1970, *Solar Phys.* **13**, 401.

Dunn, R. B.: 1971, in C. Macris (ed.), *Physics of the Solar Corona*, D. Reidel Publ. Co., Dordrecht, Holland, p. 114.

Evans, J. W.: 1957, *Publ. Astron. Soc. Pacific* **69**, 421.

Fortini, T. and Torelli, M.: 1968, in K. O. Kiepenheuer (ed.), 'Structure and Development of Solar Active Regions', *IAU Symp.* **35**, 50.
Giovanelli, R. G. and McCabe, M. K.: 1958, *Australian J. Phys.* **11**, 191.
Gold, T.: 1968, in Y. Öhman (ed.), 'Mass Motions in Solar Flares and Related Phenomena', *Nobel Symp.* **9**, 153.
Gopasyuk, S. I., Ogir, M. B., and Tsap, T. T.: 1963, *Izv. Krymsk. Astrofiz. Obs.* **30**, 148.
Hansen, R. T., Hansen, S. F., and Garcia, C.: 1973, in A. J. Hundhausen (ed.), *Proc. Conf. on Flare Produced Shock Waves*, Boulder 1974, p. 109.
Harvey, J. W.: 1969, Ph.D. Dissertation, Univ. Colorado, Boulder.
Hyder, C. L.: 1967, *Solar Phys.* **2**, 49 and 267.
Kirshner, R. P. and Noyes, R. W.: 1971, *Solar Phys.* **20**, 428.
Kleczek, J. and Hansen, R. T.: 1962, *Publ. Astron. Soc. Pacific* **74**, 507.
Kleczek, J. and Kuperus, M.: 1969, *Solar Phys.* **6**, 72.
Lilliquist, C. G., Altschuler, M. D., and Nakagawa, Y.: 1971, *Solar Phys.* **20**, 348.
Livshits, E. M. and Pickel'ner, S. B.: 1964, *Soviet Astron. AJ* **8**, 368.
McCabe, M. K.: 1971, *Solar Phys.* **19**, 451.
Meyer, F.: 1968, in K. O. Kiepenheuer (ed.), 'Structure and Developments of Solar Active Regions', *IAU Symp.* **35**, 485.
Orrall, J. Q. and Smith, H. J.: 1961, *Sky Telesc.* **22**, 330.
Pickel'ner, S. B.: 1969, *Soviet Astron. AJ* **13**, 259.
Riddle, A. C., Tandberg-Hanssen, E., and Hansen, R. T.: 1974, this volume, p. 335.
Roy, J. R.: 1973a, *Solar Phys.* **28**, 95.
Roy, J. R.: 1973b, Ph.D. Dissertation, Univ. Western Ontario, London, Canada.
Roy, J. R.: 1973c, *Solar Phys.* **32**, 139.
Rust, D.: 1968, in K. O. Kiepenheuer (ed.), 'Structure and Developments of Solar Active Regions', *IAU Symp.* **35**, 77.
Smith, S. F.: Ramsey, H. E., 1966, *Astron. J.* **71**, 197.
Smith, S. F. and Harvey, K. L.: 1971, in C. Macris (ed.), *Physics of the Solar Corona*, D. Reidel Publ. Co., Dordrecht, Holland, p. 156.
Smith, E.v.P.: 1968, in Y. Öhman (ed.), 'Mass Motions in Solar Flares and Related Phenomena', *Nobel Symp.* **9**, 137.
Stewart, R. T., McCabe, M., Koomen, M. J., Hansen, R. T., and Dulk, G. A.: 1974, this volume, p. 337.
Uchida, Y.: 1968, *Solar Phys.* **4**, 30.
Uchida, Y.: 1969, *Publ. Astron. Soc. Japan* **21**, 128.
Uchida, Y., Altschuler, M. D., and Newkirk, G., Jr.: 1973, *Solar Phys.* **28**, 495.
Wild, P.: 1969, *Proc. Astron. Soc. Soc. Australia* **1**, 181.

DISCUSSION

Altschuler: I would like to make a comment on the interesting observation early in your talk that the flare has almost disappeared at the maximum stage of the surge. It seems to me that the visible Hα flare material cannot be the same matter as in the Hα surge, because we cannot accelerate matter once it becomes neutral. Probably there is a large amount of material in an ionized state, and it is this matter which is accelerated upward. The ionized material then cools and illuminates the surge trajectory in Hα.

Athay: The material emitting Hα is certainly very significantly ionized.

Sturrock: Some time ago, you published data on a surge-flare showing that the flare had a ring-like structure. Is this case unique or typical?

Bruzek: I have only observed that one case.

Uchida: Moreton waves can be produced by quite weak flares, contrary to what seems to be generally believed. I examined Mrs Martin's data and found this to be so. Further, the propagation velocity of Moreton waves is fairly independent of the importance of the flare. As I have shown, this can only be so with the MHD fast-mode hypothesis.

On a separate matter, could you give us the estimated mass involved in large spray events? This is quite essential in discriminating between different flare models.

Bruzek: The mass estimate would be 10^{16} to 10^{17} g.

Aller: What is the typical position of a surge with respect to that of the flare or subflare?

Bruzek: In most cases, surges are not associated with large flares; a small subflare is often found at the base of the surge.

Giovanelli: The old Hale spectrohelioscope was a flexible instrument for studying ejections on the disk since the wavelength shifter enabled observation to be made over some ± 10 Å from Hα and within 1 second. It was interesting to find ejected absorbing matter associated with the early stages of some flares, presumably of the type now called 'sprays'. This often took the form of irregular blobs seen quite separate from the bright flare. Perhaps 10 min later, irregularly-shaped and distributed falling matter would be seen quite separate from the rising material. I could never see any continuity in the rise and fall of individual blobs. We do seem to have lost a great deal of instrumental capability in going over, almost universally, to photographic observations at only one or two wavelengths in or near the core of Hα.

Jefferies: Such downward moving material is characteristic of loops which form after major flares.

Meyer: You mentioned that we have not yet a direct observation of the coronal transient events. There is one other indirect observation of the fast-mode wave propagation in the corona by the 'winking filaments'. From the amplitude that these filaments receive when hit by the coronal wave one estimates a rather small energy in the fast-mode Moreton-wave. The indications are that these are only slightly non-linear. Most of the heat energy released in the flare goes obviously into the acoustic-type shock wave which continues as an interplanetary shock. This is also physically reasonable. An explosively released energy will expand the medium in the direction of least resistance and this is upwards towards the lower density. The horizontal expansion that excites the Moreton wave is resisted by the magnetic fields, which in the higher coronal regions exert stronger pressure than does the gas before the explosion. So one should probably keep this large difference in energy between the two modes in mind, even if both forms are excited by the same event.

THE BEHAVIOUR OF THE OUTER SOLAR CORONA ($3\,R_\odot$ TO $10\,R_\odot$) DURING A LARGE SOLAR FLARE OBSERVED FROM OSO-7 IN WHITE LIGHT

G. E. BRUECKNER

Naval Research Laboratory, Washington, D.C. 20375, U.S.A.

Abstract (*Solar Phys.*). A very bright coronal streamer was observed on December 13, 1971 by the Naval Research Laboratory's coronagraph on board of OSO-7. The next day, the streamer had changed it's brightness and configuration considerably. Three subsequent coronagraph images, taken on December 14 at 0407, 0418 and 0430 UT show a large plasma cloud moving outward from the Sun between 3 and 10 solar radii. They also show distinct smaller clouds moving outward with projected velocities between 950 and 1100 km s^{-1}. Traced back in time to the lower solar corona, these clouds coincide with discrete type II radiobursts observed from Culgoora between 0241 and 0256 UT. Each single cloud shows it's signature in the radio recording between 100 and 20 MHz. The drift velocity of the radio bursts can be determined to be 1600 km s^{-1} using Newkirk's coronal streamer model. Assuming, that the plasma clouds are ejected from an active region 30° behind the east limb vertically, their true velocities close to the surface of the Sun would be approximately 1400 km s^{-1}, which is in good agreement with the drift velocities determined from the type II bursts, considering all uncertainties. Therefore, the type II burst disturbance moves with the same velocity as the driving material.

The total coronal blast contains a number of particles of more than 2.5×10^{40} with a total energy of greater than 1×10^{32} erg. At 8 solar radii the disturbance lasts approximately 2.5 h.

A considerable fraction of the total energy can be found in the discrete, high velocity clouds. They carry more than 1.6×10^{31} erg.

After the event, on December 15, only a very weak remnant of the original streamer can be found. Estimates show, that the total mass contents of the streamer configuration was more than 2×10^{40} particles prior to the event. It is therefore possible to obtain the total mass of the coronal blast by emptying the preflare streamer configuration.

DISCUSSION

Dryer: I believe that you caught the birth of at least one (probably more) shock. This seems to be attested to by the type II velocities which are close to the values measured by your clouds. Thus, as time goes on, the shock, originally formed at the piston – in theory at least – would move ahead of the driving plasma piston. In reality (laboratory and detailed analysis) some time is required for the shock to form. I hope that Dr N. Krall will discuss this point this afternoon.

Smith (to Dryer): How do you determine the stand-off distance of the shock from such a small blob? Isn't it possible that the standoff distance is quite small?

Gordon Newkirk, Jr. (ed.), Coronal Disturbances, 333–334. All Rights Reserved.

Dryer: Yes, you expect the shock and piston (blob) to be initially coincident and the shock to accelerate and move away from the piston.

Schmidt: The fast blobs you discribe can certainly drive the type II, but they do not contain sufficient mass to drive an interplanetary shock. Therefore, these two phenomena do not necessarily outline the same shockfront. An interplanetary shock may well develop from the larger masses you were describing.

Brueckner: One of my slides shows the small fast clouds and the type II bursts associated with them. However, most of the material moves at 600 to 700 km s^{-1} and this may be associated with the interplanetary shock. This interpretation agrees with the onset time of the Forbush decrease. There were about 10^{32} ergs and 10^{16}–10^{17} g involved.

Schmidt: I assume that the velocities of the larger masses are rather hard to measure. But even if these masses do not move so fast, they will produce an interplanetary shock, if they are slightly hotter than the corona before the event. There is ample time for the interplanetary shock to develop after the flare and to reach 1 AU within the observed travel time.

THE CORONAL DISTURBANCE OF 12 AUGUST 1972

ANTHONY C. RIDDLE*
University of Colorado, Boulder, Colo., U.S.A.

and

EINAR TANDBERG-HANSSEN and RICHARD T. HANSEN
High Altitude Observatory, Boulder, Colo., U.S.A.

Abstract (*Solar Phys.*). The association of flare sprays, distortions of the overlying coronal structures and moving type IV radio bursts is a reasonable one and is well accepted despite the paucity of observational evidence. On 12 August 1972 there occurred a flare spray observed optically by both flare patrol (National Oceanographic and Atmospheric Administration) and coronagraph (High Altitude Observatory) instruments. A subsequent moving type IV radio burst was recorded on two swept frequency interferometers (Universities of Colorado and Maryland). In addition distinct changes in the K-coronal brightness at 1.6 $R_\odot$ were measured (High Altitude Observatory K coronameter). These observations combine to form one of the most complete sequences of measurements yet recorded covering the range from the chromosphere to about 6 $R_\odot$. The separate measurements are discussed and we show that they can be combined to form a relatively simple physical picture of the whole event.

Material ejected from a prominence behind the limb was observed in Hα out to a radius of 2 $R_\odot$. During the event the ambient coronal magnetic field above the flare at heights of 1.2–1.6 $R_\odot$ moved sideways, carrying with it the coronal plasma, as evidenced by changes seen in the K-coronal structure. The disturbance caused by the ejection moved out through the opening carrying with it material from the low corona. Subsequently (within 24 h) material left higher regions in the corona, diminishing the electron density enhancement over the region at heights of 1.2 and 1.6 $R_\odot$. The passage of the material created a shock wave which moved out through the corona ahead of the material which created it. Synchrotron radiation within the shock resulted in a type IV radio burst.

DISCUSSION

Wild: Was there an associated type II burst?

Riddle: No; the type IV started at 70 MHz and drifted to 30 MHz. I believe the source was behind the limb by about 15°.

Pneuman: I want to comment on the K-coronameter isophotes. Before the disturbance, they were roughly gaussian in shape, and square afterwards. This implies that matter was pushed to the sides of the condensations by some action moving through the middle.

Riddle: I agree, I should have made that clearer.

Gotwols: Please clarify whether the radio burst actually began at 2100 UT, or is that merely the time when it drifted into the passband of the swept frequency interferometer?

* On leave from CSIRO, Division of Radiophysics, Sydney, Australia.

Gordon Newkirk, Jr. (ed.), Coronal Disturbances, 335–336.

Riddle: The Boulder spectrograph covers up to 80 MHz whereas the burst was observed to start up at 70 MHz so the true starting of the radio event was ~2100 UT.

McLean: Why do you say that the type IV source is a shock rather than something pushed ahead of the visible matter?

Riddle: That is certainly possible, however the large height difference between the 'piston' and the radio source suggests a shock with a large standoff distance.

Sturrock: Studies of the bow shock near the Earth's magnetosphere indicate that the standoff distance is typically 0.4 times the radius of curvature of the obstacle. Does this fit your data?

Riddle: I think these theories apply for higher Mach number (4–5) than that of interest here, which is closer to 2–3. Also, I don't know how we'd find the appropriate piston radius.

Dryer: The theory for an accelerating piston certainly needs improvement. A rule of thumb is to add 10% to the distance the piston has moved to find the stand-off distance.

Riddle: We need more data to clear up the question.

OBSERVATION OF A CORONAL DISTURBANCE FROM 1 TO 9 $R_\odot$

R. T. STEWART
Institute for Astronomy, University of Hawaii, Honolulu, HI, U.S.A.

MARIE K. McCABE
On leave from Division of Radiophysics, CSIRO, Sydney, Australia

M. J. KOOMEN
Naval Research Laboratory, Washington, D.C., U.S.A.

R. T. HANSEN
High Altitude Observatory, NCAR, Boulder, Colo., U.S.A.

and

G. A. DULK*
Institute for Astronomy, University of Hawaii, Honolulu, HI, U.S.A.

Abstract (*Solar Phys.*). On 1973 January 11, a flare near the west limb of the Sun caused a coronal disturbance which was observed with a unique variety of instruments. Radio observations of a type II and a moving type IV burst were obtained by the CSIRO Division of Radiophysics at Culgoora, Australia; white-light observations of a large, moving cloud were made by the U.S. Naval Research Laboratory coronagraph on OSO-7; *K*-corona observations of a decrease in coronal density were made by the High Altitude Observatory at Mauna Loa, Hawaii and Hα observations of a flare spray were made by the Institute for Astronomy, University of Hawaii at Haleakala (and also by H.A.O.).

The flare was observed with narrow-band – 0.5 Å and tunable 0.25 Å – Hα filter telescopes at Haleakala and the spray with broadband Hα coronagraphs at Haleakala and Mauna Loa, where the bandpasses were 7 Å and 10 Å respectively. The main features of the Hα events are outlined in Figures 1(a), (b), and (c). An importance IB flare began at 00^h36^m UT at N 12°, W 80° (x of Figure 1 (a)) in McMath region 12160, spreading to y at 00^h40^m, with further brightenings at 00^h47^m to 00^h49^m. The west limb spray began at 00^h37^m, with a minor ejection from position angle 286° (a of Figure 1(a)) followed by the major ejection at 00^h39^m from 289° (b of Figure 1(a)). The positions of various bright blobs were traced out to heights $\sim 2\ R_\odot$ (Figure 1(c)); they indicated that the spray material moved outwards with nearly constant (projected) velocities of 300 to 600 km s^{-1}. Minor surge activity was observed at c, d, and e of Figure 1(b) from 00^h50^m to 01^h12^m.

The radio spectrum recorded at Culgoora showed that the flare was accompanied – within one minute – by a group of type III bursts, a short-wave fadeout, and a

* Exchange visitor from the Dept. of Astrogeophysics, University of Colorado, Boulder, Colo., U.S.A.

Gordon Newkirk, Jr. (ed.), Coronal Disturbances, 337–341.

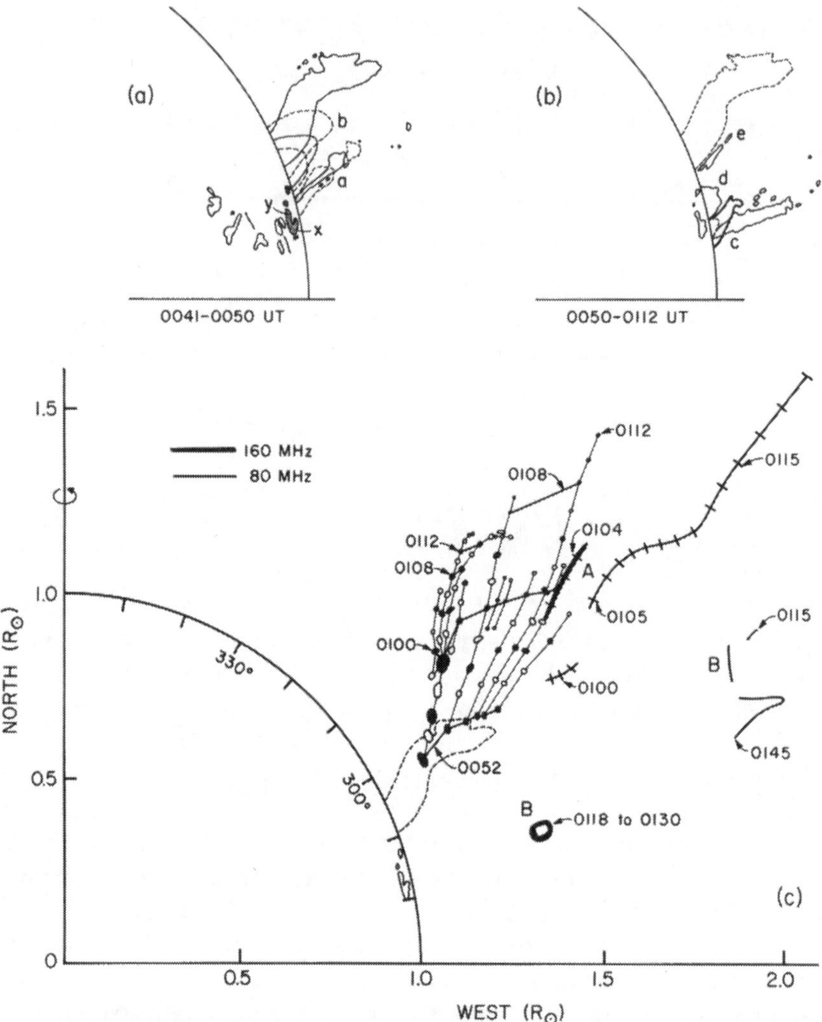

Fig. 1. Sketches of moving Hα material and positions of radio sources for the coronal disturbance of 1973 January 11. (a) Early development of the Hα spray events a and b; successive profiles are shown by full and dashed lines alternately. The flare area is hatched, with the main flare centres indicated by x and y. Plage regions are outlined and the dark points nearby are sunspots. (b) Outline of surges c, d, e at later times. For c a thick line shows its initial outline. The dashed profile outlines the spray event b at 00^h50^m. (c) Observed positions of bright blobs of the Hα spray at 2-min intervals; the light lines show the trajectories, while the heavy lines indicate the leading edge at specified times. The observed path of the moving type IV is also shown (at A) with cross marks indicating 1-min intervals, while the scatter in positions of the stationary type IV source is shown (at B) with the initial and final positions indicated by the corresponding times. In both (A) and (B) thin lines indicate 80 MHz observations, thick lines 160 MHz observations.

broadband type IV burst in the decimetre and metre wavelength ranges commencing at 00^h40^m and 01^h00^m, respectively. An intense complex type II burst started at $\sim 00^h48^m$, its beginning time being uncertain on account of the simultaneous type III bursts which accompanied a second phase of flare brightening. The frequency drifts of the various bands in the type II burst indicate that multiple shock waves travelled outwards through the corona with speeds ~ 800 to 1200 km s^{-1}.

From 00^h54^m to 01^h45^m, heliograph pictures were recorded once every second at either 80 or 160 MHz. At 80 MHz the type II source was multiple and located at 1.6 to 1.7 $R_{\odot}$ extending in position angle from 280° to 310°. The moving type IV source had a fairly constant (projected) velocity ~ 600 to 700 km s^{-1} at both 80 and 160 MHz (thin and thick lines at A in Figure 1(c)). The radio source and the leading parts of the Hα spray were at similar heights between 01^h00^m and 01^h12^m.

As the moving type IV source faded, a stationary type IV or 'storm continuum' source (B of Figure 1(c)) appeared above the flare region at ~ 1.4 $R_{\odot}$ (160 MHz) and ~ 2.0 $R_{\odot}$ (80 MHz). Both the moving and stationary sources were unpolarized.

K-corona intensities above the flare-spray region (out to 1.6 $R_{\odot}$) were measured by the coronal activity monitor at Mauna Loa. This instrument scans at a fixed height for several minutes before moving to a new height. Several fixedheight scans taken

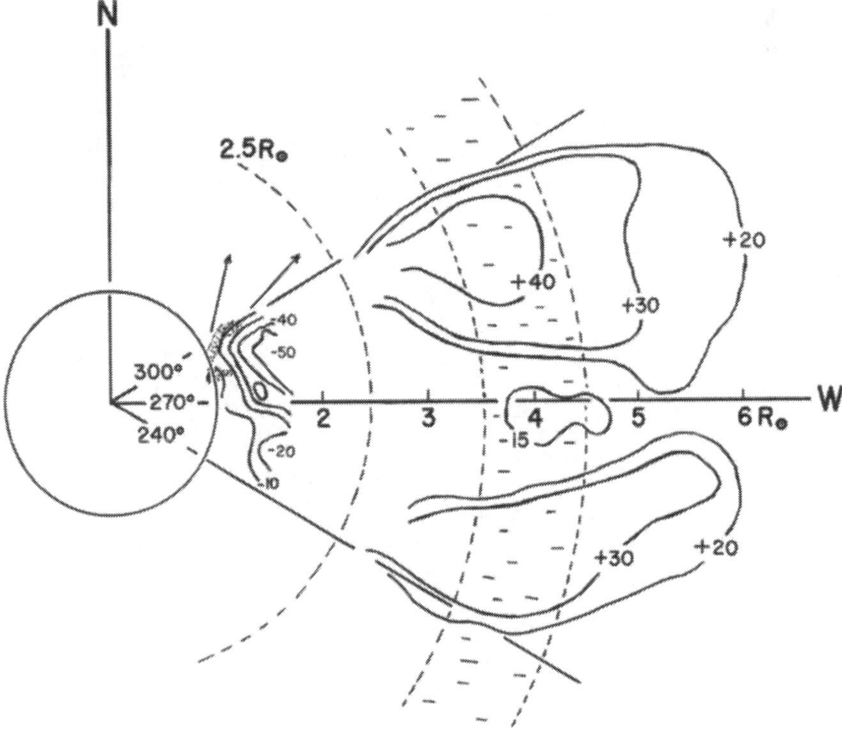

Fig. 2. Composite plot of the percentage change of the brightness of the *K*-corona (full lines) during the coronal disturbance of 1973 January 11. The isophotes below 1.6 $R_{\odot}$ were obtained by subtracting fixed-height scans taken by the coronal activity monitor at different times before and after the transient event. The isophotes between 2.5 and 6.0 $R_{\odot}$ were obtained by subtracting the OSO-7 coronagraph picture at 00^h14^m taken before the event from the one at 02^h28^m. The stippled region and the arrows indicate the Hα spray and ejection cone. The dashed arc at 2.5 $R_{\odot}$ indicates the outer edge of the OSO-7 coronagraph occulting disk. The polarization of the OSO-7 picture is tangential everywhere except in the hatched region between 3.5 and 4.5 $R_{\odot}$, where it is radial.

before and after the transient event showed that a major decrease in intensity had occurred, indicative of a depletion in the number of electrons along the observer's line of sight. The projected area of the *K*-corona depletion was extensive, ranging in height between 1.1 to 1.6 $R_\odot$ at position angles 240° to 300° (see the isophotes below 2 $R_\odot$ in Figure 2). We estimate from our limited coverage of the transient event a starting time – at 1.5 $R_\odot$ – of 00^h50^m, similar to the time of arrival of the fastest Hα spray material at that height. The depletion at 1.5 $R_\odot$ lasted until about 01^h08^m.

Instantaneous, two-dimensional pictures of the white-light corona between 3 and 9 $R_\odot$ were recorded at 00^h14^m, 01^h47^m, 02^h28^m, and 03^h30^m by the OSO-7 coronagraph. The first picture, obtained before the flare event, was subtracted from the remaining pictures to enhance their contrast. The high-contrast pictures clearly show that a cloud of enhanced electron density moved outwards from ~3 to ~9 $R_\odot$.

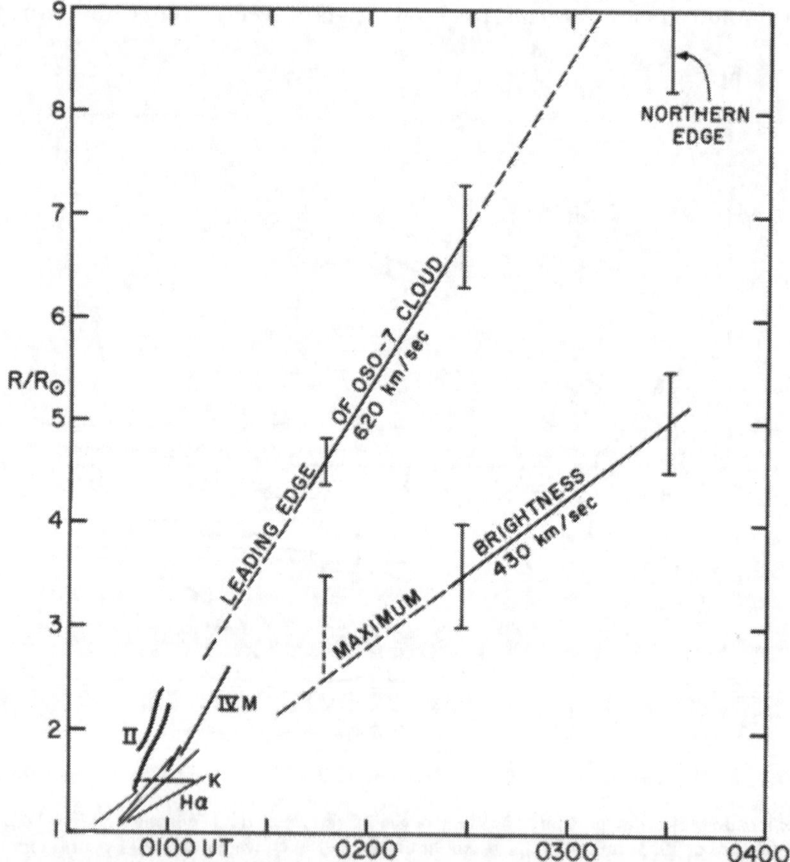

Fig. 3. Combined height-time plots of the observed ejecta in the coronal disturbance of 1973 January 11. Apart from the type II burst the projected radial velocities range from 430 km s^{-1} (the speed of the region of maximum brightness in the OSO-7 white light cloud) to 620 km s^{-1} (the speed of the leading edge of the OSO-7 cloud). The times and errors of measurement for the OSO-7 cloud are indicated. At 03^h30^m only the northern edge of the cloud was in the field of view of the coronagraph. See text for further details.

Isophotes of the percentage increase in K-corona brightness over heights 2.5 to 6 $R_\odot$ at 02^h28^m are plotted in Figure 2. The enhanced regions at position angles 250° and 300° correspond well with the two depleted regions at lower heights ($<2\ R_\odot$) observed with the coronal activity monitor.

The height of the leading edge of the white-light cloud as well as the height of the maximum excess brightness of the cloud is plotted against time in Figure 3. Also shown on this figure are the trajectories of the Hα spray (light lines) and of the type II and moving type IV bursts (labelled II and IVM, respectively). In addition, the duration of the K-corona transient at 1.5 $R_\odot$ is indicated (labelled K). We note that all the observed velocities, with the exception of that of the type II burst, lie within a velocity range of 400 to 600 km s^{-1}. Furthermore, the temporal and height agreement shown in Figure 3 suggests that a common disturbance moved outwards from the chromosphere and was manifested in various types of activity. We suggest that at the time of the flare and spray, a rearrangement of the coronal magnetic field allowed the trapped coronal gas above the flare region to expand and move outward, causing the decrease in electron density at $\lesssim 2\ R_\odot$. This material acted as a piston, driving the shock wave which caused the type II burst. The driver gas continued moving outward, producing a white-light enhancement at heights of 3 to 9 $R_\odot$. We estimate that $\sim 10^{39}$ to 10^{40} electrons (and protons) were expelled outwards to form the white-light cloud. The total energy in the expelled gas, Hα material, and radio bursts (excluding the shock wave) was $\sim 10^{31}$ erg, i.e. about one order of magnitude less than the total energy in a moderately sized flare. It will be interesting to see if an interplanetary shock wave occurred after this event.

DISCUSSION

Newkirk: The last two papers have discussed the mass content of expelled material as both cool spray material and coronal plasma. We should be very cautious in our identification of either of these with the shock material observed in interplanetary space on the basis of a mass comparison alone.

Stewart: I agree. The material in the spray is less than 10% of that in the K-corona transient or the OSO-7 clouds.

Giovanelli: Is it possible to determine whether the material is pushed out from below or pulled out from above?

Sturrock: Barnes and I have recently noticed that force-free fields may become unstable against eruption. The resulting behavior may be related to sprays. It is hard to say whether the gas should be said to be 'pulled out' or 'pushed out'.

TYPE II SOLAR RADIO BURSTS IN THE DECIMETER BAND

ALAN MAXWELL

Harvard Radio Astronomy Station, Fort Davis, Tex., U.S.A.

Abstract (*Solar Phys.*) The progress of shock waves generated by solar flares through the corona is delineated in the radio band by radio bursts of spectral type II. Previous discussion of these bursts has been mainly concerned with their characteristics at frequencies below 200 MHz. This paper discusses the characteristics of the bursts in the range 2000–200 MHz, and the information that may then be deduced about the propagation of shock waves through the lower corona. Particular attention is paid to the type II burst emitted by the flare of 1972 August 7, 1500 UT. The shock from this flare was subsequently tracked (by radio equipment on the IMP-6 satellite) through the interplanetary plasma right to the Earth.

DISCUSSION

Smith: It looks to me from the spectrum that we have the start of a type IV continuum, which may well be initiated by a shock. Isn't it a little bit dubious to call this decimetric type II?

Maxwell: The first phase appears to be a succession of type III bursts followed by type IV with pulsations. The terminology to be adopted in describing such an event is not really clear.

Wild: Did a geomagnetic storm occur at the right time?

Maxwell: Yes, even though a paper in *Nature* stated otherwise.

Dryer: Yes, on 8 August at 2354 UT. The shock was also picked up by several spacecraft. I shall discuss this a bit more this afternoon.

Gordon Newkirk, Jr. (ed.), Coronal Disturbances, 343.

OBSERVATIONS OF SPLIT-BAND HARMONIC TYPE II BURSTS WITH THE CULGOORA RADIOHELIOGRAPH AT 80 AND 160 MHz

G. J. NELSON and K. V. SHERIDAN
Division of Radiophysics, CSIRO, Sydney, Australia

Abstract (*Solar Phys.*) When the Culgoora radioheliograph started operating at both 80 MHz (the initial frequency) and 160 MHz (from May 1972) it became possible for the first time to make simultaneous observations of the fundamental and second harmonic sources in type II bursts. Three such bursts, each having harmonically related split-bands, have been observed so far. The new 80 and 160 MHz heliograph observations are illustrated in Figure 1. Since all three events had split-band structure the results are presented separately for the lower- (l) and upper- (u) frequency components and differences between the source positions in the two components will be highlighted. For the burst of 1973 May 19, which was observed at large zenith angles, calculated corrections for ionospheric refraction – approximately 1′ to 2′ for the 80 MHz sources – have been applied; the remaining two events are presented as observed since the refraction effects are thought to be small.

In the event of July 4 the fundamental bands of both the upper and lower components crossed 80 MHz, and for both components the centroid of the harmonic (at 160 MHz) was closer ($\sim 3'$) to the flare position than was the centroid of the fundamental. In the other two cases the fundamental of only the upper component crossed 80 MHz. In one case (May 19) the harmonic at 160 MHz is again (though only slightly) closer to the flare than 80F; in the other case (July 7) the two positions almost coincide and are close to the flare. Thus, while the results for the event of July 4 are in at least qualitative agreement with ray-tracing computations in refracting and scattering coronal models (Riddle, 1972a, b and private communication) the results for the other events (particularly July 7) appear to be difficult to reconcile with the computations, which show that a fundamental source is always displaced outwards, and a harmonic source inwards, from its true, projected position.

The present observations also add to the previous sample of positions of the fundamental source at 80 MHz and, at a later time, of the second harmonic of the source at 40 MHz. (No fundamentals were recorded at 160 MHz, as all fundamental bands started below 160 MHz.) Again only in the event of July 4 were both l- and u-components observed; in the l-component the second harmonic of 40 MHz is closer to the flare than the fundamental at 80 MHz, but in the u-component the relative positions are reversed. This same reverse sequence, with fundamental closer to the flare than the second harmonic, is very pronounced in the July 7 event, while the two sources nearly coincide on May 19 (in both these last cases only the u-component sources were observed).

Gordon Newkirk, Jr. (ed.), Coronal Disturbances, 345–348.

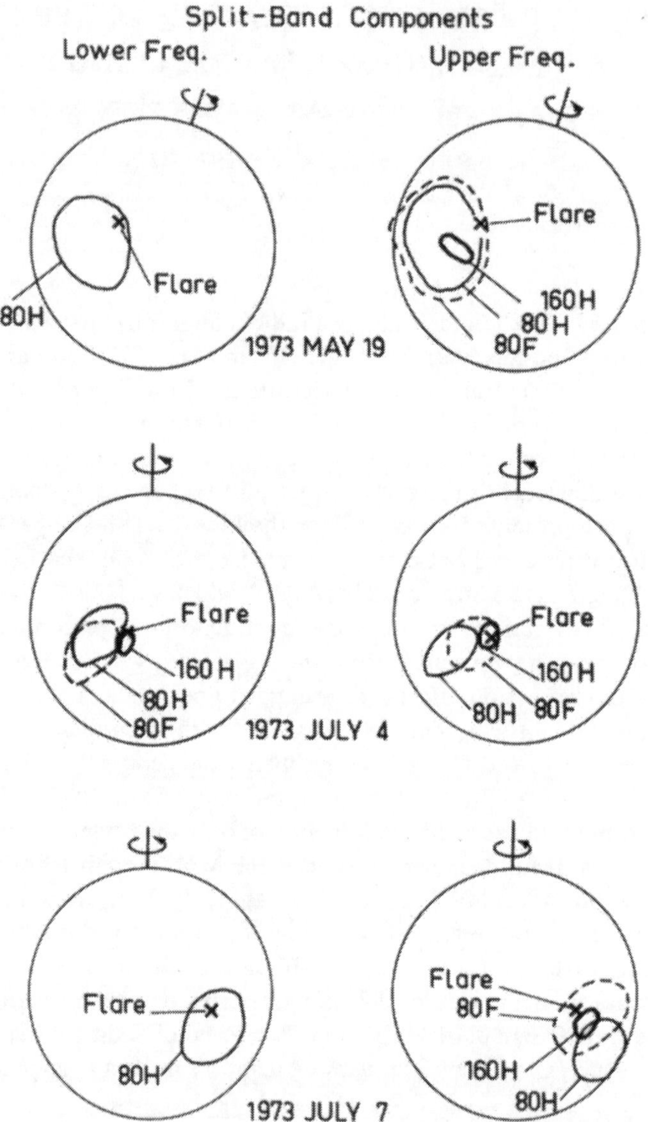

Fig. 1. Half-power contours of the sources of three splitband harmonic type II bursts observed with the Culgoora radioheliograph at 80 and 160 MHz. Full contours apply to harmonic sources and dashed ones to fundamental sources: those labelled 80F and 160H refer to simultaneous observations of the fundamental (at 80 MHz) and the second harmonic (at 160 MHz); the contours labelled 80H refer to 80 MHz observations of the same spectral feature at a later time when the fundamental emission was at 40 MHz.

The results for the *l*-component of July 4 and the *u*-component for May 19 agree with those from previous observations made without identifying split-band components (most were not split-band bursts) and are in qualitative agreement with the computations. The remaining results for the *u*-components of July 4 and 7, both showing the 80 MHz fundamental much closer ($\sim 3'$ to $4'$) to the flare than the second harmonic of 40 MHz, do not agree with our previous observations nor with the computed results for sources in a spherical corona or on the axis of a radial streamer. (Warwick (1965) reported one example where the fundamental was closer to the flare than the second harmonic for a non split-band type II burst.)

We note that all cases which appear to be incompatible with ray-tracing computations occur for sources in the *u*-component; a full explanation of these observations would probably require a better understanding of split-banding in type II bursts.

The relative positions of the sources of the two components of a split band – at the same frequency but at slightly different times – can also be deduced from the present observations, and the examples given here increase the number of such observations available by a large factor. As there were no fundamental sources observed at 160 MHz, the only new data on fundamental sources comes from the July 4 event where $(80F)_u$ is closer to the flare than $(80F)_l$, and this is opposite to the result of Dulk (1970). For the harmonic sources we find in all three cases that $(80H)_l$ is closer to the flare than $(80H)_u$; this agrees with Dulk's result but is opposite to one case reported by Labrum (1969). On the other hand, $(160H)_u$ is closer to the flare of July 4 than is $(160H)_l$, which is opposite to the only other comparable result (Dulk (1970) reported one case at 158 MHz). Thus, the small sample available so far suggests that neither the fundamental nor the harmonic bands show systematic relative positions of the lower- and upper-frequency components.

Table I summarizes the positions of the components relative to the flare position.

These, seemingly erratic, observed separations between the *l*- and *u*-components of fundamental and harmonic bands may have an explanation in existing split-band theories (McLean, 1967; Smerd *et al.*, 1974) but meaningful comparisons between theory and observation will require a knowledge of actual coronal densities and disturbance paths for particular events.

Another new and more positive result from the present observations is that the 80 MHz sources are always much larger than the corresponding 160 MHz harmonic source. The difference is much greater than expected from existing scattering calcu-

TABLE I

Projected distance from flare in minutes of arc

Date 1973		Fundamental at 80 MHz	Second harmonic at 160 MHz	Second harmonic at 80 MHz
May 19	*u*	4.3	4	4.3
	l	–	–	3.7
July 4	*u*	2.5	0	5.6
	l	5.3	1.2	3.7
July 7	*u*	1.9	1.9	6.2
	l	–	–	2.5

lations. It is concluded therefore that the harmonic emission volume is much smaller than the simultaneous fundamental emission volume and that the size of the harmonic source increases rapidly with height in the corona.

References

Dulk, G. A.: 1970, *Proc. Astron. Soc. Australia* **1**, 308.
Labrum, N. R.: 1969, *Proc. Astron. Soc. Australia* **1**, 191.
McLean, D. J.: 1967, *Proc. Astron. Soc. Australia* **1**, 47.
Riddle, A. C.: 1972a, *Proc. Astron. Soc. Australia* **2**, 98.
Riddle, A. C.: 1972b, *Proc. Astron. Soc. Australia* **2**, 148.
Smerd, S. F., Sheridan, K. V., and Stewart, R. T.: 1974, this volume, p. 389.
Warwick, J. W.: 1965, in J. Aarons (ed.), *Solar System Radio Astronomy*, Plenum Press, New York, Chapter 8, p. 131.

DISCUSSION

Sturrock: If a type II burst is caused by plasma excitation at a shock front, one would expect radiation at two frequencies originating in the shocked and unshocked gas. Can this be reconciled with the data?

Nelson: This question is the subject of Smerd's contribution later in the Symposium so I'd prefer not to comment at this time.

Smith: Isn't it possible that radiation at the second harmonic is emitted isotropically?

Nelson: For the first event I cannot tell, I have not considered the other two bursts yet from that point of view.

Smith: Are your observations consistent with radiation being produced ahead of and behind the shock to explain the split bands?

Nelson: Yes.

TYPE II BURSTS AT HECTOMETRIC AND KILOMETRIC WAVELENGTHS FROM INTERPLANETARY SHOCKS*

H. H. MALITSON, J. FAINBERG, and R. G. STONE
NASA-Goddard Space Flight Center, Greenbelt, Md. 20771, U.S.A.

The first type II radio burst to be observed at frequencies below the ionospheric cutoff was recorded by the Goddard radio astronomy experiment on the Interplanetary Monitoring Platform, IMP-6. It occurred on June 30, 1971; a tracing made from the intensity contour plot of the radio data is shown in Figure 1. On the original, intensity contours were plotted in dB above background as a function of frequency and time. The figure shows 32 discrete frequencies from 4.9 MHz to 30 kHz received by the IMP-6 experiment. The time scale covers a period of almost five hours. During this time, five individual type III bursts and one group of type III's occurred; they are shown by their outlines to be drifting rapidly from high to low frequencies. The relatively narrow-band, slowly-drifting radiation with clearly defined fundamental and second harmonic is the type II. The average drift rate, represented by the straight lines drawn through the burst, is approximately 2.5 kHz min^{-1} in this frequency range, and is roughly 100 times slower than that of the type III bursts. The ovals defining the type II emission were traced from the record and show modulation of the signal caused by the spin of the satellite and the resultant sweeping of the antenna pattern of the dipole across the source direction. On the original record, the outlines of the type III's are also filled with a pattern of modulation ovals that were omitted from the figure for clarity. The shaded portion of the record at the lower frequencies represents magnetospheric noise.

Figure 2 shows data from the IMP-6 experiment for August 7 and 8, 1972. The 3B flare at about 1500 UT on August 7 is represented by a dot at the upper left. Its associated shock front produced the sudden-commencement geomagnetic storm beginning at 2354 UT on August 8, the dot at the lower right corner of the figure. The straight line drawn between them represents the average velocity of the shock front over the 1 AU distance, about 1270 km s^{-1}. The horizontal lines in areas B, C, and D represent the times during which type II radiation at various frequencies was observed. This radiation is not observed all the time, but is sporadic – a characteristic of many type II bursts observed from the ground. Apparently, special physical conditions are necessary to produce type II emission which are not always present as the shock front proceeds outward.

* Summary of the following two papers:
'Observation of a Type II Solar Radio Burst to 37 $R_{\odot}$', H. H. Malitson, J. Fainberg, and R. G. Stone, *Astrophys. Letters* **14**, 111–114, 1973.
'A Density Scale for the Interplanetary Medium from Observations of a Type II Solar Radio Burst out to 1 Astronomical Unit', H. H. Malitson, J. Fainberg, and R. G. Stone, *Astrophys. J.* **183**, L35–L38, 1973.

Gordon Newkirk, Jr. (ed.), Coronal Disturbances, 349–353.

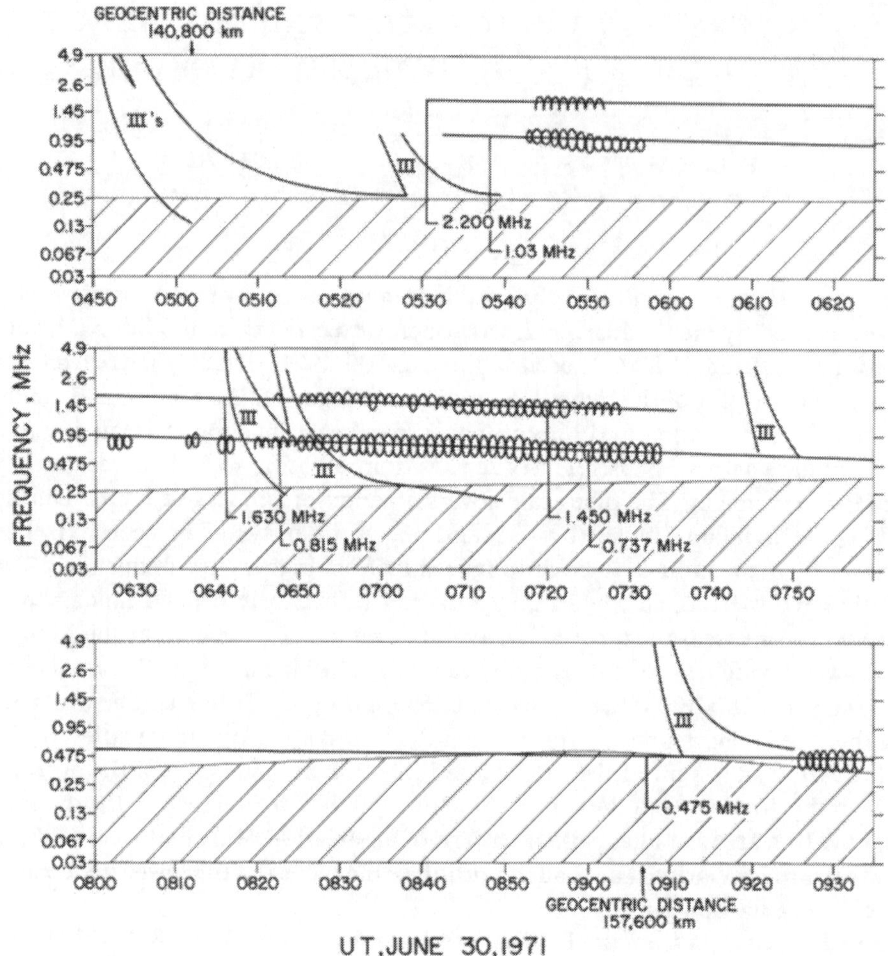

Fig. 1. [Courtesy of *Astrophys. Letters*, Gordon and Breach.]

Type II emission is commonly observed in all its wavelength ranges, both at the plasma frequency characteristic of its point of origin in the corona and at the second harmonic of that frequency. This is also true of the present data, at least in regions B and C where the burst is quite strong and well defined. Only the observations of the fundamental have been used in the figure, since those at the harmonic would tend to overlap the others.

In placing the lines for the different frequencies at certain points along the vertical distance scale, the coronal emission levels derived by Fainberg and Stone (1971) were used. These investigators developed a method, using a hectometric storm of type III bursts observed with the satellite RAE-1 over a half solar rotation, to derive a density scale in the range 10–40 $R_{\odot}$. In addition, the assumption was made that the observed type III's in the hectometric storm were all at the plasma frequency. The

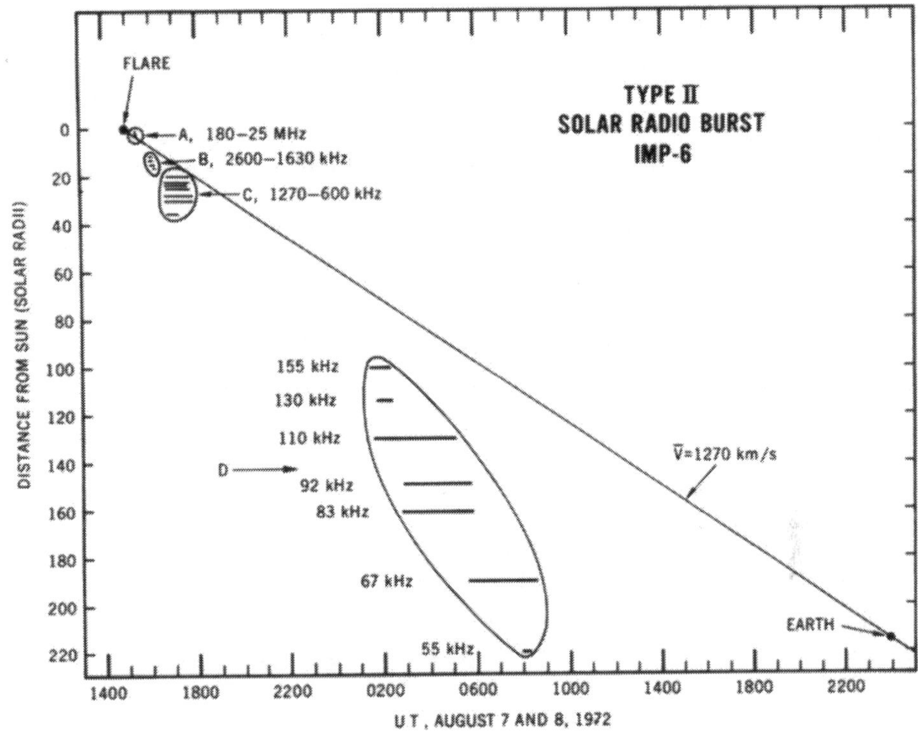

Fig. 2. [Courtesy of *Astrophys. J.*, Univ. of Chicago Press.]

lines having been arranged vertically, it is easy to see that the corresponding velocity out from the Sun is about twice the average velocity of 1270 km s^{-1}, and would result in the premature arrival of the shock front at the Earth.

For Figure 3, the same analysis was used except this time the assumption was made that type III bursts at hectometric and kilometric wavelengths are seen at twice the plasma frequency. In this case, the IMP-6 data fall in regions B′, C′, D′, and E′. Note in particular E′, the observation at 30 kHz, just before the sudden commencement of the geomagnetic storm. The measured position of the burst at this frequency was 20° east of the Sun. The 30-kHz data were omitted from Figure 2, where they would have fallen considerably below the bottom edge of the scale, at about 350 $R_\odot$. At this distance, the burst probably could not have been observed through the intervening plasma.

In Figure 3, the fit of the velocity line to the IMP-6 data is much better than in Figure 2. The agreement is remarkably good, considering the inhomogeneities in electron density along the path, the overall variation in density with time, the possible variations in velocity of the shock front, and the sporadic nature of the type II emission. This agreement, we feel, is strong evidence for the adopted density scale and strongly implies that hectometric and kilometric type III bursts are observed at twice the source plasma frequency.

We are in the process of studying more type II bursts, and expect the data to provide important contributions to the study of the interplanetary medium in two ways:

(1) As the type II bursts travel outward, the changes in intensity and variations in velocity should provide valuable information on the propagation of shock waves through the medium. With the remote sensing techniques of radio astronomy, the investigation of shock waves is not limited to those traveling only in the ecliptic plane near 1 AU.

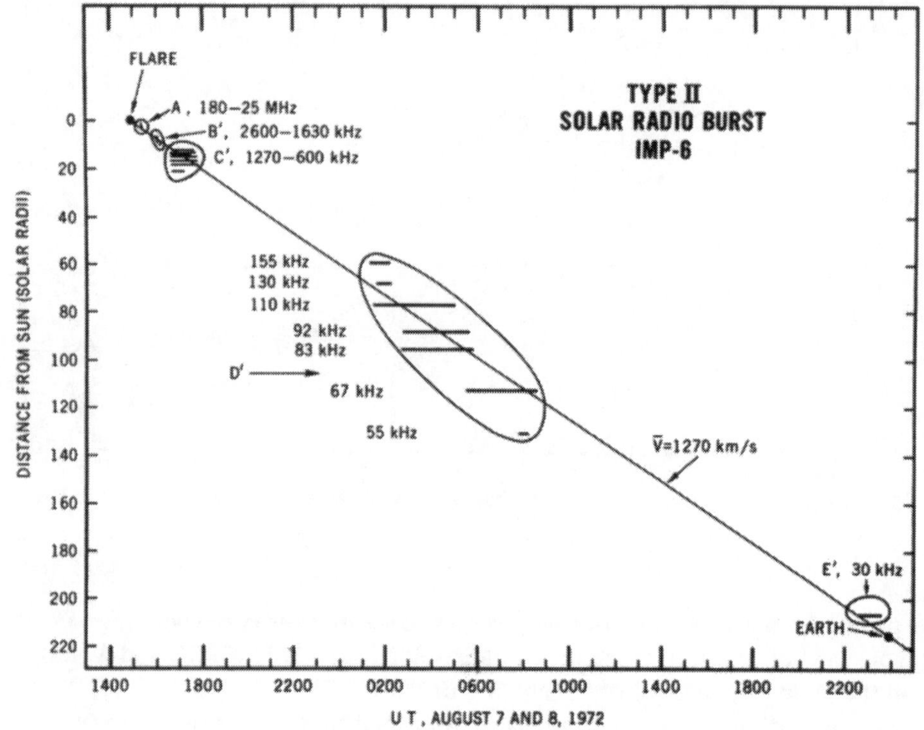

Fig. 3. [Courtesy of *Astrophys. J.*, Univ. of Chicago Press.]

(2) With type II bursts having well-defined first and second harmonic components as probes, the electron-density scale of the interplanetary medium can be determined unambiguously from regions close to the Sun out to regions close to 1 AU.

Reference

Fainberg, J. and Stone, R. G.: 1971, *Solar Phys.* **17**, 392.

DISCUSSION

Maxwell: Does the Earth noise arise in the magnetosphere, and is it wiped out when the shock reaches the Earth?

Malitson: Yes, it is magnetosphere noise but it is not wiped out; indeed, if anything, it seems to be enhanced.

Fainberg: This noise is always variable and we cannot clearly tell whether it is stronger or weaker after the shock.

Zirin: Is there any comparison between the trajectory for this shock and those of August 2 and 4?

Malitson: The corresponding records are being searched, but isolating a type II from other types of strong emission is difficult.

Smith: I see that type II burst observations combined with type III observations can provide a density model. What does a type II look like at low frequencies? Is it better or worse than a type III burst for making a density model?

Malitson: We only see the centroid of the burst so we can tell nothing about its shape or size.

Kai: I wonder why type II bursts are so rarely observed in the interplanetary medium, since the close relationship between type II bursts and the Earth's magnetic storms is well established. Is this due simply to technical problems?

Malitson: It may be that special physical conditions are necessary for a shock front to produce radio emission – e.g., the crossing of the front by a type III. Also, type II's are very sporadic and broken and it is hard to find even the strongest bursts in the records.

Dryer: In partial answer to Dr Kai's comment about other measurements of interplanetary type II's and correlation with geomagnetic storms, I might add that the first such measurement (if I recall correctly) was made by Slysh. There was a geomagnetic storm correlated with his observations on the Soviet space probe.

Malitson: That measurement was made at only two frequencies and there is some doubt that it really was a type II.

Dryer: You may be right; I do not recall if there were two or more.

Sakurai: When we compare SSC's with interplanetary shock waves as observed by satellites, we find that some SSC's are produced by interplanetary shocks, whereas the others are produced by discontinuous, or sharp, variations of interplanetary magnetic fields associated with enhanced solar winds. Therefore, there is no one-to-one correspondence between an SSC and an interplanetary shock wave. This means that there is not much chance for type II bursts to be excited in interplanetary space.

McLean: Was there any position difference between the fundamental and second harmonic component?

Malitson: There was none found within our accuracy of 1° to 2°.

Smerd: Ms Malitson noted that the interplanetary electron densities derived from interplanetary observations of type III and type II bursts could be made to agree on the assumption that the type III radiation was at the second harmonic. The same agreement can be achieved if we assume that the type III radiation is at the fundamental plasma frequency, f_p, while the fundamental of the type II burst was radiated at the enhanced plasma frequency, $f_p(\text{shock}) \sim 2f_p$, behind a strong type II shock.

EAST-WEST ASYMMETRY OF MAGNETIC BOTTLE EXPANSION AND ITS RELATION TO SHOCK WAVES PROPAGATING IN THE SOLAR ATMOSPHERE

KUNITOMO SAKURAI*

Radio Astronomy Branch, Laboratory for Extraterrestrial Physics, NASA, Goddard Space Flight Center, Greenbelt, Md., U.S.A.

Abstract (*Solar Phys.*). It is known that the sources of metric type IV radio bursts generally move outwards with speed of a few to several hundred km s^{-1} in the solar atmosphere and envelope (e.g., Kundu, 1965; Smerd and Dulk, 1971; Wild and Smerd, 1972). These sources consist of magnetic lines of force being stretched out from the flare regions and relativistic electrons being trapped by these field lines. They are often observed as expanding magnetic bottles (e.g., Smerd and Dulk, 1971). As a result of this expanding motion, the frequency range of type IV burst extends with time to lower frequencies, because of the decrease of both magnetic field intensity and energy of radiating electrons (Dulk, 1970; Sakurai, 1973a). By considering the observed expansion rate of emission frequency range, it seems possible to estimate the pattern of expansion of these radio sources: in doing so, we have analyzed the onset time differences between microwave and metric emissions of type IV bursts by referring to the longitudinal positions of parent solar flares on the solar disk, since the dependence of these differences on parent flare positions in the solar longitude seems to give a clue to estimate a general pattern of the expansion of magnetic bottles. The result thus analyzed is shown in Figure 1. Furthermore, peak flux intensities at metric frequencies for these type IV bursts have been also analyzed as a function of the longitude positions of parent flares (Figure 2). The results shown in Figures 1 and 2 are explained by assuming that, while moving outwards, the magnetic bottle tends to expand a few 10 deg east of the meridian plane which crosses the flare region (Sakurai, 1973a).

The expansion speed of these type IV sources is always much slower than that of type II radio sources, except for a few cases (see, Sakurai and Chao, 1974; Kai, 1973). Recently, Dulk *et al.* (1971) have shown that shock waves responsible for type II radio bursts usually propagate along the magnetic field lines above and near the flare regions. With the results shown in Figures 1 and 2, their analysis suggests that these shock waves tend to expand independently of the magnetic bottles in the solar atmosphere (e.g., Sakurai, 1973a).

The slow speed expansion of type IV radio sources strongly suggests that flare-ejecta observed behind interplanetary shock waves are completely different from magnetic bottles identified initially as the sources of metric type IV bursts (Sakurai

* NASA Associate with University of Maryland, Astronomy Program.

Gordon Newkirk, Jr. (ed.), Coronal Disturbances, 355–359. All Rights Reserved.

and Chao, 1974). However, these interplanetary shock waves seem to have been formed as a result of the latter expansion of shock waves responsible for type II bursts (e.g., Burlaga, 1974; Sakurai, 1973b).

As has been mentioned above, the magnetic bottles as metric type IV radio sources

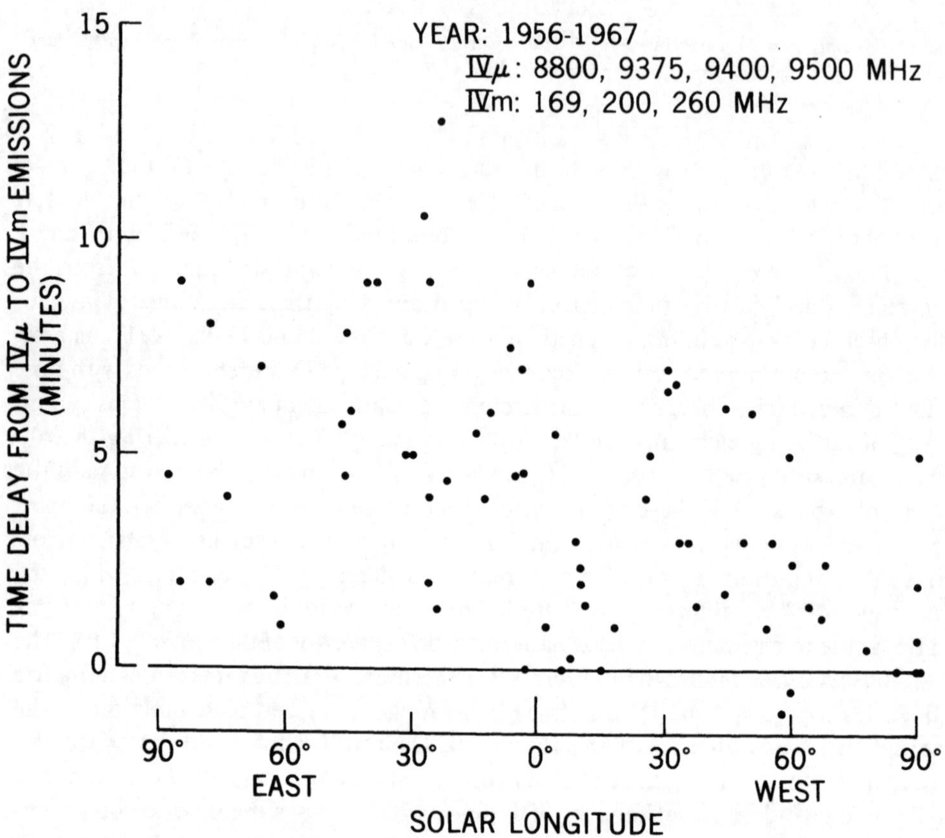

Fig. 1. Relation of the onset time differences between microwave and metric type IV radio emissions to the longitudinal positions of parent flares.

usually expand with slow speed into the solar envelope. Furthermore, they seem to push away the atmospheric gases overlaid on themselves into outer space. If this is the case, it is likely that these bottles are responsible for the ejection of the helium-enriched shells, which are observed at the Earth's orbit. These shells have been detected by Hirshberg *et al.* (1972) based on the analysis of satellite plasma data at the Earth's orbit.

The thickness of these shells is dependent on the longitudinal positions of parent flares (Figure 3). The origin of this property seems to be strongly related to the ex-

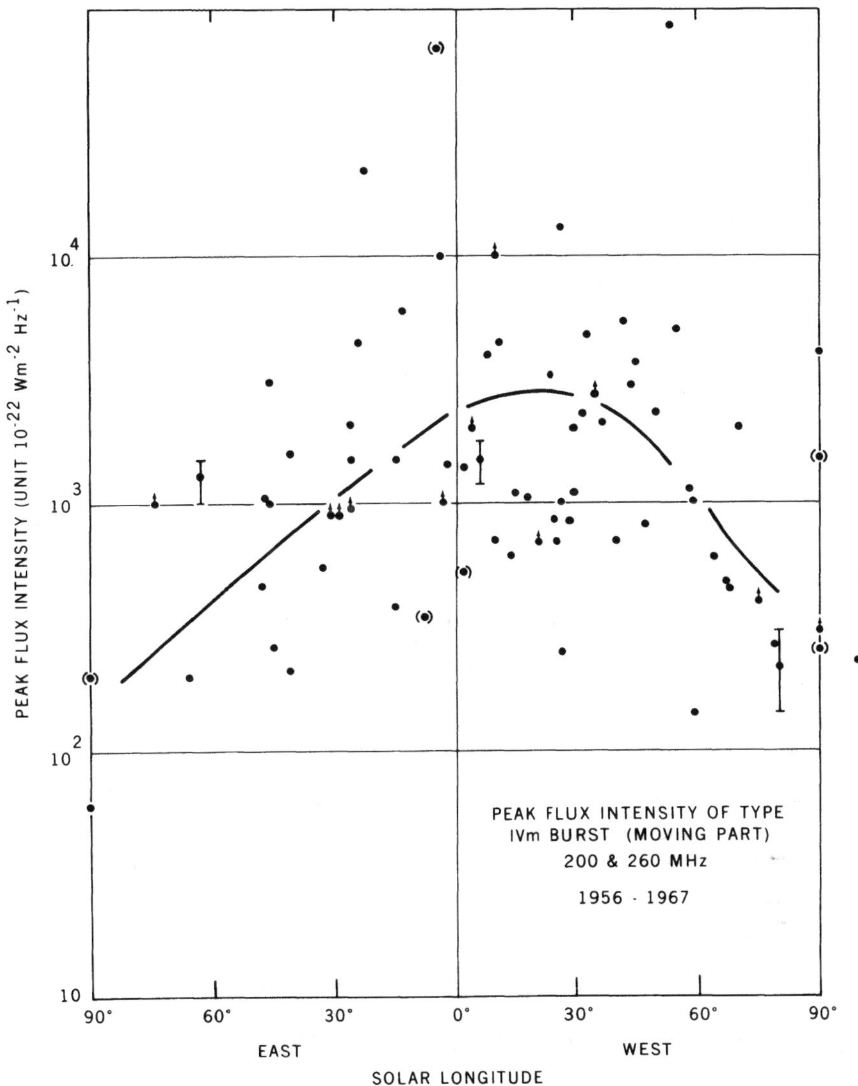

Fig. 2. The peak flux intensity of type IV metric emissions with respect to the longitudinal position of parent flares.

panding pattern of the magnetic bottles near the Sun, because the helium-enriched region is formed in the upper chromosphere and the lower corona, as shown by Brandt (1966) and Nakada (1969, 1970). The eastward expansion of these bottles seems to have produced the east-west assymmetry as seen on the helium-enriched shell. Non-spherical expansion of interplanetary shock waves seems to be also related to the field-aligned propagation of the shock waves near the Sun.

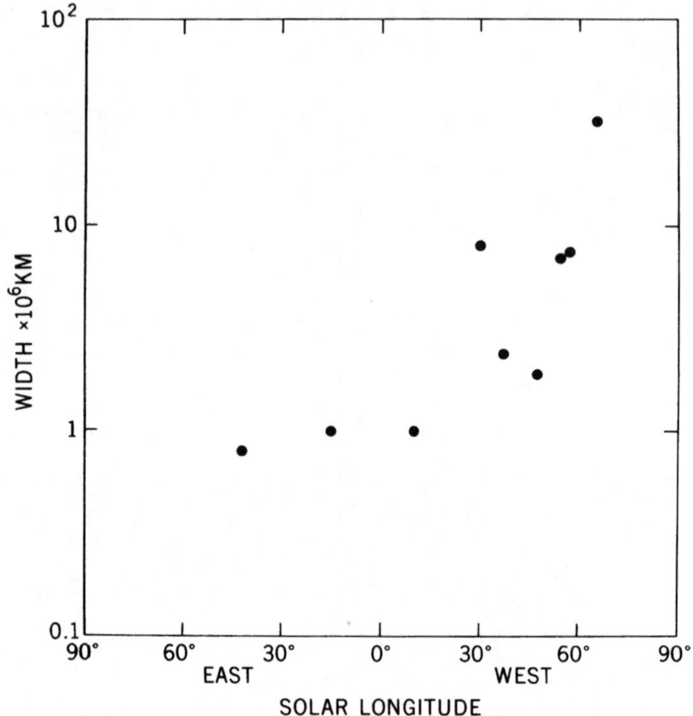

Fig. 3. Relation between the thickness of the helium-enriched shells observed at the Earth's orbit and the longitudinal positions of associated flares.

References

Brandt, J. C.: 1966, *Astrophys. J.* **143**, 205.

Burlaga, L. F.: 1974, NASA, GSFC Note X-692-72-395. *Proceedings of Boulder Conf. on Shock Waves from the Sun*, p. 123.

Dulk, G. A.: 1970, *Proc. ASA* **1**, 372.

Dulk, G. A., Altschuler, M. D., and Smerd, S. F.: 1971, *Astrophys. Letters* **8**, 235.

Hirshberg, J., Bame, S., and Robbins, D. E.: 1972, *Solar Phys.* **23**, 487.

Kai, K.: 1973, private communication.

Kundu, M. R.: 1965, *Solar Radio Astronomy*, Wiley, New York.

Nakada, M. P.: 1969, *Solar Phys.* **7**, 302.

Nakada, M. P.: 1970, *Solar Phys.* **14**, 457.

Sakurai, K.: 1973a, *Solar Phys.* **31**, 483.

Sakurai, K.: 1973b, *Physics of Solar Cosmic Rays*, Univ. of Tokyo Press, Tokyo.

Sakurai, K. and Chao, J. K.: 1974, *J. Geophys. Res.* **79**, 661.

Smerd, S. F. and Dulk, G. A.: 1971, in R. Howard (ed.), 'Solar Magnetic Fields', *IAU Symp.* **43**, 616.

Wild, J. P. and Smerd, S. F.: 1972, *Ann. Rev. Astron. Astrophys.* **10**, 159.

DISCUSSION

Newkirk: Your results critically depend upon identifying which flares are to be associated with which radio events. Of all the flares, which might be candidates for a particular event, how did you choose the ones used in your study?

Sakurai: The approach was basically statistical.

Dryer: Newkirk put his finger on the essential problem. The usual approach is to look in the 'yellow book' for an interesting flare and to pick a 'reasonable' one with large optical importance with, hopefully, type II's and/or IV's. I believe that this was done here.

Sakurai: Yes.

Pick: Caroubalos showed that the microwave energy was a good indicator of the interplanetary importance. This could be used to determine which flare.

Sakurai: Even so, it still does not lead to a clear decision.

Gotwols: I wonder if the first slide, which is a scatter plot of time delay vs longitude, really shows a statistically significant trend. It appeares to me that with a mere handful of points added at Western longitudes, no trend would remain in the data.

Sakurai: The E–W asymmetry is evident. No statistical test was applied.

SHOCK WAVES AND PLASMA EJECTION: CORPUSCULAR AND INTERPLANETARY EVIDENCE

A. J. HUNDHAUSEN

High Altitude Observatory, National Center for Atmospheric Research, Boulder, Colo., U.S.A.*

(Presented by G. Newkirk)

Abstract. The ejection of rapidly-moving solar material into interplanetary space in association with solar flares has been discussed since 1859, when geomagnetic disturbances and auroral displays followed shortly after the first observation of a flare by Carrington and Hodgson. Until the advent of *in situ* interplanetary observations in the early 1960's, such discussions were based upon the *indirect* information regarding interplanetary space that could be inferred from geomagnetic or cosmic ray data. The past decade of space exploration has provided a great deal of *direct* information regarding the interplanetary effects of solar flares and some quantitative implications regarding the nature of transient coronal disturbances.

This empirical information and related theoretical models have been reviewed by this author in several recent publications (Hundhausen, 1972a, b), to which the reader is referred for extended discussions. The most important implications of this work in the present context can be summarized as follows:

(1) Shock waves are observed to propagate past the orbit of the Earth, moving away from the Sun, after *some* solar flares. The typical shock wave sweeps past a stationary observer at ~ 500 km s^{-1}, and is thus moving at ~ 100 km s^{-1} relative to the ambient solar wind.

(2) Shock wave observations reported in the literature imply an average rate of occurrence of several shocks per 27-day solar rotation. This rate of occurrence does not change *drastically* over the solar cycle.

(3) The rate given above directly implies that most flares do not produce shock waves that have been detected near the orbit of the Earth. Although it remains far from clear why some flares produce such disturbances while other flares, though at times of greater optical importance, do not, there is some comfort in the high correlation of interplanetary shock waves with the occurrence of related (i.e., both) type II and type IV radio bursts, the probable signature of plasma ejection *and* shock propagation in the corona.

(4) Integration of the excess (relative to ambient conditions) mass and energy fluxes in interplanetary shock waves gives the following estimates of the average mass M and energy W (at one solar radius) added to the solar wind

$$M \approx 5 \times 10^{16} \text{ gm}$$
$$W \approx 2 \times 10^{32} \text{ erg.}$$

* The National Center for Atmospheric Research is sponsored by the National Science Foundation.

Gordon Newkirk, Jr. (ed.), Coronal Disturbances, 361–363.

The former is an appreciable fraction of a coronal mass, while the latter is as large or larger than any other known energy release in a large solar flare.

(5) The implication of mass ejection in association with interplanetary shock waves is consistent with the observation of post-shock solar wind with highly abnormal chemical (high helium content) and thermodynamic (low temperature) characteristics.

(6) Comparison of the observed post-shock variations in solar wind density and speed with theoretical models suggests that the mass and energy release occurs on a time scale of many hours. The lack of observational evidence for large mass motions in the corona associated with most solar flares, coupled with these long time scales, suggests to this author that the release of mass and energy into the solar wind is more extensive than that in the primary flare process. Perhaps the flare 'opens' a previously closed magnetic region, leading to emission of solar wind from a previously untapped coronal source region.

References

Hundhausen, A. J.: 1972a, in C. P. Sonnet, P. J. Coleman, and J. M. Wilcox (eds.), *Solar Wind*, NASA SP-308, Washington, pp. 393–421.

Hundhausen, A. J.: 1972b, *Coronal Expansion and Solar Wind*, Springer-Verlag, New York.

DISCUSSION

Brown: Some comments on the mass and energy figures you mention – first, though there are some uncertainties I will discuss later, it is not true that the 10^{32} erg in the blast greatly exceeds the energetic particle energy. Recent X-ray data indicate an electron energy entirely comparable with the blast energy.

If this is so and one assumes that the electrons are responsible for heating the thermal flare, then one correctly predicts a mass of around 10^{16} gm in the high temperature flare, possible escaping. Also the fact that the 10^{32} erg in the blast compares closely with the energy in the thermal flare is then not at all surprising on simple equipartition arguments.

Wild: The duration of the shock in interplanetary space suggest something similar to that of the radio storm continuum. Does anyone know if there is any correlation?

Newkirk: I am not sure anyone has looked into this.

Athay: Is it clear that what is attributed to the duration of the input is really that and not simply a dispersion in the paths followed by particles or some similar effect? In other words, is the conclusion that the input lasts for several hours the only possible one or could one construct reasonable alternatives?

Newkirk: But such dispersions would not modify the energy and mass estimates.

Sturrock: The blast wave may be the bow shock of a narrow plasmoid. If so, a 5 to 10 h duration for the piston is not required and the required mass would also be smaller.

Newkirk: If so, can you explain the observed broad angle covered by the shock? Also, unless the plasmoid were propelled to IAU, the narrow shock present close to the Sun would soon become nearly spherical.

Sturrock: This would depend upon the nature of the plasmoid. A great variety of shapes are observed in non-solar radio sources.

Dryer: The text makes the statement that additional theoretical consideration including the field are not needed at this time. I consider the incorporation of the magnetic field as an essential feature necessary to consider the ionized aspects of the interplanetary disturbances (including field compressions at the piston and anomalous transport properties – such as resistivity – indicated by theories presently in existence (such as Lee and Chen's) and presently being extended.

Newkirk: Do you refer to the effect of the field on the shock-disturbed flow or the effect of the field on the shock itself?

Dryer: The effect of the field on the disturbed flow.

Stewart: You did not discuss energy loss in the shocks. If white light observations near the Sun give sufficient energy, there is no energy loss.

Newkirk: The comparison on an event by event basis is difficult since coronal events at the limb seldom reach the Earth. The calculations already include the gravitational energy loss.

Schmidt: The determination of the duration of the initiating disturbance from a fit of a model to the observed flow-profiles may not be so bad. In a diagram published by Hundhausen *et al.* indicating total energy vs total mass content of the disturbance, the observed shocks fall on a straight line of almost constant energy per mass. In this diagram the driven shock with increasing mass flow behind the shock fall nicely into the upper end of the diagram, i.e. they have larger mass – and energy content. So they should be caused by an initiating disturbance of longer duration. The shock waves with a blast wave profile fall into the lower part of the diagram and the intermediate profiles into the middle.

Zirin: Regarding the energy in different forms in a flare, Ramaty found for protons above 5 MeV about 2×10^{30} erg for the 7 August 1972 event. Tanaka and Zirin found 2×10^{30} erg in Hα for the same flare.

Lin: The estimate (of 5×10^{30} erg) by Ramaty of energy in >5 MeV protons in the August 4 flare is a gross underestimate of the total energetic particle energy since (1) the spectrum extends well below 1 MeV, and (2) the low energy protons lose some, perhaps most, of their energy in adiabatic deceleration in propagation outward in the interplanetary medium. These two effects can easily increase the energy content in energetic particles by orders of magnitude.

Brandt: The importance of the non-spherical geometry particularly near the sun can be illustrated with a simple example. Increasing the coronal temperature in the spherically symmetric case does not always produce an enhanced solar wind. If the coronal temperature exceeds about 4×10^6 K, the gravitational nozzle vanishes, and no supersonic expansion can occur. For such high temperature regimes, the magnetic field may be needed to produce the nozzle which is necessary for a supersonic solar wind.

Brown (to Zirin): You quote 5×10^{30} erg in a flare but the point is that most of the electromagnetic radiation is in the EUV range and most of the thermal energy in the soft X-ray source. I don't know if data are available in these bands for the event you mention.

THEORY OF SHOCK WAVES AND PLASMA WAVE-EMISSION

NICHOLAS A. KRALL

Laboratory for Applied Plasma Studies
Science Applications, Inc., La Jolla, Calif. 92037, U.S.A.

Abstract. The theory of shock waves in a hot plasma is reviewed, and the types of shock waves which can exist are summarized. The detailed application of shock wave theory to the explanation of laboratory experiments is described; a comparison is made between the results of these experiments and the observations made of type II radio bursts from the Sun.

1. Introduction

The papers presented at this meeting have made it clear that there is ample evidence for the existence of shock waves in flare-produced plasma, and suggests that those are responsible for a portion of the phenomena observed in type II solar burst activity.

The model usually cited (Wild and Smerd, 1972) is the following: a magnetic wave, moving across the coronal magnetic field, reaches a region where its speed exceeds the speed of magnetic sound. At that point it steepens, and forms a magnetic shock wave. This wave produces a density compression and an electron current, and, most important, is unstable to the production of electron plasma oscillations. These plasma oscillations (frequency ω_p) scatter off ion density fluctuations to produce the observed electromagnetic radiation at frequency ω_p, and scatter off each other to produce the radiations observed at frequency $2\omega_p$. Evidence for this model includes the fact that sources of type II bursts move at Alfvén speeds, and that the emission is too strong to be thermal.

In this review I will attempt to do three things. First, review the theory of how (and what kind of) shock waves can exist in a hot plasma, second show what sorts of detailed calculations are now possible in shock wave theory (and what phenomena they have uncovered) and third describe controlled laboratory experiments which coincide well enough with the theory to convince us that we are still on the right track; in large amplitude wave theory the effects are so strongly nonlinear that only such comparison allows us to proceed with much hope that we haven't forgotten some basic effects or made some hopelessly bad approximations. In the course of this talk I will try to convince you that laboratory experiments are seeing many of the same effects that solar physicists see, and that our ability to calculate these effects in detail for laboratory applications gives promise of similar success in coronal shock theory.

2. Classical Theory of Shock Waves in Collisionless Plasma

There is a well established theory of macroscopically stable shock waves in plasma (Tidman and Krall, 1971a), with mechanisms and classification schemes which are

Gordon Newkirk, Jr. (ed.), Coronal Disturbances, 365–375. All Rights Reserved.

widely accepted. In this section I will briefly review these ideas, to set up a framework in which more recent theoretical and experimental results can be viewed.

We use the word shock wave in a very loose sense, meaning simply a situation in which the plasma changes state during the passage of a transition region. The change of state may be smooth (laminar) or turbulent. The change of state across the shock front in most cases of solar interest does not take place as a result of collisions, but rather because of collective effects. That is, plasma instabilities take energy from the ordered streams and fields of the shock and dump it into more nearly random energy. I resist saying that the energy goes into temperature. In fact, the energy often goes into motion on a shorter length scale and higher frequency scale than that of the shock itself. On the shock scale this may in fact look like temperature, but on a microscopic scale there has been no increase in entropy; rather, a large number of high frequency plasma waves have been generated, with the particles sloshing about in the troughs of these waves. Thus the conservation laws (Tidman and Krall, 1971b), (Rankine-Hugoniot relations) which connect the unshocked plasma state with the shock state, must include the electrical energy of the plasma waves in the shocked plasma, and must take account of the fact that the ions and electrons respond differently to the waves.

The greatest progress in classifying and studying shock waves has been to study under what conditions laminar solutions are possible (Sagdeev, 1966). In these cases, the plasma properties can be divided into two camps. First are the slowly and smoothly varying (i.e. laminar) quantities, such as the density $N(x, t)$, the magnetic field $B(x, t)$, the electric field $E(x, t)$ etc. Second are the more rapidly fluctuating quantities $\delta N(x, t)$, δB, δE, which are treated as perturbations on the laminar parameters. Thus,

$$N_{\text{Total}} = N(x, t) + \delta N(x, t)$$

and so on.

The laminar quantities are governed by the fluid equations of continuity and momentum transfer and by Maxwell's equations. It is generally necessary to treat the electrons and ions as separate fluids.

Laminar shock waves come out of the theory as follows: One possible motion of an electron-ion fluid is a low amplitude wave. Of the various wave motions of a plasma, many (for example: magnetosonic waves, ion acoustic waves, whistler waves) are sound-like over some frequency range, as shown in Figure 1. If the wave is of finite amplitude, nonlinear terms $(\mathbf{V} \cdot \nabla \mathbf{V})$ in the fluid equations will couple a wave (ω, k) to a higher harmonic $(2\omega, 2k)$. In the soundlike portions of the wave spectrum the harmonics travel at the same speed as the original wave. Since they are of shorter wavelength, they represent wave steepening. A shock forms when the steepening is balanced by some other mechanism, leading to a transition with a fixed transition width. One way to balance nonlinear steepening is by dispersion. When a harmonic is generated at a sufficiently short wavelength (high k) it no longer propagates with the same speed as the rest of the pulse; this phenomenon is called dispersion. It comes about because in general soundlike waves $\omega = kV_s$ only occur in some appropriate limit such as a plasma with zero mass electrons, or a plasma in which $N_e = N_i$ everywhere. Short wavelength or high frequency harmonics sense the actual electron mass, or the actual charge separation, and fall behind the longer wavelength lower frequency modes (in some cases they speed ahead of the main pulse, for a wave whose dispersion curve is of the type b in Figure 1). Dispersion will not give a shock wave by itself, but

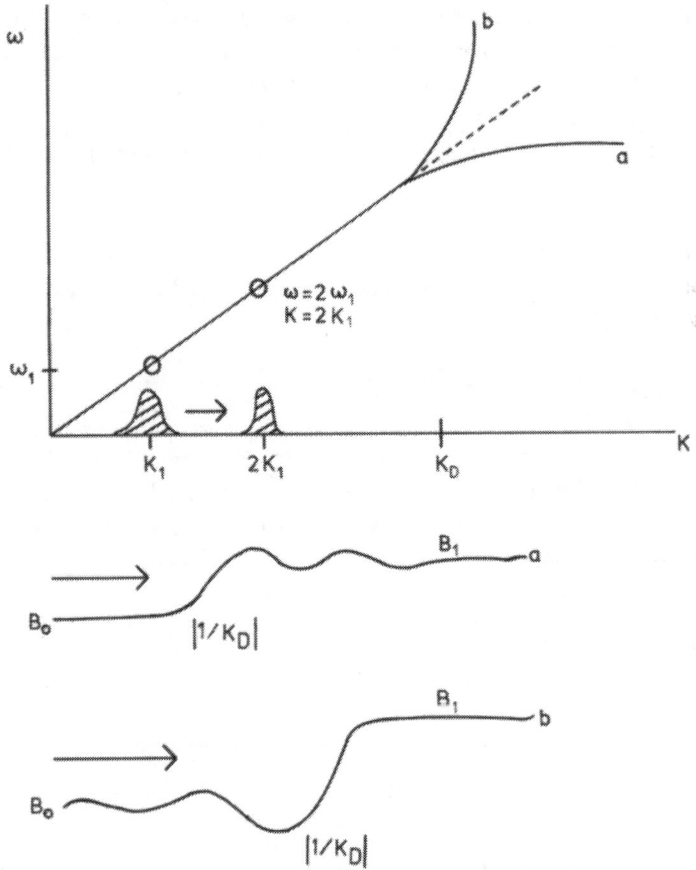

Fig. 1. Typical dispersion curves for sound-like waves in plasma.

will limit the steepening and form a large amplitude fixed profile wave, often called a soliton, or solitary wave. To convert this into a shock requires a dissipation mechanism, such as Landau damping or a plasma instability driven by currents in the soliton. Such a shock wave will have a thickness $L_s \simeq k_D^{-1}$ (see Figure 1). If the instabilities are strong enough, they can themselves provide enough dissipation to limit the steepening, by acting as a drain on the currents and fields which would build up in a very steep shock. In this case the dispersive width would never be reached, $L > 1/k_D$. Experimentally, this has proved to be the rule rather than the exception.

A third possibility is that the steepening is so rapid that it cannot be limited by either dispersion or dissipation. In this case the wave 'breaks', in much the same way that a water wave breaks (in fact, the whole shock picture described above has an exact mathematical analogy to a water wave approaching a beach). The subsequent fate of the wave cannot be calculated, and depends on the coupling of the spilling wave to the background medium. Because the steepening rate increases with wave amplitude and Mach number $\mathcal{M}$ = (shock speed)/(linear sound wave speed), the breaking point defines a critical Mach number $\mathcal{M}_c$ for each type of shock wave. In practice, $\mathcal{M}_c$ can only be calculated when dissipation is neglected, or when some very specific and simplified model is adopted to describe the dissipation.

With that preamble I will describe laminar shock waves most likely to be involved in burst phenomena, and their stability, and give a general diagram of possible laminar shocks.

2.1. Magnetosonic shock waves

These waves propagate across a magnetic field, $B = B_z(x) + B_{z0}$, and have a linear dispersion relation given by

$$\omega = \frac{kV_A}{(1 + k^2c^2/\omega_p^2)^{1/2}},$$

where ω and k are the frequency and wave number, $V_A = B/(4\pi NM)^{1/2}$ is the Alfvén speed, $\omega_p = (4\pi Ne^2/m)^{1/2}$ is the plasma frequency, c is the speed of light, N the plasma density and (m, M) the (electron, ions) mass. Thus their dispersive thickness is

$$L_s \simeq (c/\omega_p)/(2\sqrt{\mathcal{M} - 1}),$$

their Mach number is (with V the shock speed)

$$\mathcal{M} = V/V_A.$$

It can be shown that the shock amplitude and compression are

$$\frac{B}{B_0} = 2(\mathcal{M} - 1)$$

$$\frac{\Delta N}{N} = \frac{\Delta B}{B_0}\left[1 - \frac{\Delta B}{2B}\right]^{-1}.$$

The shock breaks at $\Delta B = 2B$. Note that $\Delta N/N \to \infty$ at that point (B is so strong that all the plasma is about to be reflected forward, ahead of the shock). Adding a resistivity ν to the fluid equation may allow a higher amplitude shock, depending on the model used for ν. Constant resistivity, for example, doesn't help.

Because the front of this shock contains a current carried primarily by electrons,

$$\mathbf{J} = -Ne\,\mathbf{V}_e = \frac{c}{4\pi}\,\nabla \times \mathbf{B}$$

it can be unstable to a variety of plasma oscillations if it is steep enough. In particular, the conditions to drive a strong two-stream instability is that

$$\mathcal{M} > 1 + \left(\frac{8\pi N T_e}{B^2}\right)^{1/3}.$$

This shows that low $\beta(=8\pi NT/B^2)$ shocks are easily unstable to the emission of plasma waves (Krall and Trivelpiece, 1973).

2.2. Shock waves parallel to B_0

Because existing models of shock wave produced type II bursts often assume shock waves propagating across B_0, as described above, I choose as a second wave example the whistler soliton which propagates parallel to the ambient magnetic field. This mode could, for example, propagate parallel to a neutral sheet, or along a magnetic streamer.

This wave propagates in the direction of B_0, has an electric field vector in the same direction, while the shock magnetic fields $B_\perp$ are perpendicular to B_0. The characteristic width of the shock is determined by electron-inertia produced dispersion,

$$L_s \approx \frac{cB_0}{\omega_{pi} B_\perp},$$

where ω_{pi} is the ion plasma frequency. The range of amplitudes and speeds over which this soliton propagates is

$$0 < B_\perp < B_0 (M/m)^{1/2}$$
$$V_A (M/4m)^{1/2} < V < V_A (M/2m)^{1/2}.$$

A further restriction on the amplitude of the shock comes from a quasi-neutrality assumption used to derive it, requiring that

$$B_\perp < B_0 (M/m)^{1/2} \left(\frac{m}{M} \frac{\pi N mc^2}{B_0^2/8\pi}\right)^{1/4}.$$

The shock electric field E induces electron currents in the shock front, which can drive electron plasma waves unstable, just as for the magnetosonic shock wave. The shock strength needed to drive this instability is

$$\frac{B_\perp}{B_0} > \left(\frac{4\pi N T_e}{B_0^2}\right)^{1/4} \left(\frac{M}{m}\right)^{1/4}.$$

For a low β region, or a region of relatively small B_0, this is a condition that can realistically be met, and gives a mechanism for motion of type II bursts along the magnetic lines.

The two examples given here are merely indicative of the wide variety of possible laminar shock structures. Many more are summarized and tabulated in Wild and Smerd (1972), and I refer the audience to that monograph for further details.

3. Laboratory Results Related to Flare-Produced Shock Waves

As I mentioned several times already, type II burst models are usually based on shock waves traveling across a magnetic field. There is a class of laboratory experiments which uniquely produce large amplitude magnetic waves traveling across B_0, and it is useful to see to what extent they produce the same phenomena found in the corona, and to what extent they can be used to sort out various burst models. This experiment is called a θ-pinch, because of its geometry (Hintz, 1970). A cylinder of plasma confined by a magnetic field B_z is subjected to a pulsed field δB_z at its boundary (this field is produced by a current carrying coil wrapped around the cylinder. The current is in the θ direction, hence the name of the experiment). The pulsed field creates a magnetic disturbance which propagates across the magnetic field, at speeds similar to the Alfvén speed. A magnetic shock often forms. The parameters of this experiment which make it relevant to the corona are

$$\beta = \frac{NT}{B^2/8\pi} < 1$$

$$4\pi N m c^2 > B^2$$

$$a_i > L_s > a_e > \lambda_D$$

$$V_S \simeq B_0/(4\pi N M)^{1/2},$$

where λ_D is the Debye length and (a_e, a_i) are the (electron, ion) gyroradii. This is the same ordering of the relevant parameters found in the corona.

Figure 2 shows a plot of the magnetic profile at one instant of time in the experiment. The case shown had δB antiparallel to B_0; both parallel and antiparallel cases have been thoroughly studied. The figure lists some of the plasma processes which we believe are going on during the magnetic field implosion. These are clearly the same sort of features mentioned in burst models.

Some of the results of the most recent θ-pinch experiments (Davis *et al.*, 1970; Chin-Fatt, private communication; Chin-Fatt and Griem, 1970) are the following:

(1) Emission of radiation of ω_p and $2\omega_p$
(2) Proton acceleration to $V \sim 2V_{\text{shock}}$
(3) Plasma Wave Generation
(4) Electron Heating to 10^7–10^8 °C
(5) Ion Wave Generation

In particular, the characteristics of the ω_p, $2\omega_p$ emission were (Chin-Fatt, private communication; Chin-Fatt and Griem, 1970).

(1) Emission was consistent with a bi-Maxwellian electron distribution, with the hot component 100 times as hot as the cold component

(2) Emission peaked late in time, and presumably came from a region behind the shock

(3) The emission level was about 10^3 above thermal

(4) The strengths of the two components were of the same order of magnitude

(5) The presence of ion waves was also inferred

Clearly these, particulatly the ω_p and $2\omega_p$ emission, are some of the main features of

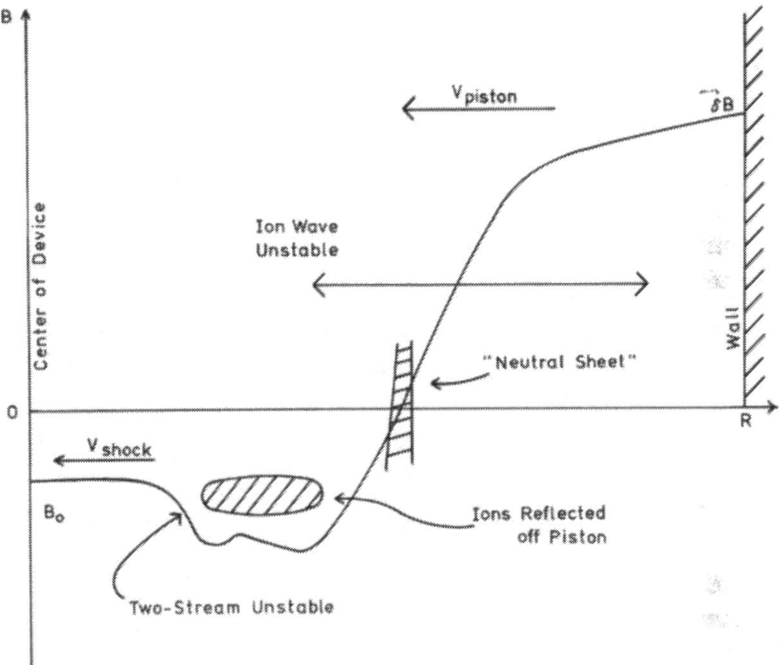

Fig. 2. Magnetic profile during magnetic implosion in a θ-pinch experiment.

the radio burst observations, and include as well direct measurement of other properties which have so far been merely the subject of speculation.

For example, the long time duration and late peaking of both the fundamental and the harmonic radiation argues against models (Saitsev, 1969) in which a coherent set of solitary waves in adjoining plasma layers give forward and backward plasma waves which scatter to give $2\omega_p$. A more plausible idea is that the plasma wave generated by instability in the shock front have isotropized by various diffusion processes in k-space (this includes the idea (Tidman and Krall, 1971a; Papadopoulos, 1969) that the particle distribution behind the shock interacts with the turbulence to develop a non-Maxwellian number of energetic particles leading to a superthermal level of plasma wave emission).

4. Recent Theoretical Advances

The ability to explain theoretically the results of θ-pinch experiments cited above offers hope for similar success in explaining coronal phenomena. The things which are well understood are the laminar shock wave structure as discussed in Section 2, the conditions under which linear instabilities can be excited, the initial rates at which instabilities take energy from the electron current, and the equipartition of that energy among electrons, ions and plasma waves. These ideas and the details of their calculation are listed in a variety of plasma texts, e.g. Krall and Trivelpiece (1973), and in previous reviews of radio burst phenomena (Wild and Smerd, 1972; Smith, 1971). The most recent theoretical work on this problem has been to apply the known stability theory in a practical and selfconsistent way. It is not often fair to assume that a magnetic pulse will develop into a shock which has a time-stationary profile determined from laminar theory, and then use this profile to calculate instabilities and plasma wave emissions. Clearly the instabilities will act on the plasma reducing the current, heating the plasma, and altering the profile which caused the instability in the first place. One recent approach to this problem has been to use the two-fluid equations to describe a plasma in which a magnetic pulse is applied at a boundary. The two fluid equations include self-consistently the effect of a variety of instabilities, which appear in the form of time and space dependent turbulence-determined transport coefficients. These include, for example, the drag on the current due to plasma wave emission and the effective heating of electrons by the plasma waves. 'Self consistent' means that, for example, if the plasma conditions say that the plasma is unstable, the resistivity begins to grow at the rate given by linear stability theory; if the instability heats the plasma and quenches itself, the resistivity returns to its stable-plasma value. In this framework, the fluid equation for electron momentum transport is

$$N\left(\frac{\partial}{\partial t}+\mathbf{V}\cdot\nabla\right)\mathbf{V}=\frac{-Ne}{m}\left(\mathbf{E}+\frac{\mathbf{V}\times\mathbf{B}}{c}\right)-\frac{\nabla NT}{m}-\frac{e}{m}\langle\delta N\delta\mathbf{E}\rangle$$

$$\delta N=N_0\, e^{i\omega t}\, e^{\gamma t}$$

$$\delta\mathbf{E}=\mathbf{E}_0\, e^{i\omega t}\, e^{\gamma t}$$

$$\gamma=\gamma(x,\, t,\, \mathbf{V},\, T),$$

where $\langle\ \rangle$ means an average over the rapidly varying instability variables and the growth of the instability γ is determined by the conditions actually present locally in the plasma. The reason this calculation is possible is that computer techniques now allow numerical integration of the nonlinear two fluid equations.

This approach has had marked success in predicting the rate of penetration of magnetic pulses into a plasma, the magnetic profile (which may or may not resemble laminar solutions), the plasma heating, and the rate and regions of production of plasma waves. Figure 3 shows our calculation of the penetration time of a magnetic pulse into a field free plasma, as a function of plasma density, and plasma species. The lines are experimental results (Bengston *et al.*, 1972), the points are calculated by numerical integration of the turbulent-transport fluid equations. Similar success is achieved for scaling with the pulse field B_{max}. Note that since this is penetration into a *field free* plasma, the Mach number would be infinite. Laminar theory is helpless in this situation. In fact, a shock does not develop in the lifetime of the experiment, but neither does the pulse steepen and break. Turbulent resistivity and viscosity allow

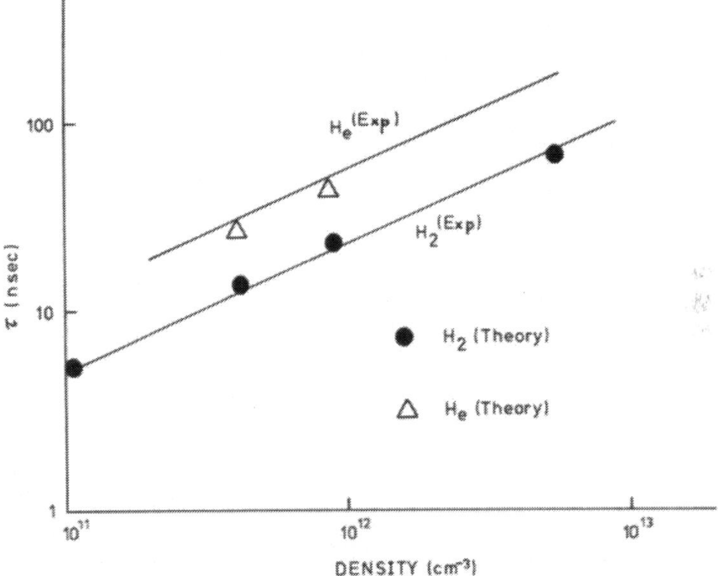

Fig. 3. Theory vs experiment for the penetration time τ of a magnetic pulse into a field free plasma.

the pulse to penetrate in a well organized fashion to the center of the machine. Figure 4 shows a typical magnetic profile during the run, and shows the spatial region in which plasma waves are excited. Note that this is a well defined region. Further into the pulse electron heating has quenched the instability. Radio emissions from these plasma waves would be more characteristic of the plasma density in the unshocked region than of the shocked plasma. This ability to perform calculations beyond the range of laminar theory and to extract real detail about heating and wave production is a hallmark of our model and offers hope that it can be extended to corona calculations.

A number of useful quantities have not been calculated, because they were not of laboratory interest, yet their calculation should be possible. For example, since the turbulent fields are known, the level of electromagnetic emission at ω_p could be calculated. Further, using the same fields, stochastic acceleration could also be calculated, either as a mechanism to accelerate particles out of the shock or as a mechanism to generate superthermal tails in the electron distribution f_e required by some models to generate more plasma waves.

Thus, it seems that many effects studied in the laboratory are directly relevant to

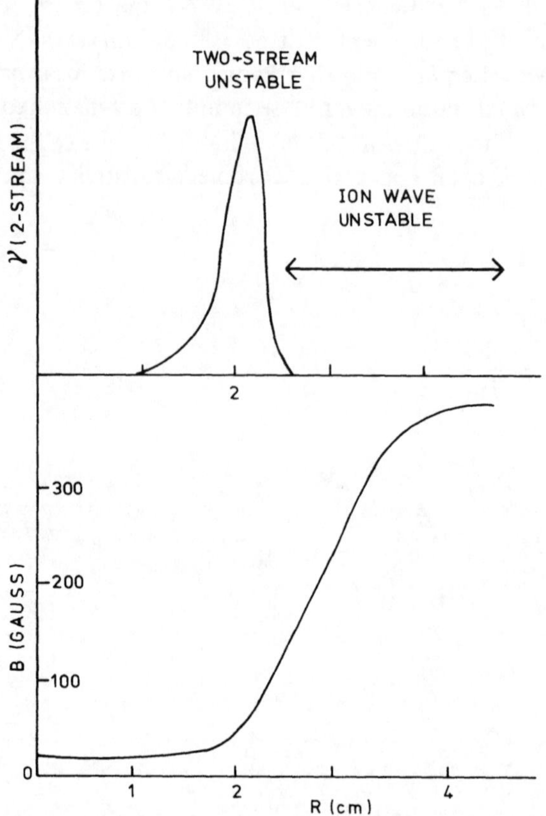

Fig. 4. Computer prediction of the two stream unstable region during penetration of a magnetic pulse.

explanation of solar corona phenomena, and collaboration between specialists of these two fields should be increasingly fruitful, particularly if the astrophysical data and laboratory theory continues to show such rapid improvement.

Acknowledgements

It is a pleasure to acknowledge an illuminating discussion with Dr D. A. Tidman on the facts and fictions of the solar burst problem. This work was performed while the author was a fellow of the John Simon Guggenheim Memorial Foundation.

References

Bengston, R. D., Marsh, S. J., Robson, A. F., and Kapetanakos, I. A.: 1972, *Phys. Rev. Letters* **29**, 1073.
Chin-Fatt, C. and Griem, H. R.: 1970, *Phys. Rev. Letters* **25**, 1644.
Davis, N., DeSilva, A. W., Dove, W., Griem, H. R., Krall, N. A., and Liewer, P. C.: 1971, in *Proc. 4th International Conference on Plasma Physics and Controlled Nuclear Fusion Research*, IAEA Vienna.
Hintz, E.: 1970, in H. Griem and R. Lovberg (eds.), *Methods of Experimental Physics* **9A**, 213.

Krall, N. A. and Trivelpiece, A. W.: 1973, *Principles of Plasma Physics*, McGraw-Hill, New York.
Liewer, P. C. and Krall, N. A.: 1974, *Phys. Fluids*, in press.
Papadopoulis, K.: 1969, *Phys. Fluids* **12**, 2185.
Sagdeev, R. Z.: 1966, in M. A. Leontovich (ed.), *Reviews of Plasma Physics*, Vol. 4, Consultants Bureau, New York.
Smith, D. F.: 1971, *Astrophys. J.* **170**, 559.
Tidman, D. A.: 1965, *Planetary Space Sci.* **13**, 781.
Tidman, D. A. and Krall, N. A.: 1971a, *Shock Waves in Collisionless Plasmas*, Wiley, New York, and references therein.
Tidman, D. A. and Krall, N. A.: 1971b, *Shock Waves in Collisionless Plasmas*, Wiley, New York, pp. 8–12.
Wild, J. P. and Smerd, S. F.: 1972, *Ann. Rev. Astron. Astrophys.* **10**, 159.
Zaitsev, V. V.: 1969, *Soviet Astron. AJ* **12**, 610.

DISCUSSION

Dryer: It has been precisely for the reasons discussed by Dr Krall that two years ago we initiated a detailed study of interplanetary structures which appeared to be associated with either flares or stream interactions. We wanted to define the shocks in terms of the parameter discussed by Dr Krall as part of the overall study of the structure itself (piston and any other discontinuities). We have started at the lowest level of sophistication, namely use of the fluid Rankine-Hugoniot equations, together with Maxwell's equation (i.e., the De-Hoffman-Teller level). Using least square fitting to suspected shocks, we have computed all parameters (shock, normal direction and velocity, etc.). The two slides show the magnetic data for a reverse shock (normal pointed back toward the sun) which moved first past Pioneer 9 – at 0.13 AU upstream of Earth – and then past OGO-5 (at Earth). The slides show the preservation if its identity; the fits to the data show that it is a supercritical, highly oblique, high-beta shock. Its signature is (in the magnetic field) characteristic of a perpendicular shock as indicated by collisionless shock theory.

Sturrock: One of the puzzles regarding type II and III bursts is that II's radiate at both fundamental and harmonic while III's radiate predominately at the harmonic. Shocks produce ion-acoustic waves, and thus we should except type II's to be dominant in the fundamental.

Krall: Ion-acoustic waves are delicate and not always seen even in the lab. They are an important source of heating in the lab situations.

Smith (to Sturrock): This is a reasonable possibility because we don't generate ion-acoustic waves in type III bursts.

Smith (to Krall): The reason we worry about shocks going along field lines is that we don't know how to generate plasma radiation from such shocks. Do you have any idea how to produce such radiation from a parallel ion-acoustic shock?

Krall: There are differences in velocity between electrons and ions in such a shock and thus there may be such a possibility, but it has not been analyzed.

Smith: What is the minimum Alfvén Mach number M_A for production of plasma radiation in the Maryland experiment you described?

Krall: $M_A = 2$–3, but this may not necessarily be a lower limit because of experimental conditions.

INTERPLANETARY SHOCK WAVES FROM McMATH REGION 11976 DURING ITS PASSAGE IN AUGUST 1972

M. DRYER

Space Environment Laboratory, National Oceanic and Atmospheric Administration, Boulder, Colo. 80302, U.S.A.

A. EVIATAR, A. FROHLICH, A. JACOBS, and J. H. JOSEPH

Tel-Aviv University, Ramat-Aviv, Israel

and

E. J. WEBER

Kitt Peak National Observatory, Tucson, Ariz. 85717, U.S.A.

Abstract (*J. Geophys. Res.*). The August 1972 events provided an excellent opportunity for synthesizing a variety of observations in a coordinated fashion for the purpose of improving flare-shock associations, and our understanding of interplanetary shock propagation and solar wind interaction with planets and comets. These observations included the usual sudden commencements of magnetic storms at Earth; preliminary shock data from Heos-2, Prognoz-1 and Prognoz-2 (at Earth) and the radially-aligned Pioneers 9 (0.77 AU) and 10 (2.2 AU) located about 45° east of the

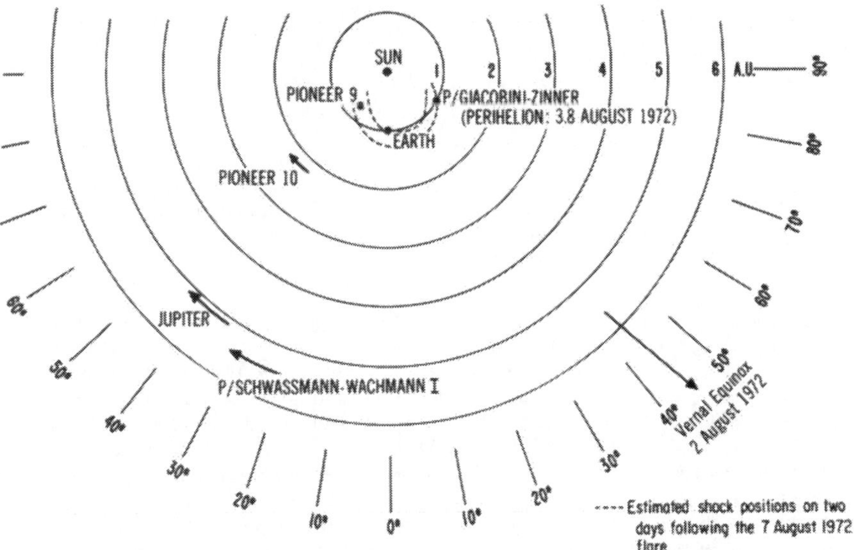

Fig. 1. Positions of Pioneers 9 and 10, Jupiter, and Comets *P*/Giacobini-Zinner and *P*/Schwassmann-Wachmann I relative to a fixed Sun–Earth axis during the epoch 1972 August 2–11.

Gordon Newkirk, Jr. (ed.), Coronal Disturbances, 377–381.

Sun–Earth axis; solar radio types II and IV (as reported in World Data Center A's UAG Report 28, 1973, and this Symposium); discrete radio source scintillations in the solar wind; and the more speculative ideas regarding the solar wind's interaction with planets and comets. In the last case, Jupiter and Comet *P*/Schwassmann-Wachmann I exhibited non-Io-associated radio emission and a brightness increase, respectively, as a possible result of shock waves from the flare and/or coronal ejection activity initiated on 1972 June 15. During the August events, Comet *P*/Giacobini-Zinner exhibited statistically-significant sudden brightness decrease following its perihelion on 1972 August 4 at 1 AU approximately 57° west of the Sun–Earth axis.

Figure 1 shows the various 'stations' in space. Shocks at Pioneers 9 and 10 were identified by detailed analysis of plasma and magnetic field data. SSCs at Earth and

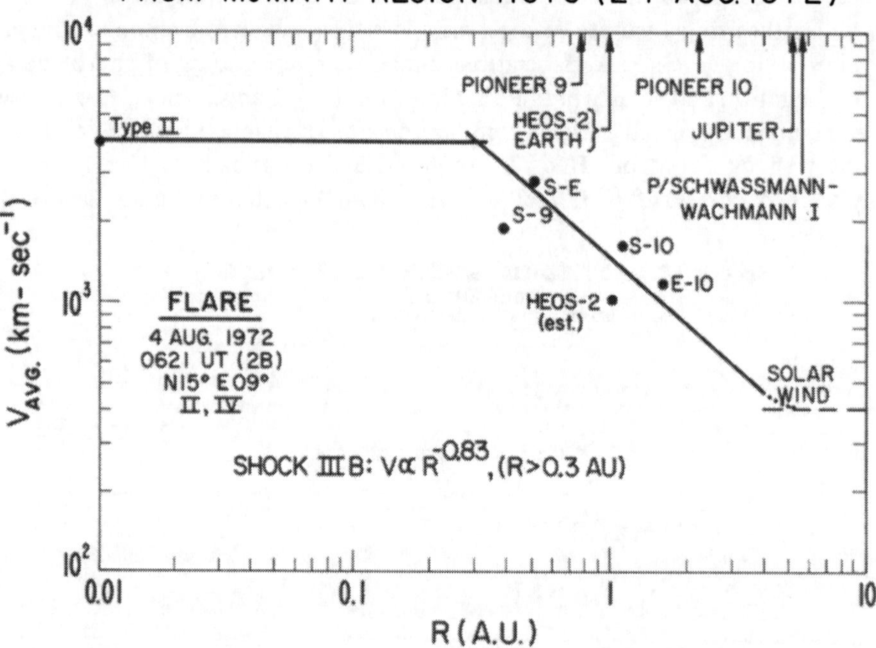

Fig. 2. Approximate shock trajectory from the flare on 1972 August 4 as inferred from the sudden commencement of a geomagnetic storm at Earth and *in situ* shock measurements observed at Pioneer 9 and Pioneer 10. The rapid deceleration suggests that (1) the energy input during the flare was impulsive thereby producing a 'blast' wave, or (2) the energy input produced a 'piston-driven' shock wave which decelerated rapidly after the cessation of the flare. The trajectory is 'approximate' in the sense that spherical symmetry is assumed. Thus, $V_{Avg}=R_i/(t_i-t_f)$, where R_i is the heliocentric radius to the measurement station, 'i'; t_i is the time of shock arrival at station, i; and t_f is the time of the flare noted on the figure. Possible non-spherical collimation of the flare's shock wave and/or large-scale deformation of the shock by ambient solar wind inhomogenities are neglected. The various points, such as S–E, S-9,... refer to the average shock velocities between the Sun and Earth, Sun and Pioneer 9, etc. In several cases, the points refer, for example, to the average velocities between Pioneer 9 and Pioneer 10. In this case, of course, $V_{Avg}=\Delta R/(t_{10}-t_9)$, etc.

a brightness decrease of *P*/Giacobini-Zinner on 1972 August 9 or 10 (followed by an equally precipitous intensity increase in the visible) were also used as timing devices. There was no response at either Jupiter or *P*/Schwassmann-Wachmann I for reasons suggested on the basis of shock deceleration noted below.

The first interplanetary shock, denoted as Shock I, was ejected following the August 2 flare (14° N, 35° E, optical 1B, types II and IV, 0316 UT maximum). Its deceleration was fairly fast with a power law index of −1.1 for its variation of average velocity as a function of heliocentric distance. We have speculated that it was deviated from a collision course with Jupiter and *P*/Schwassmann-Wachmann I by deceleration into an Alfvén wave.

Shock II originated in the flare at 1958 UT on August 2 (13° N, 28° E, optical 2B, types II and IV radio bursts). It, too, decelerated rapidly following a piston-driven effect from the Sun to ~0.3 AU.

The third interplanetary shock (Shock III) was produced by the August 4 flare (15° N, 09° E, optical 3B, types II and IV, 0621 UT maximum). Its strong deceleration has a power index of −0.8 and is shown in Figure 2. A high initial velocity is implied

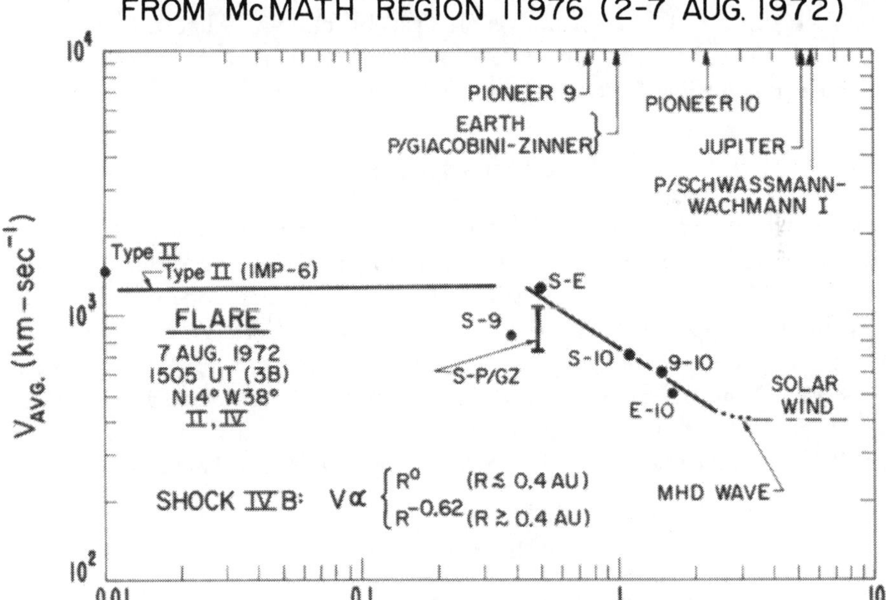

Fig. 3. Approximate shock trajectory from the flare on 1972 August 7 as inferred from the sudden commencement of a geomagnetic storm at Earth, low-frequency type II radio drifts observed by IMP 6, *in situ* shock measurements observed at Pioneers 9 and 10, and brightness diminution of Comet *P*/Giacobini-Zinner. The piston-driven character of the shock wave with fairly rapid deceleration beyond 0.5 to 1.0 AU is likely for this case. Note that shock wave decay to an MHD wave is indicated prior to 5 AU if the ambient solar wind velocity is assumed to be ~400 km s^{-1}.

at the Sun. More significantly, it appears to have decayed to a harmless Alfvén wave before reaching 5 AU in a blast-wave fashion, following a similar piston motion.

Shock IV was produced by the August 7 flare (14° N, 38° W, optical 3B, types II and IV, 1516 UT maximum) which probably produced a visual brightness decrease of *P*/Giacobini-Zinner. We believe that this is possible when the usual solar wind flux of $\sim 10^8$ cm^{-2} s^{-1} is impulsively increased behind the shock (near the piston) to $\sim 10^9$ cm^{-2} s^{-1}. Lifetime against impact dissociation for C_2 (for example) is thereby decreased to about three days which is competitive with photodissociative processes. Figure 3 shows that the shock velocity was initially fairly constant (implying continuous energy outflow at the Sun) with a more blast-like deceleration thereafter. Again, this shock also appears to have decayed prior to reaching the 5 AU position.

Detailed studies of these events will hopefully contribute to improved theories of interplanetary shock origin and propagation as well as solar wind interaction with cometary atmospheres. We do not wish to imply that the shocks noted above were the only ones produced by this spectacular display of flares. Additional shocks may indeed have been produced, and shock-shock interactions could have occurred in interplanetary space. Pending detailed analysis of all spacecraft data, including the various plasma and magnetic fluxes within the transient shock ensembles, we suggest that the shocks noted above are the essential discontinuities during the passage of McMath Region No. 11976 on the visible disk during 1972 August. Such analysis will also, hopefully, assist in the correlation of interplanetary events and the solar disturbances responsible for their existence.

Acknowledgements

We wish to express our sincere and deepest appreciation for comments, suggestions, and access to preliminary data from the following scientists: J. H. Wolfe, H. Collard, J. D. Mihalov, F. L. Scarf, D. S. Colburn, B. F. Smith, C. P. Sonett, A. Frosolone, J. McKinnon, G. Heckman, E. Roemer, G. McCorkle, J. W. Warwick, S. Zimny, H. H. Malitson, S. Cuperman, W. Bernstein, H. Rosenbauer, Š. Pintér, J. Dodge and P. Hedgecock. A more complete report of this work has been submitted under the title 'Interplanetary Shock Waves and Comet Brightness Fluctuations during June-August 1972', for possible publication in *Journal of Geophysical Research*.

DISCUSSION

Brandt: I would give the comet data very low weight in your study. F. D. Miller and B. Donn and J. Rahe have studied comet heads hit by interplanetary shocks and find no discernible effect on the comet. Hence, changes in cometary brightness or form may not be produced by shocks.

Dryer: We were looking for all possible effects of the shock. Scintillation data suggest that there were effects near the comet. The studies you cited were based upon a paucity of solar wind data.

Brandt: But comet workers searched for such effects and the evidence is negative.

Maxwell: I believe that 10^4 km s^{-1} velocity deduced from the radio data by Dodge was based on a misidentification. We observed two components – one at 1000 km s^{-1} and one at 3000 km s^{-1}.

Riddle (to Maxwell): The identification of the disturbance stated by Dodge to be a type II burst with a velocity of 9000 km s^{-1} is indeed in doubt. However, from the Boulder records there is no doubt that a disturbance with velocity 9000 km s^{-1} did exist, whatever its spectral type, as deduced by irregularities in the fringe lines on the interferometer record of a preexisting burst.

Sakurai: I would like to ask you what is the source of the piston which you say maintains the velocity before the deceleration? Is this related to the propagation of a moving type IV source in the solar atmosphere?

Dryer: I would like to give an affirmative answer, but no evidence exists between the observation at the sun and the still-sketchy plasma and field data observed by spacecraft. Regarding the latter, we have been speculating that anomalously-high alpha/proton abundance in the solar wind behind shocks (~0.15, compared with the usual 0.04) is the tracer for the plasma piston or ejecta from the flare. We need additional information provided by theory to help us identify these pistons. The piston's velocity (as presently 'identified' as just noted) is about 80% of the shock velocity which preceded it by some 5 h or so. We would like to 'connect' these observations to the type IV moving source. As spacecraft get closer to the Sun we will get more information on this question.

TYPE II BURST-SOURCES AS LOW-V_A REGIONS IN THE CORONA 'ILLUMINATED' BY FLARE-INDUCED MHD SHOCKS

YUTAKA UCHIDA

Tokyo Astronomical Observatory, University of Tokyo, Mitaka, Tokyo, Japan

and

High Altitude Observatory, National Center for Atmospheric Research, Boulder, Colo., U.S.A.

Abstract (*Solar Phys.*). The author has proposed a hypothesis that the type II burst-sources may be due to the local enhancement of the flare-produced MHD fast-mode shock wave on encountering the pre-existing low Alfvén velocity regions in the corona (Uchida, 1973). The purpose of the present paper is to show that this may actually be the case by comparing the behavior of the computor-simulated propagation and strengthening of MHD fast-mode (weak) shock with the observed characteristics of type II burst-sources.

The computation was performed as an extention of a previous treatment which interpreted the Moreton's wave phenomenon as a sweeping skirt of the flare-produced coronal MHD wavefront (Uchida, 1968; Uchida *et al.*, 1973). The behavior of the wavefront and the shock strength over the wave surface were computed by using eiconal characteristic approximation in the HAO model coronal magnetic field (Altschuler and Newkirk, 1969) and model coronal density (Perry and Altschuler, 1972) for the days for which these models as well as the type II burst data (Culgoora or Boulder) without severe refraction-effect are available (Figure 1).

It is actually seen in the calculated events (May 23, 1967; October 3, 1969; and October 13, 1969) that the projection on the plane of the sky of the calculated distribution of low Alfvén velocity regions in the corona in the vicinity of the flare coincide roughly with the observed position and shape of type II burst-sources. Furthermore, as the result of the computation of the wave behavior in the three-dimensional distribution of the propagation velocity in the corona in these cases, it is clearly seen that the wavefront tends to converge to the low Alfvén velocity regions lying in the corona. This converging effect, together with the local propagation effect of low Alfvén velocity in such regions itself, causes a considerably sharp build up in the Alfvén Mach number of the wave at such particular locations. The enhanced effect of the shock wave, and therefore such phenomena as the electron acceleration in the shock front, may be expected preferentially at these locations, since the Alfvén Mach number of the material flow behind the front (whose initial value is the ratio of the flare explosive velocity of the order of two or three hundred km s^{-1} at most, over the Alfvén velocity in the active region corona of the order of several thousand km s^{-1}) otherwise stays

* Sponsored by the National Science Foundation.

small. This explains, therefore, the observed rather irregular appearance of type II burst sources in which, for example, a distant point is brightened up in some events without any brightening at plasma level closer to the flare. It may be remarked that this anisotropic appearance of the burst sources is expected in our hypothesis even for an assumption of an isotropic emission of the wavefront from the flare explosion,

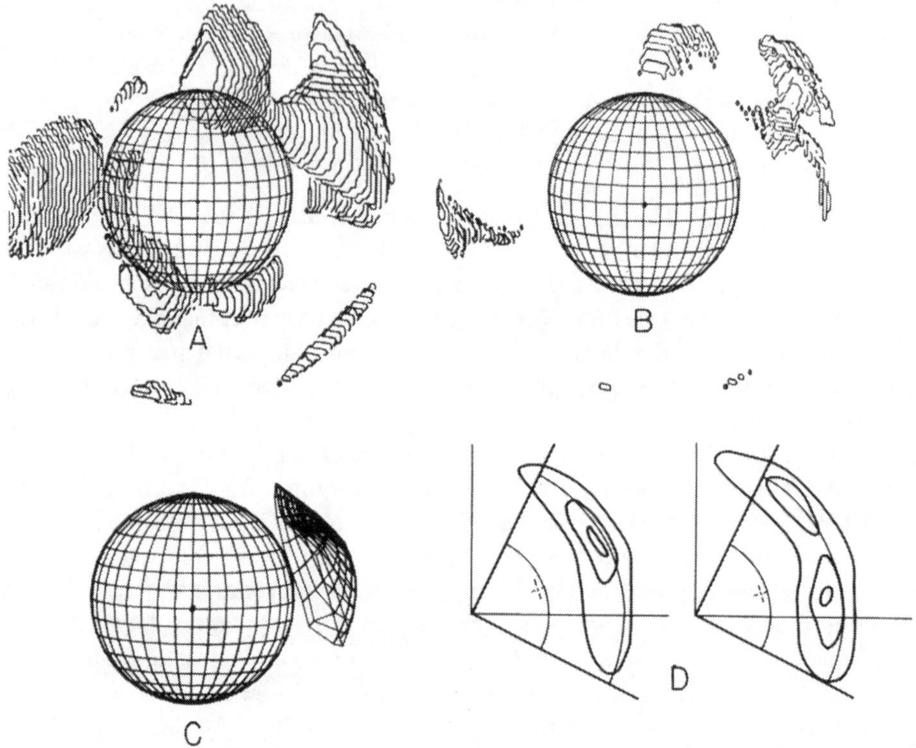

Fig. 1. Several features of the October 13, 1969 event as an example: (A) Three-dimensional representation of the regions where V_A is less than a_0. (B) Region where V_A is less than a_0 on the shell of 40 MHz plasma-level. (C) The wavefront after emission from the flare position in the corona. (D) Equal intensity contours of the observed type II burst from Dulk and Smerd (1971). The maps are rotated by the inclination angle of the solar axis: *Left* 23ʰ41ᵐ51ˢ UT; *Right* 23ʰ43ᵐ41ˢ UT.

which is a reasonable first order guess for a short rise time explosion in a medium with $(V_A/a_0)^2 \gg 1$, where the propagation of the MHD fast mode wave is isotropic in its nature (V_A = Alfvén speed and a_0 = sound speed). Even if there is an intrinsic directivity in the emission of the wavefront, the situation is similar except that some of the pre-existing low Alfvén velocity regions falling outside of the wave cone are not 'illuminated'.

One of the predictions to be made in the present hypothesis is that such properties as the drift velocity, or the time lag of the first appearance of the burst after the impulsive phase are expected to be fairly independent of the flare importance. Even the chance of association of type II bursts with a flare depends on such circumstantial situations as the existence of the low Alfvén velocity region within the reach of the wave. It is, thus, understandable that some large flares are not accompanied by a type II burst while much smaller flares may be accompanied by one. The probability of the association may increase with the flare importance because a more complex structure of the magnetic field and density is associated for the occurrence of a large flare. A similar explanation applies for the association and non-association of type II bursts with Moreton wave events, although we believe both are due to MHD wave emitted by the flare explosive phase (Uchida, 1973).

References

Altschuler, M. D. and Newkirk, G.: 1969, *Solar Phys.* **9**, 131.
Dulk, G. A. and Smerd, S. F.: 1971, *Australian J. Phys.* **24**, 185.
Perry, M. and Altschuler, M. D.: 1973, *Solar Phys.* **28**, 435.
Uchida, Y.: 1968, *Solar Phys.* **4**, 30.
Uchida, Y.: 1973, in R. Ramaty and R. G. Stone (eds.), *Proc. NASA Symposium on High Energy Phenomena on the Sun*, p. 577.
Uchida, Y., Altschuler, M. D., and Newkirk, G.: 1973, *Solar Phys.* **28**, 495.

DISCUSSION

Rosenberg: Do you think that the type II radiation is due to the focussing effect to the low V_A region, or to the fact that in such a region the shock attains a higher Mach number?

Uchida: Due to both. The shock attains a high Alfvén Mach number exclusively at such locations, and produces accelerated electrons and therefore the radio emission there.

Sturrock: What kind of disturbance is needed to set this off?

Uchida: An impulsive disturbance. I assumed a simple pressure pulse.

Smerd: What shock strengths do you find in the region of the type II's?

Uchida: This depends upon the initial shock strength. For a v_0 of 200 to 300 km s^{-1}, the Mach number will be 2 to 5.

ON THE THEORY OF MOVING TYPE IV RADIO BURSTS

A. MANGENEY

Observatoire de Meudon, France

Abstract (*Astron. Astrophys.*). In this paper it is shown that the Alfvén wave turbulence created behind a shock wave propagating along the ambiant magnetic field in a collision-less plasma, accelerates electrons of the high energy tail of the distribution up to mildly relativistic energies. The efficiency of the process depends upon the ratio of the acceleration time by the turbulence to the scattering time of the same turbulence. At high energies the accelerated particles have a power law spectrum, the steepness of which depends upon the turbulent energy.

It is suggested that moving type IV radio bursts are due to the synchrotron emission of the accelerated electrons produced when a flare induced shock wave propagates along open field lines in the solar corona.

DISCUSSION

Smith: You say that your model satisfactorily explains the association of type IV to type II bursts. What do you think determines whether a shock produces a type II or type IV burst?

Mangeney: The configuration of shock and piston.

Gordon Newkirk, Jr. (ed.), Coronal Disturbances, 387. *All Rights Reserved.*

ON SPLIT-BAND STRUCTURE IN TYPE II RADIO BURSTS FROM THE SUN

S. F. SMERD, K. V. SHERIDAN, and R. T. STEWART
Division of Radiophysics, CSIRO, Sydney, Australia

Abstract (*Astrophys. Letters*). The measured amount of band-splitting, Δf, in the spectra of nine harmonic type II bursts is illustrated in Figure 1. Here, as in previous, smaller samples (Roberts, 1959; Maxwell and Thompson, 1962; Weiss, 1965) Δf is found to increase with frequency, f.

Three kinds of interpretation of band-splitting have been offered:

(1) The splitting is due to a magnetic field; this interpretation predicts frequency-splitting in the range $f_H^2/2f_p \lesssim \Delta f \lesssim f_H$, where f_H and f_p are the electron gyro and

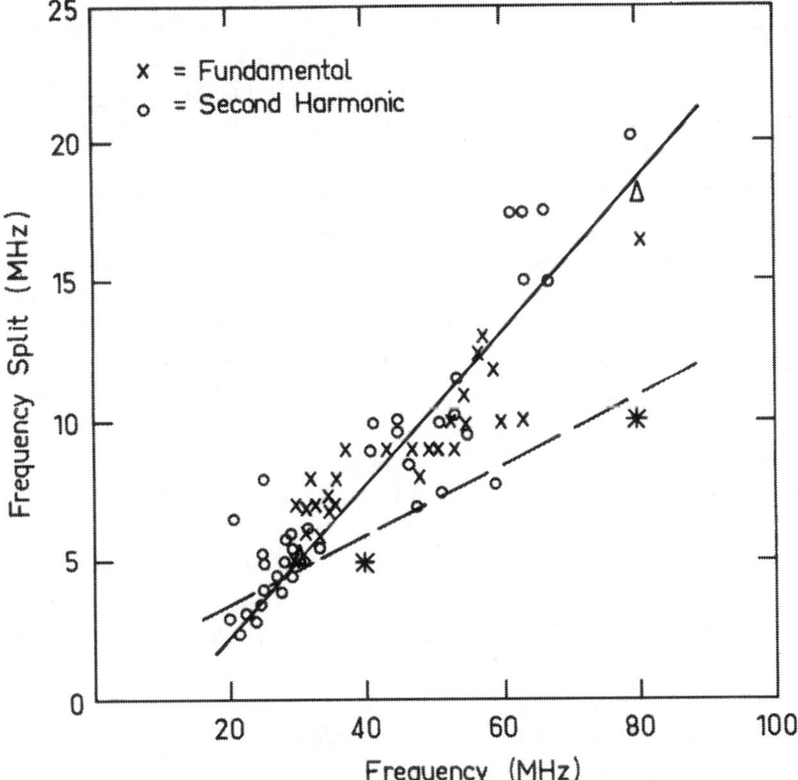

Fig. 1. The frequency separation of the two components in split-band type II bursts as a function of the mean fundamental frequency. Measured second-harmonic frequencies have been halved. The full line is a linear approximation to the present results, the broken line a similar approximation given by Weiss (1965). The points marked are mean values quoted by Maxwell and Thompson (1962), those marked △ are mean values quoted by Roberts (1959).

Gordon Newkirk, Jr. (ed.), Coronal Disturbances, 389–393.

plasma frequencies respectively and $f_H^2/f_p^2 \ll 1$. Weiss (1965) has shown that the derived magnetic fields would make the Alfvén velocity in the path of the type II disturbance larger than the speed of the disturbance. The latter could then not be a magneto-hydrodynamic shock, as is generally assumed.

(2) The splitting is due to a Doppler shift because the radiating electrons drift in opposite directions within the rising and falling branches of the shock wave. Wild and Smerd (1972) have suggested that there could be no such Doppler shift in the fundamental radiation since the latter results from the scattering of plasma waves on 'stationary' ions.

(3) The two components of a split band correspond to two maxima in the spectrum of the plasma radiation from a type II shock front. In McLean's (1967) model they originate near the axis and around the skirt of a coronal streamer because in both regions the shock front is nearly parallel to the plasma levels (though at different plasma frequencies).

Here we add another interpretation which, if correct, allows the determination of the shock strength of the type II disturbance and of the magnetic field along the path of the disturbance. According to the present hypothesis the frequency (f_l) of the lower-frequency component of a split band is identified with the plasma frequency (f_{p1}) just ahead of the shock front, while the frequency (f_u) of the upper-frequency component is identified with the plasma frequency (f_{p2}) just behind the shock; thus: $f_l = f_{p1}$, $f_u = f_{p2}$ and $\Delta f = f_u - f_l$.

This assumes (a) that radiating electrons exist at the front and the back of the shock front and (b) that, since the shock front is thin, the two components of a split band are emitted essentially from a common source. The latter assumption is supported by the occurrence at similar times of similar spectral features in the two components, a characteristic that may be hard to explain by McLean's (1967) theory. The assumption (a) may be supported by the occasional observation of 'herringbone' structure, which we takes as evidence for the forward ejection of electrons from the front, and backward ejection from the back, of a type II shock. The frequency interval that constitutes the 'backbone' can be interpreted in the same way as proposed here for band-splitting. This interpretation is made more plausible by the observation that in some cases the 'backbone' is itself split. It seems possible that the electrons exciting the plasma waves may be confined near the front and the back of a type II shock (the split-band situation) or ejected forwards and backwards (the herringbone situation), or both. The ejection of the electrons may depend on their having access to open magnetic-field lines.

With the present interpretation the band-splitting can be related to the Mach number through the Rankine-Hugoniot 'jump' condition

$$N_2/N_1 = f_{p2}^2/f_{p1}^2 = 4M^2/(3+M^2),$$

which connects the post-shock, N_2, and pre-shock, N_1, electron densities. Here the magnetic Mach number $M = v/v_A$, where v is the shock speed and $v_A = 7 \times 10^3 f_H/f_p$ is the Alfvén velocity. The measured frequency-splitting then determines the strength of a type II shock as

$$M = \frac{\sqrt{3}(f_u/f_l)}{\sqrt{4-(f_u/f_l)^2}}.$$

The derived shock strengths corresponding to the frequency-splitting of Figure 1 are shown in Figure 2; the frequency range of Figure 1 has been converted to a height range using Newkirk's (1961) streamer model. We note that the derived shock

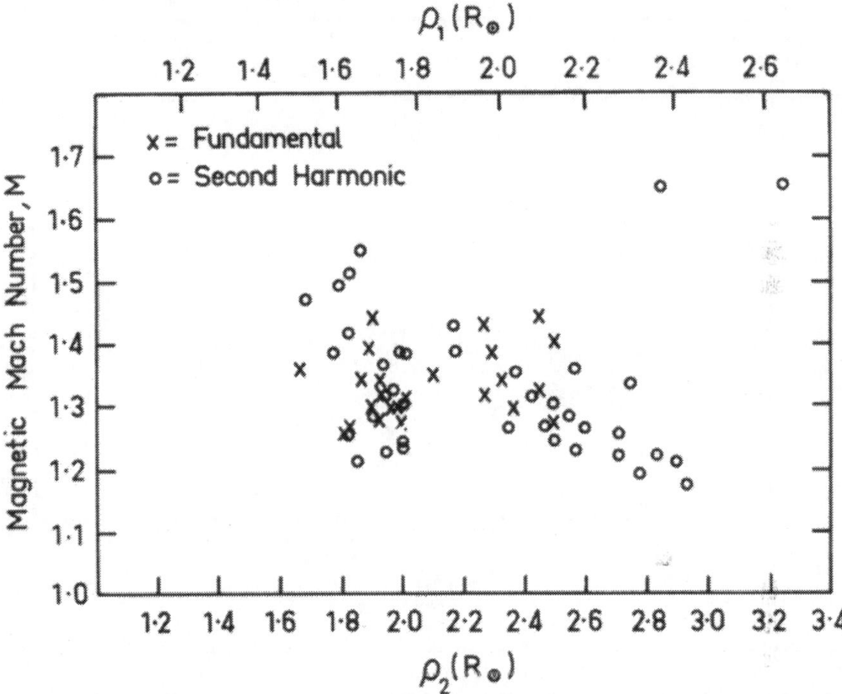

Fig. 2. The magnetic Mach number $M = v/v_A$, a measure of the shock strength of type II disturbances, as derived from the frequency-splitting of Figure 1. In converting from a frequency, f, scale to one of distance, ϱ, from the Sun's centre it has been assumed that f is the local plasma frequency along the axis of a Newkirk streamer (ϱ_1), or along a streamer with twice those densities (ϱ_2).

strengths ($1.2 \lesssim M \lesssim 1.7$) are compatible with a laminar shock structure, but are below hose ($2.0 \lesssim M \lesssim 2.9$) required in Smith's (1971, 1972) theory of turbulent type II shocks.

It is clear from the above that a knowledge of the shock strength, M, and the shock speed, v, yields the magnetic-field strength in the plasma ahead of the shock front as

$$H = 5.1 \times 10^{-5}\, v f_l / M \quad \text{G},$$

where v is in kilometres per second and f_l in megahertz. As in previous type II analyses, we derive the shock speed from the measured frequency-drift rates, assuming a radial path and a coronal density model. The magnetic fields derived in this way, again using Newkirk's streamer model, are shown in Figure 3. The points are widely scattered in the range 0.4 to 4.0 G. The main uncertainty is probably due to uncertainty in v resulting from a lack of knowledge of the actual path of the disturbance and of the densities along this path.

In a few cases the positions of both components of a split band have been observed

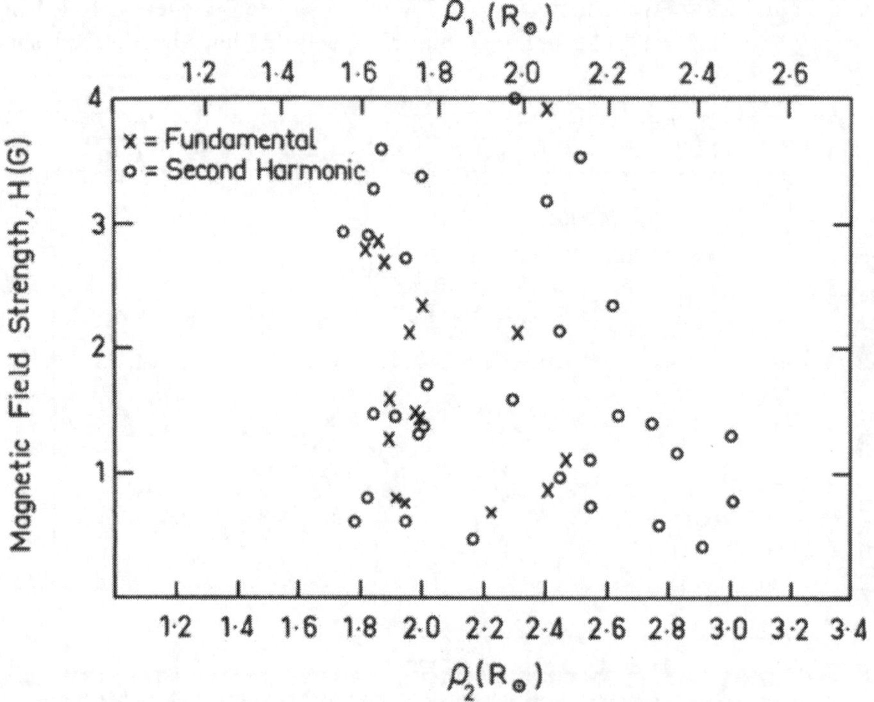

Fig. 3. The magnetic-field strength, H, as a function of distance ϱ from the Sun's centre derived from the frequency-splitting shown in Figure 1 and the radial velocities of the type II disturbances deduced from the spectral drift rates and the same two coronal streamer models (ϱ_1, ϱ_2) as in Figure 2.

with the Culgoora radioheliograph: the sources are separated by 1′ to 4′. This has been taken in support of McLean's (1967) theory (see (3) above). Some of the source separation, say $\sim 1'$, may simply reflect the distance covered by the type II disturbance during the interval (~ 1 min) between the appearance of the lower-frequency and the upper-frequency components at the heliograph frequency. However, the present theory suggests two further causes of apparent source separation.

(1) In the fundamental band the f_l-component is emitted in the undisturbed corona. Its apparent position is expected to be well beyond the true projected position, largely because of the strong, refractive outward-beaming near the plasma level (e.g. Riddle, 1972). Meanwhile the f_u-component has only to cross the shock front before emerging from the strongly refracting region; its position should be nearer the true position.

(2) While fundamental radiation can escape only outwards, harmonic radiation

can reach the observer by 'direct' or 'reflected' rays. This effect could be responsible for well-separated $2f_l$ and $2f_u$ sources if, as suggested by the comparison between split-band and herringbone structure, the radiating electrons ahead of the shock front are beamed forwards, those behind the shock front backwards.

However, the present sample of observed split-band positions is too small to decide between different split-band theories.

References

McLean, D. J.: 1967, *Proc. Astron. Soc. Australia* **1**, 47.
Maxwell, A. and Thompson, A. R.: 1962, *Astrophys. J.* **135**, 138.
Newkirk, G., Jr.: 1961, *Astrophys. J.* **133**, 983.
Riddle, A. C.: 1972, *Proc. Astron. Soc. Australia* **2**, 98, 148.
Roberts, J. A.: 1959, *Australian J. Phys.* **12**, 327.
Smith, D. F.: 1971, *Astrophys. J.* **170**, 599.
Smith, D. F.: 1972, *Astrophys. J.* **174**, 121, 643.
Weiss, A. A.: 1965, *Australian J. Phys.* **18**, 167.
Wild, J. P. and Smerd, S. F.: 1972, *Ann. Rev. Astron. Astrophys.* **10**, 159.

DISCUSSION

Smith: One of the reasons for accepting Weiss's idea that herringbone is due to a shock crossing a magnetic field was that there was often no frequency drift when herringbone was observed. What do you think about this now?

Smerd: Weiss had a very small data sample.

ON THE THERMAL INTERPRETATION OF HARD X-RAY BURSTS FROM SOLAR FLARES

JOHN C. BROWN
Dept. of Astronomy, University of Glasgow, Glasgow, Scotland

Abstract. The possible validity of thermal bremsstrahlung models of flare hard X-ray bursts is investigated quantitatively. In particular, the problem of rapid thermal conduction in 'multi-temperature' models is adequately examined for the first time by using a continuous temperature distribution consistent with the observed X-ray spectrum. This distribution is obtained from a general analytic solution for the temperature structure required to mimick *any* 'non-thermal' spectrum, the method being equally applicable to cosmic sources.

It is concluded that the thermal interpretation might extend to X-rays of hundreds of keV, a result with important consequences for flare energetics. The relationship of such a model to observations of X-ray polarization and rapid time variations is also considered.

1. Introduction

There is a general consensus in the literature that flare X-ray emission is dominated at low energies by thermal bremsstrahlung, as evidenced by continuum and line spectra (e.g. Neupert, 1969), and at high energies by non-thermal emission from the energetic particles observed directly in space (e.g. Fichtel and McDonald, 1967) and by means of their radio (Takakura, 1967) and γ-ray emissions (Chupp *et al.*, 1973). However, there is still considerable uncertainty as to the photon energies at which the transition occurs between these two types of emission. The general trend is to accept that thermal bremsstrahlung does not extend to more than a few tens of keV, and probably less than 10 keV, emission harder than this arising by bremsstrahlung of non-thermal electron streams (e.g. De Jager, 1967). Chubb (1971) has, however, recently reiterated the possibility of a multi-temperature thermal source, previously advocated by Chubb *et al.* (1966), which might fit the observations up to hundreds of keV.

Resolution of this question is crucial to considerations of flare models, particularly in respect of energy distribution in the flare, since the steep spectrum of non-thermal flare particles implies that the bulk of their energy resides at the low energy end of their distribution (e.g. Neupert, 1968; Brown, 1971). If thermal X-ray emission is dominant only up to ten keV or so, the non-thermal electron spectrum must extend down to this energy implying a large total energy amongst these particles, comparable with the total flare energy release in fact (Brown, 1971, 1972a, 1973a). This situation bears heavily on particle acceleration mechanisms, requiring that much of the magnetic energy released go directly into particles, and permitting the possibility that it is these which subsequently heat the thermal flare or at least part of it (e.g. Neupert, 1968; Kane and Donnelly, 1971; Brown, 1973b). Alternatively, if thermal X-rays dominate the spectrum to around 100 keV, the energetic electrons contributing the high energy tail must carry only a relatively small part of the flare energy,

Gordon Newkirk, Jr. (ed.), Coronal Disturbances, 395–412.

particle acceleration being then only a minor artefact of the magnetic energy release.

The tendency for most authors to accept the dominance of non-thermal electron bremsstrahlung above about 10 keV can be attributed to a number of objections raised against the multi-temperature model proposed by Chubb (1971). Firstly, there are the observations of polarization (Tindo *et al.*, 1972) and rapid time variation (e.g. Frost, 1969), in the hard X-ray emission, thought to be uncharacteristic of thermal emission (Tindo *et al.*, 1970; Kahler, 1971a). Secondly, the high temperatures ($\gtrsim 10^8$ K) needed in a multi-temperature model to produce the observed hard X-ray spectrum imply the existence of steep temperature gradients in the source and consequent rapid energy redistribution by thermal conduction, by electron escape, and by radiation (Culhane *et al.*, 1970). On these grounds, Kahler (1971a, b) calculates that the multi-temperature source would not survive long enough to sustain the observed hard X-ray burst.

It is the aim of this paper to demonstrate that these objections have not yet been adequately formulated physically and that when the formulation is complete, a multi-temperature source is not at present precluded from explaining currently available X-ray data to over 100 keV. In particular it is pointed out that conduction in the multi-temperature source can only be correctly described on the basis of a temperature structure determined by the requirement of fitting the observed hard X-ray spectrum and not by the simple empirical models of the temperature structure used by Kahler (1971a) (cf. Culhane *et al.*, 1970). Such a self-consistent description of the multi-temperature X-ray flare model is developed here from a general analytic solution for the temperature structure required to mimic *any* non-thermal X-ray spectrum. This solution may also be of interest in the field of cosmic X-ray source models, some of which involve a multi-temperature formulation (e.g. Greisen, 1971; Sunyaev and Illarionov, 1972). In the last section of the paper, possibilities of explaining also the observed hard X-ray polarization and time variations in terms of the multi-temperature model are considered.

2. X-Ray Continuum Spectra from Distributed Temperature Sources

For a thermal bremsstrahlung source at distance R, the continuum photon energy flux at the Earth is

$$F(\varepsilon)=\frac{8.1\times 10^{-39}}{R^2}\int\limits_V \frac{e^{-\varepsilon/kT}}{T^{1/2}}\,n^2\,\mathrm{d}V\,(\mathrm{keV\ cm^{-2}\ s^{-1}\ keV^{-1}}), \tag{1}$$

where $n(\mathbf{r})$, $T(\mathbf{r})$ are the electron density (cm^{-3}) and temperature at position $\mathbf{r}$ in the source volume V and ε is the photon energy (e.g. Culhane, 1969), variation of the Gaunt factor from unity being neglected. The simplest treatment of the non-isothermal model to fit photon spectra which deviate from the isothermal (exponential) form is merely by trial summing of several contributions of isothermal type, each with its temperature T and emission measure $n^2\,\Delta V$ as adjustable parameters – i.e. the integral (1) is approximated by a few term sum (e.g. Cline *et al.*, 1969; Herring and Craig, 1973; Batestone *et al.*, 1970).

A more sophisticated approach is to insert trial fitting functions for $n(\mathbf{r})$ and/or $T(\mathbf{r})$ into (1) until the required $F(\varepsilon)$ is mimicked (similarly to the fitting of trial power-law spectra in the case of non-thermal bremsstrahlung e.g. Kane and Anderson, 1970). Milkey (1971) has briefly considered this approach in connection with flares while Chambe (1971) applied it to soft X-rays from coronal condensations. Illarionov and Sunyaev (1972) (see also Felten and Rees, 1972) have investigated the case of power-law forms for n, T as functions of depth in a hot radiating atmosphere and shown how this can produce a power law for $F(\varepsilon)$. Their investigation is primarily concerned with cosmic sources but also mentions possible relevance of this result to the characteristic power law of solar flare hard X-rays.

As will now be demonstrated, however, the problem of fitting an *arbitrary* observed X-ray spectrum $F(\varepsilon)$ by a distributed temperature source can in fact be solved in an entirely general way.

Let the source volume V be divided into elements $\mathrm{d}V$ such that $\mathrm{d}V$ contains all the plasma for which the electron temperature lies in the interval T, $\mathrm{d}T$. It is then natural to define an *emission measure* $\mu(T)$ *per unit temperature* ($\mathrm{cm}^{-3}\ \mathrm{K}^{-1}$) such that

$$\mu(T)\,\mathrm{d}T = n^2\,\mathrm{d}V. \tag{2}$$

The significance of this definition is straightforward so long as a unique value of n may be associated with each T, as will be the case in a stratified source (see Section 3). If this condition is not met, V may be subdivided into elements in each of which it is valid and the definition of $\mu(T)$ modified by replacing the right side of Equation (2) by a discrete sum over these elements. For the present purpose it will be convenient to express T and μ in more appropriate units, viz.

$$T = 10^6\ T_6$$

and

$$\mu(T)\,\mathrm{d}T = 10^{41}\ \mu^*(T_6)\,\mathrm{d}T_6$$

where T_6 is now in millions of °K and $\mu^*(T_6)$ is the emission measure per 10^6 K in units of $10^{41}\ \mathrm{cm}^{-3}$. Then (1) becomes

$$F(\varepsilon) = \frac{0.81}{R^2} \int\limits_{T_6=0}^{\infty} \frac{\mu^*(T_6)}{T_6^{1/2}}\, e^{-11.6\varepsilon/T_6}\,\mathrm{d}T_6, \tag{4}$$

where ε is in keV.

Changing the variable to the inverse temperature parameter

$$t = 11.6/T_6 \tag{5}$$

and defining

$$G(t) = \mu^*(T_6)\ T_6^{3/2} \tag{6}$$

then Equation (4) becomes simply

$$F(\varepsilon)=\frac{7\times 10^{-2}}{R^2}\int_{t=0}^{\infty} G(t)\, e^{-\varepsilon t}\,\mathrm{d}t \tag{7}$$

This equation shows at once that the non-isothermal bremsstrahlung spectrum $F(\varepsilon)$ is essentially the Laplace Transform of the temperature distribution G – i.e.

$$F(\varepsilon)=\frac{7\times 10^{-2}}{R^2}\,\mathscr{L}\{G(t);\varepsilon\} \tag{8}$$

Thus, in an entirely general way, the temperature distribution required to mimic *any* observed 'non-thermal' spectrum (i.e. non-isothermal) is obtained immediately on simply inverting integral Equation (8), i.e.

$$G(t)=14.3\,R^2\,\mathscr{L}^{-1}\{F(\varepsilon);t\} \tag{9}$$

which, on re-expression in the original variables, yields

$$\mu^*(T_6)=\frac{14.3\,R^2}{T_6^{3/2}}\,\mathscr{L}^{-1}\left\{F(\varepsilon);\frac{11.6}{T_6}\right\}. \tag{10}$$

Since the Inverse Transform $\mathscr{L}^{-1}$ may be obtained for a very wide range of functions (e.g. Sneddon, 1972) this clearly shows the inadequacy of spectral data alone for establishing the existence of non-thermal processes in X-ray sources. Two particular examples of spectra may be used as illustrations. Firstly, the commonly used power-law fit to 'non-thermal' spectra, i.e. $F(\varepsilon)=A\varepsilon^{-\gamma+1}$ (where γ is the photon *number* flux index), is mimicked by a thermal emission measure distribution given by

$$\mu^*(T_6)=\frac{14.3\,R^2}{T_6^{3/2}}\,\mathscr{L}^{-1}\left\{A\varepsilon^{-\gamma+1};\frac{11.6}{T_6}\right\}=$$
$$=\frac{14.3\,R^2}{T_6^{3/2}}\,A\left(\frac{11.6}{T_6}\right)^{\gamma-2}\frac{1}{\Gamma(\gamma-1)},$$

where Γ is the gamma function (e.g. Abramowitz and Segun, 1965), i.e.

$$\mu^*(T_6)=\frac{0.36\,AR^2}{\Gamma(\gamma-1)}\left(\frac{T_6}{11.6}\right)^{-\gamma+1/2} \tag{11}$$

which is the same result as obtained by Illarionov and Sunyaev (1972) in a less general way. The form of μ^* given by (11) is illustrated in Figure 1 for the burst recorded by Cline *et al.* (1969) – see also Section 5.

A second example of some physical interest, which happens to have an analytic solution, arises in connection with flares which sometimes exhibit a power law spectrum with high energy cut-off (e.g. Cline *et al.*, 1968; Frost, 1969) above some energy ε_1 (of order 100 keV). This may be approximately represented as $F(\varepsilon)=$

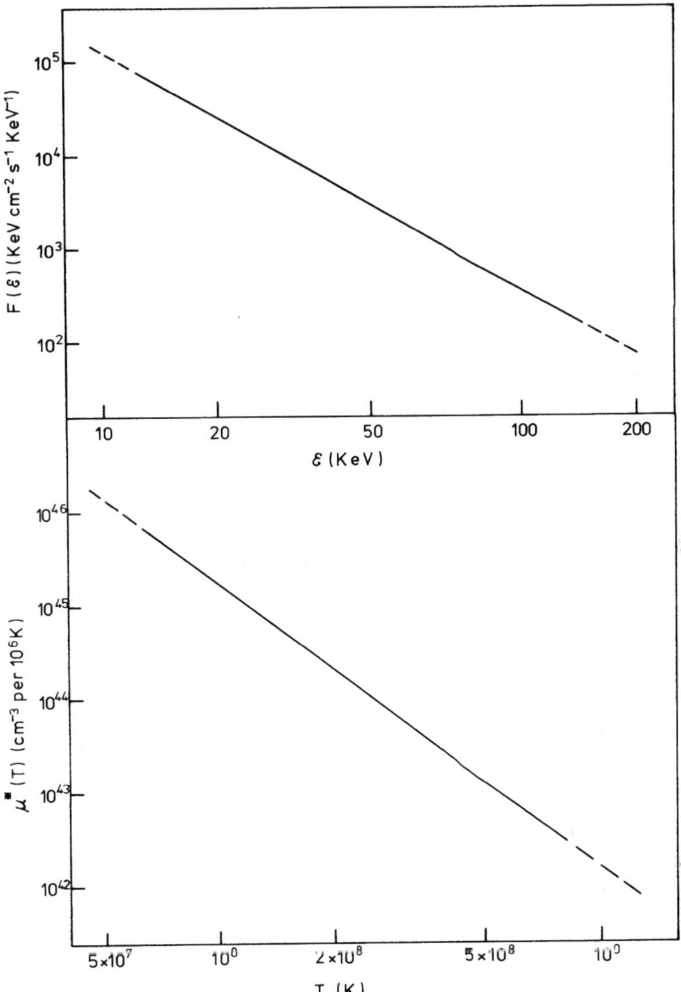

Fig. 1. Photon spectrum $I(\varepsilon)$ for a hard X-ray burst with $\gamma=3.5$ and a peak flux $I_0=300$ photons cm^{-2} s^{-1} above 80 keV (Cline *et al.*, 1969), together with the distribution of emission $\mu(T)$ required to produce it thermally.

$=A\varepsilon^{-\gamma+1}\,e^{-\sqrt{\varepsilon/\varepsilon_1}}$ (cf. Elwert and Haug, 1971). The temperature structure required to fit this is then, by (10),

$$\mu^*(T_6)=\frac{14.3\,R^2}{T_6^{3/2}}\,A\mathscr{L}^{-1}\left\{\varepsilon^{-\gamma+1}\,e^{-\sqrt{\varepsilon/\varepsilon_1}};\ \frac{11.6}{T_6}\right\}=$$

$$=3.7\times 10^{-2}\,AR^2\left(\frac{T_6}{46.4}\right)^{-\gamma+1/2} i^{(2\gamma-4)}\,\mathrm{erfc}\left(\sqrt{\frac{T_6}{46.4\varepsilon_1}}\right), \tag{12}$$

where $i^n\,\mathrm{erfc}\,x$ is the nth repeated integral of the error function (Ambramowitz and Segun, 1965).

Spectra which, unlike the above, cannot be readily fitted by analytic forms or for which the Inverse Transform is not expressible in terms of common analytic functions may nevertheless be treated by numerical transform methods (see e.g. Sneddon, 1972). The only factor not included in the above analysis is the Gaunt factor. Its inclusion would, however, add only a weighting function to $\mu(T)$.

It may be apt to emphasize that solution (10) is only meaningfully applicable to the high temperature range of $\mu(T)$ since the assumptions involved in Equation (1) refer only to hard photon energies (i.e. lines are omitted, etc.). This is clearly so since, for example, distribution (11) implies infinite amounts of plasma at zero temperature because the photon spectrum $A\varepsilon^{-\gamma+1}$ has been formally extrapolated to small ε in the integrand. In fact, however, such formal extrapolation will cause little error in solution (11) since, on the one hand, material of $kT \ll \varepsilon$ cannot significantly contribute to $F(\varepsilon)$ due to the exponential in (1) while, on the other, the steepness of typical spectra $F(\varepsilon)$ (and hence $\mu(T)$) ensures that low energy photons emanate predominantly from low temperature material. That is, as is already known, X-ray photons of energy ε can be quite closely identified with material at $kT \simeq \varepsilon$, so that solution (10) at temperature T has an accuracy comparable with that of $F(\varepsilon = kT)$. To be completely general, of course, (10) should have been solved for the entire observed spectrum $F(\varepsilon)$ – i.e. a power-law merging with an exponential at energies around 1–20 keV, corresponding approximately to a power-law temperature structure $\mu(T)$ of type (11) at high T merging into an isothermal region (μ *a* δ-function) at a few tens of millions °K. For the higher temperature regime investigated here, however, this refinement will not be necessary.

The most general use of solution (10) must lie in the accurate inference of the temperature distribution in sources (cosmic and solar) believed, from other considerations, to be of thermal type. Section 3 of this paper relates to the interpretation of $\mu(T)$ deduced in this way in the construction of models of such sources. In cases where the non-isothermal interpretation is in doubt, as it is in flares, recourse must be had to physical, and particularly energetic, considerations to assess the true nature of the source by eliminating either the thermal or non-thermal alternatives c.f. Kahler (1971a). Solution (10) provides accurately the temperature structure, demanded by the X-ray spectral data, required for such investigations. In Section 4, this method is applied to the problem of cooling of hot multi-temperature plasmas in flares.

3. Physical Data on Sources from the Differential Emission Measure

To be useful beyond the immediate context of interpreting the X-ray spectrum, the differential emission measure function $\mu^*(T_6)$ obtained in Section 2 must be more explicitly expressed in terms of the source structure. In particular, the variables n and T_6 must be separately extracted from their combination in μ^*. In the presence of the typically large temperature gradients resulting from the solution of (10) (e.g. (11)), approximately constant pressure is likely to prevail through the source (cf. the transition region in the quiet Sun) unless a highly randomised magnetic field isolates individual plasma elements. In their cooling calculations for high temperature coronal flare plasmas, both Culhane *et al.* (1970) and Kahler (1971a, b) have characterized the source by a single (constant) density – a state of affairs which is highly unlikely due to the strong pressure gradients resulting from it. This distinction is important since variations in n modify the temperature gradient deduced from $\mu^*(T_6)$ (see below) and so affect the inferred conductive flux (see Section 4). For the remainder of this paper, therefore, it will be supposed that constant pressure prevails in the source so that

$$n(\mathbf{r})\ T_6(\mathbf{r}) = 10 \times n_{10}, \tag{13}$$

where n_{10} is the density (cm^{-3}) where the temperature is 10^7 K (i.e. $T_6 = 10$) – roughly where the source becomes near isothermal as determined from soft X-ray data (e.g. Neupert, 1969).

Relation (13), together with (2) immediately yields the *volume* $\mathrm{d}V$ of plasma in a given temperature range $\mathrm{d}T_6$, viz.

$$\mathrm{d}V(\text{cm}^3) = 10^{39}\,\frac{\mu^*(T_6)}{n_{10}^2}\,T_6^2\,\mathrm{d}T_6 \tag{14}$$

so that the total volume $\Delta V(T_1, T_2)$ lying between temperatures T_1 and T_2

$$\Delta V(T_1, T_2) = \frac{10^{39}}{n_{10}^2}\int_{T_1}^{T_2} \mu^*(T_6)\,T_6^2\,\mathrm{d}T_6. \tag{15}$$

If a one-dimensional model is adopted for the source, with n and T varying along the Z coordinate, and with cross sectional area Σ, then (14) and (15) can be written

$$\mathrm{d}Z(\text{cm}) = 10^{39}\,\frac{\mu^*(T_6)\,T_6^2}{n_{10}^2\Sigma}\,\mathrm{d}T_6 \tag{16}$$

and

$$\Delta Z(T_1, T_2) = \frac{10^{39}}{n_{10}^2\Sigma}\int_{T_1}^{T_2} \mu^*(T_6)\,T_6^2\,\mathrm{d}T_6, \tag{17}$$

where $\mathrm{d}Z$ is the 'length' of the source lying in temperature interval $\mathrm{d}T_6$ and $\Delta Z(T_1, T_2)$ is the 'length' between $T_6 = T_1$ and $T_6 = T_2$.

This one-dimensional geometry is appropriate to the hot stratified atmosphere of some cosmic X-ray source models (e.g. Illarionov and Sunyaev, 1972) and also describes sources of filamentary type comprising trapped plasma in a uniform magnetic flux tube such as appears to occur in the solar corona (e.g. Culhane *et al.*, 1970). In this latter case, Z is measured along the field.

Equation (16) additionally defines the temperature gradient at each point in the source, viz.

$$\frac{dT_6}{dZ}=10^{-39}\frac{n_{10}^2\Sigma}{\mu^*(T_6)\,T_6^2} \tag{18}$$

which will be needed in calculating the thermal conductive flux in the source.

4. Energy Flow in the High Temperature Plasma

Culhane *et al.* (1970) and Kahler (1971a, b) have already discussed energy loss processes in high temperature coronal flare plasmas. Their calculations are, however, based on very simple models of the temperature structure – characterised by either a constant conductive flux (Kahler, 1971a) or a constant temperature gradient (Culhane *et al.*, 1970) along some characteristic length. These representations are unsatisfactory since the temperature structures are not compatible with the observed X-ray spectrum. Only by incorporating the X-ray spectral data into the temperature model is it possible to see how much the temperature gradient varies in the hot source, and, likewise, the conductive flux – i.e. to see the influence of energy sources on sinks.

The temperature profile derived as in Section 2 is by definition compatible with the X-ray spectrum and provides the necessary basis for conduction calculations. As the most important example, this investigation will be carried out here only for the case of a power-law spectrum $F(\varepsilon)\sim\varepsilon^{-\gamma+1}$ as considered in Section 2, but rewritten here as

$$F(\varepsilon)=(\gamma-1)\,I_0\left(\frac{\varepsilon}{\varepsilon_0}\right)^{-\gamma+1}, \tag{19}$$

where I_0 is the total flux ($\mathrm{cm}^{-2}\,\mathrm{s}^{-1}$) at the Earth of all photons of $\varepsilon\geqslant\varepsilon_0$ (keV) for some ε_0. In this case Equations (11), (14), (15), (16), (17) and (18) reduce (with $R=1$ AU) to

$$\mu^*(T_6)=\frac{0.36(\gamma-1)}{\Gamma(\gamma-1)}\frac{I_0}{\varepsilon_0^{-\gamma+1}}\left(\frac{T_6}{11.6}\right)^{-\gamma+1/2} \tag{20}$$

$$dV(\mathrm{cm}^3)=\frac{4.8\times10^{40}}{n_{10}^2}\frac{(\gamma-1)}{\Gamma(\gamma-1)}\frac{I_0}{\varepsilon_0^{-\gamma+1}}\left(\frac{T_6}{11.6}\right)^{-\gamma+5/2}dT_6 \tag{21}$$

if

$$\gamma\neq3.5,\qquad \Delta V(T_1,T_2)=\frac{5.6\times10^{41}}{n_{10}^2}\frac{(\gamma-1)}{(\gamma-\frac{7}{2})\,\Gamma(\gamma-1)}\frac{I_0}{\varepsilon_0^{-\gamma+1}}\times\\ \times\left[\left(\frac{T_1}{11.6}\right)^{-\gamma+7/2}-\left(\frac{T_2}{11.6}\right)^{-\gamma+7/2}\right]$$

while if

$$\gamma=3.5,\qquad \Delta V(T_1,T_2)=\frac{5.6\times10^{41}}{n_{10}^2}\frac{(\gamma-1)}{\Gamma(\gamma-1)}\frac{I_0}{\varepsilon_0^{-\gamma+1}}\log_e\left(\frac{T_2}{T_1}\right) \tag{22}$$

$$dZ=\frac{4.8\times10^{40}}{n_{10}^2\Sigma}\frac{(\gamma-1)}{\Gamma(\gamma-1)}\frac{I_0}{\varepsilon_0^{-\gamma+1}}\left(\frac{T_6}{11.6}\right)^{-\gamma+5/2}dT_6 \tag{23}$$

if

$$\gamma \neq 3.5, \qquad \Delta Z(T_1, T_2) = \frac{5.6 \times 10^{41}}{n_{10}^2 \Sigma} \frac{(\gamma - 1)}{(\gamma - \frac{7}{2})\, \Gamma(\gamma - 1)} \frac{I_0}{\varepsilon_0^{-\gamma+1}} \times \\ \times \left[\left(\frac{T_1}{11.6} \right)^{-\gamma+7/2} - \left(\frac{T_2}{11.6} \right)^{-\gamma+7/2} \right]$$

while if

$$\gamma = 3.5, \qquad \Delta Z(T_1, T_2) = \frac{5.6 \times 10^{41}}{n_{10}^2 \Sigma} \frac{(\gamma - 1)}{\Gamma(\gamma - 1)} \frac{I_0}{\varepsilon_0^{-\gamma+1}} \log_e \left(\frac{T_2}{T_1} \right) \tag{24}$$

and

$$\frac{\mathrm{d}T_6}{\mathrm{d}Z} = 2.1 \times 10^{-41}\, n_{10}^2 \Sigma \frac{\Gamma(\gamma - 1)}{(\gamma - 1)} \frac{\varepsilon_0^{-\gamma+1}}{I_0} \left(\frac{T_6}{11.6} \right)^{\gamma - 5/2} \tag{25}$$

This can be solved to give the Z dependence of the temperature structure, viz.:

for

$$\gamma \neq \tfrac{7}{2}, \qquad T_6 = [(\tfrac{7}{2} - \gamma)\, \beta Z + 10^{7/2 - \gamma}]^{1/(7/2 - \gamma)} \qquad \text{with } Z = 0$$

and for

$$\text{at } T_6 = 10$$

$$\gamma = \tfrac{7}{2}, \qquad T_6 = 10 \exp(\beta Z), \qquad \text{i.e. at } 10^7\,\mathrm{K}$$

where

$$\beta = \frac{2.1 \times 10^{-41}}{11.6^{\gamma - 5/2}}\, n_{10}^2 \Sigma \frac{\Gamma(\gamma - 1)}{\gamma - 1} \frac{\varepsilon_0^{-\gamma+1}}{I_0}$$

Culhane *et al.* (1970) have demonstrated that, in the high temperature regime concerned here, energy transfer in the plasma is predominantly by the flow of thermal energy down the temperature gradient. They point out, however, that this flow can only be described as a thermal conduction process provided that the collision mean free path is shorter than the scale length of the temperature gradient. In an erratum to his paper (1971a), Kahler (1971b), using again a constant thermal flux model, found that this condition is not met at the high temperatures ($\sim 10^8$ K) required for hard X-ray emission. This can be better assessed by means of (25) which shows the scale distance for change of T_6 to be

$$l \simeq \frac{\mathrm{d}Z}{\mathrm{d} \log T} = \frac{5.6 \times 10^{41}}{n_{10}^2 \Sigma} \frac{\gamma - 1}{\Gamma(\gamma - 1)} \frac{I_0}{\varepsilon_0^{-\gamma+1}} \left(\frac{T_6}{11.6} \right)^{-\gamma + 7/2}$$

while the collisional mean free path is (Billings, 1966)

$$L \simeq \frac{2.3 \times 10^{16}}{n} \times T_6 = \frac{2.3 \times 10^{15}}{n_{10}} T_6^2$$

so that

$$\alpha = \frac{L}{l} = 5.3 \times 10^{-25}\, n_{10} \Sigma \frac{\Gamma(\gamma - 1)}{\gamma - 1} \frac{\varepsilon_0^{-\gamma+1}}{I_0} \left(\frac{T_6}{11.6} \right)^{-\gamma + 11/2}, \tag{26}$$

where the regime $\alpha \ll 1$ corresponds to thermal conductive flow while with $\alpha \gg 1$ energy transfer will be essentially a matter of mass flow of hot material.

Taking typical parameters from the large burst observed by Cline *et al.* (1968), viz. $\gamma \simeq 3.5$ and $I_0 \simeq 300\ \text{cm}^{-2}\ \text{s}^{-1}$ for $\varepsilon_0 = 80$ keV then

$$\alpha = 1.6 \times 10^{-30}\ n_{10}\Sigma\left(\frac{T_6}{11.6}\right)^2. \tag{27}$$

The value of n_{10} is somewhat uncertain but is probably in excess of $10^{10}\ \text{cm}^3$ (e.g. Neupert, 1969; Culhane *et al.*, 1970) as determined by soft X-ray data. The area Σ of the hard X-ray source is entirely unknown though Takakura *et al.* (1971) have given some evidence for its being very small – certainly smaller than the soft X-ray sources photographed by e.g. Vaiana *et al.* (1968) who put their characteristic size at around 10^9 cm. Thus the area Σ is unlikely to exceed $10^{18}\ \text{cm}^2$ (cf. Culhane *et al.*, 1970) so that

$$\alpha \lesssim 1.6 \times 10^{-2}\left(\frac{T_6}{11.6}\right)^2. \tag{28}$$

Hence even at temperatures $\gtrsim 10^8$ K $(T_6 \gtrsim 100)$, α is only of order unity and conductive conditions might exist in the source. In view of the sensitivity of α on the various parameters in (26) and of the uncertainty in the values of these, however, it is appropriate to calculate the energy flow rates on both the thermal conduction approximation and on the mass flow basis.

Provided the conduction coefficient is not anomalously reduced by Plasma turbulence, then the conductive flux F_{cond} will be as given in Spitzer (1962), viz.

$$F_{\text{cond}} \simeq -7.6 \times 10^{14}\ T_6^{5/2}\ \frac{\mathrm{d}T_6}{\mathrm{d}Z}$$

which, with power-law gradient (25) reduces to

$$F_{\text{cond}} = -7.4 \times 10^{-24}\ n_{10}^2\Sigma\ \frac{\Gamma(\gamma-1)}{(\gamma-1)}\ \frac{\varepsilon_0^{-\gamma+1}}{I_0}\left(\frac{T_6}{11.6}\right)^\gamma \tag{30}$$

which illustrates how the true temperature dependence of F_{cond} differs here from the constant flux adopted by Kahler (1971a).

Following Kahler (1971b), the energy flux for a mass flow situation may be written as

$$F_{\text{mass}} \simeq \tfrac{1}{4} n\bar{v}\bar{E}.$$

If this mass flow followed the sudden creation of the high temperature situation, the density distribution $n(\mathbf{r})$ could initially be arbitrary since the configuration would be transient and constant pressure condition (25) would not apply. In fact the plasma distribution would readjust to near constant pressure on a time scale $\simeq l/\bar{v}$ which will always be short compared to typical durations of hard X-ray bursts (10–100 s). Subsequently the cooling of the hot plasma will approximate to the exchange of hot and cool material by nett flow at approximately constant pressure. Therefore during the sustained part of the burst, the mass energy flow can be written, using (25), as

$$F_{\text{mass}} \simeq -0.35\ n_{10}\ T_6^{1/2}. \tag{31}$$

comparing (28) and (29), then

$$\frac{F_{\text{cond}}}{F_{\text{mass}}} \simeq 2 \times 10^{-24}\, n_{10} \Sigma \frac{\Gamma(\gamma-1)}{\gamma-1} \frac{\varepsilon_0^{-\gamma+1}}{I_0} \left(\frac{T_6}{11.6}\right)^{\gamma-1/2} \tag{32}$$

which, for the burst observed by Cline *et al.* (1968), gives

$$\frac{F_{\text{cond}}}{F_{\text{mass}}} \simeq 10^{-3} \times \left(\frac{T_6}{11.6}\right)^3, \tag{33}$$

where $n_{10} \simeq 10^{10}$ and $\Sigma \simeq 10^{18}$ have been again assumed.

Equation (28) showed that, for this burst, the conductive cooling description breaks down for $T_6 \geqslant 100$ (i.e. $T \geqslant 10^8$ K). For these high temperatures, however, (33) shows that use of the conduction approximation description in fact *overestimates* the heat flow rate since F_{cond} exceeds F_{mass}. (This situation differs from the results of Kahler, 1971a, b due to the dependence of F_{cond} on $\mathrm{d}T_6/\mathrm{d}Z$ which is correctly given as a function of T_6 by Equation (25) and not by Kahler's empirical model.) Thus to investigate the feasibility of sustaining a multitemperature flare plasma to $T \geqslant 10^8$ it will be conservative to adopt the conductive cooling approximation throughout, as will be done from this point on.

The actual effect of the conductive flow on the temperature is given by considering the characteristic time $\tau_c = \mathscr{E}/\nabla \cdot \mathbf{F}$ where $\mathscr{E}$ is the internal energy (erg cm^{-3}) and $\mathbf{F}$ is the vector thermal energy flux. Differentiating (30) in the present one-dimensional case, and using (25), gives

$$\nabla \cdot \mathbf{F} = \frac{\mathrm{d}F_{\text{cond}}}{\mathrm{d}Z} = \frac{\mathrm{d}F_{\text{cond}}}{\mathrm{d}T_6} \times \frac{\mathrm{d}T_6}{\mathrm{d}Z} =$$

$$= -1.3 \times 10^{-65}\, \gamma \left[n_{10}^2 \Sigma \frac{\Gamma(\gamma-1)}{(\gamma-1)} \frac{\varepsilon_0^{-\gamma+1}}{I_0} \right]^2 \left(\frac{T_6}{11.6}\right)^{2\gamma-7/2} \tag{34}$$

whilc

$$\mathscr{E} = 3\, nkT_6 \times 10^6 = 3 \times 10^7\, n_{10} k = 4.2 \times 10^{-9}\, n_{10}, \tag{35}$$

i.e. the energy density is constant through the source (due to constant pressure conditions).

Combining (34) and (35), then

$$\tau_c \simeq -\frac{3.1 \times 10^{56}}{n_{10}^3 \Sigma^2} \frac{(\gamma-1)^2}{\gamma(\Gamma(\gamma-1))^2} \left(\frac{I_0}{\varepsilon_0^{-\gamma+1}}\right)^2 \left(\frac{11.6}{T_6}\right)^{2\gamma-7/2}. \tag{36}$$

This is the key result required and contains all the information needed about τ_c as a function of the observed flare parameters. If differs from the corresponding result from Kahler (1971a, b) and also from Culhane *et al.* (1970) in a number of respects, of which two are crucial.

Firstly, the negative sign of τ_c indicates that, in order to fit an X-ray power-law spectrum, the plasma is (locally) *not cooled* by conduction at all but rather *heated*. This is a consequence of the concavity of the function $F_{\text{cond}}(Z)$ (as against the convex form arising in previous empirical models) similar to the situation obtaining at the base of the quiet chromosphere transition region (cf. for example Kuperus, 1969). At high enough temperatures this state of affairs must break down, the temperature gradient inflecting into the region of maximum temperature (and appearing as a break in the hard X-ray spectrum such as is observed in fact – e.g. Frost, 1969) from which the energy flow required by (36) would ultimately emanate. Thus, though τ_c is in fact a heating rather than a cooling time, its significance in the present context is unchanged since it still represents the characteristic time for conductive *redistribution* of thermal energy in the source, and hence the time scale on which temperature structure (25), could survive in the absence of sustained energy supply (a factor considered in the next section).

The second important aspect of (36) is that it shows τ_c to be very sensitive to the adopted values of n_{10} and Σ as well as to the local temperature T_6 (depending on index γ). Any substantial uncertainty in assessing either n_{10} or Σ will therefore have a critical effect on the feasibility of the multi-thermal model for hard X-ray emission. Detailed results for the large event of Cline *et al.* (1968) are given in the next section but, bearing in mind the uncertainties involved, a tentative value of τ_c can be assigned for this event using the values for n_{10} and Σ mentioned above. These give

$$\tau_c(\mathrm{s}) \lesssim 1.5 \times 10^5 \times \left(\frac{11.6}{T_6}\right)^{3.5}$$

so that even at $T_6 = 116$ (i.e. 1.16×10^8 K) the time $\tau_c \simeq 50$ s which is within a factor of 2 of the observed decay time of this burst at 80 keV. Thus, with these parameters, the conductive temperature redistribution could actually be identified with the burst decay, contrary to the results and conclusions of Kahler (1971a, b). However, since τ_c is so sensitive to n_{10} and Σ (e.g. increasing n_{10} threefold decreases τ_c from 50 to 2 s), this identification is probably fortuitous. As discussed in the following section, further considerations indicate that τ_c is indeed probably smaller than the observed burst decay time. This would merely mean, however, that a continuing energy input was required through the burst duration and the above expression for τ_c allows evaluation of the required input which is shown below to be feasible.

5. Detailed Results for a Large Event

Some of the analytical results of the previous sections may be clarified by their representation in numerical form for a particular event, which will be chosen once again as that observed by Cline *et al.* (1968), for which, as already stated, $I_0 = 300$ cm^{-2} s^{-1} for $\varepsilon_0 = 80$ keV and $\gamma = 3.5$. Considering first the differential emission measure $\mu^*(T_6)$, this may be tabulated using (20) for a variety of temperatures, (see also Figure 1), see below.

These figures may be compared with those suggested by Chubb (1970), viz. an emission measure of around 10^{46} cm^{-3} for a 10^8 K plasma. The corresponding figure here is

$$\mu_{tot}(T>10^8)=10^{41}\int_{100}^{\infty}\mu^*(T_6)\,dT_6$$

which by (20) $\simeq 10^{47}$ cm^{-3} – a somewhat greater value than Chubb's but in adequate agreement when it is noted that the event recorded by Cline *et al.* is one of the largest and hardest ever recorded.

T_6	10	20	40	100	200	400
T(K)	10^7	2×10^7	4×10^7	10^8	2×10^8	4×10^8
μ^*(10^{41} cm^{-3} per 10^6 K)	2×10^7	2.5×10^6	3×10^5	2×10^4	2.5×10^3	3×10^2
$\mu(T)$ (cm^{-3} per 10^6 K)	2×10^{48}	2.5×10^{47}	3×10^{46}	2×10^{45}	2.5×10^{44}	3×10^{43}

Equation (22) provides the volume of the emitting plasma when evaluated between the probable temperature T_1 ($\simeq 4\times10^7$ K) at which the region becomes isothermal and an upper limit T_2 of around 10^8 K. Then

$$\Delta V \lesssim 10^{29}\ \text{cm}^3$$

for $n_{10}\gtrsim 10^{10}$ cm^3. (This also follows approximately from $\mu_{tot}=10^{47}$ cm^{-3} and a value of 10^9 cm^{-3} for n at $T=10^8$ K.) This volume is rather large since for a source filament cross-section $\Sigma\simeq 10^{18}$ cm^2, the corresponding length $\Delta Z\simeq 10^{11}$ cm exceeds a solar radius. Though no high resolution hard X-ray data exist to test the existence of such extensive sources, a single source of this size seems improbable. The problem may be eluded however, if the source consists of two filaments (c.f. Culhane *et al.*, 1970) like the soft X-ray flare (Vaiana *et al.*, 1968) each of $\Delta Z\simeq 5\times10^{10}$ cm (i.e. a solar radius) in the form of arched coronal plasma traps. Kahler (1971a) expresses the view that such an extended source is unlikely since the energy source would have to be very extended also. This is not so since, with the correct expression for $T(Z)$, conduction has been shown here to *supply* energy along the filament from a source at the high temperature end where the basic flare energy release may occur – i.e. for example, the required thermal X-ray source temperature structure could be sustained by conduction from an extremely hot coronal region heated in a plasma turbulent neutral sheet configuration.

However, such lengths are not in any case essential to the model considered here since they can be reduced by increase of n_{10} or of Σ, though this at the expense of a decreased decay time τ_c (Equation (36)) so that energy would have to be continuously supplied to sustain the temperature structure required by the burst. A twofold increase of both n_{10} and Σ would reduce ΔV by a factor of 4 and ΔZ by a factor of 8 (Equations (22) and (24)) but would reduce the time τ_c for the event of Cline *et al.* (1968) to about 2 s (Equation (36)). The total input energy flux required to sustain the burst thermally would then be the value of F_{cond} (Equation (30)) near the maximum temperature (T_6^{max})

present – i.e. at the inflection point in dT/dZ, viz.

$$F_{in} = F_{cond} \simeq 1.6 \times 10^8 \left(\frac{T_6}{11.6}\right)^{3.5}. \tag{37}$$

At $T = 10^8$ K, this gives $F_{in} \simeq 3 \times 10^{11}$ erg cm^{-2} s^{-1} at 10^8 K which is a large energy flux but over the filament area of 2×10^{18} cm^2 for the 100 s necessary, the total energy input required is about 6×10^{31} erg which is easily available in an event this large. This particular burst, was, however, so hard that the temperature needed to produce it *entirely* thermally would be nearer 10^9 K than 10^8 K (c.f. Figure 1) whereas Equation (37) shows that a reasonable maximum for F_{in} restricts T to $\lesssim 2 \times 10^8$ K. Within the terms of the preceding analysis, this would require that the thermal component of hard X-ray bursts give way continuously to non-thermal emission at the highest photon energies. Such a smooth transition between the two components is entirely reasonable, especially if the hot plasma and the high energy tail of particle streams are generated in one and the same plasma turbulent region. (Potential methods of observing this transition energy are discussed briefly in Section 6, while its importance as regards the entire question of flare energetics has already been emphasized in Section 1.) On the other hand, the above results are based purely on the assumption of normal thermal conductivity. It is entirely conceivable that the conductivity throughout the source may be reduced by the presence of plasma turbulence and by factors of up to 10^4. This would of course totally eliminate any objections to the thermal model based on the argument of rapid conductive cooling, even up to the hardest photons observed by Cline *et al.* (1969).

Fuller consideration of these latter features of the problem must, however, incorporate more specific models of the X-ray emitting plasma and of its energy source, and is therefore postponed till a future paper. The aim here has been to emphasize that the possibility of high temperature plasmas as the source of hard X-ray flares cannot be excluded on energy grounds within the limits of presently available observations. The main observations needed to resolve the issue are some limits on the dimensions of the hard X-ray flare, presumably from occultation or collimator experiments. Lacking these, there are two other observational data to which appeal may be made. viz. the polarization and the time variations of hard X-ray bursts. The relationship of these to the source model described above is discussed briefly below.

6. Polarization and Temporal Fine Structure in Hard X-Ray Flares

It has been frequently supposed (e.g. Brown, 1971; Tindo *et al.*, 1970, 1973) that the observation of substantial polarization (up to around 40%) in flare hard X-rays (Tindo *et al.*, 1970, 1972, 1973) and also perhaps of directivity (Ohki, 1969; Pintér, 1969; Drake, 1971) is indicative of non-thermal processes, and particularly of electron streams. Recently, however, Tomblin (1972) drew attention to the contribution in observed soft X-ray fluxes of source photons reflected by the photospheric albedo.

Santangelo *et al.* (1973) have shown this contribution to be even more important at high photon energies, being a large proportion of the total observed flux in the hard X-ray range. In common with all such scattered radiation, as pointed out by Santangelo *et al.* (1973), these albedo X-rays will be partially linearly polarized, thus modifying previous calculations of the polarization and directivity of non-thermal hard X-rays (e.g. Haug, 1972; Brown, 1972b). More important in the present context, however, is that photospheric scattering will *introduce* polarization and anisotropy into the X-rays observed from a thermal source above the photosphere, this source being itself unpolarised and isotropic. Thus the *presence* of polarization and anisotropy in hard X-ray flares does not after all preclude a multi-temperature thermal source. There is therefore an urgent need for quantitative work on the expected extent of these effects. Beigman and Vainshtein (1974) have calculated about 2% polarization introduced by albedo photons in the case of thermal soft X-rays but do not investigate the hard X-ray case. Considering the much greater albedo contribution at hard energies, however, it seems likely that the polarization introduced will also be considerable, perhaps comparable with that observed.

As regards the *plane* of albedo polarization, it would be expected that the maximum intensity occurs normal to the scattering plane – i.e. approximately normal to the solar radius through the source region (Santangelo *et al.*, 1973), unless the scattering is in some way anomalous as in the case of lunar polarization near full phase (Dollfus, 1966). This prediction does not in fact agree with the sparse observational data available (Tindo *et al.*, 1972) which indicate that for a few bursts the emission is partially polarized approximately *along* the solar radius through the flare – a result which may be interpreted in terms of vertically streaming electrons (Tindo *et al.*, 1972; Brown, 1972b) though this interpretation may be in doubt (Brown *et al.*, 1974). With the reservations expressed at the end of Section 5, it seems likely that the X-ray emission mechanism does become non-thermal at high enough energies but the energy at which the transition occurs is the all important question and may vary from flare to flare. If a 'radial' polarization plane *is* the rule rather than the exception for the non-thermal component then observations of the variation of the polarization plane with photon energy may provide the key to determining where thermal emission gives way to non-thermal. For at low enough energies, the plane *must* go over to the tangential case characteristic of scattering polarization, the transition occurring where thermal processes become dominant. A plea is therefore in order for theoretical work on albedo hard X-ray polarization and for observations of the energy dependence of the polarization.

Turning finally to the time dependence of the observed hard X-ray flux from flares, rapid time variations ($\simeq$seconds), well correlated with (non-thermal) microwave emission, are typical features (Frost, 1969; Parks and Winckler, 1969, 1971) which Kahler (1971a) for example, regards as uncharacteristic of thermal emission. Though the microwave burst (gyro-synchrotron) mechanism certainly involves non-thermal electrons (Takakura, 1967), it does depend on the flare field strength H in the emitting region. This field may vary rapidly with time due to magnetohydrodynamic waves with concomitant variations in the density of the frozen-in ambient plasma (e.g. Parks and Winckler, 1969, 1971; Wilson, 1972; Brown, 1973c). Such a situation could produce the observed time structure of microwave emission from trapped high energy (non-thermal) electrons and closely correlated variations in the ambient plasma density n and hence in any *thermal* X-ray emission arising from it (Equation (1)). Though this proposal is certainly subject to detailed investigation, it should serve to show that temporal fine structure in hard X-ray flares may not be taken out of hand as precluding a multi-temperature thermal emission mechanism. (The definite presence of non-thermal electrons required for microwave emission is not contrary to this view since the electron numbers involved are only 10^{-3} of those needed for the X-ray bremsstrahlung.) The factor which would ultimately limit the range of validity of such a description is that, at high enough temperatures, the electron mean free path (and hence the Maxwell relaxation time) becomes too great (Equation (26)) for description of the electron distribution as thermal at all within the observed rapid time structures (Kahler, 1971b). Above this limit the usual distinction between ambient and non-thermal electrons will be come meaningless. However, as shown by the equation preceding (26), for a density $n_{10} \simeq 10^{10}$ cm^{-3}, and time structure of a few seconds this limitation does not set in until $T \gtrsim 5 \times 10^8$ K – i.e. well into the regime normally regarded as non-thermal. Furthermore, the reduction of conductivity by turbulence, would permit much higher values of n_{10} in the model described here and consequent Maxwellian relaxation times much shorter than any observed hard X-ray fine structure. Hence it must be concluded that the important question of the relative importance of non-thermal and high temperature electrons in flare hard X-ray bursts remains open to fuller observational and theoretical investigation.

Acknowledgements

My thanks are due to Dr Steve Kahler for constructive criticisms of the preprint of this paper, and to the IAU, the Australian Academy of Sciences, and to Glasgow University whose support enabled me to attend this Symposium.

References

Abramowitz, M. and Segun, I. A.: 1965, *Handbook of Mathematical Functions*, Dover.
Batestone, R. M., Evans, K., Parkinson, J. H., and Pounds, K. A.: 1970, *Solar Phys.* **13**, 389.
Beigman, I. L. and Vainshtein, L. A.: 1974, *Astron. J. (USSR)*, in press.
Billings, D. E.: 1966, *A Guide to the Solar Corona*, Academic Press, N.Y.
Brown, J. C.: 1971, *Solar Phys.* **18**, 489.
Brown, J. C.: 1972a, *Solar Phys.* **25**, 158.
Brown, J. C.: 1972b, *Solar Phys.* **26**, 441.
Brown, J. C.: 1973a, *Solar Phys.* **28**, 151.
Brown, J. C.: 1973b, *Solar Phys.* **31**, 143.
Brown, J. C.: 1973c, *Solar Phys.* **32**, 227.
Brown, J. C., McClymont, A. N., and McLean, I. S.: 1974, *Nature* **247**, 448.
Chambe, G.: 1971, *Astron. Astrophys.* **12**, 210.
Chubb, T. A., Kreplin, R. W., and Friedman, H.: 1966, *J. Geophys. Res.* **71**, 3611.

Chubb, T. A.: 1970, in E. R. Dyer (ed.), *Proc. Leningrad Symp. Solar Terrestrial Phys.*, Vol. 1, D. Reidel Publ. Co., Dordrecht, Holland.
Chupp, E. L., Forrest, D. J., Higbie, P. R., Suri, A. N., C., and Dunphy, P. P.: 1973, *Nature* **241**, 333.
Cline, T. L., Holt, S. S., and Hones, E. W.: 1968, *J. Geophys. Res.* **73**, 434.
Culhane, J. L.: 1969, *Monthly Notices Roy. Astron. Soc.* **144**, 375.
Culhane, J. L., Vesecky, J. F., and Phillips, K. J. H.: 1970, *Solar Phys.* **15**, 394.
Dollfus, A.: 1966, in Hess, Menzel, and O'Keefe (eds.), *The Nature of the Lunar Surface*, John Hopkins, Press.
Drake, J. F.: 1971, *Solar Phys.* **16**, 152.
Elwert, G. and Haug, E.: 1971, *Solar Phys.* **20**, 413.
Felten, J. E. and Rees, M. J.: 1972, *Astron. Astrophys.* **17**, 226.
Fichtel, C. E. and McDonald, F. B.: 1967, *Ann. Rev. Astron. Astrophys.* **5**, 351.
Frost, K. J.: 1969, *Astrophys. J.* **158**, L.159.
Greisen, E.: 1971, *Physics of Cosmic X-Ray, γ-Ray and Particle Sources*, Gordon and Breach.
Herring, J. R. H. and Craig, I. J. D.: 1973, *Solar Phys.* **28**, 169.
Haug, E.: 1972, *Solar Phys.* **25**, 425.
Illarionov, A. F. and Sunyaev, R. A.: 1973, *Astrophys. Space Sci.* **19**, 61.
De Jager, C.: 1967, *Solar Phys.* **2**, 327.
Kahler, S. W.: 1971a, *Astrophys. J.* **164**, 365.
Kahler, S. W.: 1971b, *Astrophys. J.* **168**, 319.
Kane, S. R. and Anderson, K. A.: 1970, *Astrophys. J.* **162**, 1003.
Kane, S. R. and Donnelly, J. E.: 1971, *Astrophys. J.* **164**, 151.
Kuperus, M.: 1969, *Space Sci. Rev.* **9**, 713.
Milkey, R. W.: 1971, *Solar Phys.* **16**, 465.
Neupert, W. M.: 1968, *Astrophys. J.* **153**, L.59.
Neupert, W. M.: 1969, *Ann. Rev. Astron. Astrophys.* **7**, 121.
Ohki, K.: 1969, *Solar Phys.* **7**, 260.
Parks, G. K. and Winckler, J. R.: 1969, *Astrophys. J.* **155**, L117.
Parks, G. K. and Winckler, J. R.: 1971, *Solar Phys.* **16**, 186.
Pintér, Š.: 1969, *Solar Phys.* **8**, 142.
Santangelo, N., Horstman, H., and Horstman-Moretti, E.: 1973, *Solar Phys.* **29**, 143.
Sneddon, I. N.: 1972, *The Use of Integral Transforms*, McGraw Hill.
Spitzer, L.: 1962, *Physics of Fully Ionised Gases*, Interscience, New York.
Takakura, T.: 1967, *Solar Phys.* **1**, 304.
Takakura, T., Ohki, K., Shibuya, N., Fujii, M., Matsuoka, M., Miyamoto, S., Nishimura, J., Oda, M., Ogawara, Y., and Ota, S.: 1971, *Solar Phys.* **16**, 454.
Tindo, I. P., Ivanov, V. D., Mandelstam, S. L., and Shuryghin, A. I.: 1970, *Solar Phys.* **14**, 204.
Tindo, I. P., Ivanov, V. D., Mandelstam, S. L., and Shuryghin, A. I.: 1972, *Solar Phys.* **27**, 426.
Tindo, I. P., Mandelstam, S. L., and Shuryghin, A. H.: 1973, *Solar Phys.* **32**, 469.
Tomblin, F. F.: 1972, *Astrophys. J.* **171**, 377.
Vaiana, G. S., Reidy, W. P., Zehnpfennig, T., Van Speybröck, L., and Giacconi, R.: 1968, *Science* **161**, 564.
Wilson, P. R.: 1972, *Solar Phys.* **22**, 434.

DISCUSSION

I suggest that if such large amounts of very hot plasma are assumed to account for hard, fast X-ray bursts then it is difficult to account for the following flare observations:

(1) Observed weakness of Lα radiation from Fe XXVI.
(2) Small (< 1′) hard X-ray source size measured by Oda *et al.*
(3) Close time profile correlation between μ-wave, EUV and hard X-ray bursts.

Brown: The Lα of iron at 1.8 Å would be gone at those temperatures.

Acton: But it should appear during the cooling phase.

Smith (to Acton): It is completely possible to heat plasma very rapidly by dumping the energy into wave modes which are heavily damped such as ion-acoustic waves.

Kane: I think that the objection No. 3 mentioned by you, viz. the difficulty in explaining the rapid time variations, should be taken seriously. To illustrate how difficult this problem can get, I would like to show a slide. This slide shows at the top 10–580 MHz spectral radio data obtained by Dr Maxwell and at the bottom OGO-5 X-ray data above 20 keV energy. Corresponding to the 5 groups of type III bursts in the radio data, there are 5 peaks in the X-ray data. This strongly suggests that, like the type III radio emission, the time variation in the X-ray emission may have a time constant of ~ 0.1 s which is very difficult to explain by a thermal model of the X-ray source.

Meyer: Your assumed conditions appear pretty far from pressure equilibrium with the surroundings.

Sturrock: In connection with the thick/thin X-ray target controversey, Petrosian has recently analyzed the problem of a vertical electron beam and finds the polar diagram concentrated downward and a strong spectral variation with direction. He concludes that this makes the deduction of the electron spectrum from the X-ray spectrum unreliable.

Brown: I have not seen this paper but my own analysis of that problem showed that the X-ray spectral index varied by less than 0.6 with direction, which is not very significant. On the other hand, the intensity can vary by about 5. It may be that Petrosian neglected scattering and so over-estimated the beam collimation.

PART IV

ACCELERATION, CONTAINMENT AND EMISSION OF HIGH-ENERGY FLARE PARTICLES

THE CHARGE AND ISOTOPIC COMPOSITION OF SOLAR COSMIC RAYS

FRANK B. McDONALD

NASA Goddard Space Flight Center, Greenbelt, Md., U.S.A.

Abstract. The charge and isotopic composition of solar cosmic rays potentially contain a wealth of information on the acceleration and confinment of energetic particles at the Sun. As the experimental techniques have improved and with the increasing number of measurements, this potential is now being realized. It is convenient to divide the composition studies into three areas: (a) helium-iron, with energies >10 MeV nuc^{-1}; (b) helium-iron, and the trans-iron elements, with energies <2 MeV nuc^{-1}; (c) the isotopic composition. In the first area at energies above 10 MeV nuc^{-1} the solar cosmic rays appear to provide a representative sample of the solar corona. At lower energies (b) complex enhancement effects are noted. These increase with Z and decrease with energy. This result as well as the high abundance of ^{3}He suggest that the acceleration process is not a simple one and probably several stages are required.

1. Introduction

The original title for this paper as defined by the Conference Organizing Committee was the 'Acceleration, Containment and Emission of High Energy Particles'. The question of acceleration remains one of the major unsolved problems in particle astrophysics. The idea of prolonged particle containment at the Sun has increasingly become part of the accepted picture of solar cosmic rays without having been clearly established experimentally. In part this evolved from the observation of energetic particles associated with an active center (Fan *et al.*, 1968; McDonald and Desai, 1971). These display almost no velocity dispersion and appear to be co-rotating with an active center. They often persist for up to 14 days and have steep (E^{-3} to E^{-6}) energy spectra. On one occasion these persisted over 14 solar rotations. However, with increasing sensitivity it is observed (McDonald and Van Hollebeke, 1973) that there are frequently many small particle injections associated with these streams of MeV particles. Furthermore, the particle propagation is diffusive and it still has not been possible to untangle coronal and interplanetary diffusion processes or to determine the time profile of the particle injection from the active region. One channel that is beginning to reveal a significant amount of information on both the acceleration and storage processes is the charge and mass composition of solar cosmic rays. The charge composition will be a function of both the acceleration process and the properties of the source region. The isotopes such as ^{3}He should also give information on the dynamics of particle acceleration and storage in this source region. At low energies (i.e. ≤ 2 MeV nuc^{-1}) and high charges ($Z>$iron) the charge composition is strongly energy dependent. However, above ~ 10 MeV nuc^{-1} a representative sample of the solar corona is accelerated in flare-associated events for the elements helium-iron. The numerous measurements of charge composition and isotopes can best be understood by dividing the data into three regions:

(A) Helium-iron with energies >10 MeV nuc^{-1}.

Gordon Newkirk, Jr. (ed.), Coronal Disturbances, 415–420. *All Rights Reserved.*

(B) Helium-iron with energies <2 MeV nuc^{-1} and the trans-iron elements.
(C) The isotopic composition of solar cosmic rays.

2. The Charge Composition of Helium-Iron with Energies >10 MeV Nucleon

The pioneering work in this region was carried out by Fichtel and his co-workers at the Goddard Space Flight Center using rocket borne nuclear emulsion (Fichtel and Guss, 1961; Biswas *et al.*, 1963, 1965, 1966; Durgaprasad *et al.*, 1968). Over a period extending from September 1961 to August 1972 rockets were successfully recovered for 8 events. These studies indicated that the composition for $Z=2$–26 does not change markedly from event to event and is in strong agreement with the spectroscopically determined abundances in the solar atmosphere. Further observations in this range has been done with particle detectors on the OGO-V satellite by the University of Chicago group (Mogro-Campero and Simpson, 1972) and on IMP VI by the Goddard cosmic ray group (Teegarden *et al.*, 1973). In addition plastic track detectors on rocket flights and on Apollo missions have provided data for a wide variety of events. In general these data sets are in good agreement. The only notable exception to this are the University of Chicago results in the argon-iron regions.

As the measurements have become more precise it is found that the ratio of He/C–N–O can vary by a factor of ~2 from event to event. This was first noted at lower energies by Armstrong *et al.* (1971) and confirmed by Teegarden *et al.* (1973). In the iron region Bertsch *et al.*, find that when the He/C–N–O ratio is low then the He/Fe ratio is also systematically lower. However, these variations from event to event are still less than that encountered between different sets of coronal abundance

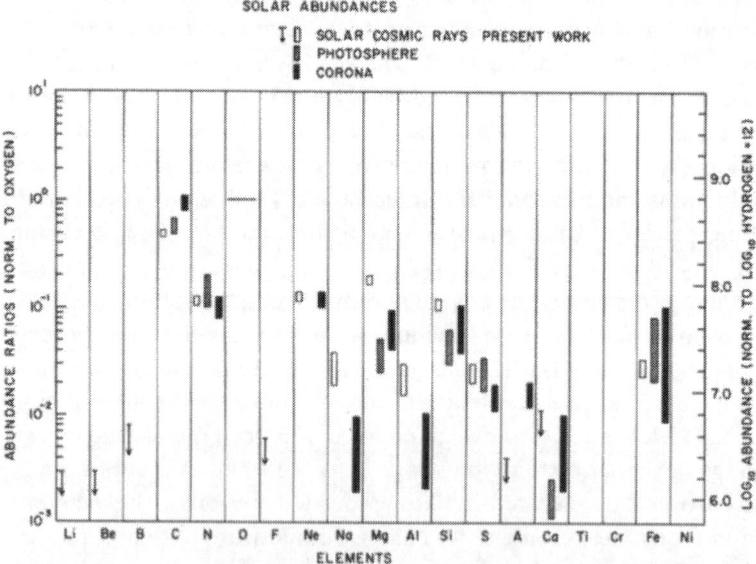

Fig. 1. A comparison of representative solar cosmic ray, coronal and photospheric abundances. All have been normalized at oxygen = 1.

data. A typical comparison between solar cosmic rays and coronal abundances is shown in Figure 1 (Teegarden *et al.*, 1973). The various measurements have all been normalized to oxygen. The spread in the particle and abundance data sets is an indication of the variation between various studies. Of particular note is the detection of Na and Al in the large flare events of August 1972 (Webber, 1973).

3. He–Fe, $E<2$ MeV nuc^{-1} and the Trans-Iron Elements

The energy interval between 2 and 10 MeV is one of transition. Above 10 MeV, electron pick-up is not important except for iron. By 2 MeV, it has become important in the oxygen region. Up to the present time, most of the experimental work has been done using plastic track detectors on the Apollo missions or the 'Electrostatis Deflection Analyzer' flown on IMP VII by G. Gloeckler and his co-workers. All measurements show there is a systematic enhancement of heavy nuclei at low energies. This enhancement increases with Z and decreases with energy. Following the example of Price (1973), it is useful to define an enrichment factor $Q(Z_1, Z_2)$ for the enrichment of Z_1, with respect to Z_2

$$Q = \frac{\dfrac{J(Z_1, E, t)\text{ solar cosmic rays}}{J(Z_2, E, t)\ S.P.}}{\dfrac{\text{solar abundance of } Z_1}{\text{solar abundance of } Z_2}},$$

where $J(Z, E, t)$ is the flux of element Z, of energy E at time t. Since at higher energies there is good agreement between the relative abundance of solar cosmic rays and coronal abundance, an equivalent definition of Q would be to replace the ratio of solar abundances with the ratio of fluxes measured at higher energies.

The different measurements presently available in the low energy range are listed in Table I.

TABLE I

Measurements of energy dependent composition of solar heavy ions

Event	Time interval studied	Detector	Reference
Strong flare	1510 to 1514 UT, 25 Jan. 1971	Lexan stack on rocket	Crawford *et al.* (1972)
Strong flare	0756 to 0760 UT, 2 Sept. 1971	Lexan stack on rocket	Price *et al.* (1973)
Small flare	Integrated over 17 to 19 Apr. 1972	Lexan + SiO_2 glass on Apollo 16	Braddy *et al.* (1973)
Very strong flare	1914 to 1918 UT, 4 Aug. 1972	Lexan stack on rocket	Price *et al.* (1973)
Almost quiet Sun	Integrated over 11 to 13 Dec. 1972	Lexan on Apollo 17	Price and Chan (1973b)
Small flare	Oct. 17–19, 1972	Electrostatic deflection analyzer	Gloeckler *et al.* (1973)

Price (1973) has summarized the systematics of heavy ion enhancement as measured by many different groups (Table I) as follows:

(1) At sufficiently low energies, the heavy elements in solar flare particles are always enriched relative to their abundance at high energies.

(2) The enhancement factor Q is an increasing function of Z from helium up to $Z \approx 44$ and probably higher. The available data is not adequate to determine any possible structure in $Q(Z, E)$.

(3) Q decreases with energy. At higher energies ($\gtrsim 2$ MeV, it starts to approach a constant value.

An example of the systematic increase of Q with Z is shown in Figure 2 for a small

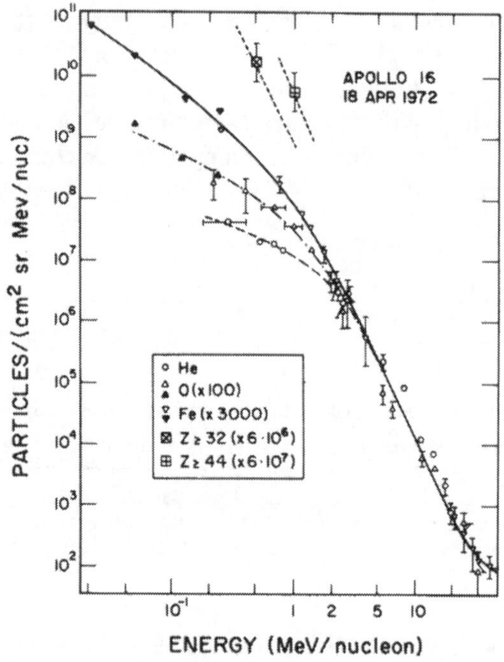

Fig. 2. Fe and heavier elements measured with $Si–O_2$ glass; O and He measured with Lexan stack for small event on 18 April 1972. The scaling factors are chosen so that the curves converge at higher energies (Price *et al.*, 1973).

event on 18 April 1972 using one of the Apollo 16 command module windows as the track detector. Oxygen, iron and the trans-iron elements have been normalized to helium above ~5 MeV nuc^{-1}. At lower energies the systematic enhancement of the heavier element is very evident. Shink *et al.* (1973) find that $Q(Z/\text{He}) \propto \exp Z/Z_1(E)$ where $Z_1(E)$ increases with energy.

Using the electrostatic deflection technique Gloeckler *et al.* (1973) have also been able to determine the charge states of oxygen and carbon. They find in average value at 100 keV nuc^{-1} of 5.4 for carbon and 7.4 for oxygen. The expected equilibrium charge states are 2.5 and 2.9. This suggests these low energy particles may have traversed a hot coronal region after acceleration where they lose further electrons. The other possibility is that the ions have undergone considerable adiabatic deceleration during their interplanetary propagation.

This low energy enhancement when fully understood should give valuable insight into the acceleration process. One possible mechanism which involved two acceleration stages has been suggested by Cartwright and Mogro-Campero (1973). Initially the ions are fully stripped and are accelerated to suprathermal energies (~ 1 MeV nuc^{-1}). They then pass through the relatively cool chromosphere where they reach charge equilibrium in their transport to the Fermi acceleration region. Since injection into Fermi acceleration is rigidity dependent, an enhancement of heavy nuclei can occur. This is probably indicative to the type of modification that will have to be made to conventional acceleration process.

4. Isotopic Composition of Solar Cosmic Rays

The satellite particle detector technology has now advanced to the point where it should be possible during the late '70's to resolve most of the individual isotopes region for C–Fe. This will be an important new tool in comparing coronal abundances with the 'Universal Abundances'. Webber *et al.* (1973) have measured the isotopes of Ne and Mg as well as O^{18} during the large August 1972 events. These are in good agreement with the expected values.

It is possible with presently available systems to resolve the isotopes of hydrogen and helium. Thus far both the University of Chicago Group (Anglin *et al.*, 1973) and Caltech (Garrard *et al.*, 1973) have been able to measure ^{3}He in a number of events. They find that the measured ratio of He^3/He^4 can be as large as 500 times the abundance ratio of 5×10^{-4} measured in the solar wind (Geiss, 1972). Furthermore, the ratio of He^3/He^4 can vary by more than an order of magnitude from event to event. Since the ratio is so much greater than the expected value, the He^3 mist be produced by nuclear interaction. For example above 10 MeV nuc^{-1} this results mainly from energetic He^4, C, N and O interacting with the nuclei in the coronal gas. The presence of these nuclear interactions products mean that solar γ-rays, neutrons and positrons would also be produced at the same time. There is a significant absence of deuterium which is not fully understood at this time.

The large values of 0.26 ± 0.08 reported for the He^3 rich event of October 13, 1969 (Gerrard *et al.*, 1973) would require the passage through at least 6 g cm^{-2} of gas for a simple slab model. This is quite unrealistic and again indicates that the basic acceleration process must be re-examined. For example, the He^3 could be built up over a long period of time by particles stored in the active region. At this time there is no evidence for such storage. On the other hand, the He^3 could be produced by 'dumping' most of the protons onto the photosphere and reaccelerating the fragmentation products. The proton 'dumping' would actually be similiar to that experienced by the electrons.

While the origin of the He^3 is not understood at this time, its presence in such significant amounts indicates it is going to be a very important factor in understanding the source region dynamics.

References

Anglin, J. D., Dietrich, W. F., and Simpson, J. A.: 1973, in R. Ramaty and R. G. Stone (eds.), *High Energy Phenomena on the Sun Symposium Proceedings*, NASA Goddard Preprint, p. 315.

Armstrong, T. P. and Krimigis, S. M.: 1971, *J. Geophys. Res.* **76**, 4230.

Bertsch, D. L., Fichtel, C. E., Pellerin, C. J., and Reames, D. V.: 1973, *Astrophys. J.* **180**, 583.

Biswas, S. and Fichtel, C. E.: 1963, *Astrophys. J.* **139**, 941.

Biswas, S. and Fichtel, C. E.: 1965, *Space Sci. Rev.* **4**, 709–36.

Biswas, S., Fichtel, C. E., and Guss, D. E.: 1966, *J. Geophys. Res.* **71**, 4071.

Braddy, D., Chan, J., and Price, P. B.: 1973, *Phys. Rev. Letters* **30**, 669.

Crawford, H. J., Price, P. B., and Sullivan, J. D.: 1972, *Astrophys. J.* **175**, L149.

Cartwright, B. G. and Mogro-Campero, A.: 1973, in R. Ramaty and R. G. Stone (eds.), *High Energy Phenomena on the Sun Symposium Proceedings*, NASA Goddard Preprint, p. 393.

Durgaprasad, N., Fichtel, C. E., Guss, D. E., and Reames, D. V.: 1968, *Astrophys. J.* **154**, 307.

Fan, C. Y., Pick, M., Pyle, R., Simpson, J. A., and Smith, D. R.: 1968, *J. Geophys. Res.* **73**, 1555.

Fichtel, C. E. and Guss, D. E.: 1961, *Phys. Rev. Letters* **6**, 495.

Geiss, J.: 1972, in C. P. Sonett, P. J. Coleman, and J. M. Wilcox (eds.), *Proc. Asilomar Conf. on the Solar Wind*, NASA, Washington, p. 559.

Gloeckler, G., Fan, C. Y., and Hovestadt, D.: 1973, *13th International Cosmic Ray Conf. Proceedings*, University of Denver, 2, p. 1492.

McDonald, F. B. and Desai, U. D.: 1971, *J. Geophys. Res.* **76**, 808.

McDonald, F. B. and Van Hollebeke, M. A.: 1973, in R. Ramaty and R. G. Stone (eds.), *High Energy Phenomena on the Sun Symposium Proceedings*, NASA Goddard Preprint, p. 404.

Mogro-Campero, A. and Simpson, J. A.: 1972, *Astrophys. J. Letters* **6**, 495.

Price, P. B., Chan, J. H., Crawford, H. J., and Sullivan, J. D.: 1973, *13th International Cosmic Ray Conf. Proceedings*, Univ. of Denver, Vol. 2, p. 1479.

Price, P. B. and Chan, J. H.: 1973b, 'Apollo 17 Preliminary Science Report', NASA Special Publication, p. 19.

Teegarden, B. J., von Rosenvinge, T. T., and McDonald, F. B.: 1973, *Astrophys. J.* **180**, 571.

Webber, W. R., Roelof, E. C., McDonald, F. B., Teegarden, B. J., and Trainor, J.: 1973, *13th International Cosmic Ray Conf. Proceedings*, Vol. 2, p. 1516.

DISCUSSION

Athay: What type of model is required to provide the 0.1 to 1 gm of matter that is needed for the ^{3}He abundance. In a normal solar model this amount of matter is reached only for depths in the very low chromosphere or photosphere.

McDonald: We used a slab model.

Sturrock: Is it known whether the ions are all fully stripped?

McDonald: The carbon and oxygen ions at 0.1 MeV are almost completely stripped, e.g. we expect C III but measured C V. That implies that they were accelerated in a hot region or that they were originally in a hot region.

X-RAY EVIDENCE FOR THE ACCELERATION, CONTAINMENT, AND EMISSION OF HIGH ENERGY FLARE PARTICLES

KENNETH J. FROST

Solar Plasmas Branch, Laboratory for Solar Physics, NASA-Goddard Space Flight Center, Greenbelt, Md. 20771, U.S.A.

Abstract. An instrument aboard the Fifth Orbiting Solar Observatory has observed hard solar X-rays from January 1969 to May 1972. A large number of X-ray bursts generated by solar cosmic ray flares have been observed. The X-ray bursts consist, in general, of two non-thermal components. The earliest occurring non-thermal component, coincident with the explosive phase, consists of a group of one to about ten X-ray bursts that are, for each burst, approximately 10 s duration and symmetrical in rise and decay. The time structure and multiplicity of these bursts is remarkably similar to that found in type III radio bursts in the meterwave band. The spectra of these bursts steepens sharply at energies greater than 100 keV indicating a limit at this energy for electron acceleration during the explosive or flash phase of the flare. For several flares these multiple X-ray bursts have occurred in coincidence with a group of type III bursts.

The second non-thermal X-ray component usually begins almost immediately after the first and in the same time interval during which type II and type IV radio emission is detected. For several events there is a remarkable similarity between the time profile of the type IV burst and the X-ray burst. The rise time of this component is of the order of a minute or two but the decay time is of the order of tens of minutes. The spectral character is power law over the whole 14 to 250 keV range. The index of the power varies from -6 to -2 and seems to track the hardness of the interplanetary solar electron spectral index.

We interpret these data as evidence that acceleration of solar particles occurs in two stages and that the X-ray components are the equivalents in the X-ray range of type III and type IV bursts in the radio range.

DISCUSSION

Smith: Has anyone tried to check the consistency of the hypothesis that the X-ray burst is associated with a moving type IV burst using the radio data? If so, what density is required?

Frost: We have not yet made a quantitative check on this speculation.

Dulk: The number of electrons in a plasmoid needed to get the radio emission is 10^{33} electrons with energy greater than 10 keV. There are about 10^{26} to 10^{27} erg involved in these electrons.

Frost: That number may be compatible with the number needed to explain the X-ray burst.

Smerd: On the evidence presented by Mr Frost it seems feasible that those hard X-ray bursts which he tentatively associated with subsequent moving type IV bursts (at metre wavelengths) may be associated with those early, concurrent stationary type IV bursts at metre wavelengths which are referred to as flare continua. The latter occur characteristically at the time of the microwave type IV burst.

Gordon Newkirk, Jr. (ed.), Coronal Disturbances, 421–422.

Note added after discussion. The sources of flare continua are high in the corona, those of the microwave IV are low; it may require position determination of hard X-ray sources to determine whether they are radiated from near the microwave source, or the stationary or moving metre-wave source.

Kundu: I am glad that Dr Smerd mentioned that the 80 MHz type IV source of March 21, 1969 was stationary and not moving as referred to by Mr Frost. It fits in nicely with the well known association of hard X-ray bursts with centimetre-wave bursts. This phase of metre-wave type IV was referred to by us as type IVA, which we interpreted as being due to the same source as centrimetre type IV. Consequently, I suggest that you should not rule out the possibility that the centimetre-continuum burst is as closely associated as the metre-wave type IV.

Regarding the lack of exact correspondence of the time profile of the hard X-ray burst with the centimetre burst, I would like to point out that the centimetre burst decreases in intensity and temperature during the post-burst phase; probably a better fit could be obtained if you compared it with another part of the X-ray spectrum. Further, the profile of a centimetre burst takes different shapes depending on the angular resolution of the beam used to track the region which flares up. In support of my contention, I'd like to show an interferometric record of a 3.7 cm burst of gradual rise and fall (g.r.f.) type, as obtained on June 9, 1973 with 3″ resolution. The slide shows fringe amplitude as a function of time; the burst started around 1806 UT, had its maximum around 1816 UT and continued beyond 1830 UT. The g.r.f. had the same time of start and maximum. However, unlike the g.r.f. which is obtained with beams covering the entire Sun, the interferometric profile reveals the impulsive nature of the burst around its peak. A significant fraction of the burst energy is emitted within a spherical volume of diameter 2″, with a corresponding brightness temperature of about 10^9 K.

Sturrock: If the moving type IV burst plasmoid produces X-rays it presumably contains a substantial mass of cool gas. Would you agree?

Frost: I intuitively expect the gas to be hot.

ACCELERATION, CONTAINMENT AND EMISSION OF HIGH ENERGY FLARE PARTICLES: RADIO EVIDENCE

A. BOISCHOT

Observatoire de Meudon, France

Abstract. The existence of non thermal radio bursts provide evidences for the acceleration of electrons in the solar atmosphere.

It is shown, from the characteristics of the bursts, that the electrons are accelerated in at least four different phases:

(1) An impulsive phase which gives μib and III bursts.
(2) A gradual phase which gives μIV and S_1IV bursts.
(3) A quasi-continuous phase which gives S_2IV bursts and noise storms.
(4) An acceleration by shock waves gives type II bursts.
(5) Eventually, another shock-wave acceleration giving the moIV burst.

1. Introduction

The large intensity of most of the radiobursts (brightness temperature between 10^7 and 10^{14} K) reveals that they are due to non-thermal mechanisms, i.e., that they take their energy from the kinetic energy of non-thermal particles.

In the following we shall assume that these particles are only electrons.

Most of the radiobursts are associated with some flare activity observed at chromospheric levels. We can divide them into two classes, somewhat arbitrarily: the bursts which are 'directly' related to the flare, begin at the same time or within a few minutes of the beginning of the Hα flare, and those the origin of which is also probably due to the flare activity, but which are observed after a longer delay, or are only statistically related to flares.

In the first group are the microwave impulsive burst (μib) the type III and type V bursts, the type II and the type IV bursts the latter being made of different components that we shall discuss later on.

In the second group, we place the many types of bursts and emissions which form the noise storm activity: type I bursts on metric wavelengths, storm type III in decametric and longer waves, storm continuums, etc....

We shall not discuss the details of the characteristics of these bursts, which may be found in Kundu (1965), Wild *et al.* (1963), Wild and Smerd (1972), or Krüger (1972), but shall insist only on a few points which are important for our study of the high energy electrons; some other characteristics are given in Table I.

The observation of the radiobursts gives us the intensity vs time and frequency, the state of polarization and some informations on the source of emission: apparent size, position, motion, etc.... From that, we wish to determine as many characteristics of the energetic electrons as possible, in particular their total number and energy distribution, the pitch angle distribution, the duration of the acceleration process, the storage and release of the electrons, etc....

Gordon Newkirk, Jr. (ed.), Coronal Disturbances, 423–435.

TABLE I

Burst	T_b (max) °K	Duration	Source	Mechanism
μib	10^9	1 to 60 mn	fixed. Low corona Double with opposite polarizations	gyroemission $B \sim 500$ G
μIV	10^9	10 mn to 2 h	fixed. Low corona Double with opposite polarizations	gyro- and synchrotron
S_1IV	?	10 mn to 1 h (?)	fixed 0.1 to 1 $R_\odot$	synchrotron (amplification?)
moIV	10^{12}	10 mn to 1 h	moves upward to several $R_\odot$ $v \sim 100$–1000 km s^{-1}	synchrotron
S_2IV	10^{12}	several hours	fixed 0.1 to 1 $R_\odot$	?
III	10^{12}	0.2 s (dm λ) to 1 h (km λ)	at height of plasma resonance Exciter: $v \sim 0.3c$	plasma oscillations Electron stream
V	?	1 to 10 mn	?	?
II	10^{12}	10 to 30 mn	at height of plasma resonance Exciter: $v \sim 1000$ km s^{-1}	plasma oscillations Shock waves
Noise storm				
continuum	10^{10}	hours	fixed at a given frequency	?
m type I	10^{12}	0.5 s	fixed at a given frequency	?
dam type III	10^{12}	few seconds	Higher the source, lower the frequency	plasma oscillations Electron stream

This can be done, even partially, only if we know the mechanisms by which the radio-waves are excited. These mechanisms are often poorly understood, and quantitative studies are generally impossible.

Moreover, there is a limitation in the determination of the electron characteristics from the observation of radiobursts: propagation effects (refraction, cut-off, scattering, etc.…) which are very difficult to take into account, can modify in a very sensitive manner the apparent intensity of the bursts and the characteristics of the sources, and then increase the difficulty of deriving the conditions within the sources from the observation of radiowaves at 1 AU. In particular, high energy electrons can exist in the solar atmosphere, which do not give any detectable radio emission.

2. Identification of the Bursts

μib and type III are very clearly defined (Kundu, 1965). The problem of μIV is more complex. μib and μIV are distinguished from the duration and shape of their time

Electrons		Associated events			
Number	Energy	Hard X-ray	Electrons (1 AU)	Protons (1 AU)	Flare
10^{30}–10^{32}	10–100 keV	+	+	–	small or large
10^{31}–10^{33}	10 keV to 5 MeV	–	+	+	large
10^{32}–10^{33}	100 keV to 5 MeV	–	+	+	large
?	100 keV to 5 MeV (?)	–	?	?	large
?	?	–	–	–	large
10^{32}–10^{35}	10–100 keV	+	+	–	small or large
?	?	–	–	–	small or large
?	?	–	–	–	large
?	?	–	–	–	small or no flare
?	?	–	–	–	
10^{32}–10^{35}	10–100 keV	–	–	–	

profile. They can be explained by electrons of similar characteristics, have the same brightness temperatures, similar spectra, similar intensity over duration ratios, etc.... This leads several authors to the conclusion that μIV is more likely the juxtaposition of several μib (Takakura, 1967).

However, there is an important difference between the two types of bursts in their association with hard X-ray bursts.

The μib is always associated with an impulsive hard X-ray burst, the two emissions being coincident in time and having the same duration (Peterson *et al.*, 1973). On the other hand, the long during μIV is accompanied only by a much shorter impulsive hard X-ray burst, similar to the above one, at the beginning of the radio emission.

It is then likely that the microwave emission in large flares is the superposition of a μib and a longer and irregular emission corresponding to the μIV. Then this requires two different processes for accelerating and/or trapping the high energy electrons.

There are even more discussions about the identification of the different components of the type IV event, and we cannot here discuss all the arguments presented by the different parties. In the author's opinion, three components can be distinguished.

The first component groups the μIV and the 'first stationary source' S_1IV. The latter is made of the dmIV (for a discussion of this dmIV see Wild *et al.*, 1963) and the mIVA (Pick, 1961), or the mIVA$_1$ of Akinyan *et al.* (1971). The two have in common to come from fixed sources in the low corona, and to begin very shortly (a few minutes at most) after the flash phase of the flare.

The second component is the moving type IV, moIV (mIVB of Pick, 1961, or mIVA$_2$ of Akinyan *et al.*, 1971) which is impossible to separate from the above S_1IV and from the third component with only single frequency or spectral observations, but which is very clearly defined with position observations. Different structures have been recognized in the moIV bursts (Wild and Smerd, 1972, and references herein).

The third component, 'second stationary source' or S_2IV corresponds to the mIVB (Wild *et al.*, 1963) and possibly a part of the dmIV. It is observed on metric and longer wavelengths from a fixed source located low in the corona and close to the flare region. (It is not clear whether the source of S_1IV is different from the source of S_2IV or not.) This component is distinguished from S_1IV by its polarization characteristics and its duration but, as it overlaps with S_1IV and moIV, these three components cannot be separated on single frequency records or dynamic spectra.

Among the different characteristics of the bursts directly associated with flares, a very important one for our problem is their probability of occurrence with a given flare. Very good statistics are still lacking, but it has been shown that:

– μib and III are observed with flares of any importance, and their intensity is not related to the importance of the flare.

– type IV bursts are only observed with large flares. Then they are accompanied by μib and some type III bursts in the early phase of the event.

– for small flares, the relation between μib and III is not clear. A flare can give either a μib only, or some III only, or both types. But type III bursts are more frequent that μib.

– many flares, even of large importance, are not accompanied by any radio-emission.

The location of the sources of emission is also very important to know; better observations are needed, specially on high frequencies.

μib and μIV come from small regions very close to the flaring site, and more precisely – in a few observations – from small sources located above photospheric regions of different magnetic polarities, giving radiowaves of opposite circular polarizations. The source is fixed, within the accuracy of the determination. This accuracy is not good enough to show the variation of position with frequency, if any. It is not possible from the observation to distinguish between the sources of μib and of μIV.

S_1IV bursts also come from a region associated with the Hα source, but more extended. The observation of the emission up to long metric wavelengths (if not in the decameter range) means that the source extends at least to 0.5 or 1.0 $R_\odot$ above the solar surface; microwave emission is no more observed from this upper part of the source. The average position of the source seems to change with frequency (lower the frequency, higher the source). This can be due to the effect of the coronal plasma on the low frequency limit of the gyroemission (Razin effect), to the variation of the energy distribution of the electrons and of the magnetic field inside the source, and also to the mechanism of emission.

For moIV burst, all the frequencies come from the same source, simple or complex (Boischot and Clavelier, 1968; Smerd and Dulk, 1971) which moves upward in the corona with a velocity of several hundreds of kilometers per second.

The source of the type III burst, on the contrary, varies with frequency, according to the plasma hypothesis, but the exact path of the source, relative to the position of the flare and to the coronal structures is not yet known exactly (Leblanc *et al.*, 1974).

A last remark about the interpretation of the bursts on microwaves. It has been often emphasized that μib and μIV show a variety of spectral shapes. It is very likely that this reflects, rather than some differences in the characteristics of the electrons and/or in the mechanisms of emission, a difference in the coronal plasma in and around the source. It is well known than in the low corona the optical depth for free-free absorption is large and, due to large density gradients, very sensitive to the frequency. Then the free-free absorption of the radio waves, which depends on the structure of the source (height and size) and of the surrounding plasma can vary with frequency in a different manner for different bursts (Ramaty and Petrosian, 1972).

3. μib and Type III Bursts

We shall first consider these two kinds of bursts together, for we can see, from Table I, that their emission can be explained by electrons of similar energies, between 10 keV and 100 keV and that they are observed at the same time.

The μib is explained by the gyroemission of the electrons in a ~500 to 1000 G magnetic field, and the type III by plasma oscillations excited by an ejected stream of such electrons.

The problem is to know if there is only one or two different acceleration processes to give the two kinds of bursts. It has been said sometimes that two accelerations are necessary because their duration must be different for the two bursts.

For μib, the life time of the electrons of energy E radiating a gyroemission in a field of 1000 G, and a thermal density of n electrons per cubic centimeter, is shown to be limited by collisions on thermal electrons.

$$\tau_c \sim 2.2 \times 10^9 \frac{E^{3/2}}{n \, \mathrm{Ln}},$$

where Ln is the Coulomb logarithm ~20.

For a density $\sim 10^{10}$ cm^{-3} which has been claimed necessary to explain the hard X-ray bursts, we find $\tau_c = 0.25$ s and 7.5 s for 10 keV and 100 keV electrons respectively (de Feiter, 1972).

Then the observed duration of the μib, several minutes, requires that there must be a continuous acceleration of electrons during this time. This conclusion has been discussed (Takakura, 1972). It depends very much on the assumed density in the source, which is not known. If we take $n = 10^8$ cm^{-3}, which is not unlikely, we can reconcile the life time of the electrons with the duration of the bursts. However, we must note that the coincidence between the positions of the sources of μib and that of the source of the slowly varying component (coronal condensations) and the increased probability to observe μib when the slowly varying component is intense (Kundu, 1959) point to large values of the density in the source.

The duration of the acceleration process can neither be determined accurately from the characteristics of the type III bursts. It has often been said that the high frequency duration of a type III burst, 0.2 to 0.5 s, gives a limit of the duration of the acceleration process, which has to be repeated to give the several type III bursts observed in a group. But we cannot disregard the possibility that a type III burst corresponds only to the head of an electron stream, or to a change of density in a quasi-continuous stream, and that the acceleration is more or less continuous during the duration of the entire group of type III, i.e. between 10 s and a few minutes.

Then there is nothing against assuming that μib and type III are due to electrons accelerated simultaneously (Kane, 1972). In this line, the difference between the two kinds of bursts is to be explained by the conditions of trapping and release of these electrons.

For μib the electrons are certainly trapped in the low corona over active centers. The double source structure, with different polarities, and the high magnetic field required for the emission mean that the trapping region has closed lines of force and is located above bipolar regions of strong magnetic fields, i.e., directly above the active regions.

On the contrary, the type III electrons must escape along an open line of force as indicated by the extension of the emission to very low frequencies. It has been said that the constancy of the exciter velocity derived from the frequency drift rate means that the electrons have a small pitch angle (the velocity parallel to $\mathbf{B}$ would increase with decreasing B for electrons of large pitch angles). But this also can be discussed as it is now thought that the observed frequency drift corresponds to the velocity of the front of the stream, i.e. of the fastest electrons, and not to the average velocity of the stream.

The theory of type V burst is not yet clear, but it is possible that this burst is the emission of some of the electrons of the stream, scattered in the surrounding corona by density and magnetic field inhomogeneities.

In the above model the electrons, accelerated in the region of the Hα flare or close to it, must encounter either closed lines of force, to give μib, or open lines to give type III or both to give simultaneously the two kinds of bursts. For Sturrock (1972) this difference is due to the magnetic configuration associated with the flare.

An important result in this respect has been obtained recently by Axisa (1974). Studying the location of the flare (Flare Productive Sites) inside three active regions, he was able to show that there is a difference in the magnetic configurations near the FSP at the photospheric levels (Hα magnetograms) between flares which are accompanied predominantly by either μib or by type III. In the former case, the flare takes place along an inversion line ($B_{\parallel}=0$) separating two compact magnetic regions of opposite polarities and large magnetic fields ($15<B<100$ G); with this configuration, only closed magnetic lines of force are present around the flare site and most of the electrons remained trapped. Type III bursts are inhibited and a μib observed (and also a hard X-ray burst).

When, on the contrary, on one side of the $B_{\parallel}=0$ line where the flare occurs there is an extensive region of weak magnetic fields, the particles can find open lines of force and the production of type III is favored.

When, in this last case the flares also affect a region of intense magnetic field such as the penumbral part of a spot, both type III and μib are observed (and also a hard X-ray burst).

Finally, one remark must be made about the escape of the type III electrons. The presence of a magnetic open structure is certainly not sufficient to explain the trajectory of the electrons. At the chromospheric level, many electrons would then leave the flare region on a line of force which turns down to the Sun at some altitude over the surface. These electrons would give emission of type *U* bursts. But it is very remarkable that the proportion of type *U* bursts to the normal type III is very small. This means that the electrons can 'recognize' even at the chromospheric level, the lines which are really open in the high corona.

This is possible probably because there is a change of density of the plasma related to the magnetic structure, which favors the propagation of the high energy electrons along or near these open lines (which are perhaps neutral lines). This requires that the roots of the magnetic structure lie very deep in the solar atmosphere, in the photospheric active region. This is an important conclusion for the structure and origin of coronal streamers, if we assume that the type III at large distance from the Sun propagate along such streamers.

This problem is also related to the propagation of the electrons in dense structures (Smith and Pneuman, 1972).

4. μIV and S_1IV Bursts

These bursts are received only at the time of large flares and last much longer than the μib or the group of type III bursts. Moreover, they are very often accompanied by an emission of high energy protons and relativistic electrons observed in the interplanetary medium, unlikely to the μib and type III, which correspond only to medium energy electron events, without protons. The acceleration process for μIV and S_1IV on the one hand, and for μib and type III on the other must then be different.

It follows that the differences between the two classes of bursts pose a serious problem.

We have, for the flash phase, electrons of medium energy which can, in many cases, escape along open lines of force to give type III bursts. Later on, we get electrons of even larger energy, in at least the same number – if the determinations of Table I are correct – which are also observed in the interplanetary medium, but which do not give rise to type III bursts.

An explanation of this phenomenon can be found in the nature of the acceleration process. If, as indicated by space observations of electron and proton events, the energy spectrum of the accelerated particles follows approximately a power law in both cases, it has only negative slopes, and the electrons cannot excite any plasma instability and type III burst. This can be done only if a positive slope in the velocity distribution function is obtained due to the dispersion of the particles propagating with different velocities. But in turn this requires that the acceleration process and/or the liberation of the electrons are impulsive enough to give this dispersion. This is certainly the case for the flash phase acceleration. But the second acceleration, which gives the μIV and S_1IV bursts, starts probably more gradually, so that the electrons are unable to excite type III bursts. This model is in agreement with some new observations of the energy distribution in the interplanetary space (Lin, 1972; Lin *et al.*, 1973).

At the beginning of an electron event, the energy distribution function presents a maximum in the 100 keV range. As time evolves, this maximum drifts toward lower and lower energies, up to the time when the true energy spectrum of the accelerated electrons, with only a negative slope, is observed. This spectrum is then observed as long as there is no rapid variation in the number of electrons accelerated.

If this stationary state, with only a negative slope, is attained before the number of electrons is large enough to excite plasma waves, type III bursts will not be observed.

The difference in impulsiveness of the acceleration process (or release) also explains the difference of the starting frequency of different type III bursts.

In this second phase, the electrons are accelerated up to energies higher than in the case of the flash phase, and are accompanied by high energy protons. They are first trapped in the same region that those giving the μib i.e. in closed lines of force of intense magnetic fields over the active region where the flare is observed. Then, they diffuse more or less rapidly to higher regions of the corona where they radiate the S_1IV burst. This diffusion is more important than for the flash phase electrons (which do not give metric emission) because the energy of the electrons is larger and probably, because there is more turbulence in the trapping region due to the larger importance of the flare.

The problem of the distinction between μIV and S_1IV has been discussed by Böhme (1972) who invokes a separate acceleration process for the latter. However, the minimum in the intensity often observed in the decimeter range can found another explanation:

The type IV on decimetric and metric wavelengths can attain very large intensities soon after the beginning of the emission (i.e. at times corresponding to the dmIV and mIVA components) with flux densities up to 10^{-18} W m^{-2} Hz^{-1} and a source size about 10′ (which is probably an overestimation). The brightness temperature of the source is then sometimes larger than 10^{12} K at, say, $\lambda = 3m$. This high brightness temperatures can be reached by an incoherent mechanism only if the electrons have an average energy of 80 MeV. This is certainly too large to fit the other observations.

Then we must conclude that the mechanism giving the decimeter and meter emission of type IV is, at least partially, coherent, or in other words, that the synchrotron waves are coherently emitted or amplified in the corona. This amplification is very likely a function of frequency and of the local conditions. Then it can explain at the same time:

– the minimum observed between μIV and the emission at longer wavelengths.

– the intensity observed in the metric range which is sometimes much larger than that of the μIV burst.

– the high variability often observed in the dmIV, which we have no reason to distinguish from the mIVA.

– the many structures of 'modulation' observed in the metric range (Slottje, 1972).

5. S_2IV Bursts and Noise Storms

S_2IV bursts (i.e. mIVB and perhaps a part of the dmIV) are observed after large flares, and last often for several hours. Most of their characteristics are similar to those of the continuum of the noise storms, and sometimes we can observe a S_2IV burst turning into a normal noise storm after a few hours. This is the reason why we shall study them together.

The mechanisms of the emission which are at the origin of S_2IV bursts and noise storm activity (continuum and fast bursts) are still unknown. Therefore we have no idea of the amount and the energy of the electrons required, and all our conclusions about the acceleration process will be very speculative. However, we can show that these electrons are certainly accelerated in a process different from the two previous ones.

An important difference between S_2IV burst and noise storm on the one hand, and the bursts studied above on the other, is that the former are accompanied neither by microwave emissions, although they can be as intense as the S_1IV burst in the metric range, nor by interplanetary electrons. (Increases of high energy electrons during the passage of large active centers on the disc have been observed but with an intensity much smaller than those accompanying the flash phase of the flare) (McDonald, 1970).

Another noteworthy point is the difference of nature between the metric and decametric noise storm bursts (Boischot *et al.*, 1970, 1971). The former are essentially type I bursts, the mechanism of which being unknown, and the latter mainly type III bursts beginning at a frequency between 60 and 30 MHz, and extending down to much longer wavelengths. These type III bursts have the same average characteristics than the type III observed at the flash phase of a flare, except their starting frequency.

The high intensity of the metric continuum, and of type I bursts, requires certainly non thermal electrons in the source for many hours or days, i.e. in regions between 0.2 and 1.0 $R_\odot$ (but not below to explain the absence of cm and dm emission).

The type III storm bursts reveal on another hand that high energy electrons – between 10 to 100 keV if we believe the theory of type III bursts – escape from this source into the interplanetary medium. During an average noise storm as many as 10^4 type III burst can be observed in a day. If, as assumed for the isolated type III, each burst requires 10^{32} to 10^{35} electrons, we arrive at 10^{36} to 10^{39} electrons of 10 to 100 keV every day. This number is certainly too large to be attributed to one single or even a few flares, and requires a quasi-continuous acceleration during the whole life time of the storm.

The above figure is derived from the observation of type III bursts only, and is probably underestimated, as it has been shown that the so called type III_b bursts (De la Noë and Boischot, 1972), nearly as frequent as the normal type III burst in the decameter range, must also be attributed to the escape of electrons of similar energy, and reveal that many other electrons can escape the storm region without giving any radio emission.

Then we arrive at the conclusion that the noise storm and the S_2IV bursts require another source of acceleration quasi-continuous during the life time of the noise storm.

We have for the moment very little idea of what this process is, and where it occurs. It can take place in the active center at chromospheric or low coronal levels. However to explain in the absence of microwave emission, it is more likely that it takes place in current sheets or near neutral points higher in the corona, at an altitude of about 1 $R_\odot$ (Steward and Labrum, 1972).

The relation between the noise storms and the direct observation of high energy electrons in the interplanetary space poses another problem. It seems from the published observations that the increase of the number of these electrons at the time of noise storms is much smaller, if it even exists (Anderson, 1969) than the increase after an isolated group of flare associated type III bursts. This requires an explanation which could be that the number and energy of the electrons claimed to be necessary to excite a type III burst are much overestimated, and that smaller fluxes can as well give storm type III bursts, with the same average characteristics. This, at the same time, would help solving the problem of the total number of electrons escaping during a noise storm, discussed above.

If true, this conclusion would be very important for the general theory of the type III burst.

6. Moving Type IV Bursts (moIV)

The acceleration process leading to the emission of the moIV is not known. Two possibilities exist, both as likely.

The radiating electrons could be some of the electrons accelerated in the second phase (which gives μIV and S_1IV bursts) which remained trapped in a special magnetic configuration (plasmoïd) ejected from the flaring region in the high corona (Schmahl, 1972) or which propagate along open lines of force and encounter some local re-inforcements of magnetic field such as may be observed in a solitary wave in parallel propagation (Warwick, 1969).

– The electrons can be either accelerated locally in the wake of a shock wave moving up in the corona. Such an acceleration process has been studied by Lacombe and Mangeney (1969) and more recently by Mangeney (1974).

Even though the problems of confinment of high energy electrons in a plasmoïd, and of their acceleration by a shock wave is very interesting, we have too little information on the nature of the electrons, and too little place in this paper to discuss them in more details.

7. Type II Bursts

We shall be brief also about type II bursts. It is generally agreed that the electrons required to excite plasma waves are accelerated either in the front or in the wake of a shock wave. We have for the moment too few informations on these electrons (number, energy distribution,...) to be able to discuss the details of the acceleration process. The recent observation of type II bursts in the interplanetary medium down to 500 kHz (Stone and Fainberg, 1972) means that the electrons are certainly accelerated by the shock wave and are not flare electrons trapped in or near the shock.

8. Summary

To sum up, we propose the following classification of the acceleration phenomena, based on the study of the radio bursts.

(1) Coincident with the flash phase of the flare, an 'impulsive' acceleration with a duration of a few seconds to a few minutes, giving electrons of medium energy (1 to 100 keV) without protons. These electrons are either trapped near the flare center in a high magnetic field closed to the surface of the Sun to give the μib burst (and the impulsive hard X-ray burst) or they escape along open lines of force to give the type III bursts. Their energy is not sufficient to allow their diffusion from the trapping region into the higher corona where they would have given metric emission, but some of the type III electrons might be scattered to give the type V burst.

(2) In the case of large flares only a second phase of acceleration more gradual and lasting longer (minutes to hours) giving electrons of higher energy (or a flatter spectrum) (100 keV to a few MeV) and also protons. These electrons are first trapped in the same closed loops region which gives the μib, where they radiate the μIV burst; then they diffuse to larger heights to give the S_1IV burst.

(3) A third phase which is enhanced just after large flares but works also in the absence of flare. We have no information on the number and energy of the accelerated electrons. They give the S_2IV bursts and the noise storms. This acceleration might take place near the flare itself, in the active center, but more likely at high altitude in the corona.

(4) Another kind of acceleration, in the wake of a shock wave will give the moIV, but a possibility remains that this burst is due to the electrons accelerated in phase 2 and trapped in special moving structures such as plasmoïds.

(5) Also in the front or in the wake of a shock wave, an acceleration process at the origin of the type II burst.

References

Akinyan, S. T., Mogilevsky, E. I., Bohme, A., and Krüger, A.: 1971, *Solar Phys.* **20**, 112.
Anderson, K. A.: 1969, *Solar Phys.* **6**, 111.
Axisa, F.: 1974, *Solar Phys.* **35**, 207.
Bohme, A.: 1972, *Solar Phys.* **25**, 478.
Boischot, A. and Clavelier, B.: 1968, *Ann. Astrophys.* **31**, 445.
Boischot, A., De la Noë, J., Du Chaffaut, M., and Rosolen, C.: 1971, *Compt. Rend. Acad. Sci.* **272**, 166.
Boischot, A., De la Noë, J., and Pedersen, B. M.: 1970, *Astron. Astrophys.* **4**, 159.
De Feiter, L. D.: 1972, *Space Sci. Rev.* **13**, 827.
De la Noë, J. and Boischot, A.: 1972, *Astron. Astrophys.* **20**, 55.
Kane, S. R.: 1972, *Solar Phys.* **27**, 174.
Krüger, A.: 1972, *Physics of Solar Continuum Radiobursts*, Academic Verlag, Berlin.
Kundu, M. R.: 1959, *Ann. Astrophys.* **22**, 1.
Kundu, M. R.: 1965, *Solar Radio Astronomy*, Interscience Publ., New York.
Lacombe, C. and Mangeney, A.: 1969, *Astron. Astrophys.* **1**, 325.
Leblanc, Y., Kuiper, T. B. H., and Hansen, S.: 1974, this volume, p. 227.
Lin, R. P.: 1972, in R. Ramaty and R. G. Stone (eds.), *Proc. of High Energy Phenomena on the Sun*, GSFC Preprint X-693-73-193, p. 439.
Lin, R. P., Evans, L. G., and Fainberg, J.: 1973, *Astrophys. Letters* **14**, 191.
Mangeney, A.: 1974, this volume, p. 387.
McDonald, F. B.: 1970, in Manno and Page (eds.), *Intercorrelated Satellite Observations Related to Solar Events*, D. Reidel Publ. Co., Dordrecht, Holland, p. 34.
Peterson, L. E., Datlowe, D. W., and McKenzie, D. L.: 1973, in R. Ramaty and R. G. Stone (eds.), *Proc. of High Energy on the Sun*, GSFC Preprint X-693-73-193, p. 132.
Pick, M.: 1961, *Ann. Astrophys.* **24**, 183.
Ramaty, R. and Petrosian, D.: 1972, *Astrophys. J.* **178**, 241.
Schmahl, E. J.: 1972, *Proc. Astron. Soc. Australia* **2**, 95.
Slottje, C.: 1972, *Solar Phys.* **25**, 210.
Smerd, S. F. and Dulk, G. A.: 1971, in R. Howard (ed.), 'Solar Magnetic Fields', *IAU Symp.* **49**, 616.
Smith, D. F. and Pneuman, G. W.: 1972, *Solar Phys.* **25**, 461.
Stewart, R. T. and Labrum, N. R.: 1972, *Solar Phys.* **27**, 192.
Stone, R. G. and Fainberg, J.: 1973, in R. Ramaty and R. G. Stone (eds.), *Proc. of High Energy Phenomena on the Sun*, GSFC Preprint X-693-73-193, p. 519.
Sturrock, P. A.: 1972, *Solar Phys.* **23**, 438.
Takakura, T.: 1967, *Solar Phys.* **1**, 304.
Takakura, T.: 1972, *Solar Phys.* **26**, 151.
Takakura, T.: 1973, in R. Ramaty and R. G. Stone (eds.), *Proc. of High Energy Phenomena on the Sun*, GSFC Preprint X-693-73-193, p. 179.
Warwick, J. W.: 1968, *Solar Phys.* **5**, 111.
Wild, J. P. and Smerd, S. F.: 1972, *Ann. Rev. Astron. Astrophys.* **10**, 169.
Wild, J. P., Smerd, S. F., and Weiss, A. A.: 1963, *Ann. Rev. Astron. Astrophys.* **1**, 291.

DISCUSSION

Pick (on Boischot paper): I would like to comment on the old classification of type IV bursts and the new one. (a) What we called type IVMA occurred at the time of the type IVμ. Its main characteristic was a similar behavior in the intensity evolution to that of the microwave burst. This phase very likely corresponds to the 'flare continuum' which is a stationary source as revealed by the Culgoora observations. The electron energy would be less than 1 MeV. (b) The type IVMB corresponds to the moving type IV burst as described by Boischot (*Compt. Rend. Acad. Sci.*, 1959). Concerning the dm type IV burst, both moving and stationary type IV bursts are observed – the large fluctuations that you reported are in fact observed during the stationary part. The sources of the metric and decimetric type IV burst do not correspond to the same position; the intensity behaviors are not similar, large fluctuations being observed in the dm range.

Kai: Flare-continua generally coincide in time with microwave IV bursts. The example of a flare-continuum source, which I mentioned before, suggests that microwave IV bursts originate in the most inner part of the extended flare-continuum source.

Schmidt: I agree with Dr Sturrock that it is still somewhat premature to construct a good theoretical model for the phase I. But I think some points made by several speakers at this symposium point towards a phenomenological model which would take into account the effects of the low-lying current sheets along the legs of the typical helmet configuration. Dr Pneuman described these current sheets for the static configuration. Dr Boischot has mentioned these current sheets as possible source regions of type III bursts in phase I, whereas the type III bursts which come in phase III may come from the upper current sheet which is located above the two others. The time profile of the total energy release during a major flare may support this view. In phase I the major part of the flare energy is still confined in the region inside of the two lower current sheets, so that they are more compressed than usual (by dynamic effects) and that their rate of dissipation goes up. On the other hand, during phase III the major part of the flare energy has already left the whole flare region. As described in the model of Carmichael mentioned by Dr Sturrock, the upper current sheet will then reconnect the magnetic flux expanded earlier very effectively.

Pneuman: I would like to comment upon the very important point that Dr Boischot made regarding the commonness of type III bursts and the relative rarity of *U*-bursts which are supposed to have the same physical explanations. If this is true, then doesn't this argue strongly against a purely chromospheric origin of flares? If the flare occurred first in the chromosphere then one would expect electrons to be ejected in all directions – both along closed and open field lines. Why do the electrons always seem to find the open lines?

On the other hand, the explanations might not be the same. I believe that closed flux tubes contain more material than open field lines. Hence, the electrons that would normally give rise to *U*-bursts are quenched quenched due to coulomb collisions whereas the electrons on open field lines can get out of the lower corona more easily.

PARTICLE ACCELERATION IN SOLAR FLARES

P. A. STURROCK

Institute for Plasma Research, Stanford University, Stanford, Calif., U.S.A.

Abstract. A review of observational data supports the proposal that there are two distinct phases of particle acceleration in solar flares. 'Phase 1' is associated with the flash phase and is here interpreted as acceleration during field-line reconnection. 'Phase 2' is associated with type II and type IV radio bursts, and is ascribed to stochastic acceleration in the turbulent plasma behind a magnetohydrodynamic shock formed ahead of an ejected plasmoid.

1. Introduction

It is generally agreed that, in the preflare state, energy is stored in magnetic fields. More specifically, the energy released during a flare is derived from the 'free energy' associated with currents in the solar atmosphere in an active region. It is widely believed that the mechanism by which this energy is converted into other forms is a plasma instability. If the current in the preflare state is in the form of a 'current sheet', the appropriate instability is of the 'tearing mode' type analyzed by Furth *et al.* (1963). After the instability has reached its nonlinear stage, the reconnection probably proceeds substantially as described by Petschek (1964). One of the simplest geometries for the preflare magnetic field is that proposed by Carmichael (1964) and independently by Sturrock (1966) (Figure 1).

Magnetic energy released during the tearing process must somehow be transferred to the plasma. Dungey (1958) realized that, when reconnection occurs, a DC electric field will be developed in the reconnection region which could lead to particle acceleration. Friedman and Hamburger (1969) and Coppi and Friedland (1971) have proposed that micro-instabilities, of the two-stream type, are likely to occur during reconnection since intense currents will flow in small regions. These micro-instabilities will lead to high-frequency electric fields which can lead to particle acceleration. Within the context of this model, the most obvious source for particle acceleration is, therefore, the reconnection process itself.

When we come to consider observational data, we find that the question of acceleration cannot be answered so simply. There have been many reviews of the properties of solar flares, perhaps the most recent being that of Švestka (1973). Particle emission from solar flares has recently been reviewed by Lin (1974) and Simnett (1974). These recent reviews support a proposal made by de Jager (1969), that there are two stages of acceleration of particles in solar flares, which we shall refer to as 'Phase 1 Acceleration' and 'Phase 2 Acceleration'.

Phase 1 acceleration occurs during the flash phase of a flare and is of limited duration, of order 1–2 min. Spacecraft observations indicate that large numbers of electrons are accelerated to energies in the range 10–100 keV during this phase. During this phase, type III radio bursts occur, which are usually ascribed to the passage of discrete beams of electrons, with energies in the above range, through

Gordon Newkirk, Jr. (ed.), Coronal Disturbances, 437–445.

the corona and into interplanetary space. It is a common view that these particles travel through coronal streamers, possibly in the current sheets which are believed to lie within streamers. It is known that impulsive nonthermal X-ray bursts occur during this stage of a flare. These bursts may be understood as the result of electron

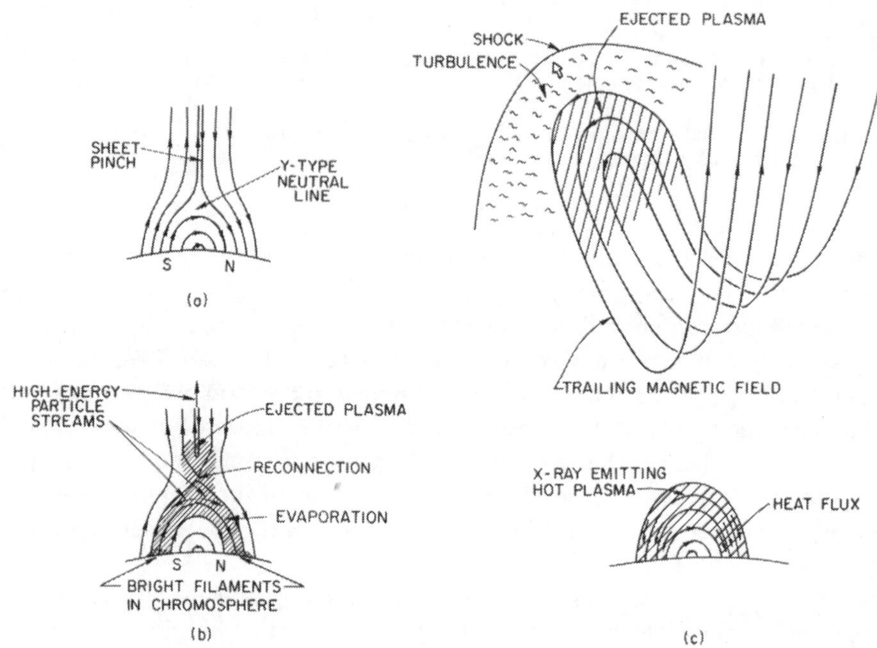

Fig. 1. (a) Example of 'open' current sheet configuration above a bipolar magnetic region. (b) Reconnection results in particle acceleration which in turn leads to evaporation. (c) Following reconnection, magnetic field near Sun's surface is in current-free state and a mass of plasma, trailing magnetic field lines, is ejected from the corona.

streams, similar to those which escape into interplanetary space, streaming down into the chromosphere. Petrosian (1973) has shown that the polar diagram of bremsstrahlung radiation becomes peaked in the 'forward' direction (pointing into the photosphere) as one considers photons of energy of order 100 keV produced by mildly relativistic electrons. This means that impulsive X-rays are not a good indicator of the electron spectrum above about 100 keV. On the other hand, spacecraft observations indicate that there is typically a cut-off in the electron spectrum between 100 keV and 500 keV. There is no indication, from observations made with particle detectors, that Phase 1 acceleration typically leads to a flux of protons or other ions. If such a flux is produced, the number of accelerated ions is much smaller than the number of accelerated electrons. It may be noted that Frost (1972) has recently introduced the term 'type III X-ray burst' to characterize the impulsive X-ray bursts occurring during this stage, to discriminate them from X-ray bursts sometimes observed later in the flare, as will be described below.

Phase 2 acceleration occurs in only a small fraction of flares, namely those flares which exhibit 'high-energy' phenomena. The most direct evidence for Phase 2 acceleration is the detection, by spacecraft, of streams of relativistic electrons and ions. These measurements show that Phase 2 acceleration typically occurs 5–10 min after the flash phase of the flare. Proton streams have power-law spectra with detectable fluxes at 100 MeV, and sometimes at 1 BeV or more. Electron fluxes are much smaller than the proton fluxes, but the energies may extend up to 10 MeV or more. The abundance distribution of the ion stream varies from flare to flare, and has a higher contribution of high-Z elements than is characteristic of the photosphere. Particle fluxes from Phase 2 acceleration seem sometimes to arrive promptly and sometimes to propagate primarily through diffusion.

The occurrence of high-energy particle streams, which we attribute to Phase 2 acceleration, correlates well with type II and type IV radio bursts (Kundu, 1965). Type II bursts are generally attributed to shock fronts moving through the corona; this may possibly be a 'blast wave', but it is more likely to be a 'bow shock' ahead of a moving plasmoid, as shown in Figure 1 (c). Type IV radiation is generally attributed to gyro-synchrotron radiation by mildly relativistic electrons either in magnetic field patterns anchored to the photosphere, or in the closed magnetic field of a moving plasmoid. The ejected plasma cloud shown in Figure 1 (c) has only open field lines, but reconnection at the neutral sheet could produce a plasmoid with a closed, toroidal field pattern, as shown in Figure 2. Such a field pattern, in a moving plasmoid, would provide a good fit to the radio source observed on March 1, 1969 (Riddle, 1970).

Impulsive X-ray bursts occur also during the time-frame of Phase 2 acceleration.

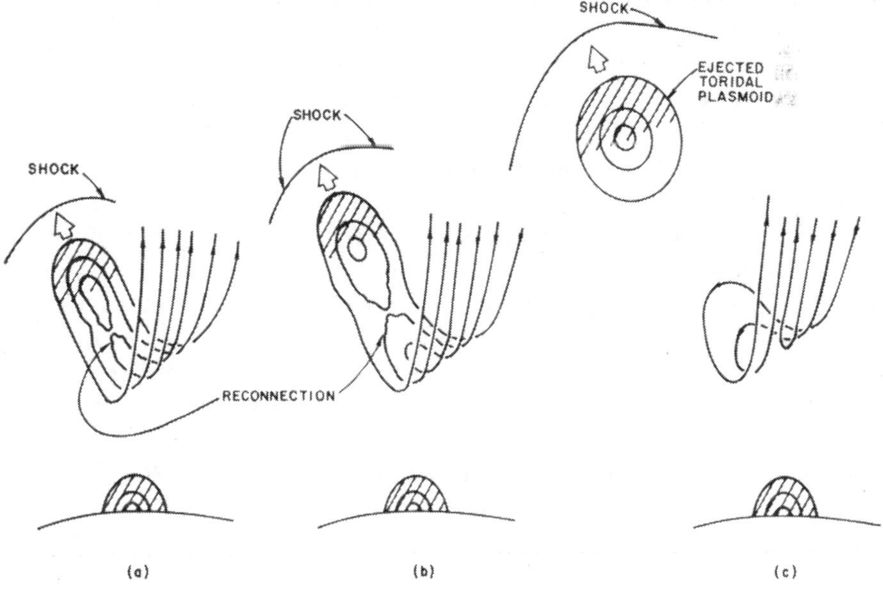

Fig. 2. (a, b) Reconnection can occur in current sheet which develops in ejected plasmoid. (c) Such reconnection could lead to an ejected plasmoid with a closed magnetic-field configuration, possibly toroidal.

Since they occur at about the same time as type IV radio bursts, Frost (1972) has introduced the term 'type IV X-ray bursts' to describe them. These X-ray bursts differ from type III X-ray bursts in that they do not show a cut-off in the X-ray spectrum (indicative of a cut-off in the electron spectrum).

2. Phase 1 Acceleration

We next discuss briefly possible mechanisms of Phase 1 acceleration. Since the reconnection process converts magnetic energy into plasma kinetic energy, this must give rise to some increase in particle energy. A simple estimate of the energy E (eV) acquired by electrons and protons is given by the energy balance equation

$$\frac{1}{8\pi} B^2 = 2n \times 10^{-11.8} E, \tag{1}$$

where n is the electron density. This becomes

$$E = 10^{10.1} B^2 n^{-1}. \tag{2}$$

Adopting a typical value $B = 10^{1.5}$ (gauss) for the coronal magnetic field of an active region, and $n = 10^8$ (cm^{-3}) for the preflare coronal density, Equation (2) leads to the estimate $E = 10^{5.1}$, in good agreement with the fact that the electron energy spectrum typically cuts off at about 100 keV. Note that, as the flare develops, the density of coronal plasma apparently increases due to evaporation from the chromosphere. This increase in n will lead to a corresponding decrease in E. Data concerning soft X-ray emission from flares indicate that, during the late stage of a flare, the coronal density may be in the range $n = 10^{10}$. This would lead to $E = 10^3$. The mean free path of particles of this energy in a plasma of this density is only 10^8 cm. Hence these particles would not escape freely to the chromosphere and into interplanetary space: rather, they would lose energy simply by heat conduction.

Acceleration of electrons during the Phase 1 stage is most probably due to the interaction of electrons with RF electric fields developed in the tearing region as the result of a two-stream electron-ion instability. Since the phase velocity of the excited waves are comparable with the electron streaming velocities, there will be little interaction between these waves and ions. These considerations are consistent with the observational fact that Phase 1 acceleration is mainly electron acceleration, with comparatively little proton acceleration. We also note that energy considerations are consistent with the fact that there is typically a sharp change in the electron energy spectrum at an energy of order 100 keV. A more extensive discussion of particle acceleration during the flash phase of solar flares is given by Smith (1974).

3. Phase 2 Acceleration

We next discuss possible mechanisms for Phase 2 acceleration. Observational evidence indicates that this probably occurs in association with a shock front. It is known, from observations of the Earth's bow shock, that a collision-free shock leads to shocked gas which is in a state of MHD turbulence. This leads to the possibility of stochastic particle acceleration (Hall and Sturrock, 1967; Newman, 1973), which comprises a class of acceleration mechanisms of which Fermi's is a particular case.

Suppose that the mean magnetic field strength in the shocked gas is B (Gauss), the mean plasma density n (cm^{-3}), and the thickness of the shocked region is L (cm). Suppose that the 'depth of modulation' of the turbulence is δ, so that the magnitude of the fluctuating magnetic field is δB. The phase velocity of waves comprising the turbulence will be comparable with the Alfvén velocity v_{A}, given by

$$v_{\mathrm{A}} = 10^{11.3}\, Bn^{-1/2}. \tag{3}$$

The frequency spectrum of the MHD turbulence will extend from low frequencies up to a maximum of approximately ω_{gp}, the radian gyro-frequency of protons, given by

$$\omega_{gp} = 10^{4.0}\, B. \tag{4}$$

This means that the turbulence loses correlation for distances larger than λ given by

$$\lambda = \frac{v_{\mathrm{A}}}{\omega_{gp}} = 10^{7.3}\, n^{-1/2}. \tag{5}$$

If a particle has sufficiently high energy, it will traverse the turbulent plasma sampling the electric field in different cells, the cell size being given by Equation (5). The magnitude of the electric field $\mathscr{E}$ (esu) is given by

$$\mathscr{E} = \frac{v_{\mathrm{A}}}{c}\, \delta B = 10^{0.8}\, \delta B^2\, n^{-1/2}. \tag{6}$$

If a particle travels a total distance L, it samples (L/λ) different cells, and gains or loses an energy $e\mathscr{E}\lambda$ (erg) per cell. Treating this as a random walk process, the expectation value of the final energy E_{av} is given by

$$E_{\mathrm{av}} = 10^{11.8} \left(\frac{L}{\lambda}\right)^{1/2} eE\lambda. \tag{7}$$

On using Equations (5) and (6), we find that this may be re-expressed as

$$E_{\mathrm{av}} = 10^{7.0}\, \delta L^{1/2} B^2 n^{-3/4}. \tag{8}$$

We may also estimate the absolute maximum energy which could be acquired by any particle by assuming that the electric field has the same sign over the total path length L. One then finds, using Equation (6), that the maximum energy is given by

$$E_{\mathrm{max}} = 10^{3.3}\, \delta L B^2\, n^{-1/2}. \tag{9}$$

Let us consider the above formulas for a large flare characterized by $L = 10^{10}$, $B = 10^{1.5}$. If the density in the shocked gas is that of the preflare coronal density in the active region, of order $n = 10^8$, we find, adopting $\delta = 10^{-1}$, that $E_{\mathrm{av}} = 10^8$ and $E_{\mathrm{max}} = 10^{11.3}$. On the other hand, if the density is that characteristic of the post-flare plasma, calculations presented elsewhere (Sturrock, 1973) would indicate that $n = 10^{9.4}$. We then find that $E_{av} = 10^7$ and $E_{max} = 10^{10.6}$.

Stochastic acceleration typically leads to a power-law energy spectrum. The above considerations indicate that there should be a change in slope of the energy spectrum in the range 10–100 MeV, for a large flare. Data reported by Price (1973) show evidence for such a change of slope. We also see that there is no difficulty in understanding that some particles can acquire energies as high as 10 BeV.

We may also note that particles will experience stochastic acceleration, as described above, only when they have gyro-radii r_g comparable with or larger than the turbulence cell size, i.e.

$$r_g \geqslant \lambda. \tag{10}$$

Electrons will typically be relativistic, for which

$$r_g = 10^{-2.5} EB^{-1}, \tag{11}$$

so that the 'injection energy' of electrons, $E_{e,\,\mathrm{inj}}$, is given by

$$E_{e,\,\mathrm{inj}} = 10^{9.8} Bn^{-1/2}. \tag{12}$$

For the two cases considered above, this energy is estimated to be $10^{7.3}$ eV or $10^{6.6}$ eV. Ions will typically be nonrelativistic at injection, so that r_g is given by

$$r_g = 10^{2.2} A^{1/2} Z^{-1} E^{1/2} B^{-1}. \tag{13}$$

Equation (10) therefore leads to the following formula for the injection energy of ions:

$$E_{i,\,\mathrm{inj}} = 10^{10.2} Z^2 A^{-1} B^2 n^{-1}. \tag{14}$$

The injection energy for protons, for the above two cases, is found to be $10^{5.2}$ eV and $10^{3.8}$ eV. Since the injection energy for ions is considerably lower than that for electrons, one can understand that Phase 2 acceleration typically produces larger fluxes of ions than of electrons.

In Table I, certain acceleration parameters are presented for three sets of primary

TABLE I

	$L = 10^{8.5}$, $B = 10^3$	$L = 10^{8.5}$, $B = 10^2$	$L = 10^{10}$, $B = 10^2$
W(erg)	$10^{29.8}$	$10^{27.8}$	$10^{32.3}$
n_{PF} (cm^{-3})	$10^{11.6}$	$10^{10.2}$	$10^{9.7}$
Phase 1			
$E_{e\,(\mathrm{knee})}$ (eV)	$10^{4.5}$	$10^{3.9}$	$10^{4.4}$
Phase 2			
$E_{p\,(\mathrm{knee})}$ (eV)	$10^{7.6}$	$10^{6.6}$	$10^{7.7}$
$E_{p\,(\mathrm{max})}$ (eV)	$10^{11.0}$	$10^{9.7}$	$10^{11.5}$
$E_{e\,(\mathrm{inj})}$ (eV)	$10^{7.0}$	$10^{6.7}$	$10^{7.0}$
$E_{p\,(\mathrm{inj})}$ (eV)	$10^{4.6}$	$10^{4.0}$	$10^{4.5}$

flare parameters B and L. The total flare energy W (erg) and the post-flare density are given, estimated on the model recently proposed (Sturrock, 1973). Using these values of B, L and n, the preceding formulas are used to calculate the energy at which the energy spectra should show a 'break' or 'knee' (for both Phase 1 and Phase 2 acceleration), the maximum energy to which protons might be accelerated, and the injection energy for electrons and protons (for Phase 2 acceleration).

Finally, we may note that the model proposed for Phase 2 acceleration meets the principle requirements set down by Cartwright and Mogro-Campero (1972), as a result of analysis of abundance data. The final acceleration is stochastic; the injection condition depends primarily on rigidity; and accelerated particles may be in a fairly dense plasma (the post-flare plasma) before injection.

A more detailed treatment of Phase 2 acceleration, based on the theory of stochastic acceleration in a turbulent plasma (Hall and Sturrock, 1967; Newman, 1973) will be presented at a later date.

Acknowledgements

This work was supported by the National Aeronautics and Space Administration under Grant NGL 05-020-272 and the Office of Naval Research under Contract N00014-67-A-0112-0062.

References

Carmichael, H.: 1964, in W. Hess (ed.), *AAS-NASA Symp. on the Physics of Solar Flares*, NASA SP-50, p. 459.
Cartwright, B. G. and Mogro-Campero, A.: 1972, *Astrophys. J. Letters* **177**, L43.
Coppi, B. and Friedland, A. B.: 1971, *Astrophys. J.* **169**, 379.
De Jager, C.: 1969, in C. de Jager and Z. Švestka (eds.), 'Solar Flares and Space Research', *COSPAR Symp.*, 1.
Dungey, J. W.: 1958, *Cosmic Electrodynamics*, Cambridge University Press, p. 98.
Friedman, M. and Hamburger, S. M.: 1969, *Solar Phys.* **8**, 104.
Frost, K. J.: 1972, private communication.
Furth, H. P., Killeen, J., and Rosenbluth, M. N.: 1963, *Phys. Fluids* **6**, 459.
Hall, D. E. and Sturrock, P. A.: 1967, *Phys. Fluids* **10**, 2620.
Kundu, M. R.: 1965, *Solar Radio Astronomy*, Interscience, New York.
Lin, R. P.: 1974, *Space Sci. Rev.* **16**, 189.
Newman, C. E.: 1973, *J. Math. Phys.* **14**, 502.
Petrosian, V.: 1973, *Astrophys. J.* **186**, 291.
Petschek, H. E.: 1964, in W. Hess (ed.), *AAS-NASA Symp. on Physics of Solar Flares*, NASA SP-59, p. 425.
Price, P. B.: 1973, in R. Ramaty and R. G. Stone (eds.), *Proc. Symposium on High Energy Phenomena on the Sun*, NASA Goddard Space Flight Center, p. 377.
Riddle, A. C.: 1970, *Solar Phys.* **13**, 448.
Simnett, G. M.: 1974, *Space Sci. Rev.* **16**, 257.
Smith, D. F.: 1974, this volume, p. 253.
Sturrock, P. A.: 1966, *Nature* **211**, 695.
Sturrock, P. A.: 1973, in R. Ramaty and R. G. Stone (eds.), *Proc. Symposium on High Energy Phenomena on the Sun*, NASA Goddard Space Flight Center, p. 3.
Švestka, Z.: 1973, *Conf. on Solar Terrestrial Relations*, Calgary, Canada, p. 23.

DISCUSSION

Uchida: What is the amount of mass involved in the process (tongue shaped mass ejection in Carmichael's model) you are talking about? That can only bring out the mass initially lying in the corona and must be small.

Sturrock: The numbers are given in my Goddard talk, but the mass ejected from the corona is evaporated from the chromosphere in the course of flare.

Uchida: Isn't it true that the closure of the field pattern in your slide by reconnection occurs before the evaporation process takes place?

Sturrock: Reconnection is not restricted to the flash phase; it lasts as long as the flare itself, i.e. for hours.

Kane: You mentioned that in Phase 2 the acceleration occurs in a high density region. However the hard X-ray observations of behind-the-limb flares indicate that the energetic electrons (in Phase 2) are located in a low density region. How do these electrons get from the acceleration region to the hard X-ray source? How much energy is lost in the transportation process?

Sturrock: Acceleration occurs in a current sheet, then the accelerated particles cause evaporation. However, the model has been not worked out in sufficient detail to answer your question.

Dryer: Your point about the term 'bow shock' with its analogy to the Earth's bow shock is appropriate. Some clarification is necessary, however. If your magnetic piston maintains its geometrical size as it moves from sub- to superalfvénic velocities, the shock will develop, at first, far in front of the barrier, quickly approaching its classical position (steady-state) given as a function of the density compression ratio across the shock (i.e. as a function of specific heat ratio and Mach number). As the blob (still the same size) changes its Mach number, the shock's position changes accordingly just as in the case of the Earth. The comments about turbulence between the shock and the driver are not appropriate with the proviso that the undisturbed magnetic field should – if the Earth's bow shock's example may be used – be highly oblique or perpendicular to the shock normal. If the field is more or less parallel to the shock normal, there should indeed be such turbulence as seen behind the Earth's shock. If, however, the blob expands, the shock adjusts by moving away from the driver piston. Its velocity will respond in accordance with the acceleration, constant velocity, or deceleration of the piston.

Sturrock: I agree.

Lin: Your acceleration mechanism takes place in the turbulent region *behind* the shock. Observations of the Earth's bow shock and interplanetary shocks show that the particles are accelerated *at* the shock or in *front* of the shock. Is there anything wrong with a simple Fermi mechanism between the shock and magnetic structures ahead of it for the second phase acceleration? Protons would naturally get more energy than electrons for a Fermi mechanism since the proton velocities are lower for the same energy.

Sturrock: I would need some numbers to make any comment.

Vrabec: Your figures of magnetic field configurations associated with flares showed essentially only their two-dimensional geometry. Would you please comment on the importance of the three-dimensional configuration of the magnetic field with respect to stability and containment.

Sturrock: I assumed the extension in the third dimension to be comparable with the other two.

Meyer: My comment refers to the stochastic particle acceleration process. You mentioned that one can compute a minimum injection energy for acceleration to occur from the requirement that the gyro radii initially should not be smaller than the smallest turbulent scale. It seems that this would be required for those induced electric fields that are perpendicular to the magnetic field. However under very turbulent conditions one might expect also a rapid shearing of magnetic fields which induce electric fields along the magnetic field. Such electric fields would also accelerate particles whose gyro radii are smaller than the turbulent scale, so that the minimum injection energy requirement would be lifted. Of course such electric fields would differentiate strongly between electrons and protons. The former, due to their mobility, will pick up most of the energy available from such fields and might thereby shortcircuit such fields before protons are able to make use of them.

Smith: A very important part of this model is boiling off part of the chromosphere. The only person who has done detailed calculations regarding this possibility is Brown. I would like to know whether he considers this possible.

Brown: On the calculations of mine that Dean Smith mentions, I find chromospheric evaporation by electron streams very favourable in respect of ejecting the necessary 10^{16} gms. However, if you are really suggesting using this upflow to feed the current sheet with material, I wouldn't believe it because the dynamical upflow time is around a minute – much larger than the time scale for rapid variation in the sheet as inferred from hard X-ray time profiles and radio data.

Sturrock: There does seem to be evidence that evaporation occurs (e.g. Hudson and Ohki) but how effective it is may be governed by thermal conduction across the magnetic field.

Leblanc: In your model of magnetic field, what is the height of the top of the magnetic arch where the acceleration of particles occurs?

Sturrock: The height is inferred from observation and built into the model.

Uchida: I have a rather exotic model in which I seek the mass and energy source loss in the photosphere or low chromosphere. People have sought the flare origin in the corona or the high chromosphere mainly because no disturbed photosphere was observed lying beneath the Hα flare area. However, recent observation with high time and spatial resolution of low lying layers as performed by Zirin and Tanaka in λ3835 has revealed the existence of perturbed photosphere, though pretty small in size. We are tempted to explain the entire flare as due to a dynamical process if we remember that the main part of the total energy of large flares, 10^{32} erg, is mainly due to the interplanetary flux with particle number of 10^{40} to 10^{41} but this is readily realised if a tiny cube (10^8 cm^3) of the photospheric or low chromospheric material is accelerated to a final velocity of the order of 500 km s^{-1}, for example by the melon seed mechanism in a highly stressed field near neutral lines in the photosphere, where photospheric or low chromospheric brightening of such a size was reported by Zirin and Tanaka. It may be added that the location of these is also the location of the kernels in Hα which have very conspicuous characteristics incomparable to the rest of the Hα area according to Zirin.

Sturrock: Without seeing your model I cannot comment.

Kundu: I'd like to ask Dr Sturrock whether he considers the decameter continuum which lasts for several days as enough evidence for trapping of particles near the Sun.

Smerd: The late stationary type IV burst at metre wavelengths, the storm continuum, is compatible with the arrival and passage over several hours of a 'slow' disturbance ($\sim$100 km s^{-1}) at the coronal plasma levels. The disturbance may well be the 'driver gas' deduced from interplanetary shock wave observations.

If the storm continuum becomes bursty and continues (for $\sim$days) as a I-III storm, this can, in my opinion, only be interpreted as continued, intermittent acceleration of $\sim$10 to $\sim$100 keV electrons (and perhaps also sub-relativistic protons). The radio storm evidence suggests that the acceleration of the type I-radiating electrons is due to mhd-turbulence in a closed, strong-field loop, that of the storm III-radiating electrons due to mhd-triggered re-connections at the cusp or the neutral sheet above the closed-loop structure. In my opinion neither the storm continuum nor a radio storm are evidence for the trapping for hours or days respectively.

Enome: From strong proton flares electrons with energies of tens of MeV are often observed. The spectrum of these electrons is continued smoothly to the non-thermal electron spectrum, which strongly suggests the same origin for both electrons. If a hard X-ray second component is supposed to be accelerated in the plasmoid, strong synchrotron radiation will be expected from the relativistic electrons in the magnetic field of order of 5 G. This is not observed.

Kai: We have recorded a complex burst (1973 June 26) with the 80 and 160 MHz radioheliograph which could clarify a rather confusing concept on flare-continuum, moving type IV and storm continuum. The *flare-continuum* appeared in a very extended source which moved outward slowly ($\sim$400 km s^{-1}). It is noted that the 160 MHz source faded out more quickly. Later a *polarized moving IV source* appeared just on the top of the flare-continuum source and continued to move outwards, whereas the flare-continuum source shrank toward the disk, and showed a bipolar structure. This source remained active for a few hours and it was identified with a *storm-continuum*.

Smerd: Smerd and Dulk (*IAU Symp.* **43**, 1970) suggested that a flare continuum may be evidence for plasma radiation from sub-relativistic electrons injected into closed-field loops during the flash phase. The duration of the flare continuum ($\sim$10 min) is taken as evidence for further acceleration of the injected electrons, probably due to the interaction of the loop with a shock wave.

Note added after discussion. The difference between this and the same process for the second-phase acceleration of the microwave IV-radiating electrons (and protons) is probably that the flare-continuum electrons are in a 'high', active-region spanning loop, those of the microwave IV in 'low' loops near the flare explosion.

Rosenberg: One should be careful to attribute everything to the flare energy. Due to the flux, the magnetic configuration in the corona can change considerably, leading to different heating conditions (e.g. difference between closed and open field lines). This helps to understand also the type I continua, which are not at all flare related, but more related to active regions, strong fields, specific magnetic structures. Changes in heating, or excess heating via Alfvénic or magnetosonic flux also creates turbulence in the coronal structures, which in turn can lead to fast particles. The latter are in that case *not* directly connected to the flare.

MILLIMETER RADIO EVIDENCE FOR CONTAINMENT MECHANISMS IN SOLAR FLARES

E. B. MAYFIELD, K. P. WHITE III, and F. I. SHIMABUKURO
Aerospace Corp., El Segundo, Calif., U.S.A.

Abstract. Recent theories of solar flares are reviewed with emphasis on the aspects of pre-flare heating. The heating evident at 3.3-mm wavelength is analyzed in the form of daily maps of the solar disk and synoptic maps compiled from the daily maps. It is found that isotherms defining antenna temperature enhancements of 340 K correspond in shape and location to facular areas reported by Waldmeier. Maximum enhancements occur over sunspots or near neutral lines of the longitudinal magnetic fields which indicates heating associated with chromospheric currents. These enhancements are correlated with flare importance number and are observed to increase during several days preceding flaring. This evidence for a containment mechanism in the chromosphere is collated with current theories of solar flares.

1. Introduction

Most of the published theories of solar flares have been based on the presumption that the magnetic field of an active region is the source of energy for a flare. Although the exact process of field annihilation is not certain, it is believed that the conversion occurs at the neutral line or sheet separating fields of the opposite polarity. The principal difficulty for these theories as Parker (1963) showed, is the excessive time required for the process to occur. Whereas flares are observed to reach flash phase in typically 10^2 to 10^3 s, calculations yield times of 10^5 s or longer to provide the 10^{32} erg expected to be released by a major flare. Despite this difficulty, the premise is regarded as correct and subsequent work has examined various models to identify a rapid mode of field diffusion and annihilation. Recent theoretical work based on magnetohydrodynamic models has been successful in identifying processes of field annihilation which are sufficiently fast and compatible with experimental observations.

One of the first of these was that of Jaggi (1963) who considered three types of hydromagnetic instabilities in a plasma; namely, infinite conductivity, finite conductivity, and magnetic field gradient. His results, which were based on work by Furth *et al.* (1963), showed that only the finite conductivity case leads to an instability with growth times of the order of seconds or minutes. Finite conductivity instabilities in a plasma have been investigated in detail by Furth *et al.* (1963) for the case of a current sheet between fields of opposite polarity. For this model they find two unstable modes, a rippling and a tearing mode. Applied to conditions in the chromosphere, Jaggi showed that the rippling mode gave excessive times except for very short wavelength instabilities, but that the tearing mode was unstable for long wavelengths and gave typical times of 10^2 to 10^3 s. One of the characteristics of this model is that the temperature of the current sheet should rise prior to the onset of the instability.

Petschek (1964) proposed standing magnetohydrodynamic shock waves in the neutral region separating fields of the opposite polarity as the mechanism for con-

Gordon Newkirk, Jr. (ed.), Coronal Disturbances, 447–460.

verting magnetic to plasma energy. In this model, which is based on the Sweet (1958) X-type magnetic field configuration, the plasma and fields flow into the neutral region and the fields are annihilated by Alfvén waves. Assuming typical flare volume values of 10^9 cm for the scale length, 500 G average field and 2×10^{11} cm^{-3} density, the annihilation time is 10^2 to 10^3 s. The inflow velocity determines the rate of energy conversion and is insensitive to conductivity in the shock region, but a conductivity corresponding to a temperature of 10^6 K is assumed for the initial condition. Based on this high initial value of temperature at the neutral point, heating should be observed prior to a flare.

Syrovatskii (1969) considered collisionless dissipation of magnetic energy in the neutral region separating fields of opposite polarity as the annihilation mechanism. In this model, currents in the neutral sheet can increase very rapidly, leading to plasma turbulence and magnetic field reconnection in times of the order of 10^{-1} s. This also results in charged particle acceleration to very high energies observed to accompany flares. Part of this energy is dissipated in heating the chromosphere and corona and is responsible for the observed optical radiation. The onset of the instability in this model is controlled by conditions in the neutral region and turbulent resistance in the plasma becomes important when the ion thermal velocity exceeds the directed (current) electron velocity. This indicates, for chromospheric conditions, temperatures of about 10^4 K prior to a flare.

Coppi and Friedland (1971) have proposed a model which does not require shock or Alfvén waves in the neutral field region for the dissipation of field energy. Flow velocities into the neutral sheet are subsonic but sufficiently large to generate strong electric fields and to initiate turbulent electrical resistivity. This results in field annihilation in the neutral region from a tearing mode instability. They assume that the lower corona is the region of flare onset and show that the model is very sensitive to temperature variations. Although they presume that initially temperatures in the neutral region are not high, values of 10^5–10^6 K are indicated as typical for flare conditions. This value is shown to be reached in phases. When instability occurs, the dissipation of about 10^{32} erg can be released in about 10^2 to 10^3 s.

Experimental evidence in support of these theories has been previously reported by Mayfield *et al.* (1970). Results of observations at 3.3 mm wavelength showed enhanced emission associated with magnetic fields, especially near the neutral lines separating fields of the opposite polarity. For regions in which antenna temperatures were enhanced by more than about 8% over a quiet area temperature, importance class 2 or greater flares were observed to occur usually within about 24 h. Although the results of the investigation were limited to larger flares during a short period, it verified heating associated with flaring regions and particularly at or near the neutral region of magnetic fields.

The present study is based on an analysis of extensive observations of solar emission at 3.3 mm wavelength obtained with the Aerospace radio telescope. These data cover the years 1967 through 1972 and, with a few missing periods, provide a continuous record of solar emission. Flare data for the same period, obtained from *Solar-*

Geophysical Data published by the National Oceanic and Atmospheric Administration, have been compared with the radio data. These results show a close association between regions of enhanced 3.3 mm radio emission and flares and statistical evidence for flare occurrence proportional to enhancement for importance 1N and 1B flares. The heating at millimeter wavelengths is observed to occur about 24 h prior to flare onset and to be near the neutral magnetic field region.

2. Observations

The observations were made with the 15-ft (4.57 m) diameter antenna at Aerospace in El Segundo, Calif. Although the instrument has been described by Jacobs and King (1965) and by King *et al.* (1966), a brief description is given. The antenna is a Cassegrain in a polar mount with surface tolerance for 400 GHz (0.75 mm) operation. For this study it was operated at 90 GHz (3.3 mm). Beamwidth at half-power points is about 2.8′ and pointing accuracy is about 20″. The antenna is controlled by means of an on-line computer which permits various tracking modes. For this investigation, the antenna was programmed to obtain a temperature matrix of 19 × 19 points centered on the disk. The angular separation of the readings was 99″. Emission temperatures are normalized to an undisturbed region near the center of the disk which was selected from daily maps and the Fraunhofer Institut maps. By normalizing to a quiet region, problems of absolute calibration and atmospheric attenuation are avoided.

Observations near the limb are inaccurate because of the antenna beamwidth which limits the useful area to less than 0.7 $R_{\odot}$. Analytical methods to restore the data beyond this radius have not been successful.

Daily maps are made from the normalized temperature matrix by constructing isotherms of the antenna temperature at 2% increments for 4% or greater enhancement. A typical isothermal map is shown in Figure 1 for 30 June 1971. This shows two regions which are enhanced 7% and 10%. These maxima are closely associated with magnetic field regions near sunspots at these locations. For presentation of longer term data, the daily maps are plotted in synoptic format to show a solar rotation. These are organized after Waldmeier (1972) so that comparison can be readily made with the photospheric maps of the Sun. Each solar rotation shows maps of the emission as it appeared near central meridian passage for latitudes to $\pm 50°$. In addition to the date which is identified by a dot, the time at which maps were made is identified by the short vertical line. Crosses shown on the figure mark locations used for normalizing emission temperatures. Location and maximum enhancement for each region are shown by the filled circle and given by the number next to it in percent of enhancement. Values of the isotherms are also given.

Examples of comparisons of the radio synoptic maps with the Waldmeier heliographic maps of the photosphere are presented in Figures 2, 3, 4 and 5. During these four solar rotations, one can see at a glance how well the 4% enhancement contours correspond to the extensions of the facular regions. Differences in shape and extent

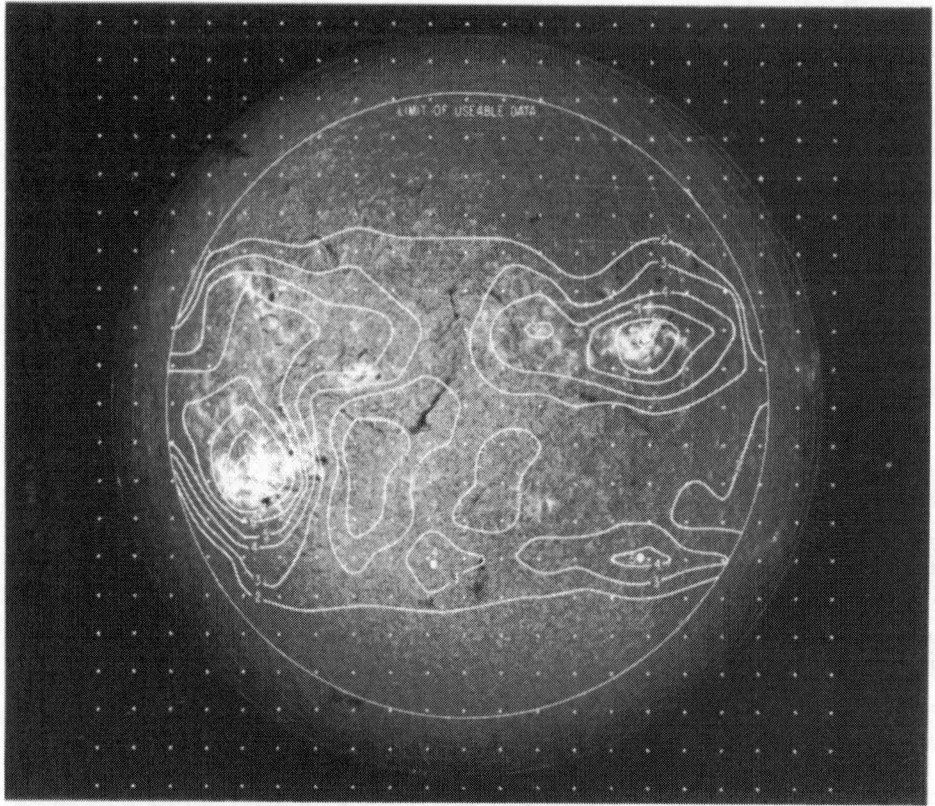

Fig. 1. The 3.3-mm solar radio temperature contour map made at The Aerospace Corp. on 30 June 1971 is superimposed on a concurrent Hα photograph. The contours are labeled in percent enhancement relative to the temperature of the undisturbed region denoted by the hatched contours.

can be attributed to a number of causes, which include: (1) slow, evolutionary changes in the active regions, since the photospheric maps pertain to the maximum development of the regions, while the millimeter maps reflect the appearance closest to central meridian passage; (2) rapid changes in the active regions, as caused by flares, whose residual effects can influence the radio contours (although this has been allowed for and corrected when possible); (3) selection of an invalid normalization point for a particular daily radio map (e.g., a dark absorption filament), which would cause an apparent 'growth' or 'contraction' of the enhancement levels, and; (4) the lack of a physical connection between the two phenomena.

The first three points are really just problems of observation and have been accounted for as described previously. The fourth point implies that, to the extent that the two phenomena are physically distinct, one would not expect a correspondence in shape and location. We feel that the striking correspondence evident in comparisons between the two types of maps for all of the solar rotations we have been able to analyze shows there is an apparent physical connection between the two phenomena. On the basis of these data, we will adopt the view that the facular regions and

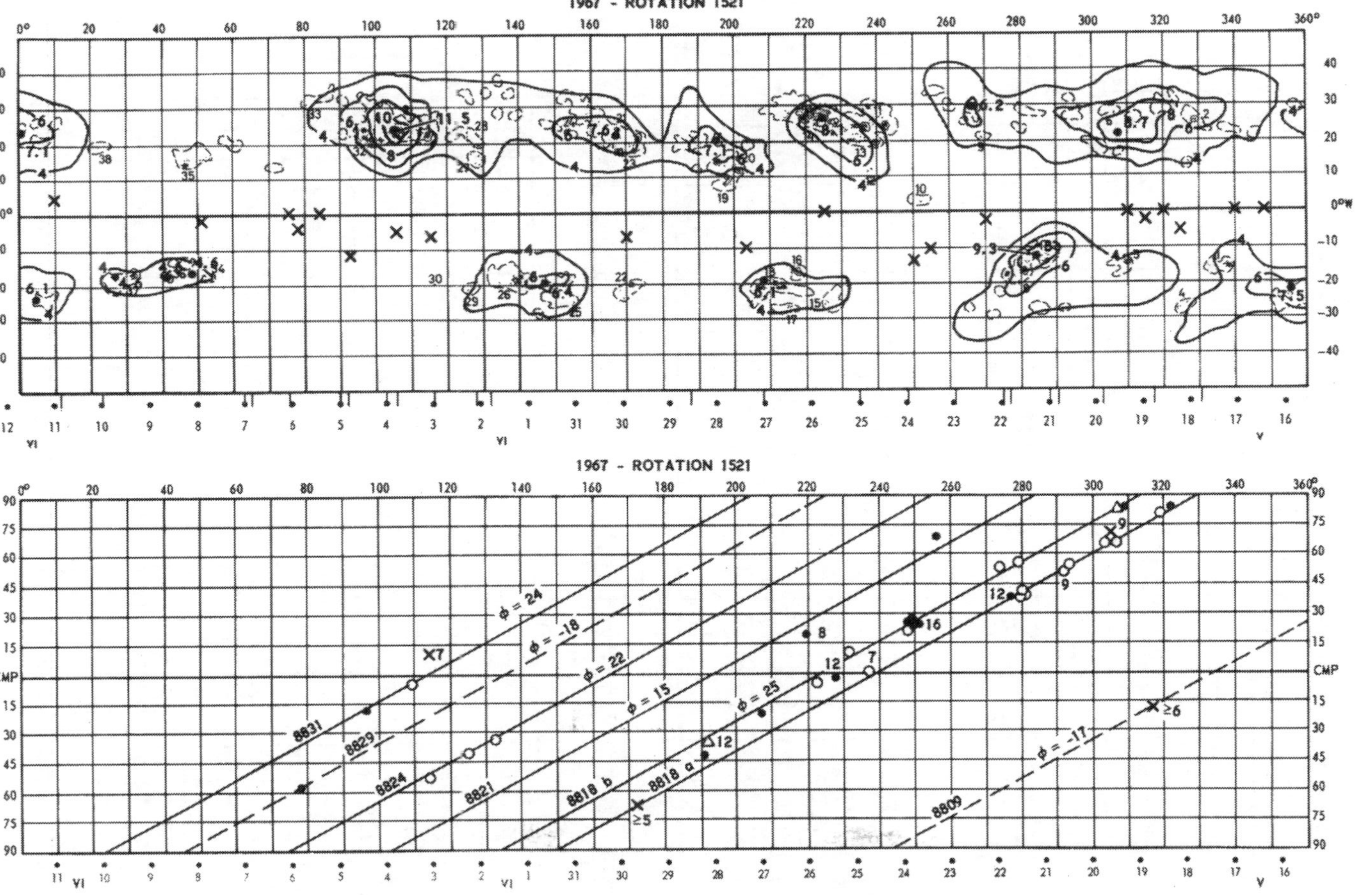

Fig. 2. *Upper:* Synoptic 3.3-mm radio temperature contour map for Carrington rotation 1521 has been superimposed on the corresponding heliographic map of the photosphere by Waldmeier. The contours are at levels of 4, 6, 8%, etc., enhancement above the quiet regions indicated by crosses. Peak enhancements are also shown by the filled circles. Note the close correspondence between the 4% contour levels and facular areas and between the peak enhancements and sunspots. *Lower:* The synoptic flare chart for Carrington rotation 1521. Each center of activity is designated by a McMath calcium plage number, latitude, and Carrington longitude. The flare designations are explained in the text.

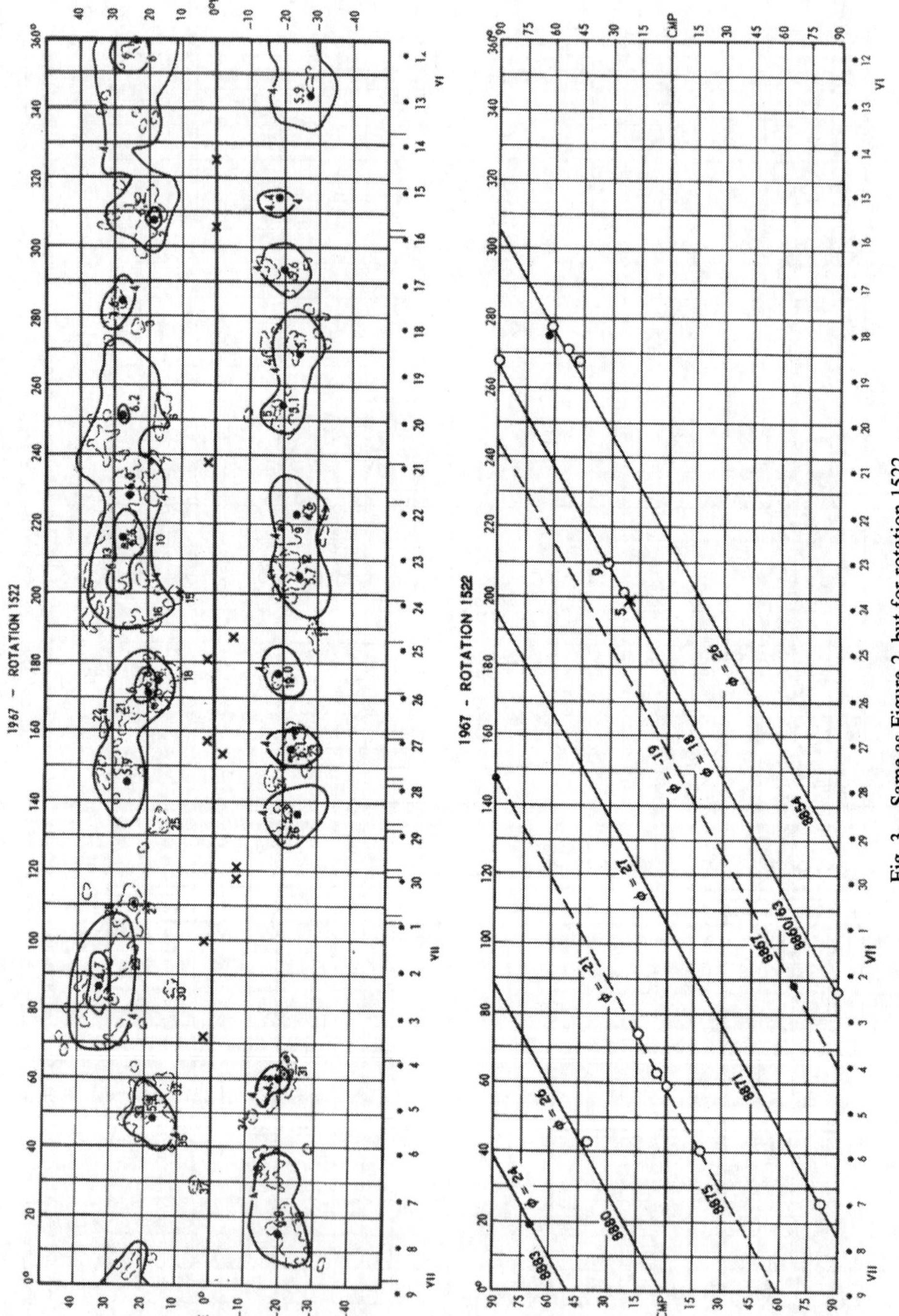

Fig. 3. Same as Figure 2, but for rotation 1522.

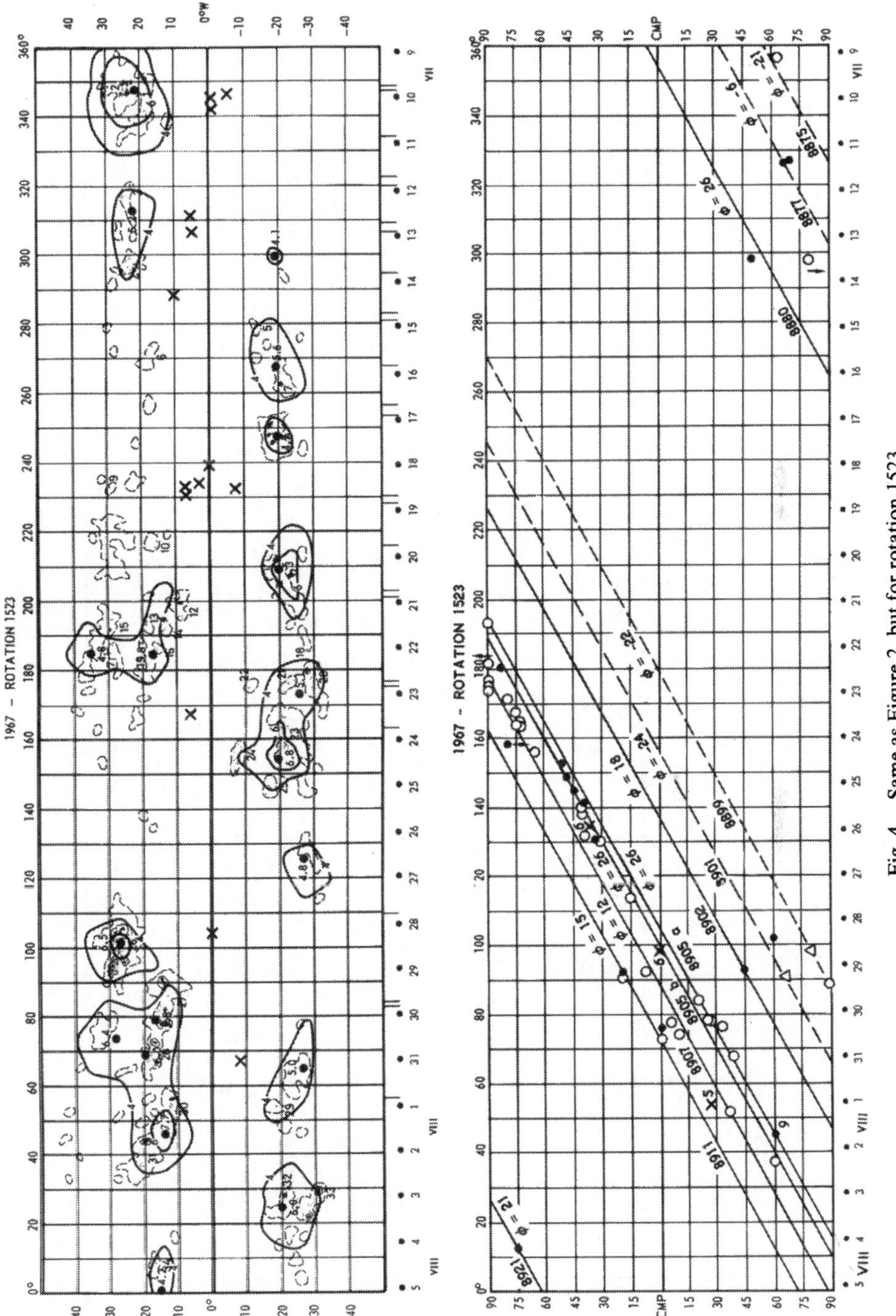

Fig. 4. Same as Figure 2, but for rotation 1523.

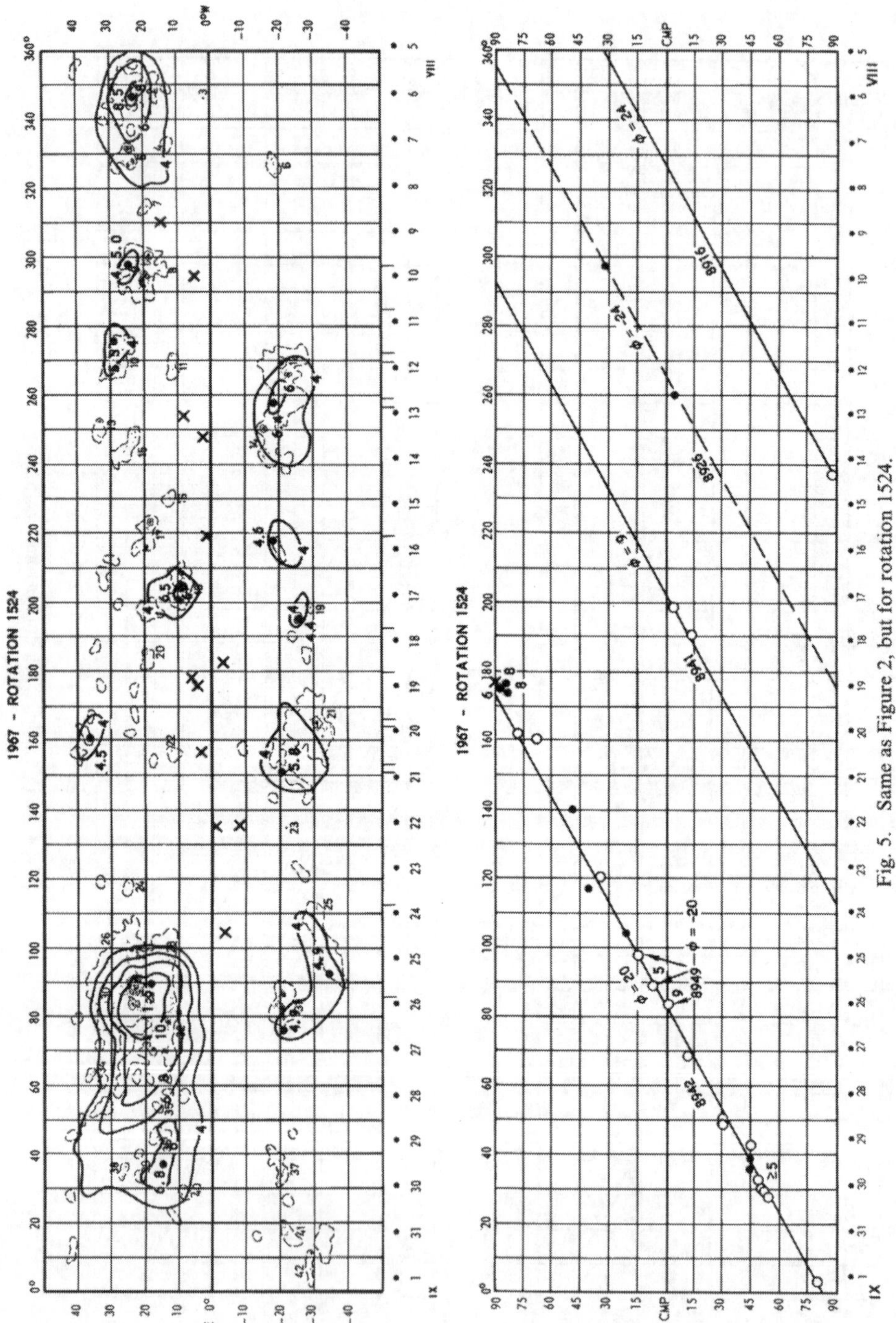

Fig. 5. Same as Figure 2, but for rotation 1524.

the millimeter enhancements are manifestations of the same phenomenon in the solar atmosphere; namely, magnetic fields which are responsible for the increased emissions in both cases. Figures 2 through 5 show that scattered facular regions, with no prominent sunspots, rarely exceed about 5% enhancement. If they do, the cause can usually be attributed to a scattering of small sunspots or pores, not all of which appear on Waldmeier's maps. Furthermore, the peak enhancements more likely than not, will correspond in location to the sunspots, even in active regions not exhibiting much flaring. These observations suggest regarding the measured millimeter enhancements as being made up of two components; one component can be attributed to heating which also manifests itself in the faculae; the other component should be associated with the sunspot magnetic fields and some sort of containment of an excess-heated plasma. The latter component will be discussed in more detail in a later section dealing with flares. The former component will be discussed next.

By inspection, the contours in greatest agreement with the facular areas are the 4% contours. In some cases, a slightly greater contour, perhaps 5%, would produce a cleaner correspondence. In either case, it is clear that most outlying facular regions, uncontaminated by the sunspot component, would lie between the 4% and 5% contours. Thus, we shall adopt a mean enhancement of 4.5%$\pm$0.5 as representative of the excess heating at 3.3 mm measured by the radio telescope. With the adoption of 7500 K (Linsky, 1973) as the temperature of the quiet sun, the enhancement above facular regions corresponds to about 340 K. Such a measurement must be corrected for the fraction of the radio beam which is actually being filled by the heated structures. An estimate of such a fraction involves many assumptions about the fine structure of the low chromosphere and an indefensible extrapolation of models of heating in faculae, which models do not extend sufficiently high into the solar atmosphere. Future work could attempt to achieve consonance between the physical parameters indicated by the mm measurements and a facular model, such as that by Chapman (1970), but such is beyond the scope of the present work. For now the correspondence between the 4.5% enhancement at 3.3 mm and faculae can only be substantiated on the basis of spatial coincidence.

The question of the longevity of active regions can be investigated by inspection of successive synoptic maps to note the recurrence of given active regions. In this respect it is important to point out the sensitivity and precision of the data contained in the synoptic maps, because these parameters will directly influence the visibility of the enhancements. As stated before, the lowest contour level plotted corresponds to a 4% enhancement, so chosen because at this level individual active regions are usually discernable, whereas a 3% contour usually will run nearly the whole length of the map, providing very little information. Additionally, the repeatability of measurements made on different daily maps either on the same day or different days can be as large as $\pm$0.5%, though the average is probably close to $\pm$0.3%, with some outstanding examples showing variations of no more than 0.2% over a five or six-day period. To maintain a precision of about 10% (e.g. 4$\pm$0.4%), we felt it was safest to display the 4% contour as the lowest level.

Therefore, the longevity of an active region is defined as the time from when it first exceeded 4% enhancement to when it first disappeared below the 4% level. In this context, some regions are found to last less than a full solar rotation, while others can persist and be followed over five or six rotations. This result is entirely consistent with the result that the 4% contour level coincides in shape and location with the facular areas, which have lifetimes that can be as short as a few days or as long as a few solar rotations.

Another feature of the synoptic maps whose longevity can be gauged is the quiet regions, as indicated by the clustering of crosses. The nature of these regions is still in question, but certain examples of them can be followed over three or four rotations.

Solar flare data for the period 1967 through 1972 were plotted in synoptic format for comparison with the mm maps. Examples of these are shown in Figures 2–5 for rotations 1521–1524. The longitude and date are given at the bottom of the plots and central meridian distance is given vertically. This permits displaying all flares for an active region during a disk passage as well as its time of occurrence and location. Heliographic longitude for each region is determined from time of central meridian passage, and latitude as well as McMath-Hulbert calcium plage number is indicated with each region. These plots show a consistency for the flaring locations to remain fixed in position during a disk passage. In complex regions two or more centers may exist and these show only small excursions from a fixed location. Although the reports from individual observatories show significant dispersion in location and time, the mean values reported in the Comprehensive Part of *Solar-Geophysical Data* of the flare centroids are clearly very consistent.

The flares reported here are for importance number 1B or greater. Those of number 1B are shown by open circles, importance 2 by filled circles and importance 3 by triangles. In addition we have included flares with an index number of 5 or greater based on the classification method of Dodson and Hedeman (1971). These are shown by crosses with the index number next to them. The Dodson-Hedeman classification method is based on five parameters: (1) X-ray emission, (2) 10 cm radio flux, (3) optical importance number, (4) meter wave flux and (5) dynamic radio spectrum.

3. Discussion

Active regions plotted in this synoptic fashion show a close association with a fixed location during a disk transit. Some regions have more than one active part but these typically remain distinct. These results show that flares are produced by unique areas of centers of activity which retain close association with the magnetic fields of the underlying sunspots. This is an important result and is in disagreement with the widely accepted opinion that flare data are subject to large errors and that both location and times of occurrence are very uncertain. Although this may be true for reports from individual observatories, the mean values reported in *Solar-Geophysical Data* are highly consistent.

Since flares occur in active regions near sunspot groups and the 3.3-mm emission is associated with the facular regions of magnetic fields, there is a correlation of the flares with enhanced mm regions. This correlation, however, is complex and no simple rules can be stated. Evidence for two components of the mm emission can be seen in the maps. One of these is associated with the general magnetic fields in the faculae and accounts for the enhancements of 4 or 5%. This is the slowly-varying component and has been shown by Shimabukuro *et al.* (1973) to be due to magnetoionic processes in fields. The second component which accounts for the larger, rapidly-changing emission is associated with strong fields and occurs near the neutral lines or sheets of these fields. This strong emission is probably caused by currents in the neutral sheet separating fields of the opposite polarity as has been postulated by the theories discussed above.

Centers of activity which produce the greatest number of flares occur in regions of enhanced 3.3-mm emission, typically with enhancements greater than 8%. A distinctive feature of the mm-emission in these centers is that the radio plage is usually small and the thermal gradient is large. The 4% isotherm tends to occur near the maximum of emission. In other centers which may have a large enhancement but produce few or small flares, the 4% isotherm encloses a large area. This indicates that major flare-producing regions have concentrated magnetic fields with necessarily large magnetic flux density and associated $\nabla \times H$ current densities. These results are consistent with previous reports that magnetically complex γ-type sunspot groups produce the greatest number of flares.

In the most active regions, the location of the flare centroids is usually near the maximum of the mm emission. Since the longitude of each active region is very stable during a single transit and the mm regions change slowly, the association of flare centroid and maximum enhancement is close. For less active regions, however, the flare center is frequently located several degrees from the maximum and may be as great as 10 deg. The mm enhancements in these cases are usually part of a complex region or two regions joined by a common 4% isotherm. This indicates that the stronger magnetic flux of the original sunspot group has been dissipated into the region marked by faculae and associated with the slowly-carying component of the mm emission. Location of the flare centers for these regions is usually within the 5 or 6% isotherm but rarely a flare may occur at the 4% level of the mm maps.

Although all of the centers of activity which flare are associated with an enhanced region of mm emission, the reverse is not true and a few regions with up to 8 or 9% enhancement may not flare. These are typically part of a large multiple region similar to those which produce a few small flares during a transit. They provide additional information on the kinetic processes in chromospheric magnetic fields which lead to flares and are valuable in evaluating flare theories.

On the basis of these results, it is not possible to distinguish critically between the proposed theoretical models of flares. The model of Furth *et al.* (1963) is specific, however, in discussing two instability modes with differing time constants, i.e. non-local and local. The pre-flare millimeter heating supports this model and indicates that the large-scale instability has a time constant of about 24 h. Further the data show that a critical enhancement is reached at about 8% and for heating greater than that, a region will always produce a flare of importance 1N or greater. This is shown in Figure 6 which summarizes the data for 1N flares. Similar results are observed for 1B flares and also for importance 2 or greater flares as has been previously reported by Mayfield *et al.* (1970). This is also consistent with the Furth *et al.* model and indicates initiation of local instabilities although this does not exclude other models which require high temperatures as an initial condition to instability. Lack of antenna resolution prevents indentification of precise locations of the millimeter heating in the neutral magnetic field region which could distinguish smaller and hotter areas in this

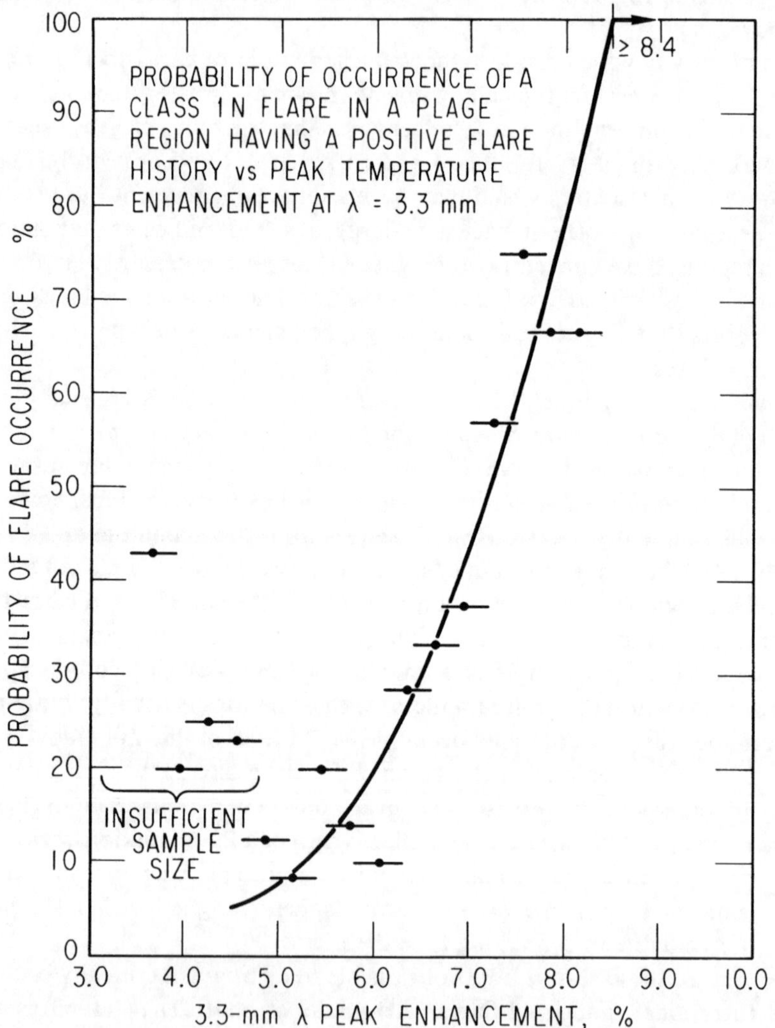

Fig. 6. The probability of occurrence of a class 1N flare-prone plage region within one to two days following measurement of the peak temperature enhancement at 3.3-mm wavelength, given as a function of the peak temperature enhancement.

transition. No apparent changes have been observed in either Hα filtergrams or magnetograms of the longitudinal component of the field associated with the 8% enhancement of the 3.3 mm emission.

4. Concluding Remarks

These extensive observations of solar emission at 3.3 mm wavelength show that strong enhanced regions are closely associated with the neutral region of longitudinal magnetic fields. The 4% enhanced isotherms agree closely with the facular regions and indicate that this is also due to magnetic fields. These features of the emission have been shown to be consistent with magneto-ionic theory. Regions which are enhanced greater than 8% always flare and indicate that some mechanism critical to flares occurs at this level. These results are consistent with recent hydromagnetic theories of flares which are based on magnetic field annihilation at the neutral region of opposing polarities. The observed heating is presumed caused by a current sheet in the neutral region.

Acknowledgements

The authors thank Mr P. Alailima who helped obtain the radio data and Mrs Mary Gates who helped in preparing the synoptic maps.

Part of this work was supported by NASA under contract NAS2-7292, part of it by Air Force Contract F04701-72-C-0073 and part of it by the Aerospace Company Financed Research Program.

References

Chapman, G. A.: 1970, *Solar Phys.* **14**, 315.
Coppi, B. and Friedland, A. B.: 1971, *Astrophys. J.* **169**, 379.
Dodson, H. W. and Hedeman, E. R.: 1971, *NOAA Report UAG* – **14**, World Data Center A, U.S. Dept. of Commerce.
Furth, H. P., Killeen, J., and Rosenbluth, M. N.: 1963, *Phys. Fluids* **6**, 459.
Jacobs, E. and King, H. E.: 1965, *I.E.E.E. International Convention Record Part 5.*
Jaggi, R. K.: 1963, *J. Geophys. Res.* **68**, 4429.
King, H. E., Jacobs, E., and Stacey, J. M.: 1966, *I.E.E.E. Trans. on Antennas and Propagation, A.P.* **14**, 82.
Linsky, J. L.: 1973, *Solar Phys.* **28**, 409.
Mayfield, E. B., Higman, J., and Samson, C.: 1970, *Solar Phys.* **13**, 372.
Parker, E. N.: 1963, *Astrophys. J. Suppl.* **8**, 177.
Petschek, H. E.: 1964, in W. N. Hess (ed.), *AAS-NASA Symposium on the Physics of Solar Flares*, Government Printing Office, Washington, D.C., p. 425.
Shimabukuro, F. I., Chapman, G. A., Mayfield, E. B., and Edelson, S.: 1973, *Solar Phys.* **30**, 163.
Solar-Geophysical Data: 1967–1972, U.S. Dept. of Commerce, Boulder, Colo.
Sweet, P. A.: 1958, in B. Lehnert (ed.), 'Electromagnetic Phenomena in Cosmical Physics', *IAU Symp.* **6**, 123.
Syrovatskii, S. I.: 1969, in C. de Jager and Z. Švestka (eds.), 'Solar Flares and Space Research', *COSPAR Symp.*, 346.
Waldmaier, M.: 1972, 'Heliographic Maps of the Photosphere', *Publikationen der eidgenössischen Sternwarte Zürich.*

DISCUSSION

Smith: You have used the word containment. Isn't it possible that the heating is due to slow continuous reconnection at a current sheet?

Mayfield: We have used the word containment because there is a threshold beyond which a region is certain to flare.

Smith: This is confusing terminology because most people interpret containment as containment of particles. What you are really talking about is a threshhold.

Mayfield: We interpret the enhanced 3.3 mm radio emission preceding flares as evidence for containment because of the existence of a critical level of enhancement. This does not show the presence of energetic particles but does indicate a heating mechanism that is a precursor to flares and is observed to occur about 24 h prior to onset.

Kundu: (1) I would like to point out that we have shown conclusively that Hα dark filaments and limb prominences are associated with absorption and emission features at 3 mm. In your first slide where you associate a 3 mm depression with a coronal hole, I believe I've seen a fairly large filament whose orientation is in general agreement with the orientation of your depression feature. Consequently I feel that it might as well be associated with the filament.

(2) I do not understand why you are associating the 3 mm emissive regions with the magnetic neutral line only. As you know, these regions are polarized – the emission comes from two regions of opposite polarity, separated by a line of zero polarity or of neutral line as seen by Mt. Wilson magnetograms.

(3) Regarding the preheating as an indicator of flare production, I would like to make the following comment. The phenomenon of preheating was first observed by the French group in 1957. We noticed that several hours before a burst occurs a region greatly increases in intensity and polarization within a core-size of about 1′ at 3 cm. Indeed if we plot the number of bursts as a function of the core-intensity of the region, we find a very similar linear relation to that indicated in your last slide at 3 mm. Consequently, I contend that the 3 cm preheating can be used as efficiently as an indicator of flare production as your 3 mm preheating. This result was published in *Compt. Rend. Acad. Sci.* 1957, *Ann. Astrophys.* **22**, *Paris Symposium on Radio Astronomy* and in *Solar Radio Astronomy*.

Mayfield: (1) The data we report have a resolution of only 2.8′ and we are not able to resolve filaments. We are aware of your observations which show the association of dark filaments and reduced emission. Our evidence for the association of reduced emission and coronal holes is not shown in Figure 1 but in later data and is still tentative.

(2) The polarized mm emission in a bipolar magnetic field has been observed to be right circular above the positive field and left circular above the negative field. However, the total power which we are reporting here is associated with the neutral line or region as shown in longitudinal magnetograms.

(3) Similar observations of enhancement and polarization prior to flares have been reported at 3 and 10 cm by others (particularly the Japanese). However our observations at 3.3 mm occur low in the chromosphere and show evidence of a critical temperature enhancement above which flares will occur within about 24 h.

ACCELERATION, CONTAINMENT, AND EMISSION OF VERY LOW ENERGY SOLAR FLARE PARTICLES

R. P. LIN, R. E. McGUIRE*, and K. A. ANDERSON*

Space Sciences Laboratory, University of California, Berkeley, Calif. 94720, U.S.A.

Abstract. We present the first observations of protons down to 44 keV and electrons down to $\lesssim 2$ keV emitted in an impulsive solar particle event. The observations are from the Apollo 15 Subsatellite during a flare event which began on September 1, 1971. We obtain a lower limit estimate of the energy contained in $\gtrsim 0.05$ MeV protons in the flare. The effects of adiabatic deceleration in the interplanetary medium and dE/dx energy loss in the corona are discussed. We conclude that energetic protons may constitute a major energy release in large solar flares.

1. Introduction

It has been known for some time that in many small flares (importance 1 or subflares) energetic, 20–100 keV electrons accelerated during the flash phase constitute a major portion of the total flare energy (Lin and Hudson, 1971; Lin, 1973; Kane, 1974). If the electron spectrum extends to energies as low as $\lesssim 5$ keV as indicated by observations, then the electrons would clearly contain the bulk of the flare's energy. Satellites have measured the proton component in large proton flares down to energies of ~ 0.3 MeV. These observations indicate that protons to energies of 0.3 MeV are accelerated and emitted into the interplanetary medium by these flares and that the energy spectrum, which is typically a power law in energy, is still rapidly rising with decreasing energy at ~ 0.3 MeV (Verzariu and Krimigis, 1972). If the power law proton spectrum extends to even lower energies, then the protons could contain a major portion of the energy in these large flares.

In this paper we present Apollo 15 Subsatellite observations of low energy protons and electrons during the solar flare event of 1 September 1971. These observations cover the range from ~ 44 keV to ~ 2 MeV for protons and $\lesssim 2$ keV to ~ 340 keV for electrons. At these low energies the energy change processes of dE/dx, adiabatic deceleration and Fermi acceleration play an important role in the propagation of these particles. Here we provide some estimates of the magnitude of energy changes due to these various processes. We also obtain a lower limit to the energy contained in energetic, >0.04 MeV protons emitted by this flare into the interplanetary medium.

2. Experimental Details

The observations reported here are from the University of California particle experiment aboard the Apollo 15 Subsatellite. This Subsatellite was ejected from the Apollo 15 Lunar Science Module on 4 August 1971, into a lunar orbit with initial perilune

* Also Physics Dept., University of California, Berkeley, Calif. 94720, U.S.A.

Gordon Newkirk, Jr. (ed.), Coronal Disturbances, 461–469. *All Rights Reserved.*

of 102 km, apolune of 139 km, and orbit inclination of 28.5° to the lunar equatorial plane. The Subsatellite samples particle fluxes at $\sim 60\ R_E$ distance from the Earth, and during this solar event the Subsatellite was in the distant magnetosheath and magnetotail regions.

The instrumentation in the UCal (Berkeley) experiment consisted of a pair of surface barrier semiconductor telescopes, one of which was open and the other of which was covered by a thin, low Z foil. This foil stopped protons of less than ~ 340 keV from registering in the telescope. By suitable subtraction of the fluxes observed in the two telescopes, electron and proton fluxes could be identified and measured. Details of this arrangement are given in Anderson *et al.* (1972).

In addition four large geometric factor, hemispherical plate electrostatic analyzers utilizing channel multiplier detectors provided measurements of electrons at ~ 0.5, 2, 6, and 14 keV. The ~ 6 and ~ 14 keV detectors were surrounded by plastic anticoincidence jacket to reject penetrating particles and thereby reduce the background. Further details are contained in Anderson *et al.* (1972).

3. Observations

The first major solar particle event observed by the Subsatellite began on 1 September 1971 at ~ 2000 UT. The relevant solar, interplanetary and terrestrial times are summarized in Table I. No optical flare was observed but both radio and X-ray emission was reported in the *Solar-Geophysical Data Bulletin* at ~ 1930 UT. It is very likely that this event originated in a flare in McMath plage region number 11842 which

TABLE I

Event times

Solar Events	
Radio (1 September 1971)	Time
Type II Intensity 3	1933.8–1948
Type III Intensity 2 g	1934.2–1935.4
Type IV Intensity 2	1934.5–2009.5
X-ray (Explorer 33)	
1–8 Å	1930–2009
Interplanetary Events	
Storm Sudden Commencement 4 September 1646 UT	
Terrestrial Events	
Entered magnetotail from magnetosheath 3 September ~0400 UT	

went over the west limb of the Sun August 31. The observations are complicated by several problems.

(a) The onset of this event was not observed by the Subsatellite since it was in the midst of a battery charge. The first particle data was obtained at the beginning of 2 September 1971.

(b) The Subsatellite was in the magnetosheath and crossed into the magnetotail during the event. It passed on several occasions through the magnetotail plasma sheet which contains large fluxes of protons and electrons in these low energy ranges which would mask the solar particle fluxes.

(c) The gain of the open telescope changed early in the event. Although we do not know the cause of this gain change we were able to estimate its magnitude from a comparison with the foil telescope during electron shadows. The fluxes have been corrected for the gain change.

(d) Due to an engineering oversight the Subsatellite 19 to 8 bit floating point accumulators registers 16–31 counts per accumulation period as 0 to 15 counts per accumulation period.

During the solar event we have attempted to obtain values of the solar particle fluxes which are uncontaminated by these effects. The proton points are plotted in Figure 1. Particularly in the lowest two energy channels it was difficult to separate solar fluxes from terrestrial fluxes, and from background due to penetrating particles. However, the main features of the low energy event are clear from this figure. These are:

(a) There is a distinct velocity dispersion of the peaks down to the 72–130 keV channel with the highest energy peaking earliest.

(b) The profiles are generally similar to those observed previously at higher energies.

(c) There was a storm sudden commencement at ~1700 UT on 4 September. Associated with this SSC was the characteristic increase in low energy proton fluxes observed in front of an interplanetary shock. These fluxes dominated the intensity-time profile from $\lesssim$1200 UT on 4 September. A characteristic flux decrease was observed after the SSC at the beginning of 5 September.

We thus associate the proton fluxes observed in the period ~0000 UT 2 September to ~1600 UT 4 September with an impulsive injection of protons at the time of the flare, followed by propagation in the interplanetary medium.

The energy spectrum of the proton event is shown in Figure 2. We have plotted the maximum flux in each energy channel. The spectrum below ~1 MeV is very shallow with $dJ/dE \propto E^{-0.65}$. Also shown are the 10–30 MeV and 30–60 MeV fluxes from the *Solar-Geophysical Data Bulletins.* The best fit over the entire spectral range is given by $dJ/dP = 55 \exp(-P/65)$, where P is particle rigidity in MeV/c.

During this event electrons were observed as well. Their intensity-time profiles are typical of electron events and are shown in Figure 3. Electrons were observed to $\lesssim$2 keV, and the electron energy spectrum, plotted in Figure 4, extends smoothly in a power law $dJ/dE \propto E^{-1.9}$ to $\lesssim$2 keV.

3. Discussion

In a preliminary analysis we have estimated the diffusion coefficients for each proton energy channel by comparison with the analytic solution of the diffusion-convection-

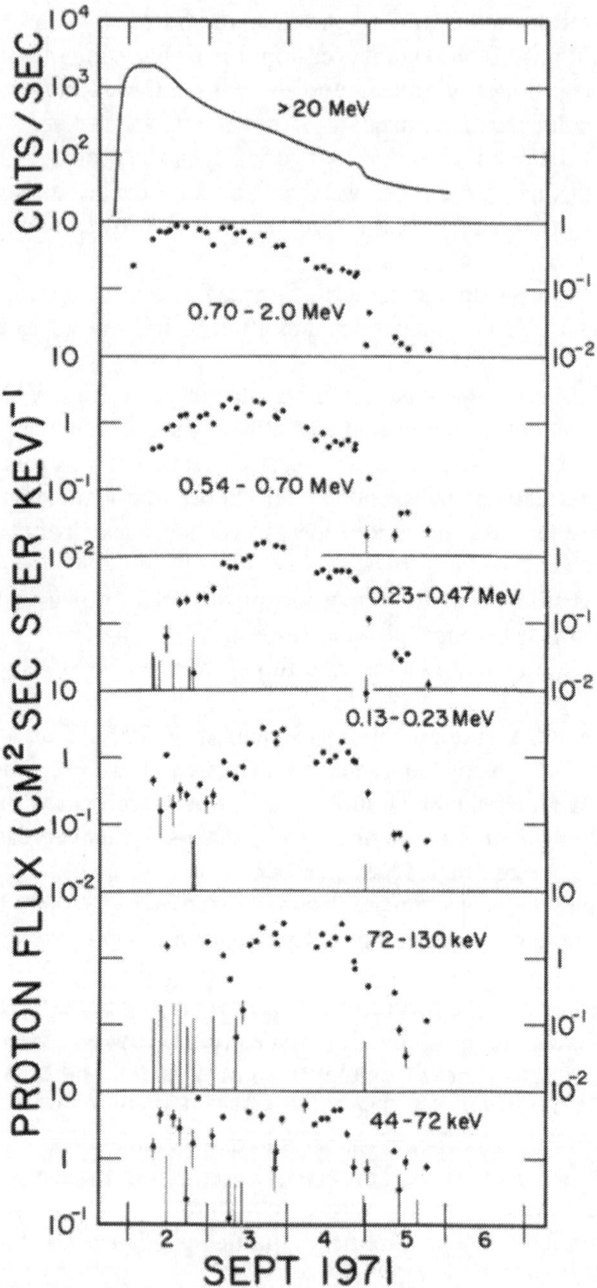

Fig. 1. The low energy proton observations for the flare event of 1 September 1971. The >20 MeV profile is from the UCal IMP-6 experiment. All the rest of the data is from the Apollo 15 Subsatellite, chosen at times when the fluxes are free from terrestrial fluxes. Early in the event the 44–72 keV and the 72–130 keV channels may be contaminated.

energy loss transport equation (Parker, 1965)

$$\frac{\partial U}{\partial t}+\frac{1}{r^2}\frac{\partial}{\partial r}\left(r^2 VU-r^2 k\frac{\partial U}{\partial r}\right)-\frac{2V}{3r}\frac{\partial}{\partial T}\left(\alpha TU\right)=0.$$

$U(r, T, t)$ is the differential number density, T is the kinetic energy, V is the solar wind velocity, assumed constant, and $\alpha \approx 2$ for non-relativistic particles; r and t are distance and time respectively. This has been solved for an impulsive injection of particles at

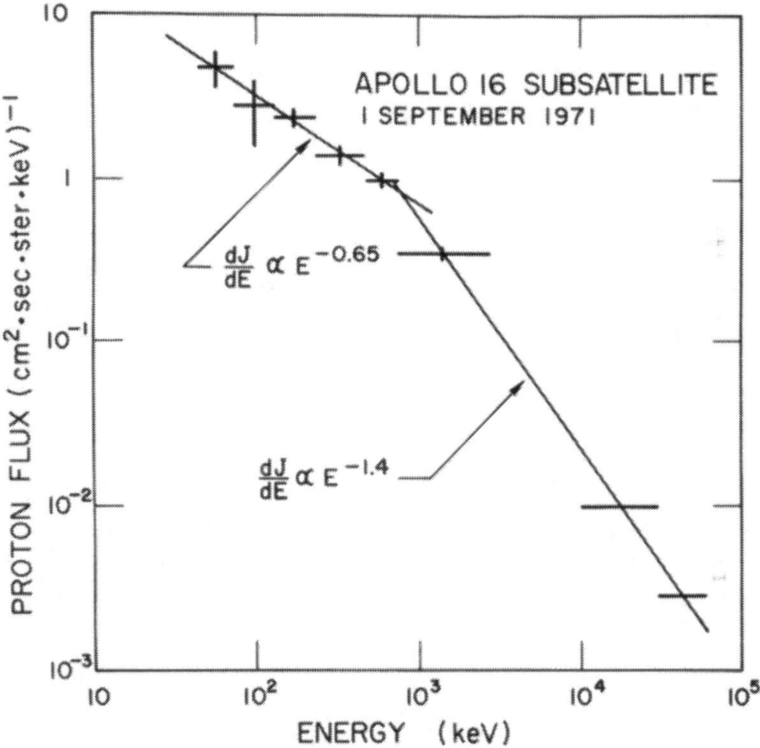

Fig. 2. The proton energy spectrum for the 1 September 1971 flare event, plotted from the maximum flux in each energy interval.

the Sun with $k=k_0 r$ and with energy dependence $U \propto T^{-\mu}$ (Fisk and Axford, 1968). Ng and Gleeson (1971) find that for times, t, such that

$$\eta k_0 t \gg 2(rr_0)^{1/2},$$

where

$\eta=[(2+V/k_0)^2+16\,V(\mu-1)/3k_0]^{1/2}$, and
r_0 = radius of injection surface,
r = radius at point of observation;

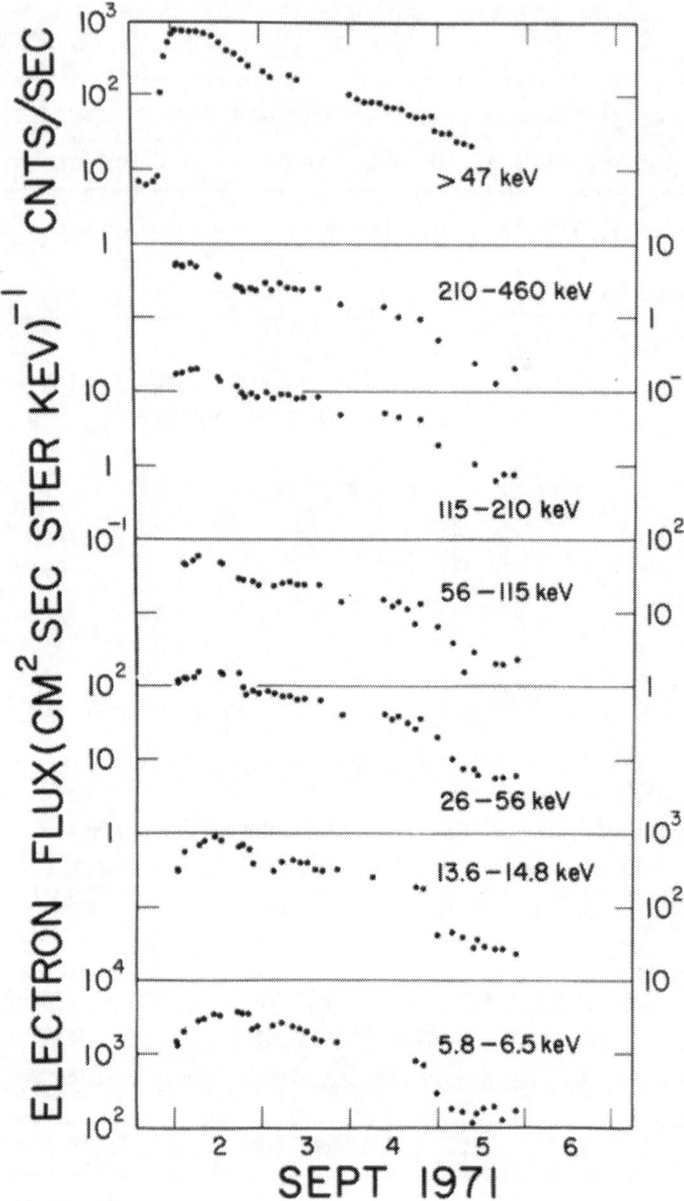

Fig. 3. The low energy electron observations for the flare event of 1 September 1971.

the peak in the particle density occurs at

$$t_{\max} = \frac{r - r_0}{(\eta + 1)\, k_0}.$$

Taking $r \approx 1$ AU and $r_0 \approx 7 \times 10^5$ km (one solar radius), we find that $\eta t_{\max} k_0 > 10^{13}$

cm $\gg 2(rr_0)^{1/2} \approx 2 \times 10^{12}$ cm for $\mu > 1$. From the time of maximum in each channel we have estimated k_0 for $V = 400$ km s^{-1}.

The diffusion coefficients range from ~ 1 to 5×10^{20} cm^2 s^{-1} at 1 AU, comparable in value to those derived earlier by Lupton and Stone (1973) and others at energies of ~ 1 to 10 MeV. There is a discernible, although weak, trend to lower values of k at lower energies. In this range of k convection and adiabatic energy loss clearly play an important role in the particle propagation. Thus, it is likely that these particles were emitted at the Sun with substantially higher energies, and then lost energy through repeated scatterings with the magnetic field in the expanding solar wind. Estimates by Palmer (1973) from Monte-Carlo simulations indicate that for these values of k typically $\gtrsim 75\%$ of the original particle energy is lost by the time of maximum intensity at 1 AU.

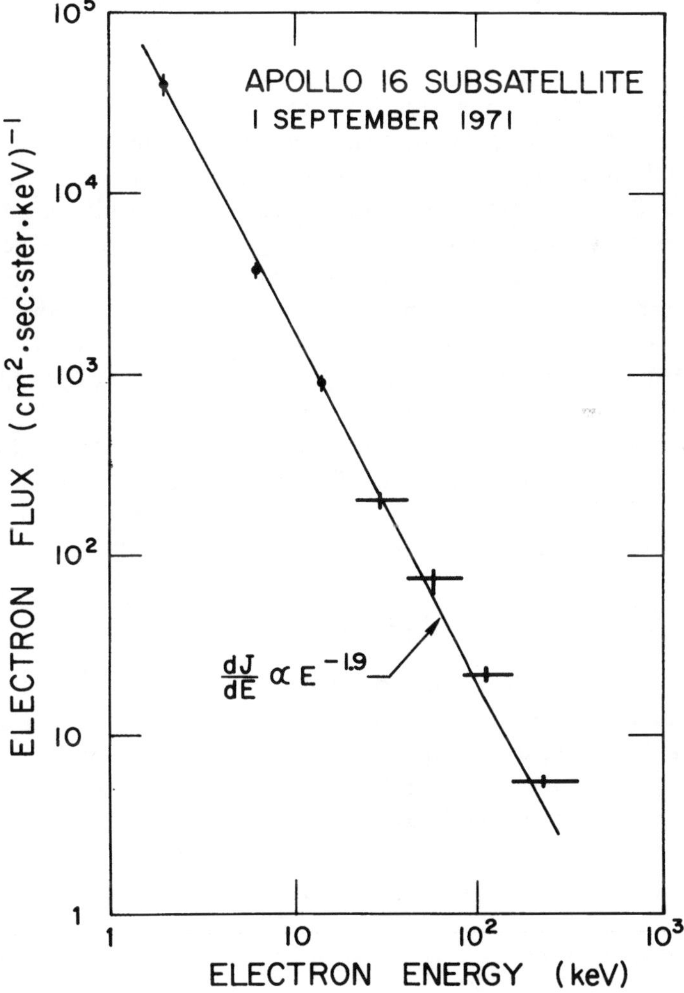

Fig. 4. The electron energy spectrum for the 1 September 1971 event.

We can obtain a lower limit to the total energy in emitted protons by assuming, unrealistically, that no energy is lost to adiabatic deceleration in the propagation of these protons from the Sun to the Earth. To estimate the number of protons, we assume the protons fill an $\sim 100°$ cone of interplanetary field lines uniformly out to ~ 1.5 AU by the time of maximum intensity at 1 AU. Then the lower limit on the total energy contained in protons above ~ 0.04 MeV is approximately

$$\mathscr{E}_{\text{protons}}(E > 0.04 \text{ MeV}) \gtrsim 5.4 \times 10^{28} \text{ erg}$$

compared with about 2×10^{28} erg above 10 MeV. The actual total energy in protons *emitted* by the flare may be an order of magnitude or more higher than these numbers since the protons most likely started with several times the energy at the Sun that they were observed to have at 1 AU, and since protons below 0.04 MeV were not included. The 1 September 1971 flare was not a large proton event at 1 AU; the August 4, 1972 event was two to three orders of magnitude more intense. We note that previous estimates of the energy release in a large solar flare only took into account protons above ~ 10 MeV (Hundhausen, 1972) and gave a figure of $\sim 2 \times 10^{31}$ erg. If a similar proton energy spectrum exists for those events as in this event, then estimates of the total energy in energetic protons must be revised upward by a minimum factor of ~ 3, to $\gtrsim 6 \times 10^{31}$ erg. If the energy losses in propagation are taken into account energetic protons may easily contain a major portion of the total flare energy.

The dE/dx energy loss in the solar atmosphere must be very small for the electrons since the spectrum extends as a smooth power law to ~ 2 keV. We obtained an upper limit to the amount of material traversed of $\sim 5\ \mu\text{g cm}^{-2}$ of ionized hydrogen (Lin, 1972). If we attribute the bending in the proton spectrum just to dE/dx energy loss, then the amount of material traversed is $\sim 20\ \mu\text{g cm}^{-2}$ of ionized hydrogen. These values are remarkably small and may indicate either that (1) the acceleration processes occur high in the solar corona or that (2) the energy loss formulae are inapplicable, as for example in a very hot plasma where the average particle energy, $\sim KT$, is on the order to tens of keV (Anderson, 1972).

4. Summary

The observations presented here show that protons of energies down to ~ 0.4 MeV at 1 AU are impulsively emitted into interplanetary medium from solar flares. The energy contained in these low energy protons is at least ~ 3 times the energy in > 10 MeV protons, and thus, these protons may constitute a major portion of the total flare energy in large proton flares.

Acknowledgements

This research was supported in part by NASA contract NAS 9-10509 and NASA grant NGL 05-003-017.

References

Anderson, K. A.: 1972, *Solar Phys.* **27**, 442.

Anderson, K. A., Chase, L. M., Lin, R. P., McCoy, J. E., and McGuire, R. E.: 1972, *J. Geophys. Res.* **77**, 4611.

Fisk, L. A. and Axford, W. I.: 1968, *J. Geophys. Res.* **73**, 4396.

Hundhausen, A. J.: 1972, in C. P. Sonett, P. J. Coleman, Jr., and J. M. Wilcox (eds.), *Solar Wind*, NASA SP-308, p. 393.

Jokipii, J. R.: 1971, *Phys. Rev. Letters* **26**, 666.

Kane, S. R.: 1974, this volume, p. 105.

Lin, R. P.: 1972, in R. Ramaty and R. G. Stone (eds.), *Proc. Symposium on High Energy Phenomena on the Sun*, Goddard Space Flight Center, Greenbelt, Maryland, Sept. 28–30.

Lin, R. P.: 1974, *Space Sci. Rev.* **16**, 189.

Lin, R. P. and Hudson, H. S.: 1971, *Solar Phys.* **17**, 412.

Lupton, J. E. and Stone, E. C.: 1973, *J. Geophys. Res.* **78**, 1007.

Murray, S. S., Stone, E. C., and Vogt, R. E.: 1971, *Phys. Rev. Letters* **26**, 663.

Ng, C. K. and Gleeson, L. J.: 1971, *Solar Phys.* **20**, 166.

Palmer, I. D.: 1973, *Solar Phys.* **30**, 235.

Parker, E. N.: 1965, *Planetary Space Sci.* **13**, 9.

Verzariu, P. and Krimigis, S. M.: 1972, *J. Geophys. Res.* **77**, 3985.

DISCUSSION

Newkirk: Could you put a number on the amount of material which has been encountered by your low energy protons?

Lin: 15 μg cm^{-2} if there is no energy loss.

Dryer: It is very interesting to see your estimate of energy for greater than 50 keV protons (for the 1972 August 4 flare) as 5×10^{32} erg. The hourly-averaged bulk velocities for Pioneer 9 show about 100 km s^{-1} for the lower energy protons. Although additional detailed study of the finer time-resolved data is forthcoming, I could hazard a guess at this time that – on the basis of estimates for flares in the past – the energy would be at least equal to the amount you presented for >50 keV protons.

Lin: The estimated energy continues to rise.

SOLAR RADIO PULSATIONS

B. L. GOTWOLS

The Johns Hopkins University, Applied Physics Laboratory, Silver Spring, Md., U.S.A.

Abstract (*Solar Phys.*). Several models for pulsating type IV radio bursts are presented based on the assumption that the pulsations are the result of fluctuations in the synchrotron emission due to small variations in the magnetic field of the source. It is shown that a source that is optically thick at low frequencies due to synchrotron self absorption exhibits pulsations that occur in two bands situated on either side of the spectral peak. The pulsations in the two bands are 180° out of phase and the band of pulsations at the higher frequencies is the more intense. In contrast, a synchrotron source that is optically thin at all frequencies and whose low frequency emission is suppressed due to the Razin effect develops only a single band of pulsations around the frequency of maximum emission. However, the flux density associated with the later model would be too small to explain the more intense pulsations that have been observed unless the source area is considerably larger than presently seems reasonable.

DISCUSSION

Dulk: I believe that you did not include the effects of betatron acceleration and deceleration on the electrons as the tube contracts and expands. Inclusion of this would greatly increase the depth of the modulations of the radiation that you calculate.

Gotwols: I agree. I considered a highly idealized homogeneous flux tube 10000 km in diameter by 100000 km in length.

Rosenberg: You used quite large gyro frequencies; would your result be much effected if you use smaller values for the magnetic field strength.

Gotwols: The calculated flux is quite sensitive to the assumed nonrelativistic gyro frequency. To some extent it is possible to compensate for a decrease in the gyro frequency by assuming a larger number of energetic electrons. I don't think the gyro frequencies I have assumed are particularly large for a spectrum with a peak in the decimetric wavelength range.

McLean: We (Sheridan and I) have recently observed a spectrum showing a very fine series of regular pulses. These show a clear sawtooth wave form with a rise time about twice the decay time for each pulse which I find difficult to provide with a standing wave oscillation as the cause of the modulation.

Gotwols: If the standing wave in the magnetic flux tube is strictly sinusoidal, then it is true that the theory presented here can not yield an asymmetrical temporal profile at a single frequency.

Gordon Newkirk, Jr. (ed.), Coronal Disturbances, 471.

CORONAL MAGNETIC FIELDS AND ENERGETIC PARTICLES

GORDON NEWKIRK, JR.

High Altitude Observatory, National Center for Atmospheric Research, Boulder, Colo., U.S.A.*

Abstract.** Magnetic fields in coronal and interplanetary space play a critical role in the propagation and storage of energetic particles from the flare site to 1 AU. The current investigation is designed to determine if the detailed configuration of the coronal magnetic fields below 2.5 $R_\odot$ has a measurable influence on the escape of particles to 1 AU, can account for the observed broad longitudinal distribution of proton flares, and can provide a realistic site for storage of energetic particles.

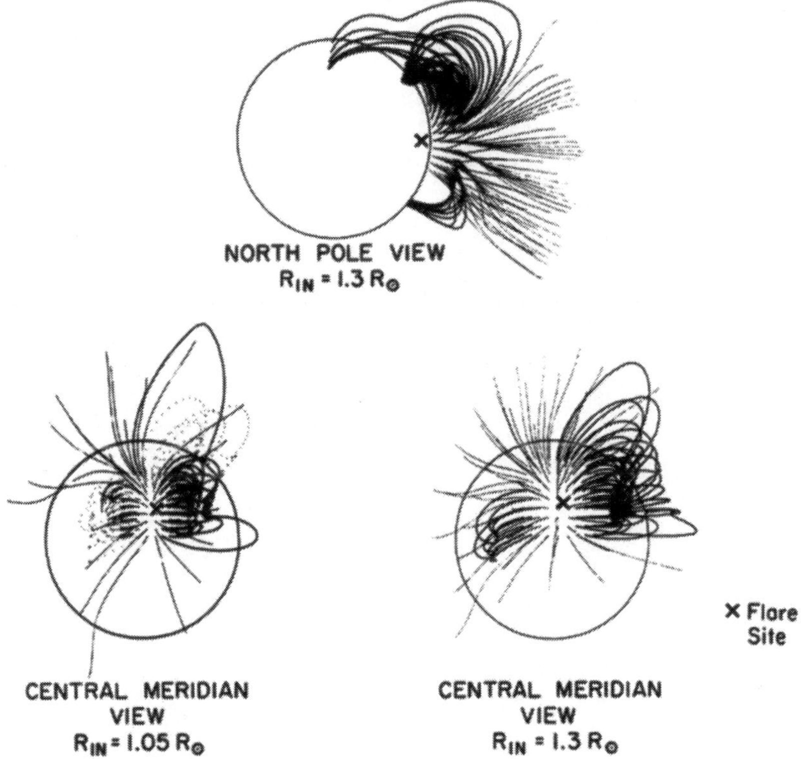

Fig. 1. A typical example of the trajectories of particles originating at a 36° × 36° injection surface centered on a flare (×) for two different values of R_{in}. Mirroring particle trajectories appear as heavy solid lines; escaping trajectories, as light solid lines; and impacting trajectories, as dotted lines. The particular event displayed is the long duration proton flare of 20 April 1971.

* The National Center for Atmospheric Research is sponsored by the National Science Foundation.

** In R. Ramaty and R. G. Stone (eds.), *Proc. of Symposium on High Energy Phenomena on the Sun.*

Gordon Newkirk, Jr. (ed.), Coronal Disturbances, 473–476.

Particles are followed in the ambient magnetic field from their injection in the mirroring condition at a surface above a flare until they either:

(1) impact on the photosphere;

(2) escape into interplanetary space;

(3) mirror at some distant location in the corona.

We use the guiding center approximation but neglect drifts, pitch angle diffusion, and scattering from magnetic inhomogeneities. Energy loss by Coulomb collisions is included for a model corona. The ambient magnetic fields are calculated using a potential model incorporating Mt. Wilson photospheric magnetic data. A typical example of such field line calculations for an injection surface placed at 1.3 $R_{\odot}$ covering a sector $36° \times 36°$ centered on the flare appears in Figure 1.

The calculations are applied to 46 flares of reported importance $>2+$ which occurred in the period 1959–1971 at times when good magnetic data were available. To minimize the influence of any systematic effect of the spiral interplanetary field, attention was concentrated on the 27 western hemisphere flares. These included:

(1) 14 flares without proton events;

(2) 6 flares with long duration (lifetimes in excess of 3 days at 100 MeV) or with recurrent proton events;

(3) 7 flares with impulsive proton events.

A statistical examination of these calculations (Table I) shows that many particles

TABLE I

Probability of injected particles having a particular fate
(36° Injection sector at 1.3 $R_{\odot}$)

(1)	(2) Escape	(3) Impact	(4) Mirror with $R_{max} < 2\ R_{\odot}$	(5) Mirror with $R_{max} > 2\ R_{\odot}$	(6) Remain at Sun
All flares $>2+$ (46)	0.28 *531*	0.002 *3*	0.68 *1287*	0.04 *79*	0.72 *1369*
Western hemisphere flares $>2+$, non-proton (14)	0.16 *222*	0.001 *1*	0.80 *1127*	0.04 *50*	0.84 *1178*
Western hemisphere flares $>2+$, long duration proton (6)	0.24 *143*	0.003 *2*	0.74 *441*	0.02 *14*	0.76 *457*
Western hemisphere flares $>2+$, impulsive proton (7)	0.39 *276*	0.004 *3*	0.58 *405*	0.02 *16*	0.61 *424*

Note: The division between impulsive and long duration proton flares is taken as a reported lifetime of 3 days for 100 MeV protons. Number of flares appear in parenthesis while number of field lines appears in italics.

($\sim$60%) find themselves initially mirroring in closed magnetic arches, that a smaller fraction (16–40%) escape immediately, and that a still smaller fraction (1–15%) impact onto the photosphere. However, only a small number ($\sim$4%) of the injected particles would find themselves mirroring on field lines which extended above 2.0 $R_\odot$. Differences in the ambient magnetic field configurations between proton and non-proton flares are also examined with the conclusion that proton flares have significantly more open field lines emerging from the injection surface than do flares of the

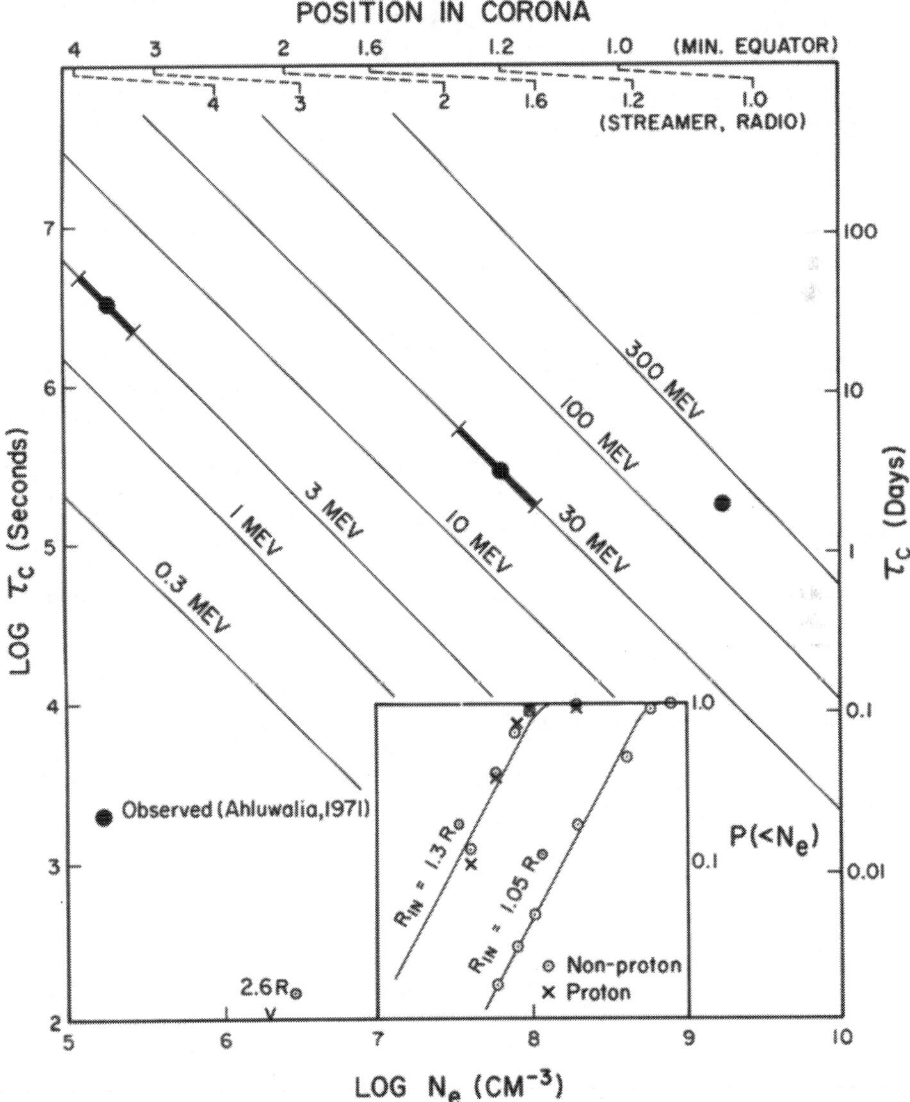

Fig. 2. The lifetimes of protons of various energies moving in a completely ionized corona of density N_e together with the equivalent position in the corona for each density. The inset shows the probability that temporarily trapped particles will encounter an average density less than N_e.

same importance which do not produce protons at 1 AU. We conclude that the ambient field configuration has a detectable influence on whether or not protons escape from a given flare.

Comparing the calculation of the average spread in longitude of field lines opened by the solar wind at 2.5 $R_{\odot}$ with observations of proton and electrons at 1 AU, we conclude that the present model cannot account for the observed longitude distribution of energetic particles.

The mean electron density encountered by particles mirroring in the magnetic field permits us to determine the probability that particles injected from the flare regions will encounter a mean electron density smaller than any given value. Mean electron densities smaller than $\sim 3 \times 10^7$ cm^{-3} occur for less than 10% of the mirroring particles (Figure 2). This implies that storage of 10 MeV particles for longer than 1 day and for 3 MeV particles for longer than a few hours is extremely unlikely. *Thus, particle storage in high magnetic arches cannot account for the long duration (1–50 days) of particles with energies* $\gtrsim$ *10 MeV at 1 AU.* Such particles must be generated by a continuously-operating acceleration mechanism.

DISCUSSION

Duncan: Particles injected at a position where the field strength is B cannot mirror until they encounter a field strength equal to or greater than B. Hence, in a symmetrical loop most of them will plunge back into the chromosphere. If only a negligible number impact, it implies that the magnetic field strength in the vicinity of the flare is less than that on other parts of the chromosphere. Is this what you are inferring?

Newkirk: No. This calculation assumes that particles are injected in the mirroring condition, and thus those particles which would strike the chromosphere immediately below the injection surface are neglected. The fraction of isotropically injected particles which would impact would be 5 to 10 times higher than I have indicated.

Wentzel: This situation inevitably leads to velocity anisotropics; these cause various plasma instabilities; the resulting waves scatter the particles and greatly increase the loss rate. Depending on particle densities, time scales of minutes are typical.

Newkirk: I agree. These calculations represent an *upper* limit to the time particles can be stored.

Smerd: The problem may be related to that of meter-wave type I storms for similar periods. If the type I-radiating electrons are accelerated intermittently throughout the storms to keV in a Fermi-like manner by MHD waves in a closed magnetic loop, the corresponding protons would reach energies of MeV. These protons could account for the observed sub-relativistic proton storms when they diffuse out of the closed-field region.

COHERENT EFFECT IN THE PREFERENTIAL ACCELERATION OF RELATIVISTIC SOLAR HEAVY COSMIC RAY NUCLEI

SHAHINAZ M. A. YOUSEF

Astronomy Dept., Cairo University, Cairo, Egypt

Abstract (*Solar Phys.*). The observed enrichment of heavy nuclei in the composition of relativistic cosmic rays including solar, may be explained on the basis of a coherent accelerating mechanism. This mechanism requires the existence of high intensity relativistic electron beams or preferably blobs (with number density $\geqslant 10^{11}$ electrons cm^{-3}) in the accelerating solar regions. The radius of each blob should be less than its plasma wavelength, i.e. of the order of cm.

If such electron blobs do exist, then each would be partially self focused by its magnetic attraction force due to the pinch effect which counteract $(V/c)^2$ of the coulomb repulsive force of the electrons. As these blobs propagate through the solar atmosphere they will electrostatically pick up a suitable number of positive ions, strip them of their electrons and capture them in their deep potential wells. This seeding of these electron blobs by positive ions would render them stable at least over short periods.

These captured positive ions would have no choice but to travel with the blobs with their relativistic velocities. Assuming the energy of each electron in the blob to be 1 MeV, then the corresponding energy of a captured proton would be 1836 MeV and that of say an iron nucleus would be 102.8 GeV.

These positive ions are actually accelerated by the internal coherent electric field of the blob's electrons. What ever happens to each blob after the acceleration of the positive ions should not in principle affect these ions.

Since the capture of positive ions is electrostatic in nature, the blobs preferentially collects high Z nuclei.

Briefly these electron blobs can serve as vehicles for the preferential acceleration of heavy cosmic ray nuclei.

The principle of this mechanism has recently been already applied in a modified way in the electron ring accelerators and its possibility, for the acceleration of cosmic rays has been briefly pointed out in 1956 by Veksler and by Budker, independently.

DISCUSSION

Meyer: The relativistic electron cloud needs only relatively few ions for charge compensation. Now usually in the solar plasma there are many ions available, which is different from the situation in the plasma accelerators. One wonders whether under such circumstances the ions really have to travel along with the relativistic electron bunch. I could imagine that the ions would basically just stay where they are and are only effected by the passing electrons through a kind of polarization field, so that no ions would travel along but only a kind of polarization wave pattern. Did you investigate this situation?

Yousef: I should expect that there must be a threshold or some critical condition that allows the electron cloud to suck only a smaller number of the surrounding ions, preferably the heavy ions, strip them

Gordon Newkirk, Jr. (ed.), Coronal Disturbances, 477–478. All Rights Reserved.

of their electrons and trap them inside the cloud or drag some ions that settle near its tail. The very short time of acceleration (which I guess is less than 10^{-5} s limits the number of trapped ions to a small percentage of the total electron number forming the cloud. I would like to stress here that the electron cloud does not have to necessarily be stable. However, I did not investigate the polarization field you mentioned.

Smith: There are two problems which I can see with your model. (1) How do you generate beams with densities greater than 10^{11} cm^{-3} in a corona where acceleration occurs in regions where the density is less than 10^{10} cm^{-3}? Your beams, especially the ones with velocities of $0.5c$ should be very unstable. Do you have any idea how to stabilize them.

Yousef: In pinching, the plasma attains higher densities.

Smith: I am familiar with the work on relativistic electron beams in the laboratory. What I am really worried about is the generation of such beams under solar conditions. All laboratory pinches with which I am familiar which lead to particle acceleration inevitably involve a current sheet and these don't give very coherent dense pulses of particles.

Krall (to Smith): There is some precedent in laboratory plasmas for propagating electron beams a distance of very many instability growth lengths. One mechanism would be to convert a small fraction of the beam energy into temperatures perpendicular to the beam. This increases the Debye length and can lead to finite size effects which stabilize the non-neutralized beam.

Laboratory experiments have shown some preferential acceleration of a small number of ions in pinch experiments. I agree that the conditions would have to be pretty special to produce this.

Another coherent acceleration mechanism is inverse Cerenkov radiation. In the rest frame of the electron beam the ions move super-fast, emit plasma waves and slow down by recoil. In the lab frame this corresponds to ion acceleration.

A CONDUCTIVE COOLING MODEL FOR CONFINED SOLAR FLARE PLASMA*

L. W. ACTON and W. T. ZAUMEN
Lockheed Palo Alto Research Laboratory, Palo Alto, Calif., U.S.A.

Abstract (*Solar Phys.*). The evolution of the X-ray emission from hot solar plasmas for which radiative cooling is insignificant is computed on the basis of a simple conduction model. This model considers a hot and a cool region separated by a sharp boundary. The plasma is assumed to be confined within a vertical column of constant cross-sectional area with the hot X-ray emitting region lying above the cool region, and both being in hydrostatic equilibrium. Because the electron temperature is high and thermal conductivity varies at $T^{5/2}$ the hot region remains approximately isothermal during the cooling process. Cooling then proceeds by raising the temperature of the cool material at the interface thereby moving the boundary into the cool region. The emission measure, $\text{E.M.} \equiv \int N_e^2 \, dV$, of the hot volume therefore increases as the cooling progresses. For this model, the emission measure may be expressed as a function of temperature, T, by $\{\text{E.M.}\} = \{\text{E.M.}_0\} \, (T_0/T) \, [(T_0 - T_f)/(T - T_f)]^2$ where E.M._0 is the emission measure when the hot region is at temperature T_0, and T_f is the temperature of the cool region. This expression is shown to satisfactorily describe observations of small fast-decaying flares. Larger, slower events tend to show a decreasing emission measure during late decay phases and cannot be fitted with this model.

* This research has been carried out under support of NASA Contract NASw-2316 and by the Lockheed Independent Research Program.

Gordon Newkirk, Jr. (ed.), Coronal Disturbances, 479.

THE GYRO-SYNCHROTRON RADIATION FROM MOVING TYPE IV SOURCES IN THE SOLAR CORONA

G. A. DULK
Division of Radiophysics, CSIRO, Sydney, Australia

Abstract (*Solar Phys.*). Calculations of the gyro-synchrotron emission are made for conditions which might be expected in moving type IV sources in the solar corona. Two simple models for an evolving source are treated: a uniform cube and an inhomogeneous sphere. The results suggest that most moving sources have the following features: (1) A rather strong magnetic field, ≈ 10 G, is carried out within the source. This is required to achieve the high degree of circular polarization often observed. (2) Synchrotron self-absorption causes the source to be optically thick at frequencies less than about 100 MHz, thus restricting the bandwidth of the radiation. The self-absorption decreases as the source moves outward and expands. The turnover frequency, which separates the optically thick and thin spectral regimes, moves rapidly to lower frequencies, accompanied by a change from low to high circular polarization. In the case of an inhomogeneous source, the source appears to be larger at the lower frequencies. (3) Razin-Tsytovich suppression cannot be an important factor in determining the characteristics of most sources.

DISCUSSION

Rosenberg: A plasma blob with such a strong magnetic field would have to be self-contained by the field; hence, the field should be quite tangled up. How would that effect the radiation and in particular the polarization?

Dulk: The degree of polarization would go down drastically. Thus, we require a well-ordered nearly homogeneous field with few meanderings or inclusions with the opposite polarity.

Smith: In the 'westward ho' event we have evidence of a very coherent field since both polarizations are seen. Have you applied your calculations to each source in this event?

Dulk: If we take a model of a toroidal magnetic field of about 10 G, the double source with opposite polarizations could be explained. By inverting a toroid, i.e. introducing a ring current with its poloidal field, we could end up with a single source in the center of the ring where the field is high and the particles radiate efficiently.

Kane: What are the ion and electron densities inside the blob?

Dulk: In the two models I have considered we have

	Model 1	Model 2	
B	3.4	13.6	G
N_T	10^{35}	10^{33}	electrons with energy > 10 keV
n	10^5	10^3	cm^{-3}
m_c	$\lesssim 0.5$	$\lesssim 0.8$	degree of circular polarization

* Exchange visitor from the Dept. of Astro-Geophysics, University of Colorado, Boulder, Colo., U.S.A.

Gordon Newkirk, Jr. (ed.), Coronal Disturbances, 481–482. *All Rights Reserved.*

Mangeney: If you consider the case of a ring current, you should also have two sources; since particles will remain in the external regions where the magnetic field is lower/longer than inside.

Dulk: It depends on the distribution of the particles. If most of them have pitch angles to allow them to thread through the center, they will radiate strongly only in the strong field near the center. On the other hand, if you put all the particles in the outer regions, with small pitch angles such as are in the Earth's radiation belts, the source will be a ring with radiation of the opposite polarization. In between these cases you might have a central source of one polarization surrounded by a ring of opposite polarization. The latter kind has not been observed.

Schmidt: I take it that you are well aware about the problem caused by a field strength of 13 G in this environment. I think the timescales are too long to balance the surplus pressure by inertia. Would you like to comment on this problem?

Dulk: It is definitely a problem but the large field seems to be required to give the large degree of polarization observed. The configuration is not stable but relaxes with a typical time of a few minutes to a few tens of minutes. Even if there were no plasma inside the source, the field could not be contained by the surrounding corona. So we must assume that forces internal to the plasmoid, such as magnetic tension in field loops, will prevent it from deteriorating too rapidly. I refer to the paper by Altschuler, Lilliequist, and Nakagawa, who investigated the stability of a toroid under photospheric conditions. It is an open question what happens to such a source in the corona.

Brown: I would just like to add that if you were to associate these sources with Frost's second phase hard X-rays, the 10^{27} erg you require would have to be increased by several orders (in the 10–100 keV particle range). This would make the energy and pressure problems much worse.

Dulk: I wouldn't suggest such as association.

PART V

REPORTS ON SPECIAL OBSERVATIONS

TEMPORAL OBSERVATIONS OF THE λ5303 EMISSION LINE PROFILE DURING THE 74 MINUTE TOTALITY FROM THE CONCORDE SST AT THE 30 JUNE 1973 TOTAL SOLAR ECLIPSE: PRELIMINARY INTENSITY VARIATIONS ABOVE AN ACTIVE REGION*

D. H. LIEBENBERG and M. M. HOFFMAN

University of California, Los Alamos Scientific Laboratory, Los Alamos, N.M., U.S.A.

Abstract (*Nature*). Apparatus was flown aboard the French-British Concorde SST to obtain high resolution emission line intensity profiles during the 30 June 1973 total solar eclipse. A prime objective was to obtain profiles that could be used to determine the presence in the Fe XIV coronal emission line, 530.3 nm, of the 300 s periodicity such as is observed in the photosphere and chromosphere. The long totality duration of 74 min on the French Concorde 001 provided a unique opportunity for this study. At the 55000 ft cruise altitude weather influences are negligible, scattered light background reduced and seeing conditions excellent. The instrumentation included a high resolution pressure scanned Fabry-Perot interferometer and high resolution recordind. This instrumentation was perfected in earlier Los Alamos Scientific Laboratory solar eclipse flights since 1965 (Liebenberg, 1965, 1967; Hoffman *et al.*, 1970). A 12 cm aperture window and 7.7 cm aperture $f/13$ telescope were designed to optimize the use of the available space as shown in Figure 1.

Results were obtained in a preliminary analysis by determining the emission line peak intensity for a single location in the solar corona near the base of a helmet structure above the NW limb. Averages over one minute intervals are shown in Figure 2 for a 15 min section of the recording. Background intensity of 0.05 on the arbitrary intensity scale has been subtracted from each average. A shift in time base by t_0+30 s was performed to determine whether the large statistical noise might cause the average over a one minute interval to change. As is seen the 6 ± 1 min periodicity of the 530.3 nm peak intensity is plainly observed at a level of 2 standard deviations. Further analysis should determine whether this is a progressive wave motion and the extent to which the line profile itself is modified in correlation with the intensity fluctuations.

Earlier observations (Noxon, 1966; Morel, 1972) have been inconclusive in showing an intensity fluctuation in Fe XIV line emission. The present data are completely free of the influence of the Fraunhofer absorption line introduced in scattered light to coronagraphic observations. Measurements in X-ray lines have shown a similar intensity periodicity with which the present preliminary results in Fe XIV line emission agree (Chapman *et al.*, 1972).

* Work performed under the auspices of the U.S. Atomic Energy Commission.

Gordon Newkirk, Jr. (ed.), Coronal Disturbances, 485–487. *All Rights Reserved.*

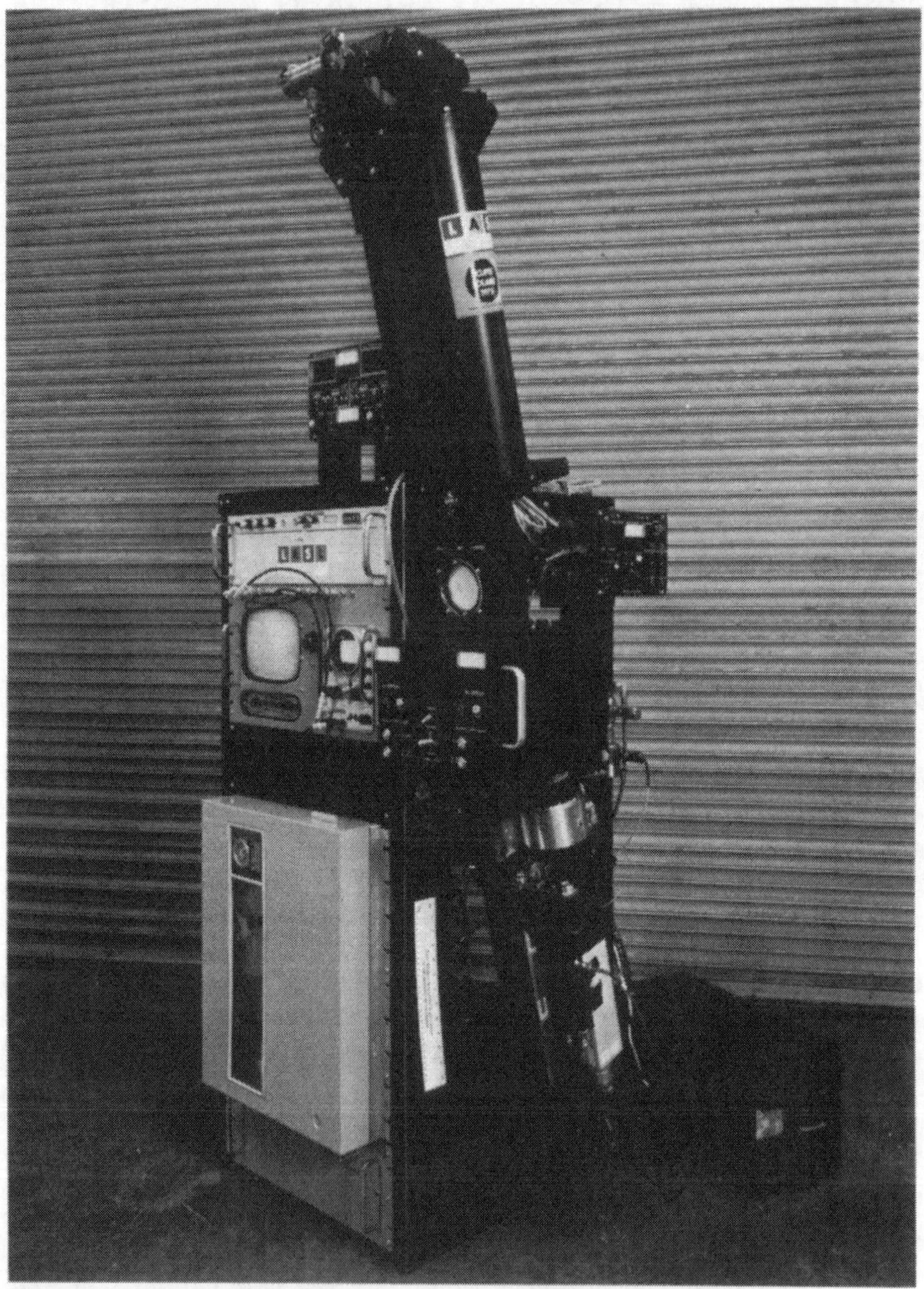

Fig. 1. The Los Alamos apparatus as installed on the Concorde SST. The optical train cover is removed to show the Fabry-Perot interferometer cell and the image intensified vidicon camera. Video data monitoring and recording equipment and image guidance controls are arranged so that one person can operate the equipment.

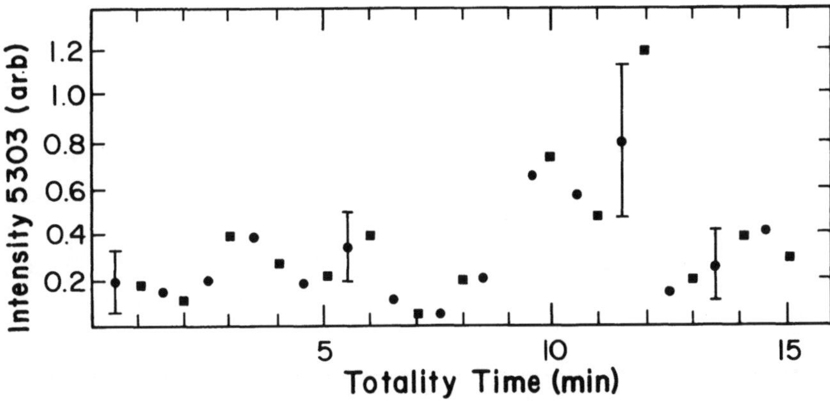

Fig. 2. Peak intensity of λ5303 in arbitrary units vs time during totality. The points ● are averages over one minute intervals and ■ are the same with a shift in time base t_0 to $t_0 + 30$ s as discussed in the text. Standard deviations are shown for a few points and are typical for all points.

Acknowledgements

We thank Mr R. Lang, Mr Ed Brown, M. Sanders, J. Calligan and Dr A. Cox of LASL for help with equipment and flight plans required for these observations. To the chief test pilot M. André Turcat, Aerospatiale, and scientific coordinator Dr P. Charvin, Institute de Astronomie et du Géophysique, we are indebted for a successful Concorde flight. Financial support was provided by U.S. Atomic Energy Commission through LASL, U.S. National Science Foundation Grant AG-433 and the Delegation General for Scientific and Technical Research of France.

References

Chapman, R. D., Jordan, S. D., Neupert, W. M., and Thomas, R. J.: 1972, *Astrophys. J.* **174**, L97.
Hoffman, M. M., Brown, E., and Liebenberg, D. H.: 1970, *Am. Astron. Soc.*, June 1970.
Liebenberg, D. H.: 1965, *Proc. NASA Solar Eclipse Symposium*, Dec. 1965.
Liebenberg, D. H.: 1967, *Astron. J.* **72**, 307.
Morel, C.: 1972, 'Detection of Waves in the Solar Corona', M. S. Thesis, Univ. Colorado.
Noxon, J. F.: 1966, *Astrophys. J.* **145**, 400.

DISCUSSION

Schmidt: Do you plan to evaluate your results to find propagation effects from time-distance correlations?

Liebenberg: We have reasonably complete data around limb and extending to nearly 2 $R_\odot$. In addition to intensity measurements, full line profiles were obtained and will be used in the more complete analysis.

Brandt: H. R. Chapman, S. Jordan, W. Neupert, and R. Thomas have observed line intensity variations with the OSO-7 spectrometer experiment. The EUV lines studied refer to a variety of excitation conditions which cover the chromosphere, transition zone, and the lower corona. In all cases, a power spectrum analysis showed intensity oscillations with periods of approximately 300 s. Thus, these oscillations extend at least to the altitude of the lower corona.

Athay: Did the window temperature cause any problems?

Liebenberg: The skin temperature was 128 °C (hence no consideration problems). Our experiment could not be affected.

SKYLAB: A PROGRESS REPORT

R. M. MacQUEEN

High Altitude Observatory, National Center for Atmospheric Research, Boulder, Colo., U.S.A.*

(Presented by G. Newkirk)

Abstract. The Skylab Apollo Telescope Mount contains six principal instruments spanning the X-ray to visible wavelength range. These experiments include an externally occulted white light coronagraph from the High Altitude Observatory which observes the outer solar corona from 1.5 to 6.0 radii from Sun center, in broadband (3500–7000 Å) white light with approximately 10″ spatial resolution. An X-ray spectrographic telescope of the American Science and Engineering, Inc. employs six filters and an objective grating to observe a 48′ field of view in the wavelength range 3.5–6.0 Å, with approximately 3″ spatial resolution. A scanning ultraviolet polychromator, spectroheliometer of the Harvard College Observatory is capable of observing 1.2 Å spectral resolution from 300–1350 Å with a 7 detector array. The field of view of the instrument is determined by its operational mode and ranges from $5' \times 5'$ to $5'' \times 5''$. An X-ray telescope from the Marshall Space Flight Center employs five metal filters to observe the 5–33 Å spectral region with 2.5″ spatial resolution over the 40′ field of view. The Naval Research Laboratory has supplied two instruments: the first, an XUV spectroheliograph, covers wavelength regions 150–335 Å and 321–625 Å, with somewhat better than 5″ spatial resolution and a spectral resolution of 0.13 Å (for a 10″ feature). The second instrument, a slit spectrograph, covers the spectral range 970–3940 Å in two bands, 970–1970 and 1940–3940, and has 0.5 and 0.1 Å spectral resolution respectively, with a $2'' \times 60''$ slit. Additionally, these principal experiments are supported by two Hα telescopes and a broadband (150–600 Å) ultraviolet monitor for astronaut use. All experiments except that of the Harvard College Observatory (which utilizes photoelectric detectors) employ film to be recovered and installed during astronaut extravehicular activity. Operating in concert in a joint observing program designed to obtain observations of certain solar phenomena, the experiments have now completed more than two months of manned operation and, in the case of the AS&E, HCO and HAO instruments, approximately two additional months of unmanned operations. Representative preliminary results are outlined from several of the experiments below.

DISCUSSION

Pneuman: One thing that perplexes me about these apparent changes in the white light corona is that, if we observe the corona in the green line, one has to look for a long time to see any changes at all. Of course, that bubble you showed was a real transient, but couldn't some of those other rather subtle changes be also due to perspective effects?

* The National Center for Atmospheric Research is sponsored by the National Science Foundation.

Gordon Newkirk, Jr. (ed.), Coronal Disturbances, 489–490.

Newkirk: The hour-by-hour changes in the fine scale ray structure appear to be quite common. At present we do not know what fraction of these changes are produced by the turning on and off of fine jets of coronal plasma, by the influence of waves, or by perspective changes.

Schmidt: In Asilomar you put forward the conjecture that one might see the current sheets in the streamers only for a short time, from the right viewing angle. In one of the early frames of the movie there was a thin radial feature. Could this be such a feature?

Newkirk: I think that feature was an artifact but other features whose appearance and disappearance are due to perspective changes have been seen.

Schmidt: Which magnetograms were used to identify the X-ray bright points with bipolar magnetic features?

Newkirk: I don't know.

Kiepenheuer: Are the knots moving with the same velocity as the expanding blob?

Newkirk: The knots are probably Hα emitting blobs and appear to be moving slower.

Vrabec: Would the bright points in the low corona correspond to the emerging flux regions described by Martin.

Martin: The lifetime, size, distribution on the disk, and approximate number (50–100 day^{-1}) of X-ray bright points agrees very well with our observations of ephemeral regions. I think they are probably the same solar features.

Smith: With the wide bandwidth of the coronagraph, will it be possible to determine the densities in transients?

Newkirk: We cannot directly distinguish Thompson scattering and line emission in say Hα but concomitant Skylab and ground observations and the observed polarization should allow these components to be separated to determine densities, masses, etc.

PRELIMINARY RESULTS FROM THE NRL/ATM INSTRUMENTS FROM SKYLAB SL/2

R. TOUSEY, J.-D. F. BARTOE, J. D. BOHLIN, G. E. BRUECKNER, J. D. PURCELL, V. E. SCHERRER, R. J. SCHUMACHER, N. R. SHEELEY, and M. E. VANHOOSIER
U.S. Naval Research Laboratory, Washington, D.C. 20375, U.S.A.

Abstract. The Naval Research Laboratory prepared three instruments for ATM and also a series of rocket payloads for purposes of calibration, under the overall direction of R. Tousey, the Principal Investigator. In this preliminary report a summary of results from operations during the first mission (SL/2) and sample results are presented.

The extreme ultraviolet (XUV) spectroheliograph, S082A, performed well, obtaining 219 exposures over the two ranges, 150–350 Å and 300–630 Å with spatial resolution reaching at least 2″ near the grating normal. Figure 1, left to right upper, is a section from 200–304 Å; below, 304–400 Å is covered. The exposures were made during an M-2 flare on June 15, 1973, which produced the conspicuous linear array

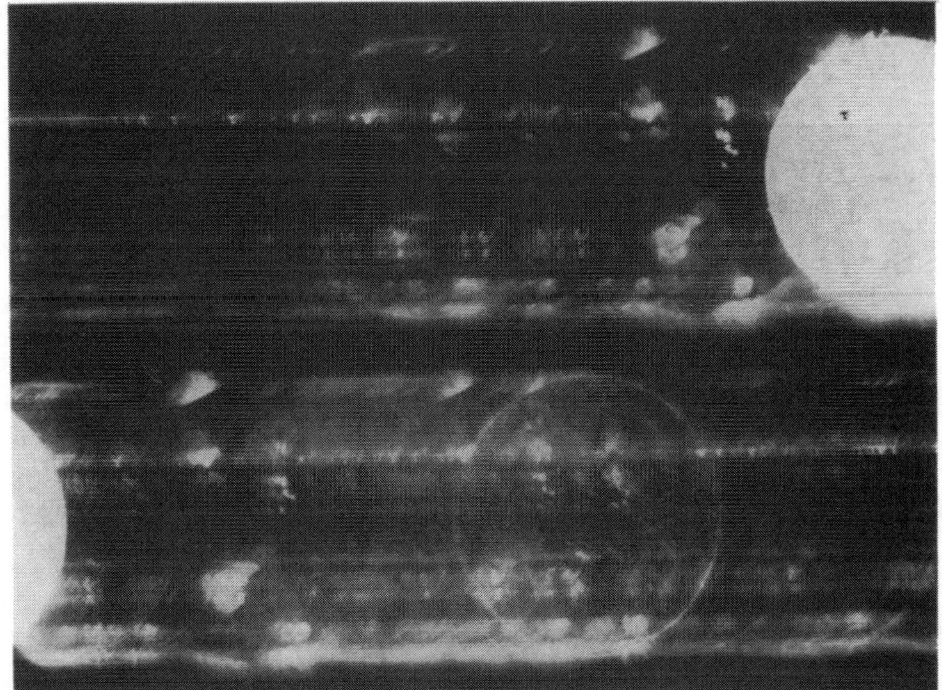

Fig. 1. Spectroheliograms of the 200–400 Å region obtained by S082A on June 15, 1973 during an M-2 flare. In the upper section the flare is seen in He II 304 Å as the solarized black feature against the disk, followed by images converging to the series limit at 228 Å and the continuum. In the lower section the flare is conspicuous in Fe XVI 335 Å and 361 Å. Solar north is left and west up. λ increases to the right.

Gordon Newkirk, Jr. (ed.), Coronal Disturbances, 491–495.

of images. In He II, the flare in 304 Å is black by solarization against the overexposed image of the disk. Thirteen series members are resolved, followed by the continuum beyond 228 Å. Present are flare images in many highly ionized species such as Fe XIV, XV, and XVI. In all these images the flare is complex in form, containing ribbons and threads. A few extremely high ionization lines were recorded, and in these the form was a nucleus without ribbons. In Figure 1 Fe XXIV 255 Å is present as the nearly circular feature about 30″ in diameter, $\frac{1}{3}$ to the right across the upper spectrum.

Many other striking features are present. The active regions show loops and swirls that appear to delineate magnetic field lines in the corona. Fine detail as well as diffuse clouds are present, seen both against the disc and above the limb. Fe XV 284 Å shows these forms beautifully.

The quiet disk in coronal lines is dark, but shows diffuse emission above the limb with a darker disc below where activity is present, and intense localized emission above active regions. In transition layer and low coronal lines the limb is a bright ring. Mg IX 368 Å produces the intense ring in the lower part of Figure 1. The limb is not uniform, but contains breaks, points of origin of loops, and varies in width, depending on the character of the surface nearby.

Spicules are conspicuous in He II 304 Å, especially over the polar regions where the surface is rather dark. In Figure 1 few quiescent prominences are present in He II, but most of the emission high above the limb is Si XI.

A properly exposed He II 304 Å disc image in Figure 2 is compared with an image

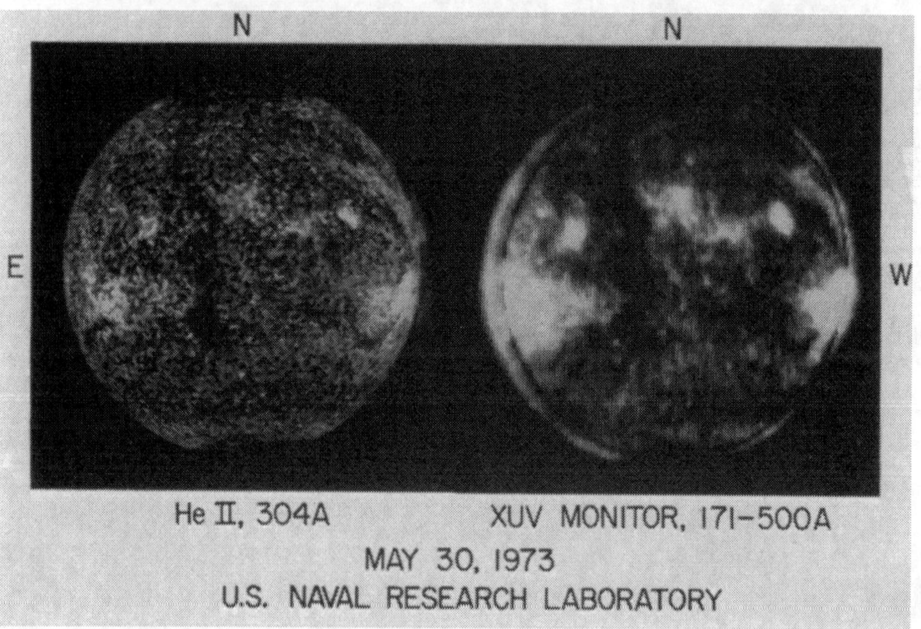

Fig. 2. An S082A spectroheliograph in He II 304 Å compared with a heliogram, (171–500 Å), produced by TV by the NRL XUV Monitor. Both images were made on May 30, 1973 and are mounted with north up and east left.

produced by the NRL 'XUV Monitor'. This projected an undispersed image, 80% coronal and 20% chromospheric, from 171–500 Å, on a phosphor conversion layer deposited on the face plate of an SEC Vidicon. This TV image, viewed by the crewman and also transmitted to ground, had about 10–15″ spatial resolution, and by means of a reticle array could be used for pointing the ATM. In He II the chromospheric network is present almost everywhere, but varies in character with boundaries crisp in some regions, and diffuse in others. The regions of diffuse network seem to correspond to regions that are mainly unipolar, such as regions on each side of a filament channel.

There are also regions where the emission is depressed. In Figure 2 there are five: one at each pole; the great rift running from the N-polar hole to 30° S; a broad, long region following latitude 45° S which is the remnant of a filament channel; and another filament channel complex in the NW. The darkest of these are now called coronal holes, but their presence in He II 304 Å, (and also in He I 584 Å) shows that they extend well down into the chromosphere.

Scattered over the He II image and seen best within the holes are many isolated bright points. These show also in undispersed XUV, and some are visible in the XUV Monitor image in spite of its low resolution. These points seem to be associated with small bipolar regions, or small unipolar regions in a region of opposite polarity.

The emission from the coronal lines shows well in the XUV Monitor, as emission high above the limb produced by intense active regions often behind the limb, and as intense emission from active regions on the disc. By contrast, in He II the emission over the limb is from prominences, plus coronal emission from the near-blend Si XI 303.32 Å. Also, the active regions are less intense in He II.

More often than expected, eruptive prominences were recorded in He II 304 Å. Figure 3 shows the largest observed on SL/2. It was followed about 8 h later by an eruption recorded by the white-light coronagraph, presumably the result of another prominence near the same location that erupted later. No corresponding emission is visible in Fe XV 284 Å, or in the Si XI corona, which is seen clearly separated from He II by its dark limb. Although a detailed comparison with Hα has not yet been made, the He II eruptive prominence seems more intense and larger than those usually seen in Hα.

The other major NRL instrument was the chromospheric UV spectrograph, S082B. The ranges covered were 970–1970 Å, and 1940–3940 Å, with spectral resolutions of 0.04 and 0.08 Å, respectively. The instrument was not stigmatic, but produced a slit spectrum from a region on the Sun 2″ × 60″. The position and angle of the slit were controlled precisely by the crewman who used the white-light S082B slit display, the Hα display, or solar coordinates. The one camera carried on SL/2 recorded nearly 1600 spectra, and the three more cameras to be flown on SL/3 and 4 are expected to bring the total close to the limit, 4800.

At this time, study of the SL/2 spectra has hardly begun. In Figure 4 short sections of two exposures are reproduced with slit tangent to the limb and 12″ inside, or 4″ above. The photospheric spectrum contains mainly Fraunhofer lines, but does show

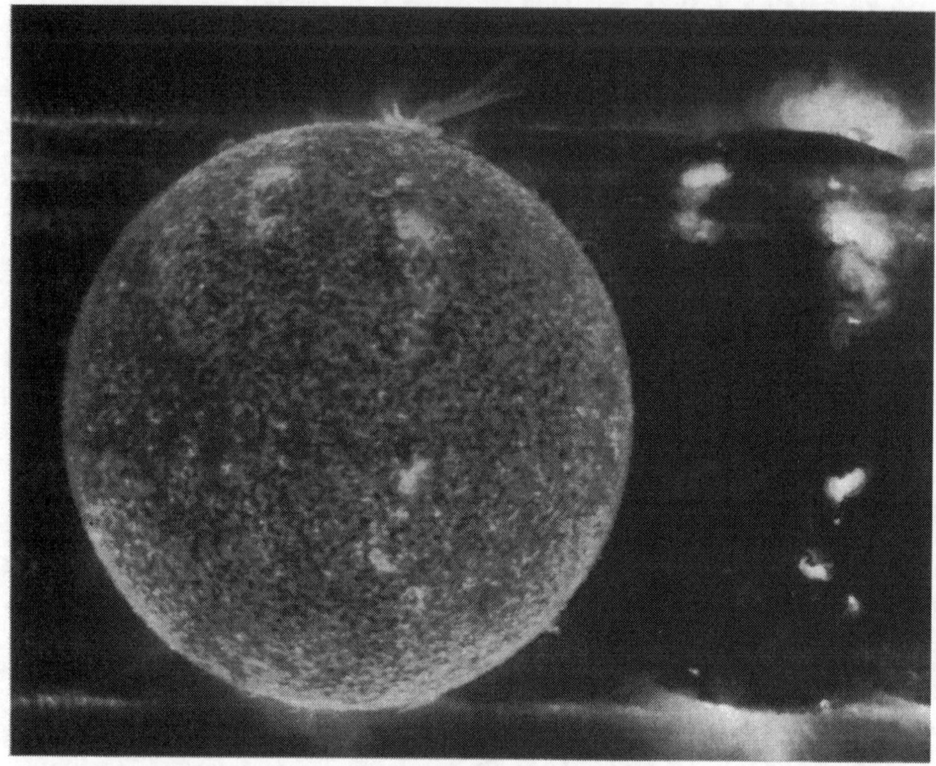

Fig. 3. A section of an S082A spectroheliogram showing He II 304 Å with an eruptive prominence at 01:46 UT on June 10, 1973. The overlapping image to the right is Fe XV 284 Å.

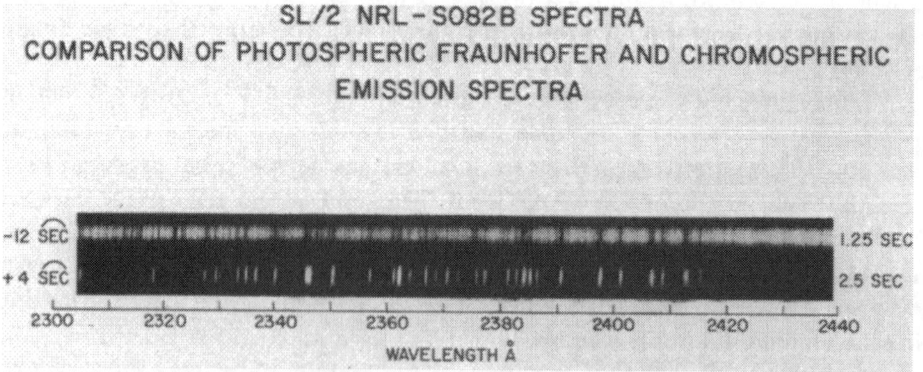

Fig. 4. A sample of spectra produced by S082B. Shown are short sections of spectra exposed with the 2″ × 60″ slit placed 12″ inside and 4″ outside the limb.

two emission lines of Si II, 2334.50, 2350.28 Å. Above the limb the photospheric continuum is absent, and the spectrum is chromospheric essentially like flash spectra recorded during a total eclipse. The instrument was used in major programs to record spectra across the limb, of prominences and filaments, active regions, the chromospheric network, and transients. Flare spectra were obtained. Another application was to photograph the setting sun to search for atmospheric contaminants and to measure the distribution of the normal atmospheric constituents to very high altitudes. The spectral resolution was sufficient to record line profiles and the results from SL/2 show a wealth of surprising shapes.

SOLAR EUV PHOTOELECTRIC OBSERVATIONS FROM SKYLAB

E. M. REEVES, P. V. FOUKAL, M. C. E. HUBER*, R. W. NOYES, E. J. SCHMAHL, J. G. TIMOTHY, J. E. VERNAZZA, and G. L. WITHBROE

Center for Astrophysics, Harvard College Observatory

and

Smithsonian Astrophysical Observatory, Cambridge, Mass., U.S.A.

Abstract. Most of the atomic species originating in the solar atmosphere between the upper chromosphere and the corona have their strong characteristic wavelengths in the extreme ultraviolet region of the spectrum. A simple normal-incidence spectrometer system with solar blind detectors such as the Harvard instrument operating between approximately 250 Å and 1350 Å is ideally suited for observing in this most interesting range of the solar atmosphere where the temperature rises outward from 10^4 to 3×10^6 K. The temperature range represented by the various atomic and ionic species in the extreme ultraviolet is associated with many types of solar structure, prominences and filaments, the supergranulation cells and network, active regions and their associated loop structures and other features. Simultaneous observations in lines of different characteristic temperatures provide a three-dimensional probe of the solar atmosphere. In the instrument, the principal polychromatic position observes the Lyman continuum, Lα, C II, C III, O IV, O VI, and Mg X with seven detectors simultaneously from the same spatial image element, 5″ in size. Approximately 60 additional polychromatic positions are used routinely to carry out specific observing programs, for example, covering several lines of a given stage of ionization, observing lines or continuum from specific species of interest such as helium in prominences, comparing combinations of lines from a given ionic species such as O V where the relative intensities give a rather direct measurement of the density at a given temperature, or measuring differing positions in the Lyman continuum providing intensity measurements which can be interpreted in terms of the departure from ionization equilibrium.

There are a diversity of experimental modes available both through remote instrument operation from the ground and through astronaut operation at the ATM control console. The modes include wavelength scans of selected solar features, and rasters covering fields of view of 5″ × 5′, 1′ × 5′, and 5′ × 5′. Time resolutions from 0.040 s to 5.5 m are provided in the range of modes.

Figure 1 shows a set of observations in the first polychromatic grating setting covering the excitation range from the Lyman series to Mg X, taken near the center of the Sun on May 29, 1973. The photographic presentations are converted to a 64-level grey scale from the numerical data. The chromospheric network can easily be seen

* Also at the Federal Institute of Technology (ETH-Z), Zürich, Switzerland.

Gordon Newkirk, Jr. (ed.), Coronal Disturbances, 497–500.

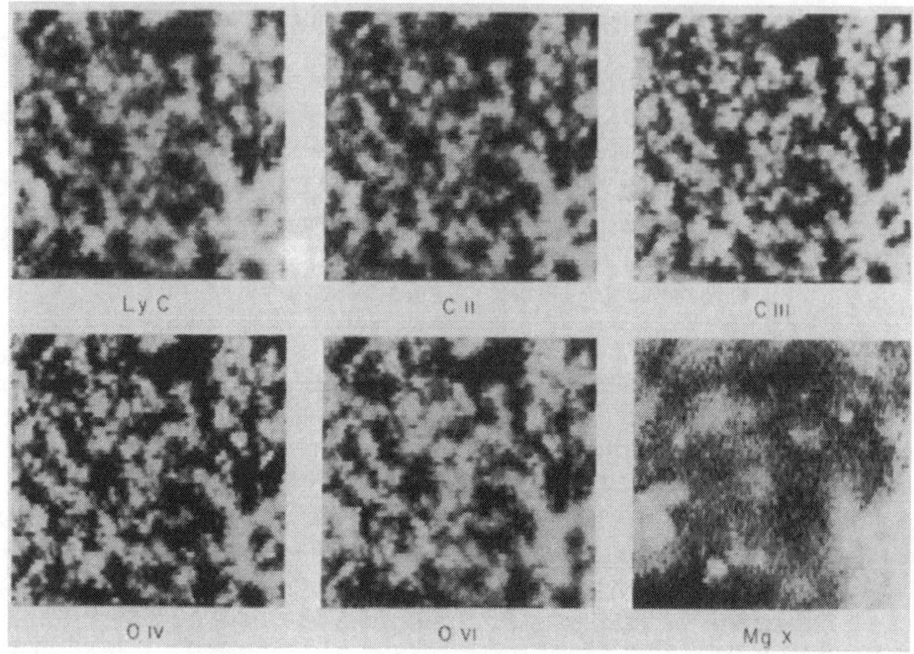

Fig. 1.

extending almost unchanged, except in contrast, from the Lyman continuum at 10^4 K through the resonance lines of C II, C III, O IV, and O VI. Between the temperature of 3×10^5 K characteristic of O VI and the 1.5×10^6 K temperature of formation of Mg X there is an abrupt change in the observed network structure with height. In Mg X the chromospheric network is almost totally unrecognizable, although in some places remnants can be seen to remain if the network is followed up the temperature sequence from below. The similarity in structure observed from the middle chromosphere through the transition zone lines suggests that the magnetic field pattern in the network must be essentially vertical through temperatures as high as 3×10^5 K. After this temperature, there is a rapid onset of some change, markedly affecting the progression of intensity with height in the solar atmosphere and smearing the structure laterally. Examination of the data shows that the network observed in the chromosphere and transition zone exhibits a strong correlation with the CaK network observed from the ground, and that the contrast between the network and the centers of the supergranulation cells increases with excitation energy, reaching a maximum in lines such as C III, O IV and O VI where the enhancement between the network and the central portion of the cells is a factor of 10 to 15.

Figure 2 shows a photographic presentation of the data obtained during a raster made at the northeast limb of the Sun on June 10, 1973, at approximately 1146 GMT in the light of Fe XV 417 Å, formed at a temperature of about 2.5×10^6 K in the solar corona. Both the solar disc and the corona some 4′ above the limb are quite low in intensity while the lower corona and the active region loop appear more intense. The active region under these loops did not appear on the limb until the following day. The loops are persistent structures, frequently interconnecting active regions or other magnetic regions of opposite polarity. While the loops can be seen most clearly on the limb from O VI to Fe XV they can also be seen on the disk in many EUV lines, perhaps the most clearly in Ne VII 465 Å, formed in the chromosphere to corona transition

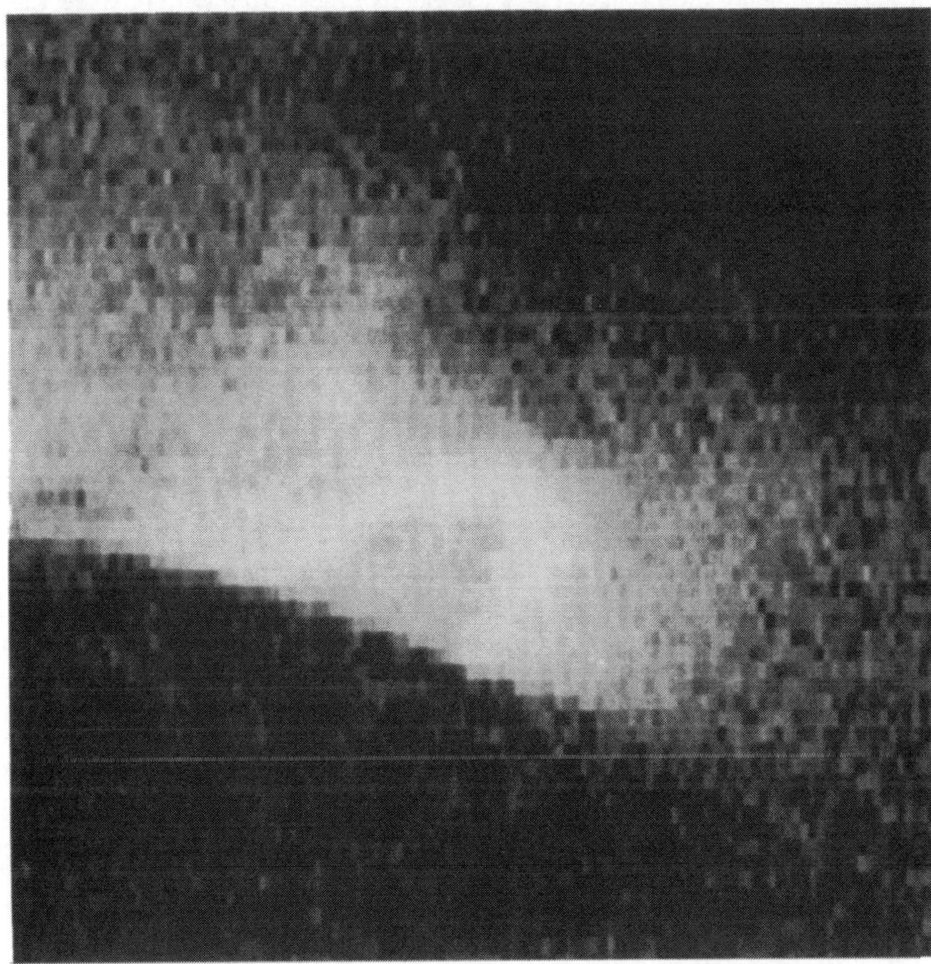

Fig. 2.

zone. In the coronal loop the enhancement in Fe XV is about 5 times the corona at the same height. Assuming that the loop has cylindrical symmetry, it is possible to imply a density increase of about an order of magnitude over the quiet corona. Even in the farthest regions of the corona at the edge of the raster, the average coronal intensity, while still a factor of 30 below the corona at the height of the loop, is at least 10 times the background scattered light and detector noise level in the instrument. The Fe XV line at 417 Å is optically thin and it can immediately be seen from the picture that the

Fe XV does not exist as a thin shell surrounding some cooler material but rather is of almost uniform density. Analyses are in progress to differentiate the density structure for different types of loops observed, determine their temperature, and observe the way in which the feet of the loops are imbedded in the active regions.

The use of photoelectric detection and the availability of accurate intensity data from a series of calibration rocket flights will enhance the capability to interpret the data in terms of physical parameters in the solar atmosphere. The resolution of the HCO instrument together with the simultaneous observations of the other ATM instruments covering a wide range of excitation energies, when coupled with the supporting ground-based observations frequently undertaken in the Coordinated Observing Program, should provide a number of new insights into some of the outstanding problems of the solar atmosphere. The Harvard instrument will also acquire data on Comet Kohoutek, absorption properties of the upper atmospheric layers of the Earth, and observations during the transit of Mercury on November 10, 1973.

DYNAMIC EVENTS IN THE X-RAY CORONA

(A Progress Report from the AS&E X-ray Telescope on Skylab)

G. S. VAIANA, A. S. KRIEGER, J. K. SILK, A. F. TIMOTHY, R. C. CHASE, J. DAVIS, M. GERASSIMENKO, L. GOLUB, S. KAHLER, and R. PETRASSO

American Science and Engineering, Cambridge, Mass., U.S.A.

Abstract. Data obtained by the AS&E X-ray Telescope Experiment during the first Skylab mission have revealed a variety of temporal changes in both the form and brightness of coronal structures. Dynamical changes have been noted in active regions, in large scale coronal structures, and in coronal bright points. The coronal activity accompanying a series of Hα flares and prominence activity between 0800 and 1600 UT on 10 June 1973 in active region 137 (NOAA) at the east limb is shown in Figure 1. It is characterized by increases in the brightness and temperature of active region loops and a dramatic change in the shape and brightness of a loop structure. Figure 2 shows the reconfiguration of an apparent polar crown filament cavity between 1923 UT on 12 June 1973 and 1537 UT on 13 June 1973. A ridge of emitting material which attains a peak brightness at least four times that of the surrounding coronal structures appears within the cavity during the course of the event. Typical X-ray photographs with filters passing relatively soft X-ray wavelengths (3–32, 44–54 Å) show 90 to 100 X-ray bright points (Vaiana *et al.*, 1973). On twelve occasions in the data from the first mission, such bright points were seen to increase in intensity by two orders of magnitude in less than 4 min. Such an event is shown in Figure 3.

Gordon Newkirk, Jr. (ed.), Coronal Disturbances, 501–504.

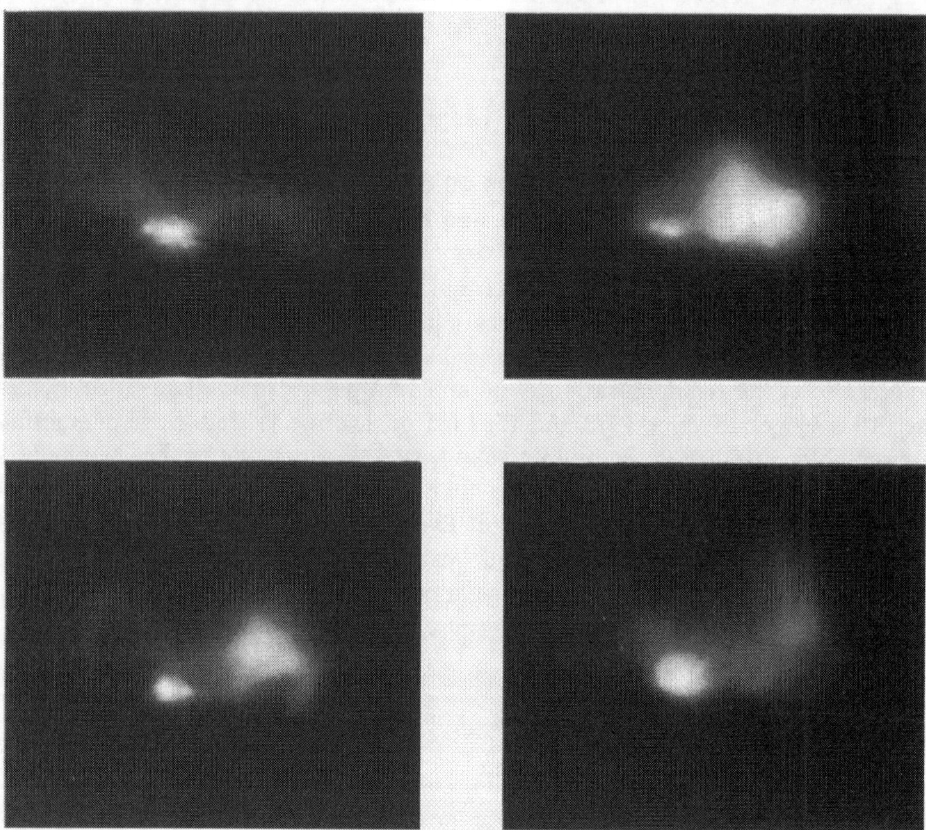

Fig. 1. The temporal behavior of active region 137 (at the east limb in 3–17 Å X-rays. *Top left:* A 16 s exposure at 0807 UT June 10, 1973. *Top right:* A 4 s exposure at 0944. The northern loop structure has changed dramatically in form and brightness; the southern loop structure is not seen on this print because of its low surface brightness. Examination of the corresponding 16 s exposure indicates that it did not change. *Bottom left:* 4 s exposure at 1135 UT. The southern loop structure and the southern bright core have increased in brightness. The northern loop has declined in brightness and the position of maximum brightness along the loop has changed. *Bottom right:* 16 s exposure at 1611 UT. The brightness of all features in the region has declined although it is still above the pre-flare value.

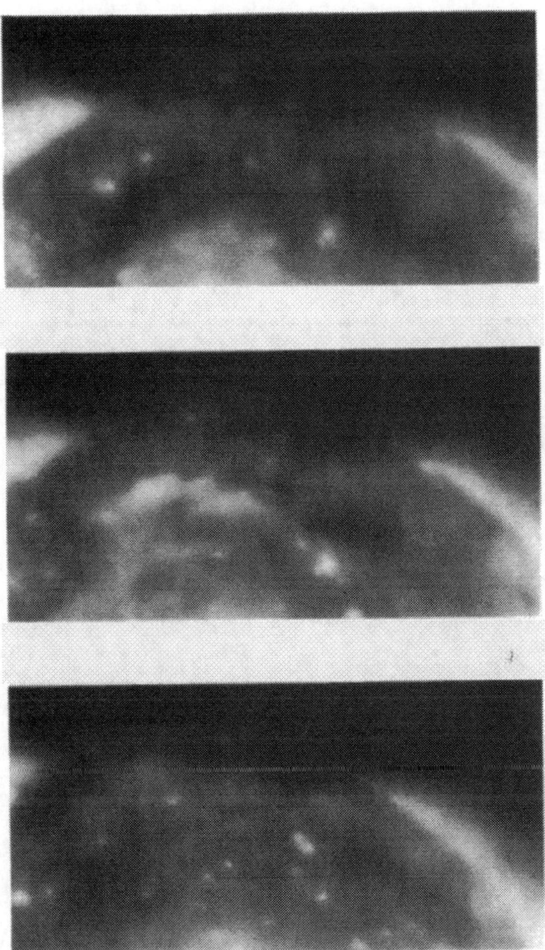

Fig. 2. The brightening and subsequent disappearance of an apparent filament cavity on June 12–13, 1973. *Top:* The appearance of the feature at 1923 UT in a 64 s exposure (3–32, 44–54 Å). *Center:* At 0118 UT, June 13, 1973, the bright structure is evident in the center of the cavity. *Bottom:* By 1537, June 13, 1973, the bright structure is no longer visible and the cavity is not distinguishable.

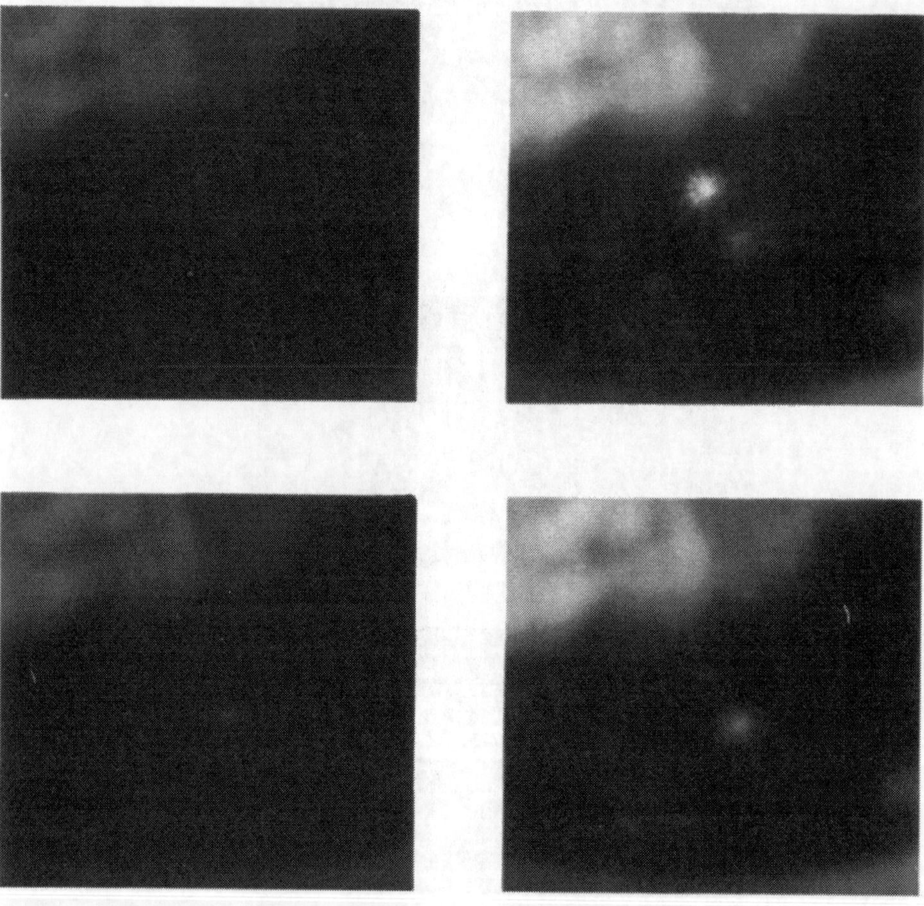

Fig. 3. A sudden increase in the X-ray intensity of a coronal bright point. *Top left:* A portion of a 64 s exposure in 3–32, 44–54 Å X-rays (June 12, 0505 UT). *Top right:* The corresponding 256 s exposure (June 12, 0506 UT). *Bottom left:* A portion of a 64 s exposure at 0641 UT. *Bottom right:* The corresponding 256 s exposure at 0642 UT.

THE HIGH ALTITUDE OBSERVATORY WHITE LIGHT CORONAGRAPH EXPERIMENT

R. M MacQUEEN, J. T. GOSLING, E. HILDNER, R. H. MUNRO, A. I. POLAND, and C. L. ROSS

High Altitude Observatory, National Center for Atmospheric Research, Boulder, Colo., U.S.A.*

Abstract. The coronagraph has obtained observations, both by astronaut and ground command, spaced periodically throughout the mission on a time center of approximately six to eight hours. Programs of more frequent operation – to examine short-term temporal variations and transient activity in the corona – have also been run at various times during the mission. The initial coronagraph observations of structures near limb passage experimentally verify that there are several time scales on which visual changes in these structures occur: (a) approximately one-half rotation, presumably accompanying major reorientations of coronal magnetic fields governing large scale coronal structures, (b) hours to days wherein changes to smaller coronal features are due either to structural changes of particular coronal features or to perspective effects and (c) less than hours – during coronal transients – which caused major reorientation of coronal structures by their passage through the coronal medium. Observations of the latter phenomena have provided some of the more spectacular results from the coronagraph (see Figure 1). The particular case illustrated was one stage of a complex series of events associated with activity in region 137, National Oceanic and Atmospheric Administration numbering, near east limb passage on 10 June 1973. The limb exhibited a major system of active prominences extended from the equator to north 25° and produced surges and sprays through the early hours of the day. From approximately 0700 to 0900 UT several Hα and small X-ray flares occurred and 0815 an eruptive prominence was observed by ground-based telescopes to ascend to 1.4 $R_{\odot}$. This latter event is presumably the source of the transient observed by the coronagraph between 0929 and 1001 by ground command. During this period the material front moved with the apparent projected velocity of 450 km s^{-1} from 3.6 to 4.8 $R_{\odot}$, while the arch-like structure expanded to a diameter greater than 2 $R_{\odot}$. Figure 1 shows the event at 0943 GMT; it is one photograph of the 144 obtained. Within the structure are numerous bright areas and extended structures apparently mapping the distended magnetic field configuration. The general appearance of this transient is indicative of a large magnetic loop, or bottle, expanded outward from the Sun, its leading edge compressed by interaction with the ambient corona. Although there apparently was not a metric radio burst associated with this event, other transients observed later in the mission have associated metric wavelength activity.

* The National Center for Atmospheric Research is sponsored by the National Science Foundation.

Gordon Newkirk, Jr. (ed.) Coronal Disturbances, 505–506.

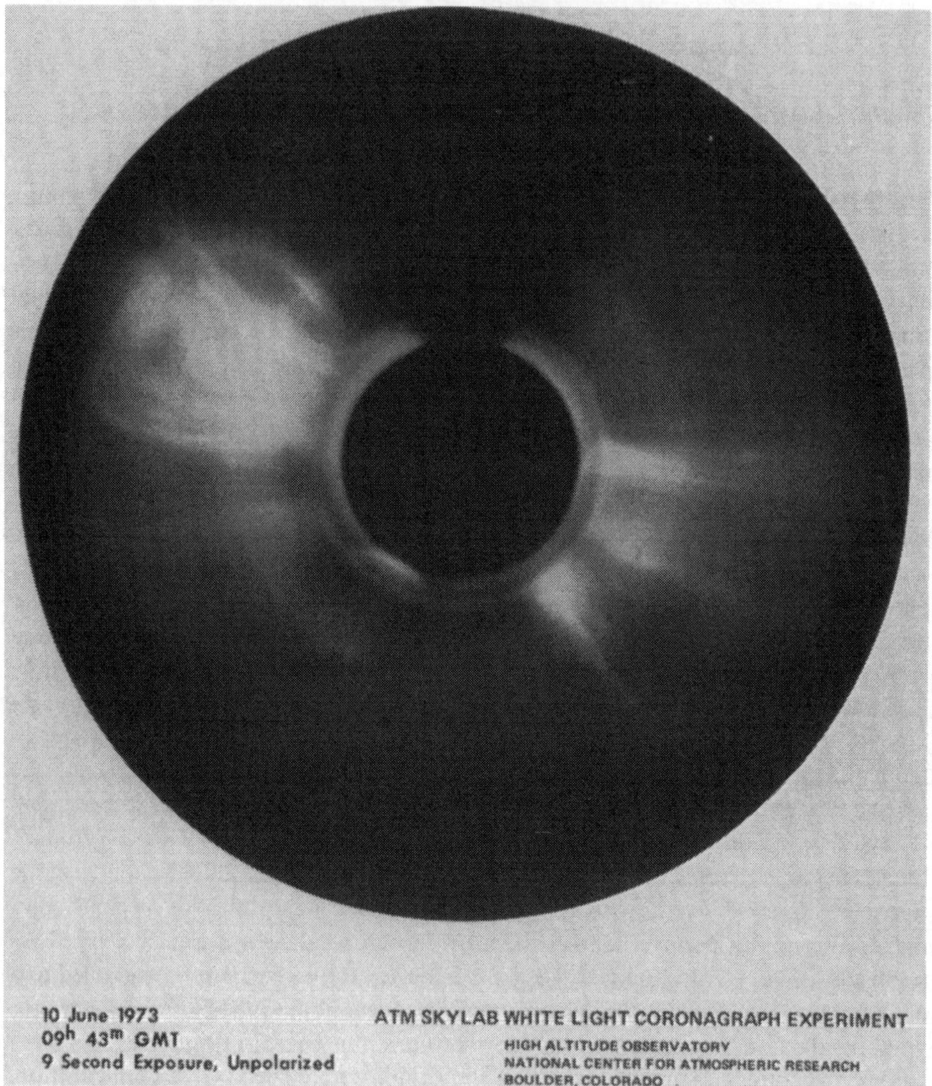

Fig. 1. The coronal transient of 10 June 1973. The photograph was made at 0943 GMT, more than one hour following ground observation of an eruptive prominence on the solar limb. Solar north is up, and east to the left. The coronagraph support pylon observes the northern solar corona, and present in the southeast is a cusp of scattered light due to contamination on the coronagraph external occulting disk.

INDEX OF SUBJECTS

Zeitfracht Medien GmbH
Ferdinand-Jühlke-Straße 7
99095 Erfurt, Deutschland
produktsicherheit@kolibri360.de